Ecological Studies, Vol. 114

Analysis and Synthesis

Edited by

O.L. Lange, Würzburg, Germany
H.A. Mooney, Stanford, USA

Ecological Studies

Volumes published since 1989 are listed at the end of this book.

Springer
Berlin
Heidelberg
New York
Barcelona
Budapest
Hong Kong
London
Milan
Paris
Santa Clara
Singapore
Tokyo

K. Winter J.A.C. Smith (Eds.)

Crassulacean Acid Metabolism

Biochemistry, Ecophysiology and Evolution

With 123 Figures and 45 Tables

Springer

Dr. Klaus Winter
Smithsonian Tropical Research Institute
P.O. Box 2072, Balboa
Republic of Panama

Dr. J. Andrew C. Smith
Department of Plant Sciences
University of Oxford
South Parks Road
Oxford OX1 3RB,
and Magdalen College
Oxford OX1 4AU
UK

Front cover: The cover motif shows CAM plants in southern Madagascar, with a simplified scheme of the CAM cycle superimposed. The "spiny forest" of Madagascar represents a high-biomass community of CAM plants, comprised principally of succulent members of the Didiereaceae up to 10 m tall (*Alluaudia ascendens, A. dumosa, A. procera*) and Euphorbiaceae (*Euphorbia leucodendron, E. oncoclada, E. plagiantha, E. stenoclada*) with species of Crassulaceae (*Kalanchoë* spp.), Asphodelaceae (*Aloë* spp.) and Cucurbitaceae (*Xerosicyos danguyi*) in the undergrowth.

ISBN 3-540-58104-9 Springer-Verlag Berlin Heidelberg New York

Library of Congress Cataloging-in-Publication Data. Crassulacean acid metabolism: biochemistry, ecophysiology, and evolution/Klaus Winter, J. Andrew C. Smith (eds.). p. cm. – (Ecological studies: vol. 114) Papers from an International Workshop on Crassulacean Acid Metabolism, held at Smithsonian Tropical Research Institute, Republic of Panama, Mar. 21–26, 1993. Includes bibliographical references and index. ISBN 3-540-58104-9 1. Crassulacean acid metabolism-Congresses. I. Winter, Klaus, 1949– . II. Smith, J.A.C. (J. Andrew C.). III International Workshop on Crassulacean Acid Metabolism (1993: Smithsonian Tropical Research Institute, Republic of Panama) IV. Series: Ecological studies: V. 114. OK881.C73 1996 581.5-dc20 95-7373

Cover design: Springer-Verlag, Design & Production

Typesetting: Thomson Press (I) Ltd., Madras

SPIN: 10100195 31/3137/SPS – 5 4 3 2 1 0 – Printed on acid-free paper

Dedicated to the memory of
Alan P. Smith
(1945–1993),
a pioneer in plant ecophysiology.

Preface

Crassulacean acid metabolism (CAM) is one of the classic examples of a metabolic adaptation to environmental stress. Well over 5% of vascular plant species engage in this water-conserving photosynthetic pathway to obtain carbon for growth and reproduction. Intensified research activity over the last few years has led to major advances in understanding the biology of CAM plants – from their biochemistry and molecular biology to fundamental aspects of their physiology, ecology and evolution. While a holistic approach characterized CAM research in the past, recent studies have become ever more specialized and analytical. CAM research is ready for a detailed review that attempts to integrate these new developments.

This book has its origins in an International Workshop on Crassulacean Acid Metabolism held at the Tupper Conference Center of the Smithsonian Tropical Research Institute (STRI), Republic of Panama, from March 21 to 26, 1993. Over 60 scientists came together to review and discuss recent developments in their research. More than a decade had elapsed since the last major meeting on CAM plants, held at the University of California at Riverside in January 1982. Since then, important new areas of research have emerged. These include studies of gene expression and the molecular basis of CAM, of the ecophysiology of tropical epiphytes and other CAM plants in their natural environments, of the productivity of several agronomically important species, and of CAM in submerged aquatic species. Today, it is no longer possible to regard CAM as "a rather eccentric phenomenon displayed by a small group of xeromorphic plants", as an eminent plant physiologist wrote as recently as 1987. We hope this book will provide a coherent overview of the rapid advances that have been made in this field and of our current understanding of CAM.

We are indebted to the Smithsonian Tropical Research Institute for making funds available for organizing the workshop in Panama, and especially to the Director, Ira Rubinoff, for his continual support and encouragement. Many members of the STRI staff, in particular Mónica Alvarado, María Luz Calderón and Gloria Maggiori, deserve special thanks for an enormous amount of assistance throughout the meeting. We all owe a great debt to our late colleague, Alan P. Smith, who did so much to establish plant ecophysiology as a research discipline at STRI. Alan was heavily involved in the planning of this workshop, but sadly was not able to participate in the meeting itself. He was an inspiration to many people at STRI and in the world of tropical ecophysiology.

The book begins with a summary of the basic principles of CAM to lay the foundations for three major sections that cover the different areas of current research. Part A deals with the biochemistry of the CAM cycle, Part B focuses on the environmental and developmental control of CAM and underlying molecular mechanisms, while Part C is devoted to the ecological diversity and evolutionary origins of CAM. The three sections are followed by a chapter that attempts to draw together some of these latest findings and to identify research directions that may be of particular importance in the years ahead. Because of the large number of CAM plants now known, we have also compiled a complete list of genera in which CAM has been documented, with which the book concludes. Overall, we hope the result is a volume that will be useful both to research scientists and to advanced students in plant biochemistry, physiology and ecology who wish to learn more about the metabolic basis of plant adaptation to environmental stress.

Panama City and Oxford, 1995

Klaus Winter
J. Andrew C. Smith

Contents

List of Contributors

W.W. Adams III

Department of Environmental, Population, and Organismic Biology, University of Colorado, Boulder, CO 80309-0334, USA

M.N. Angelov

Department of Biochemistry, Life Sciences Building, The University of Georgia, Athens, GA 30602, USA

B.J. Barkla

Department of Plant Sciences, University of Oxford, South Parks Road, Oxford OX1 3RB, UK

D.M. Bartholomew

Department of Plant Sciences, University of Oxford, South Parks Road, Oxford OX1 3RB, UK

G. Bartoli

Dipartimento di Biologia Vegetale, Università di Napoli "Federico II", Via Foria 223, I-80139 Napoli, Italy

M. Bäßler

Institut für Pflanzenphysiologie und Mikrobiologie, Freie Universität Berlin, Königin-Luise-Str. 12–16, D-14195 Berlin, Germany

M. Bettey

Department of Plant Sciences, University of Oxford, South Parks Road, Oxford OX1 3RB, UK

C.C. Black

Department of Biochemistry, Life Sciences Building, The University of Georgia, Athens, GA 30602, USA

H.J. Bohnert

Departments of Biochemistry, of Molecular and Cellular Biology, and of Plant Sciences, The University of Arizona, Tucson, AZ 85721, USA

A.M. Borland

Department of Agricultural and Environmental Science, Ridley Building, The University, Newcastle upon Tyne NE1 7RU, UK

J.V. Brederode

Department of Plant Ecology and Evolutionary Biology, University of Utrecht, Padualaan 8, 3584 CH Utrecht, The Netherlands

J. Brulfert

Laboratoire de Biochimie Fonctionelle des Membranes Végétales, CNRS, F-91198 Gif-sur-Yvette, France

P.J. Carter

Institute of Biomedical and Life Sciences, University of Glasgow, Glasgow G12 8QQ, UK

J.-Q. Chen

Department of Biochemistry, Life Sciences Building, The University of Georgia, Athens, GA 30602, USA

S.-H. Cheng

Department of Botany, Washington State University, Pullman, WA 99164-4238, USA

J.C. Cushman

Department of Biochemistry and Molecular Biology, Oklahoma State University, Stillwater, OK 74078-0454, USA

Z. Dai

Department of Botany, Washington State University, Pullman, WA 99164-4238, USA

B. Demmig-Adams

Department of Environmental, Population, and Organismic Biology, University of Colorado, Boulder, CO 80309-0334, USA

R.L. Doong

Department of Biochemistry, Life Sciences Building, The University of Georgia, Athens, GA 30602, USA

G.E. Edwards

Department of Botany, Washington State University, Pullman, WA 99164-4238, USA

C.A. Fewson

Institute of Biomedical and Life Sciences,
University of Glasgow, Glasgow G12 8QQ, UK

B. Fißlthaler

Institut für Pflanzenphysiologie und Mikrobiologie, Freie Universität Berlin, Königin-Luise-Str. 12-16, D-14195 Berlin, Germany

H. Griffiths

Department of Agricultural and Environmental Science, Ridley Building, The University, Newcastle upon Tyne NE1 7RU, UK

J. Hann

Department of Botany and Plant Sciences, University of California, Riverside, CA 92521, USA

U. Heber

Julius-von-Sachs-Institut für Biowissenschaften, Universität Würzburg, Mittlerer Dallenbergweg 64, D-97082 Würzburg, Germany

J. Ingram

Department of Plant Sciences, University of Oxford,
South Parks Road, Oxford OX1 3RB, UK

W. M. Kaiser

Julius-von-Sachs-Institut für Biowissenschaften, Universität Würzburg, Mittlerer Dallenbergweg 64, D-97082 Würzburg, Germany

S. Kaur

Department of Botany and Plant Sciences, University of California, Riverside, CA 92521, USA

J.E. Keeley

Department of Biology, Occidental College, Los Angeles, CA 90041, USA

M. Kluge

Institut für Botanik, Technische Hochschule Darmstadt,
Schnittspahnstr. 10, D-64287 Darmstadt, Germany

M.S.B. Ku

Department of Botany, Washington State University,
Pullman, WA 99164-4238, USA

B. Lenz

Institut für Pflanzenphysiologie und Mikrobiologie, Freie Universität Berlin, Königin-Luise-Str. 12–16, D-14195 Berlin, Germany

U. Lüttge

Institut für Botanik, Technische Hochschule Darmstadt, Schnittspahnstr. 3–5, D-64287 Darmstadt, Germany

C.E. Martin

Department of Botany, University of Kansas, Lawrence, KS 66045-2106, USA

G. Meyer

Institut für Pflanzenphysiologie und Mikrobiologie, Freie Universität Berlin, Königin-Luise-Str. 12–16, D-14195 Berlin, Germany

S. Neimanis

Julius-von-Sachs-Institut für Biowissenschaften, Universität Würzburg, Mittlerer Dallenbergweg 64, D-97082 Würzburg, Germany

G.A. Nimmo

Institute of Biomedical and Life Sciences, University of Glasgow, Glasgow G12 8QQ, UK

H.G. Nimmo

Institute of Biomedical and Life Sciences, University of Glasgow, Glasgow G12 8QQ, UK

P.S Nobel

Laboratory of Biomedical and Environmental Sciences, 900 Veteran Avenue, University of California, Los Angeles, CA 90024-1786, USA

G.B. North

Laboratory of Biomedical and Environmental Sciences, 900 Veteran Avenue, University of California, Los Angeles, CA 90024-1786, USA

C.B. Osmond

Research School of Biological Sciences, Institute of Advanced Studies, Australian National University, Box 475, Canberra 2601, Australia

O. Pantoja

Department of Plant Sciences, University of Oxford,
South Parks Road, Oxford OX1 3RB, UK

A. Patel

Department of Botany and Plant Sciences,
University of California, Riverside, CA 92521, USA

A.J. Pennington

Department of Pharmacology, University of Edinburgh,
1 George Square, Edinburgh EH8 9JZ, UK

E.A.H. Pilon-Smits

Department of Plant Ecology and Evolutionary Biology,
University of Utrecht, Padualaan 8, 3584 CH Utrecht,
The Netherlands

M. Popp

Institut für Pflanzenphysiologie, Universität Wien, Althanstr. 14,
A-1091 Wien, Austria

J.A. Raven

Department of Biological Sciences, University of Dundee,
Dundee DD1 4HN, UK

S.A. Robinson

Research School of Biological Sciences, Institute of Advanced
Studies, Australian National University, Box 475, Canberra 2601,
Australia

J.M. Schmitt

Institut für Pflanzenphysiologie und Mikrobiologie,
Freie Universität Berlin, Königin-Luise-Str. 12–16,
D-14195 Berlin, Germany

M. Schomburg

Institut für Botanik, Technische Hochschule Darmstadt,
Schnittspahnstr. 10, D-64287 Darmstadt, Germany

A. Sheriff

Institut für Pflanzenphysiologie und Mikrobiologie,
Freie Universität Berlin, Königin-Luise-Str. 12–16,
D-14195 Berlin, Germany

N. Smirnoff

Department of Biological Sciences, Hatherly Laboratories, University of Exeter, Prince of Wales Road, Exeter EX4 4PS, UK

J.A.C. Smith

Department of Plant Sciences, University of Oxford, South Parks Road, Oxford OX1 3RB, UK

R.A. Spicer

Department of Earth Sciences, The Open University, Milton Keynes MK7 6AA, UK

S.J.S. Sung

Department of Biochemistry, Life Sciences Building, The University of Georgia, Athens, GA 30602, USA

H. 't Hart

Department of Plant Ecology and Evolutionary Biology, University of Utrecht, Padualaan 8, 3584 CH Utrecht, The Netherlands

I.P. Ting

Department of Botany and Plant Sciences, University of California, Riverside, CA 92521, USA

M.S. Tsiantis

Department of Plant Sciences, University of Oxford, South Parks Road, Oxford OX1 3RB, UK

A. Virzo De Santo

Dipartimento di Biologia Vegetale, Università di Napoli "Federico II", Via Foria 223, I-80139 Napoli, Italy

L. Walling

Department of Botany and Plant Sciences, University of California, Riverside, CA 92521, USA

M.B. Wilkins

Institute of Biomedical and Life Sciences, University of Glasgow, Glasgow G12 8QQ, UK

K. Winter

Smithsonian Tropical Research Institute, P.O. Box 2072, Balboa, Republic of Panama

H. Ziegler

Institut für Botanik und Mikrobiologie, Technische Universität München, Arcisstr. 21, D-80290 München, Germany

G. Zotz

Smithsonian Tropical Research Institute, P.O. Box 2072, Balboa, Republic of Panama

Abbreviations

ABA	abscisic acid
ACC	acetyl CoA carboxylase (EC 6.4.1.2)
ATP	adenosine 5′-triphosphate
BA	6-benzylaminopurine
CAM	crassulacean acid metabolism
cDNA	complementary DNA
Chl	chlorophyll
CDCF	5- (and 6-)carboxy-2′,7′-dichlorofluorescein
CPS	carbamoyl phosphate synthetase (EC 6.3.5.5)
D	deuterium (^{2}H)
EPI	environmental productivity index
EPR	electron paramagnetic resonance
F_m	maximum fluorescence level (fully reduced quinone acceptor Q_A in the reaction centre of photosystemII) of a dark-adapted leaf (F_m′denotes illuminated leaf)
F_o	minimum fluorescence level (fully oxidized quinone acceptor Q_A in the reaction centre of photosystem II)
F_v	variable (flash-induced) fluorescence rise $(= F_m - F_o)$
F_s	steady-state fluorescence of illuminated leaf
F6-P	fructose 6-phosphate
FBP	fructose 1,6-bisphosphate
FW	fresh weight (fr. wt.)
GCMS	gas chromatography – mass spectrometry
GUS	ß-glucuronidase (EC 3.2.1.31)
G3PDH	glyceraldehyde 3-phosphate dehydrogenase (NAD^+-dependent, EC 1.2.1.12; $NADP^+$-dependent, EC 1.2.1.13)
G6-P	glucose 6-phosphate
JA-Me	(±)methyl jasmonate
IRGA	infra-red gas analyzer
ka	thousand years
LAI	leaf area index
Ma	million years
MDH	malate dehydrogenase (EC 1.1.1.37)

ME	malic enzyme
mRNA	messenger RNA
NAD/NADH	nicotinamide adenine dinucleotide/reduced form
NADP/NADPH	nicotinamide adenine dinucleotide phosphate/reduced form
NAD-ME	NAD^+-dependent malic enzyme (EC 1.1.1.39)
NADP-ME	$NADP^+$-dependent malic enzyme (EC 1.1.1.40)
NMR	nuclear magnetic resonance
NPQ	nonphotochemical fluorescence quenching
NPR	nonphotorespiratory
NTP	nucleoside triphosphate
PC	pyruvate carboxylase (EC 6.4.1.1)
PCO(C)	photosynthetic carbon oxidation (cycle)
PCR(C)	photosynthetic carbon reduction (cycle)
PEG	polyethylene glycol
PEP	phospho*enol*pyruvate
PEPC	phospho*enol*pyruvate carboxylase (EC 4.1.1.31)
PEPCK	phospho*enol*pyruvate carboxykinase (EC 4.1.1.49)
PFK	phosphofructokinase (EC 2.7.1.11)
PFP	pyrophosphate: fructose 6-phosphate 1-phosphotransferase (EC 2.7.1.90) (also called PP_i-PFK)
PGA	3-phosphoglycerate
p_a	ambient CO_2 partial pressure
p_i	intercellular CO_2 partial pressure
P_i	inorganic phosphate
PP_i	inorganic pyrophosphate
PPDK	pyruvate, P_i dikinase (EC 2.7.9.1)
PPFD	photosynthetic photon flux density (400 to 700 nm)
PR	photorespiratory
PSI	photosystem I
PSII	photosystem II
Q_A	primary quinone acceptor of PSII
RH	relative humidity
RUBISCO	ribulose 1,5-bisphosphate carboxylase/oxygenase (EC 4.1.1.39)
RWC	relative water content
SAI	stem area index
SD	standard deviation
SE	standard error
TCA cycle	tricarboxylic acid cycle
UTP	uridine 5′-triphosphate
VPD	leaf-to-air vapour-pressure difference
WUE	water-use efficiency
$\delta^{13}C$	carbon-isotope ratio (‰)
Δ	carbon-isotope discrimination (‰)

An Introduction to Crassulacean Acid Metabolism. Biochemical Principles and Ecological Diversity

K. WINTER[1] and J.A.C. SMITH[2]

Discovery of Dark CO_2 Fixation

Crassulacean acid metabolism (CAM) is one of three metabolic pathways found in the photosynthetic tissues of vascular plants for assimilation of atmospheric CO_2. In contrast to C_3 and C_4 plants, CAM plants take up CO_2 from the atmosphere predominantly at night, subsequently assimilating this CO_2 to the level of carbohydrate during the following light period. Elucidation of the complete pathway of carbon assimilation in CAM plants took nearly 150 years and encompassed many fundamental discoveries in plant biochemistry.

Nocturnal uptake of CO_2 by succulent plants was already the subject of investigations by de Saussure in 1804. The photosynthetic tissues of such plants were also shown to acidify at night and deacidify during the day, the first formal report of this property being that of Heyne in 1815. By the early twentieth century, it was known that carbohydrate levels vary reciprocally with organic acids, and that succulent plants exhibit an "inverse rhythm" of stomatal opening at night and closure during the day. But it was not until the late 1940s that all the connections between these observations were made and the reaction sequence of the CAM cycle finally understood (Bonner and Bonner 1948; Thurlow and Bonner 1948; Thomas 1949; Thomas and Beevers 1949). In summary, CO_2 taken up at night combines quantitatively with a product of carbohydrate breakdown to form organic acid, and decarboxylation of this organic acid during the following day releases CO_2, which is assimilated through the C_3 pathway as storage carbohydrate is regenerated. The discovery of CAM and the historical development of this first phase of CAM research has been carefully documented by Bruinsma (1958), Ranson and Thomas (1960), Wolf (1960) and Walker (1992).

Over the last five decades, CAM research has entered its second phase. A vast amount of information has been obtained on the biochemistry of the metabolic cycle, its regulation, and the biochemical variations exhibited by different groups of plants. These developments have been discussed in detail in a number of earlier reviews (e.g. Kluge and Ting 1978; Osmond 1978; Osmond and Holtum 1981; Osmond et al. 1982; Ting and Gibbs 1982; Edwards and Walker 1983; Medina

[1]Smithsonian Tropical Research Institute, P.O. Box 2072, Balboa, Republic of Panama
[2]Department of Plant Sciences, University of Oxford, South Parks Road, Oxford OX1 3RB, UK

Ecological Studies, Vol. 114
Winter/Smith (eds) Crassulacean Acid Metabolism

1984; Ting 1985; Winter 1985; Lüttge 1987, 1993; Bohnert et al. 1988; Griffiths 1988; Lüttge and Smith 1988; Leegood and Osmond 1990; Smith and Bryce 1992; Leegood 1993). In essence, we can define the CAM cycle in terms of the following sequence of metabolic steps, also shown schematically in Fig. 1:

At night: 1. Formation of the CO_2 acceptor phospho*enol*pyruvate (PEP) by glycolytic breakdown of storage carbohydrate (starch and/or soluble sugars) in chlorenchyma cells.
2. CO_2 fixation by the cytosolic enzyme PEP carboxylase (PEPC) resulting in malic-acid synthesis.
3. Storage of malic acid in the large central vacuole of the chlorenchyma cells.

During the day: 1. Release of malic acid from the cell vacuole.
2. Decarboxylation of malate in the cytoplasm, leading to liberation of CO_2 and formation of the three-carbon compounds pyruvate and/or PEP.
3. Assimilation of the liberated CO_2 via ribulose 1,5-bisphosphate carboxylase/oxygenase (RUBISCO) and the C_3 photosynthetic carbon reduction cycle, with regeneration of storage carbohydrate largely from pyruvate and PEP in gluconeogenesis.

Biochemistry

With the basic metabolic cycle understood by the late 1940s, much biochemical research was directed towards identification and characterization of the enzymes

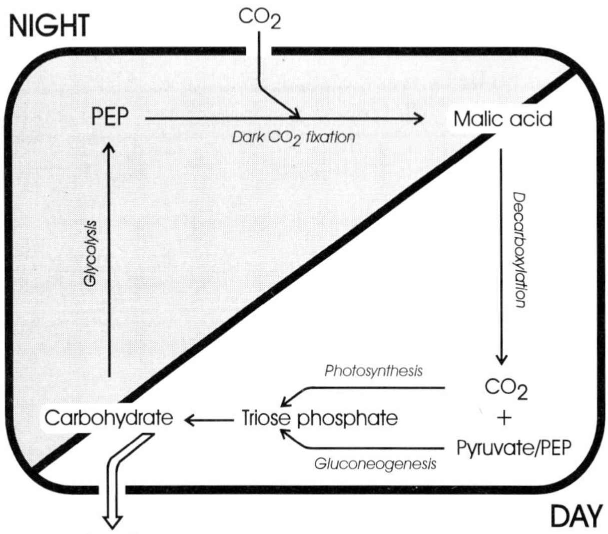

Fig. 1. Simplified schematic outline of the CAM cycle showing temporal organization of the principal metabolic steps

of the CAM pathway. During the 1970s and 1980s, this work led to an appreciation of the complexity of the CAM pathway in terms of intracellular compartmentation, coordination of metabolite fluxes, and the importance of regulatory processes in minimizing futile cycling of reaction products.

Foremost amongst these key control points in the CAM cycle is the enzyme PEPC, which is downregulated during the day by dephosphorylation, and is converted by phosphorylation at night to an "active" form relatively insensitive to feedback inhibition by malate (Winter 1982; Nimmo et al. 1984; Chap. 3). Carbon flow in CAM also involves the participation of different intracellular organelles, and subcellular fractionation techniques have been developed for CAM plants to study the properties of isolated chloroplasts, mitochondria and vacuoles. This has been particularly important in allowing direct access to the transport proteins at the vacuolar membrane (the tonoplast). In CAM plants, the vacuole commonly occupies more than 95% of cell volume, and study of the properties of the tonoplast is central to understanding how acid influx and efflux are coupled to carbon flow during the day-night cycle (Lüttge 1987; Smith and Bryce 1992; Chap. 4).

A considerable degree of biochemical diversity is seen within the standard framework of the CAM cycle. In many species a major part of the storage carbohydrate accumulates as soluble sugars rather than insoluble glucan (starch), which itself has implications for the intracellular compartmentation of metabolism (Chap. 2). Additionally, these CAM plants utilize PEP carboxykinase as their principal decarboxylating enzyme, rather than the NAD- and/or NADP-malic enzymes used in species storing solely starch. There has also been renewed interest in the biochemical significance of the nocturnal citric-acid accumulation observed in some species (Thomas 1947; Lüttge 1988). Although citric-acid accumulation does not appear to be associated with net carbon gain, under certain conditions it can exceed that of malic acid (Chap. 19). Other variants on the basic CAM pathway may, conceivably, still be discovered.

Much more information on the regulation of the CAM cycle at the molecular level promises to become available through the identification of genes encoding key components of the reaction pathway. The first genes to be isolated from a CAM plant – the *Ppc1* and *Ppc2* genes encoding PEPC in *Mesembryanthemum crystallinum* – were cloned and sequenced 6 years ago (Cushman and Bohnert 1989a,b; Cushman et al. 1989), and many other sequences are now available (Chap. 9). In particular, these approaches should allow insight into the mechanisms by which enzyme activity is regulated through transcriptional and post-transcriptional control of gene expression.

Phenotypic Plasticity

Many CAM plants engage in a certain amount of daytime CO_2 uptake directly via C_3 photosynthesis. Depending on the species concerned, its developmental state and prevailing environmental conditions, both the amplitude of the CAM rhythm and its contribution to carbon gain (relative to C_3 photosynthesis) may vary substantially. In many stem-succulent species, such as columnar cacti or platyopuntias, atmospheric CO_2 is taken up almost exclusively at night, even when plants are well watered (Nobel 1988). On the other hand, in some species the capacity for nocturnal CO_2 assimilation is so low that it suffices merely to reduce the net loss of repiratory CO_2 at night compared with typical C_3 plants; these species, which are said to display "CAM-cycling" (Ting 1985; Chap. 12), usually show net CO_2 uptake only during the daytime.

Between these two extremes, one finds a continuous spectrum of intermediate gas-exchange patterns. For example, many species with succulent leaves show quite prolonged net CO_2 uptake at the very beginning and towards the end of the light period, as well as throughout the dark period. Consequently, it is convenient to recognize four distinct phases of gas exchange in CAM plants, as shown in Fig. 2. In these plants, gas exchange is curtailed by stomatal closure during the middle of the day (Phase III) as a result of the high intercellular partial pressure of CO_2 (p_i) generated by malate decarboxylation. When malate decarboxylation nears completion and the CO_2 partial pressure has dropped, Phase IV gas exchange commences.

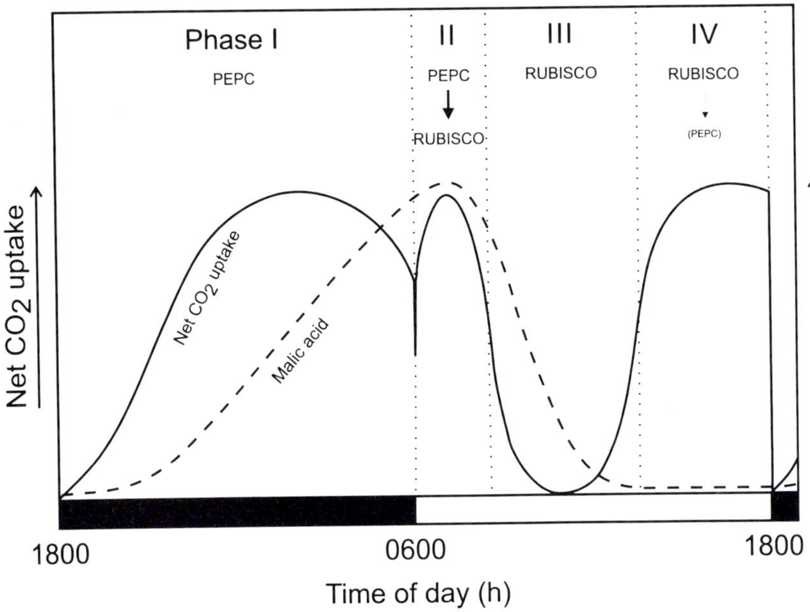

Fig. 2. Day-night cycle of CO_2 fixation and malic-acid content typically observed in well-watered succulent-leaved CAM plants of the genus *Kalanchoë* (Crassulaceae). *Phase II* is a transitional phase between PEPC-mediated and RUBISCO-mediated CO_2 fixation. PEPC may also be catalytically active towards the end of *Phase IV*, but this does not result in malic-acid accumulation. (After Osmond 1978; Leegood and Osmond 1990; Smith and Bryce 1992)

CO_2 fixation via RUBISCO is associated with a larger discrimination against $^{13}CO_2$ compared to the much more abundant $^{12}CO_2$ than is CO_2 fixation via PEPC. The mass-spectrometric monitoring of uptake of $^{13}CO_2$ versus $^{12}CO_2$ has become a powerful tool for assessing relative rates of carboxylation via PEPC and RUBISCO during Phases II and IV (Chap. 15). While the stomata are closed, photosynthesis is not readily accessible to the application of gas-exchange techniques, but novel non-intrusive approaches for measuring chlorophyll *a* fluorescence in intact leaves and stems now allow direct estimation of photosynthetic activity in the chloroplasts during Phase III (Winter et al. 1990; Keiller et al. 1994; Chaps. 6, 7 and 8).

In CAM plants, the relative abundance of ^{13}C, expressed as the $\delta^{13}C$ value[3], has proved to be a valuable index of the relative contributions of dark CO_2 fixation and light CO_2 fixation to total carbon gain over longer periods of time (Osmond et al. 1973). $\delta^{13}C$ values have been particularly useful in research on the environmental regulation and taxonomic occurrence of CAM (Chap. 21). This is illustrated in a study on species of epiphytic orchids from Australia, which were found to exhibit a wide, continuous spectrum of $\delta^{13}C$ values ranging from those typical of the C_3 pathway (−33 to −24‰) to those indicating full CAM (−15 to −12‰) (Fig. 3).

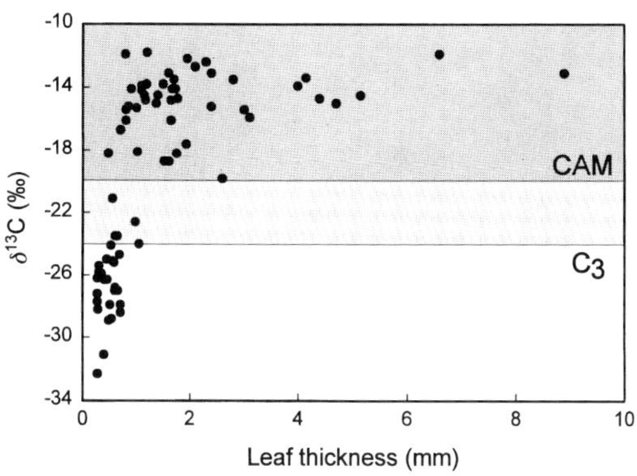

Fig. 3. Relationship between leaf thickness and $\delta^{13}C$ value for tropical and subtropical species of epiphytic Orchidaceae from Australia. Those species in which leaf succulence is mainly attributable to a water-storage tissue that lacks chloroplasts (and hence does not participate in CAM) were not included. *Darker shading* denotes $\delta^{13}C$ values indicating pronounced CAM; *lighter shading* includes $\delta^{13}C$ values of plants in which dark CO_2 fixation contributes to a slight but measurable extent to total carbon gain. (After Winter et al. 1983)

---

[3] $$\delta^{13}C(‰) = \left[\frac{^{13}C_{sample}/^{12}C_{sample}}{^{13}C_{standard}/^{12}C_{standard}} - 1\right] \times 1000$$

The generally accepted standard is CO_2 generated from carbonaceous rock from the Pee Dee belemnite (PDB) formation.

Environmental conditions have a profound effect on the balance between light and dark CO_2 fixation in numerous species, some of which can switch from the C_3 to the CAM mode of photosynthesis, for instance in response to water deficits or photoperiod. Such species have been variously described as "C_3/CAM intermediates", or as "facultative" or "inducible" CAM plants. Important examples are known from the Aizoaceae (e.g. *Mesembryanthemum*: Winter and von Willert 1972; Chaps. 8, 9 and 10), Clusiaceae (e.g. *Clusia*: Tinoco Ojanguren and Vázquez-Yánes 1983; Ting et al. 1985; Schmitt et al. 1988; Chaps. 19 and 20), Crassulaceae (e.g. species of *Kalanchoë* and *Sedum*: Chaps. 11 and 21) and Portulacaceae (e.g. *Calandrinia*: Mooney et al. 1974; Winter et al. 1981). In *Mesembryanthemum crystallinum*, the environmentally induced change from C_3 to CAM is accompanied by very large increases in the activity of PEPC and other enzymes. This species is also amenable to the methods of molecular biology, and indeed has become one of the most intensively studied higher-plant systems for analyzing the responses of gene expression to environmental stress (Chaps. 9 and 10).

Ecophysiology and Species Diversity

Most interpretations of the selective advantage of the CAM pathway attribute its evolutionary success to the highly efficient use of water per unit CO_2 assimilated. This derives from two factors. First, the stomata are open mainly at night rather than during the day. At night, the driving force for transpiration (the difference in water-vapour pressure between the plant and ambient air) is relatively low: this can reduce the amount of water lost per unit CO_2 fixed by an order of magnitude in environments such as those characteristic of desert cacti (Gibson and Nobel 1986). Second, as in C_4 plants, PEPC-mediated CO_2 fixation is associated with low values of p_i. The steeper gradient in CO_2 partial pressure between the ambient air and intercellular spaces results in higher rates of CO_2 influx for a given stomatal conductance or unit of water transpired. These factors combine to produce water-use efficiencies for CAM plants several times higher than those for C_3 species (Kluge and Ting 1978; Nobel 1991).

Additionally, the photosynthetic organs of most CAM plants have a distinctly succulent morphology and a high water-storage capacitance, features that aid survival in tropical and subtropical regions with intermittent water supply or strongly seasonal rainfall. Since there is a limit to the concentration of malic acid that can be accumulated in the vacuole of the chloroplast-containing cells, the succulent morphology also provides a very high storage capacity for nocturnally assimilated CO_2 per unit surface area of photosynthetic organ. Strong positive correlations between leaf succulence and the occurrence and magnitude of CAM have been reported for families such as the Crassulaceae (Teeri et al. 1981) and Orchidaceae (Fig. 3).

The efficiency of photosynthetic CO_2 assimilation in CAM plants is markedly improved by the high values of p_i generated during malate decarboxylation in

Phase III. Under these conditions – when p_i can at times exceed 40 mPa Pa^{-1} (4%[v/v]; Cockburn et al. 1979, Spalding et al. 1979) – the carboxylase activity of RUBISCO is maximized and its photorespiratory oxygenase activity minimized (Osmond et al. 1982). An important role has been ascribed to these high values of p_i in preventing damage to the photosystems under conditions of high photosynthetic photon flux density (PPFD) and drought stress (Osmond 1982; Griffiths et al. 1989; Osmond et al. 1991).

This CO_2-concentrating mechanism incurs an extra energetic cost associated with the diel cycle of malic-acid accumulation and degradation (Edwards et al. 1982), against which can be set the energy savings resulting from suppression of photorespiration. The net result appears to be that the energetic cost of carbon assimilation in CAM plants does not differ greatly from those in C_3 plants (Chaps. 16, 24 and 25). Indeed, although plants exhibiting CAM are most common in high-light environments, several species occur in deeply shaded parts of tropical forests (Winter et al. 1986; Medina 1987).

Much of our understanding of the ecological significance of CAM has been derived from work on desert cacti, which have been the subject of detailed study since the beginning of this century (Nobel 1988). However, ecophysiologists have been slower to investigate other types of CAM plants in their natural habitats. This has required partly the development of more portable measuring equipment, and partly an awareness of the great taxonomic diversity of CAM plants. Although the occurrence of CAM in tropical bromeliads and orchids was recognized early on by Warburg (1886), it was not until the extensive surveys of Nuernbergk (1961) and Coutinho (1963, 1969) that the ecological significance of these CAM plants became more apparent. Epiphytic bromeliads and orchids in tropical forests can be subjected to frequent short-term water deficits, since their root systems function primarily as holdfasts, and the substrates on which they grow have low water-retaining capacity compared with soil environments. With their distribution essentially unaffected by any edaphic factor, epiphytic bromeliads can show a particularly close correlation between photosynthetic pathway and habitat aridity, providing one of the clearest demonstrations of the ecological importance of CAM in water-limited environments (Griffiths and Smith 1983; Smith 1989; Fig. 4).

Some of the most extreme epiphytes possessing CAM are essentially rootless, such as the bromeliad *Tillandsia usneoides*. Even more remarkably, CAM is also found in the autotrophic roots of "shootless" epiphytic orchids. These organs do not possess stomata, yet exhibit all four phases of CAM gas exchange shown in Fig. 2 (Benzing and Ott 1981; Winter et al. 1983, 1985; Cockburn et al. 1985). Relative to the size of the family, there have been few ecophysiological studies of the Orchidaceae (Winter et al. 1983; Earnshaw et al. 1987; Goh and Kluge 1989; Benzing 1990), but increased interest in the biology of tropical epiphytes has led to the discovery of CAM in further families such as the Clusiaceae (Tinoco Ojanguren and Vásquez-Yánes 1983; Ting et al. 1985). With the range of growth forms and physiological plasticity seen in the genus *Clusia*, this group has already become a major area of CAM research (Chaps. 19 and 20).

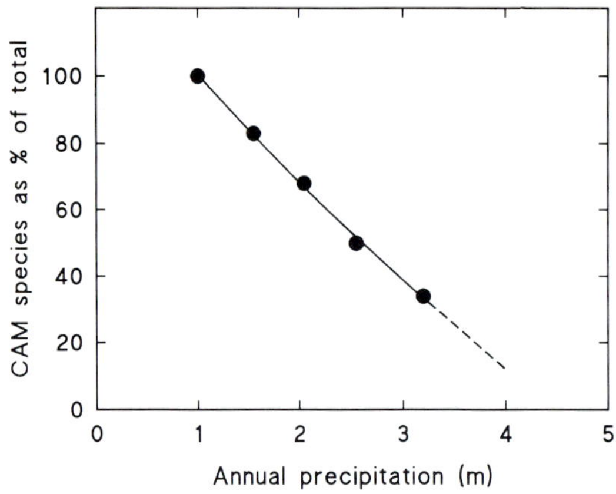

Fig. 4. Relationship between the percentage of CAM species and annual precipitation for epiphytic bromeliads along a climatic gradient in Trinidad. CAM and C_3 species of bromeliads were identified by measurements of $\delta^{13}C$ values and diel changes in titratable acidity. The *broken line* indicates extrapolation to the wettest site on the island, where 10% of the epiphytic bromeliads were CAM species, but for which only approximate meteorological data are available. (After Griffiths and Smith 1983; Smith 1989)

The fact that (functional) stomata are not required for CAM is also illustrated by its occurrence in the aquatic genus *Isoetes* (Isoetaceae; Lycopodiopsida) and certain other submerged aquatic macrophytes (Keeley 1981; Chap. 18). In these species, CAM appears to be an adaptation to low CO_2 availability during the daytime – as for example in shallow pools, where the concentration of dissolved inorganic carbon can be greatly reduced by the photosynthetic activity of co-occurring C_3 plants. Apart from its intrinsic interest for the photosynthetic ecology of aquatic plants, the occurrence of CAM in the fern ally *Isoetes* emphasizes that this photosynthetic option has probably arisen many times in evolution in taxonomically diverse groups (Ehleringer and Monson 1993; Chap. 24).

As a result of the intensive ecophysiological studies of the last 15 years, we now have a much better picture of the taxonomic occurrence and distribution of CAM plants (Table 1; a complete list of the genera in which CAM has been recorded is given in Chap. 26). CAM is displayed in 33 families of vascular plants, ranging from the aquatic fern allies of the Isoetaceae and tropical epiphytic ferns through the gymnosperm *Welwitschia* (Dittrich and Huber 1974; Winter and Schramm 1986) to eight families of monocotyledons (Liliopsida) and 21 families of dicotyledons (Magnoliopsida). A large proportion of the total number of CAM species is represented by the four monocotyledonous families Agavaceae, Asphodelaceae, Bromeliaceae and Orchidaceae, and by the dicotyledonous families Aizoaceae, Asclepiadaceae, Cactaceae, Crassulaceae and Euphorbiaceae (Chap. 26). However, a number of the families contain only a few CAM species or species in which CAM acitivity is expressed very weakly (e.g. *Welwitschia*). In addition, the CAM activity in some of the submerged angiosperms with the isoetid life-form (Raven et al. 1988) appears to be very low (e.g. Hydrocharitaceae and Apiaceae; Webb et al. 1988).

Table 1. Plant families containing species capable of crassulacean acid metabolism (CAM). Major families are those containing either a high proportion of species showing CAM or a large number of CAM species. Minor families are those with only a small number of CAM species or species showing only weak CAM activity. Nomenclature for the family names follows Dahlgren et al. (1985) for the monocotyledons and Cronquist (1981) for the dicotyledons. Within each column, families are listed in the order suggested by the phylogenetic analysis of Chase et al. (1993)

Land plants		Aquatic plants
Major families	Minor families	
Agavaceae	Polypodiaceae	Isoetaceae
Asphodelaceae	Vittariaceae	Alismataceae
Orchidaceae	Welwitschiaceae	Hydrocharitaceae
Bromeliaceae	Dracaenaceae	Crassulaceae
Aizoaceae	Commelinaceae	Plantaginaceae
Didiereaceae	Piperaceae	Apiaceae
Cactaceae	Portulacaceae	
Clusiaceae	Cucurbitaceae	
Crassulaceae	Passifloraceae	
Euphorbiaceae	Vitaceae	
Asclepiadaceae	Oxalidaceae	
	Geraniaceae	
	Asteraceae	
	Apocynaceae	
	Rubiaceae	
	Gesneriaceae	
	Lamiaceae	

Further work will be necessary to determine the functional importance and evolutionary significance of CAM in these taxa.

Excluding the Orchidaceae, an estimated 9 000 species probably perform CAM (see Chap. 26). Far greater uncertainty exists regarding the number of CAM species in the Orchidaceae (Winter et al. 1983), a family thought to contain 19 500 species (Dressler 1993), of which about 73% are believed to be tropical epiphytes (Atwood 1986). If 50% of these tropical epiphytes were CAM species, this would give about 7 000 CAM species in the Orchidaceae. Thus, in total, there could be approximately 16 000 species of CAM plants, corresponding to over 6% of the estimated 260 000 species of vascular plants (Akeroyd and Synge 1992) – a much larger percentage than is represented by C_4 species, but a value that will almost certainly require revision as more extensive studies are performed on the Orchidaceae.

Conclusions

CAM represents a photosynthetic CO_2-concentrating mechanism that improves the ability of plants to assimilate carbon in water-limited and CO_2-limited environments. This biochemical pathway is taxonomically widespread among vascular plants and is found in many succulent species in arid regions, as well as in large numbers of tropical epiphytes and in some submerged aquatic macrophytes. PEPC is a key control point for CO_2 fixation in all CAM plants, but there are important differences between species in the pathways of malate decarboxylation and regeneration of storage carbohydrate. Furthermore, in some species the extent to which CAM contributes to total carbon gain is highly variable and depends on environmental conditions. The mechanisms underlying the expression and regulation of CAM – both at the molecular/biochemical level and at the whole-plant/environmental level – are now the subject of intensive research. This book reviews recent progress in these areas, together with other studies of the ecology and diversity of CAM plants that are contributing to our understanding of the functional significance of this metabolic pathway.

References

Akeroyd J, Synge H (1992) Higher plant diversity. In: Groombridge B (ed) Global diversity: status of the earth's living resources. Chapman & Hall, London, pp 64–87

Atwood JT Jr (1986) The size of the Orchidaceae and the systematic distribution of epiphytic orchids. Selbyana 9: 171–186

Benzing DH (1990) Vascular epiphytes: general biology and related biota. Cambridge University Press, Cambridge

Benzing DH, Ott DW (1981) Vegetative reduction in epiphytic Bromeliaceae and Orchidaceae: its origin and significance. Biotropica 13: 131–140

Bohnert HJ, Ostrem JA, Cushman JC, Michalowski CB, Rickers J, Meyer G, DeRocher EJ, Vernon DM, Krueger M, Vasquez-Moreno L, Velten J, Hoefner R, Schmitt JM (1988) *Mesembryanthemum crystallinum*: a higher plant model for the study of environmentally induced changes in gene expression. Plant Mol Biol Rep 6: 10–28

Bonner W, Bonner J (1948) The role of carbon dioxide in acid formation by succulent plants. Am J Bot 35: 113–117

Bruinsma J (1958) Studies on the crassulacean acid metabolism. Acta Bot Neerl 7: 531–588

Chase MW, Soltis DE, Olmstead RG, and 39 others (1993) Phylogenetics of seed plants: an analysis of nucleotide sequences from the plastid gene *rbc*L. Ann Mo Bot Gard 80: 528–580

Cockburn W, Ting IP, Sternberg LO (1979) Relationships between stomatal behavior and internal carbon dioxide concentration in crassulacean acid metabolism plants. Plant Physiol 63: 1029–1032

Cockburn W, Goh CJ, Avadhani PN (1985) Photosynthetic carbon assimilation in a shootless orchid, *Chiloschista usneoides* (DON) LDL. Plant Physiol 77: 83–86

Coutinho LM (1963) Algumas informações sôbre a ocorrência do "Efeito de De Saussure" en epífitas e erbáceas terrestres de mata pluvial. Bol Fac Filos Ciên Letr USP No 288. Botânica 20: 81–98

Coutinho LM (1969) Novas observações sôbre a ocorrência do "Efeito de De Saussure" e suas relações com a suculência, a temperatura folhear e os movimentos estomáticos. Bol Fac Filos Ciên Letr USP No 331. Botânica 24: 77–102

Cronquist A (1981) An integrated system of classification of flowering plants. Columbia University Press, New York

Cushman JC, Bohnert HJ (1989a) Nucleotide sequence of the *Ppc2* gene encoding a housekeeping isoform of phosphoenolpyruvate carboxylase from *Mesembryanthemum crystallinum*. Nucleic Acids Res 17: 6743–6744

Cushman JC, Bohnert HJ (1989b) Nucleotide sequence of the gene encoding a CAM specific isoform of phosphoenolpyruvate carboxylase from *Mesembryanthemum crystallinum*. Nucleic Acids Res 17: 6745–6746

Cushman JC, Meyer G, Michalowsky CB, Schmitt JM, Bohnert HJ (1989) Salt stress leads to differential expression of two isogenes of phosphoenolpyruvate carboxylase during crassulacean acid metabolism induction in the common ice plant. Plant Cell 1: 715–725

Dahlgren RMT, Clifford HT, Yeo PF (1985) The families of the monocotyledons. Structure, evolution, and taxonomy. Springer, Berlin Heidelberg New York

De Saussure T (1804) Recherches chimiques sur la végétation. Chez la V.[e] Nyon, Paris

Dittrich P, Huber W (1974) Carbon dioxide metabolism in members of the Chlamydospermae. In: Avron M (ed) Proc 3rd Int Congr Photosynthesis. Elsevier, Amsterdam, pp 1573–1578

Dressler RL (1993) Phylogeny and classification of the orchid family. Cambridge University Press, Cambridge

Earnshaw MJ, Winter K, Ziegler H, Stichler W, Crutwell NEG, Kerenga K, Cribb PJ, Wood J, Croft JR, Carver KA, Gunn TC (1987) Altitudinal changes in the incidence of crassulacean acid metabolism in vascular epiphytes and related life forms in Papua New Guinea. Oecologia 73: 566–572

Edwards G, Walker DA (1983) C_3, C_4: mechanisms, and cellular and environmental regulation, of photosynthesis. Blackwell, Oxford

Edwards GE, Foster JG, Winter K (1982) Activity and intracellular compartmentation of enzymes of carbon metabolism in CAM plants. In: Ting IP, Gibbs M (eds) Crassulacean acid metabolism. American Society of Plant Physiologists, Rockville, pp 92–111

Ehleringer JR, Monson RK (1993) Evolutionary and ecological aspects of photosynthetic pathway variation. Annu Rev Ecol Syst 24: 411–439

Gibson AC, Nobel PS (1986) The cactus primer. Harvard University Press, Cambridge

Goh CJ, Kluge M (1989) Gas exchange and water relations in epiphytic orchids. In: Lüttge U (ed) Vascular plants as epiphytes: evolution and ecophysiology. Springer, Berlin Heidelberg New York, pp 139–166

Griffiths H (1988) Crassulacean acid metabolism: a re-appraisal of physiological plasticity in form and function. Adv Bot Res 15: 43–92

Griffiths H, Smith JAC (1983) Photosynthetic pathways in the Bromeliaceae of Trinidad: relations between life-forms, habitat preference and the occurrence of CAM. Oecologia 60: 176–184

Griffiths H, Ong BL, Avadhani PN, Goh CJ (1989) Recycling of respiratory CO_2 during crassulacean acid metabolism: alleviation of photoinhibition in *Pyrrosia piloselloides*. Planta 179: 115–122

Heyne B (1815) On the deoxidation of the leaves of *Cotyledon calycina*. Trans Linn Soc Lond 11: 213–215

Keeley JE (1981) *Isoetes howellii*: a submerged aquatic CAM plant? Am J Bot 68: 420–424

Keiller DR, Slocombe SP, Cockburn W (1994) Analysis of chlorophyll *a* fluorescence in C_3 and CAM forms of *Mesembryanthemum crystallinum*. J Exp Bot 272: 325–334

Kluge M, Ting IP (1978) Crassulacean acid metabolism. Analysis of an ecological adaptation. Springer, Berlin Heidelberg New York

Leegood RC (1993) Carbon dioxide-concentrating mechanisms. In: Lea PJ, Leegood RC (eds) Plant biochemistry and molecular biology. Wiley, Chichester, pp 47–72

Leegood RC, Osmond CB (1990) The flux of metabolites in C4 and CAM plants. In: Dennis DT, Turpin DH (eds) Plant physiology, biochemistry and molecular biology. Longman, Harlow, pp 274–298

Lüttge U (1987) Carbon dioxide and water demand; crassulacean acid metabolism (CAM), a versatile ecological adaptation exemplifying the need for integration in ecophysiological work. New Phytol 106: 593–629

Lüttge U (1988) Day-night changes of citric-acid levels in crassulacean acid metabolism: phenomenon and ecological significance. Plant Cell Environ 11: 445–451

Lüttge U (1993) The role of crassulacean acid metabolism (CAM) in the adaptation of plants to salinity. New Phytol 125: 59–71

Lüttge U, Smith JAC (1988) CAM plants. In: Baker DA, Hall JL (eds) Solute transport in plant cells and tissues. Longman, Harlow, pp 417–452

Medina E (ed) (1984) Physiological ecology of CAM plants. CIET (Unesco-IVIC), Caracas

Medina E (1987) Aspectos ecológicas de plantas CAM en los trópicos. Rev Biol Trop 35 (Suppl): 55–70

Mooney H, Troughton JH, Berry JA (1974) Arid climates and photosynthetic systems. Carnegie Inst Wash Year Book 73: 793–805

Nimmo GA, Nimmo HG, Fewson CA, Wilkins MB (1984) Diurnal changes in properties of phosphoenolpyruvate carboxylase in *Bryophyllum* leaves: a possible covalent modification. FEBS Lett 178: 199–203

Nobel PS (1988) Environmental biology of agaves and cacti. Cambridge University Press, Cambridge

Nobel PS (1991) Physicochemical and environmental plant physiology. Academic Press, San Diego

Nuernbergk EL (1961) Endogener Rhythmus und CO_2 Stoffwechsel bei Pflanzen mit diurnalem Säurerhythmus. Planta 56: 28–70

Osmond CB (1978) Crassulacean acid metabolism: a curiosity in context. Annu Rev Plant Physiol 29: 379–414

Osmond CB (1982) Carbon cycling and stability of the photosynthetic apparatus in CAM. In: Ting IP, Gibbs M (eds) Crassulacean acid metabolism. American Society of Plant Physiologists, Rockville, pp 112–127

Osmond CB, Holtum JAM (1981) Crassulacean acid metabolism. In: Hatch MD, Boardman NK (eds) The biochemistry of plants, vol 8, photosynthesis. Academic Press, New York, pp 283–328

Osmond CB, Allaway WG, Sutton BG, Troughton JH, Queiroz O, Lüttge U, Winter K (1973) Carbon isotope discrimination in photosynthesis of CAM plants. Nature 246: 41–42

Osmond CB, Winter K, Ziegler H (1982) Functional significance of different pathways of CO_2 fixation in photosynthesis. In: Lange OL, Nobel PS, Osmond CB, Ziegler H (eds) Physiological plant ecology II, water relations and carbon assimilation. Encyclopedia of plant physiology, new series, vol 12B. Springer, Berlin Heidelberg New York, pp 479–547

Osmond CB, Adams WW III, Smith SD (1991) Crassulacean acid metabolism. In: Pearcy RW, Ehleringer JR, Mooney HA, Rundel PW (eds) Plant physiological ecology. Chapman & Hall, London, pp 255–280

Ranson SL, Thomas M (1960) Crassulacean acid metabolism. Annu Rev Plant Physiol 11: 81–110

Raven JA, Handley LL, MacFarlane JJ, McInroy S, McKenzie L, Richards JH, Samuelsson G (1988) The role of CO_2 uptake by roots and CAM acquisition of inorganic C by plants of the isoetid life form: a review, with new data on *Eriocaulon decangulare* L. New Phytol 108: 125–148

Schmitt AK, Lee HSJ, Lüttge U (1988) The response of the C_3-CAM tree, *Clusia rosea*, to light and water stress. I. Gas exchange characteristics. J Exp Bot 39: 1581–1590

Smith JAC (1989) Epiphytic bromeliads. In: Lüttge U (ed) Vascular plants as epiphytes: evolution and ecophysiology. Springer, Berlin Heidelberg New York, pp 109–138

Smith JAC, Bryce JH (1992) Metabolite compartmentation and transport in CAM plants. In: Tobin AK (ed) Plant organelles. Cambridge University Press, Cambridge, pp 141–167

Spalding MH, Stumpf DK, Ku MSB, Burris RH, Edwards GE (1979) Crassulacean acid metabolism and diurnal variations of internal CO_2 and O_2 concentrations in *Sedum praealtum* DC. Aust J Plant Physiol 6: 557–567

Teeri JA, Tonsor SJ, Turner M (1981) Leaf thickness and carbon isotope composition in the Crassulaceae. Oecologia 50: 367–369

Thomas M (1947) Plant physiology, 3rd edn. J & A Churchill, London

Thomas M (1949) Physiological studies on acid metabolism in green plants. I. CO_2 fixation and CO_2 liberation in crassulacean acid metabolism. New Phytol 48: 390–420

Thomas M, Beevers H (1949) Physiological studies on acid metabolism in green plants. II. Evidence of CO_2 fixation in *Bryophyllum* and the study of diurnal variation of acidity in this genus. New Phytol 48: 421–447

Thurlow J, Bonner J (1948) Fixation of atmospheric CO_2 in the dark by leaves of *Bryophyllum*. Arch Biochem 19: 509–511

Ting IP (1985) Crassulacean acid metabolism. Annu Rev Plant Physiol 36: 595–622

Ting IP, Gibbs M (eds) (1982) Crassulacean acid metabolism. American Society of Plant Physiologists, Rockville

Ting IP, Lord EM, Sternberg LSL, DeNiro MJ (1985) Crassulacean acid metabolism in the strangler *Clusia rosea* Jacq. Science 229: 969–971

Tinoco Ojanguren C, Vásquez-Yánes C (1983) Especies CAM en la selva húmeda tropical de los Tuxtlas, Veracruz. Bol Soc Bot Méx 45: 150–153
Walker D (1992) Energy, plants and man. Oxygraphics Limited, Brighton
Warburg O (1886) Über die Bedeutung der organischen Säuren für den Lebensprozess der Pflanzen (speziell der sog. Fettpflanzen). Untersuchungen Bot Inst Tübingen 2: 53–150
Webb DR, Rattray MR, Brown JMA (1988) A preliminary survey for crassulacean acid metabolism (CAM) in submerged aquatic macrophytes in New Zealand. NZ J Mar Freshwater Res 22: 231–235
Winter K (1982) Properties of phosphoenolpyruvate carboxylase in rapidly prepared, desalted leaf extracts of the crassulacean acid metabolism plant *Mesembryanthemum crystallinum* L. Planta 154: 298–308
Winter K (1985) Crassulacean acid metabolism. In: Barber J, Baker NR (eds) Photosynthetic mechanisms and the environment. Elsevier, Amsterdam, pp 329–387
Winter K, von Willert DJ (1972) NaCl-induzierter Crassulaceensäurestoffwechsel bei *Mesembryanthemum crystallinum*. Z Pflanzenphysiol 67: 166–170
Winter K, Schramm MJ (1986) Analysis of stomatal and nonstomatal components in the environmental control of CO_2 exchange in leaves of *Welwitschia mirabilis*. Plant Physiol 82: 173–178
Winter K, Osmond CB, Pate JS (1981) Coping with salinity. In: Pate JS, McComb AJ (eds) The biology of Australian plants. University of Western Australia Press, Nedlands, pp 88–113
Winter K, Wallace BJ, Stocker GC, Roksandic Z (1983) Crassulacean acid metabolism in Australian vascular epiphytes and some related species. Oecologia 57: 129–141
Winter K, Medina E, Garcia V, Mayoral ML, Muniz R (1985) Crassulacean acid metabolism in roots of a leafless orchid, *Campylocentrum tyrridion* Garay & Dunsterv. J Plant Physiol 118: 73–78
Winter K, Osmond CB, Hubick KT (1986) Crassulacean acid metabolism in the shade. Studies on an epiphytic fern, *Pyrrosia longifolia*, and other rainforest species from Australia. Oecologia 68: 224–230
Winter K, Lesch M, Diaz M (1990) Changes in xanthophyll-cycle components and in fluorescence yield in leaves of a crassulacean-acid-metabolism plant, *Clusia rosea* Jacq., throughout a 12-hour photoperiod of constant irradiance. Planta 182: 181–185
Wolf J (1960) Der diurnale Säurerhythmus. In: Ruhland W (ed) Encyclopedia of plant physiology, vol 12. Springer, Berlin Göttingen Heidelberg, pp 809–889

Part A
Biochemistry of Carbon Flow During Crassulacean Acid Metabolism

Preface

Building on the earlier work of Lavoisier with animals, de Saussure carried out extensive measurements at the beginning of the nineteenth century of respiratory quotients in plants. De Saussure was unable to explain the simultaneous uptake of CO_2 and O_2 when he placed the cactus *Opuntia* in the dark in an atmosphere enriched in CO_2, but this was in fact the first direct demonstration of nocturnal CO_2 fixation by CAM plants (Edwards and Walker 1983; Walker 1992). Nearly two centuries later, O_2 and CO_2 exchanges during the different phases of the CAM cycle can be studied very precisely using gas-phase electrodes. This approach yields surprisingly high O_2: CO_2 stoichiometries associated with light-dependent CO_2 assimilation (Chap. 1). The origin of these values may only be fully understood when the polarographic techniques can be combined with simultaneous measurements of instantaneous isotope discrimination.

The substrate for dark CO_2 fixation in CAM plants is PEP, which is derived from carbohydrates synthesized during daytime photosynthesis. As a consequence, carbohydrate levels change in a reciprocal manner to malic acid during the diel rhythm. However, CAM plants differ in the principal carbohydrate stored during the light period, some accumulating starch and others soluble sugars. Chapter 2 discusses the metabolic implications of these different forms of carbohydrate storage during the CAM cycle.

The next chapters deal with two fundamental and interrelated processes in CAM: the regulation of PEPC by malate, and the regulation of malate fluxes between cytosol and vacuole. CAM is basically a single-cell phenomenon. The two carboxylating enzymes, PEPC and RUBISCO, occur in the same cell and are not restricted to different tissues as in C_4 plants. But to avoid the futile cycling of CO_2 produced by malate decarboxylation back into malate, PEPC activity must be suppressed during the daytime. Early work with *Mesembryanthemum crystallinum* demonstrated that PEPC exists in two states during the 24-h cycle (Greenway et al. 1978) and that PEPC is more sensitive to feedback inhibition by malate (and has a lower affinity for PEP) during the light period (Winter 1982). The mechanism underlying the modulation of PEPC activity is now known to be phosphorylation of the PEPC protein at night, followed by its dephosphorylation during the light period (Chap. 3).

At night, accumulation of malic acid in the large central vacuole of the chlorenchyma cells is essential for pH homeostasis and maintenance of normal cytoplasmic metabolism. During the light period, the direction of malic-acid flow is reversed to release malate for decarboxylation in the cytoplasm. Chapter 4 discusses our current knowledge of nocturnal malic-acid accumulation in the vacuole, a process that is energized by one or both of the tonoplast H^+ pumps, the H^+-ATPase and H^+-PP_iase. Malate transport into and out of the vacuole probably occurs via separate pathways, and mechanisms are presumed to exist for the regulation of influx and efflux at the appropriate times of the day–night cycle. The next chapter (Chap. 5) expands on the possible role of changes in the physicochemical properties of the tonoplast in controlling malate fluxes when daily temperature fluctuations are extreme.

Most of the information on intracellular transport during the CAM cycle has been derived from studies with isolated membranes or organelles. Very little information is available on fluxes through the CAM pathway in vivo, but Chapter 6 discusses the novel application of pH-indicating dyes to monitor real-time changes in the activity of the tonoplast H^+ pumps during the day–night cycle. Combined with measurements of net CO_2 uptake and chlorophyll *a* fluorescence, this approach is starting to reveal how malic-acid efflux from the vacuole is co-ordinated with malate decarboxylation and photosynthetic CO_2 reduction in the light. The last chapter of the first part of the book (Chap. 7) describes the use of measurements of chlorophyll *a* fluorescence and of leaf carotenoids to investigate photochemical responses in CAM plants to high levels of PPFD in excess of those required for photosynthesis. Evidence is summarized for the involvement of the so-called xanthophyll cycle (Yamamoto 1979; Hager 1980) in the protection of the photosynthetic apparatus of CAM plants from damage by excess excitation energy.

References

Edwards G, Walker DA (1983) C_3, C_4: mechanisms, and cellular and environmental regulation, of photosynthesis. Blackwell, Oxford

Greenway H, Winter K, Lüttge U (1978) Phosphoenolpyruvate carboxylase during development of crassulacean acid metabolism and during a diurnal cycle in *Mesembryanthemum crystallinum*. J Exp Bot 29: 547–559

Hager A (1980) The reversible, light-induced conversions of xanthophylls in the chloroplast. In: Czygan FC (ed) Pigments in plants. Fischer, Stuttgart, pp 57–79

Walker D (1992) Energy, plants and man. Oxygraphics Limited, Brighton

Winter K (1982) Properties of phosphoenolpyruvate carboxylase in rapidly prepared, desalted leaf extracts of the crassulacean acid metabolism plant *Mesembryanthemum crystallinum* L. Planta 154: 298–308

Yamamoto HY (1979) Biochemistry of the violaxanthin cycle in higher plants. Pure Appl Chem 51: 639–648

1 Stoichiometric Nightmares: Studies of Photosynthetic O_2 and CO_2 Exchanges in CAM Plants

C.B. Osmond[1], M. Popp[2], and S.A. Robinson[1]

1.1 Introduction

The unusual gas-exchange stoichiometries of succulent plants which engage in crassulacean acid metabolism (CAM) have been recognized since de Saussure (1804), whose observations are best accessible today via Thomas (1949). In beautifully crafted sentences, sometimes exceeding 100 words, Thomas reviewed the history of gas-exchange experimentation in CAM plants. He left his indelible mark on the dark reactions of CAM with the words "The positive contribution made by the Leipzig workers was to draw attention to the possible cardinal importance (cf. Gustafson's qualified statement) of carbon dioxide rather than oxygen in controlling diurnal fluctuations in the acidity of the green cells of plants showing crassulacean acid metabolism. The present writer felt that this possibility became a probability when he had considered the experimental results obtained at Leipzig and elsewhere in the light of Wood and Werkman's discovery". There is no clearer, and more generous, expression of the origins of the hypothesis that led us to an understanding of the mechanisms of dark CO_2 fixation and malic-acid synthesis in CAM plants. Appreciation of photosynthetic CO_2 fixation in CAM plants is less clear, historically. Bennet-Clark (1933) attributed the recognition of malate decarboxylation as an internal CO_2 source for photosynthesis to A. Mayer (1878, 1899) and to H.A. Spoehr (1913) (cited in Bennet-Clark 1933).

Quantitative studies of gas exchanges in CAM improved with the use of modified Warburg manometers (Wolf 1939), the Ambler apparatus (Thomas and Beevers 1949), and infra-red gas analyzers (Gregory et al. 1954). Subsequent applications of radioactive and stable isotopes have gone a long way to elucidating the detailed mechanisms of carboxylation reactions and oxygen-exchange processes. However, although four distinct phases of gas exchange behaviour have been identified using these techniques (Osmond 1976; Griffiths et al. 1990), difficulties remain in the measurement and interpretation of component O_2 and CO_2 exchanges.

[1]Research School of Biological Sciences, Institute of Advanced Studies, Australian National University, Box 475, Canberra 2601, Australia
[2]Institut für Pflanzenphysiologie, Universität Wien, Althanstr. 14, A-1091 Wien, Austria

Ecological Studies, Vol. 114
Winter/Smith (eds) Crassulacean Acid Metabolism

In the present paper we wish to describe contemporary approaches to the two-century-old, unresolved problems of photosynthetic gas exchange in CAM plants. We shall describe a combined O_2/CO_2 electrode system which makes simultaneous measurements possible in all phases of CAM. However, we conclude that it is still necessary to bring the full arsenal of modern technologies, especially mass spectrometry and chlorophyll fluorescence quenching analysis, to bear on gas exchange in CAM if we are to understand this all-options photosynthetic process.

1.2 Simultaneous Measurements of O_2 and CO_2 Exchange Using an O_2/CO_2 Electrode System

Net gas exchange in leaf discs (10 cm^2) of *Kalanchoë daigremontiana* was measured at different times of the day and night. The lower epidermis of the leaf discs was removed, usually in a single piece, to help facilitate speedy gas exchange. This treatment had no other effects on gas-exchange processes. Discs were placed in thermoregulated chambers at 17 or 25 °C (Model LD2) fitted with an O_2 electrode and readout device (Model CB 1D), all from Hansatech Ltd., Kings Lynn, Norfolk, UK. The chambers were modified to accept gas-phase CO_2 electrodes (Model MI-720) used with specially manufactured readout devices (Model CM-1), both from Microelectrodes Inc., Londonderry, NH, USA. The O_2 electrodes and chamber volumes were calibrated by injection of air as described by Walker (1988), and the CO_2 electrodes were calibrated by passing water-saturated standard gas mixtures containing 1% and 10% (v/v) CO_2 through the chamber. Illumination was provided by a red LED light source (Hansatech LC1-UB36) which delivered saturating light (approx. 1500 μmol m^{-2} s^{-1}) to the surface of the leaf disc. Differences in the heat-transfer characteristics of different chambers meant that tissue temperature increased by 3 to 5 °C following illumination in some cases. The CO_2 electrodes showed somewhat slower responses than the O_2 electrodes, and it was usually necessary to make observations during successive 10-min cycles of light and dark to obtain steady rates in particular treatments.

1.3 Photosynthetic O_2/CO_2 Stoichiometry During C_3 Photosynthesis in Phase IV

Well-irrigated, constitutive CAM plants such as *K. daigremontiana* engage in C_3 photosynthesis during the late light period (Phase IV: Osmond 1976; Griffiths et al. 1990). Net O_2 and CO_2 exchanges in the light and dark in Phase IV are shown in Table 1. If we assume that respiratory O_2 uptake and CO_2 evolution continue during photosynthesis, then the stoichiometry of O_2/CO_2 exchange during Phase IV is 1.4 $\pm$ 0.3 in these experiments, done with leaf discs from different plants grown in different conditions and measured with different sets of apparatus (Table 1.1).

However, it should be noted that the stoichiometry of O_2/CO_2 exchange during dark respiration in Phase IV is 1.9 ± 0.4, indicating that some CO_2 is refixed by phospho*enol*pyruvate carboxylase (PEPC) which becomes active during Phase IV (Kluge 1968; Osmond and Allaway 1974). If we correct the respiratory CO_2 output to that of O_2 evolution, and assume that recycling of CO_2 via PEPC continues during photosynthesis, then the photosynthetic stoichiometry in Phase IV becomes 1.3 ± 0.2.

This value is similar to that observed in C_3 plants at CO_2- and light-saturation in leaves which are well supplied with NO_3^- (S.A. Robinson, unpubl.). Because this stoichiometry of C_3 photosynthesis reduces to unity in the presence of inhibitors of NH_4^+ incorporation, we suspect that the excess of O_2 evolution over CO_2 fixation reflects electron transport required to generate the additional ATP consumed in light-dependent NO_3^- assimilation. Stimulation of photo-synthetic O_2 evolution by NO_3^- has been demonstrated in peas and barley (de la Torre et al. 1991). Photosynthetic CO_2 incorporation in CAM plants during Phase IV may display similar excess ATP demand, especially if CO_2 fixed via PEPC into malate enters a futile cycle via decarboxylation and resynthesis of

Table 1.1. Photosynthetic and respiratory O_2/CO_2 exchange in leaf discs of *Kalanchoë daigremontiana* during Phase IV of the diurnal cycle performed with plant material cultivated in Beaufort (USA) and Canberra (Australia). Measurements were done with 5% (v/v) CO_2

Treatment	O_2 evolution (μmol m^{-2} s^{-1})		CO_2 uptake (μmol m^{-2} s^{-1})		O_2/CO_2	
	Dark	Light	Dark	Light	Dark	Light[a]
30 °C Beaufort						
8 h light	−2.4	12.6	−1.0	10.2	2.4	1.3(1.2)
	−2.6	17.5	−0.9	12.5	2.9	1.5(1.3)
	−2.4	15.5	−1.1	10.2	2.2	1.6(1.4)
9 h light	−2.7	15.9	−2.1	10.9	1.3	1.4(1.4)
	−2.1	18.4	−1.8	11.0	1.2	1.6(1.6)
	−2.6	20.4	−1.4	11.4	1.9	1.8(1.6)
28 °C Canberra						
10 h light	−1.9	8.0	0	6.0	–	1.7(1.3)
	−1.9	8.6	0	7.3	–	1.4(1.1)
	−2.0	11.4	0	8.7	–	1.5(1.3)
11.5 h light	−1.6	6.6	−1.0	3.7	1.6	1.7(1.5)
	−2.1	6.7	−1.0	3.6	2.1	1.9(1.5)
	−2.2	7.3	−1.1	3.7	2.0	2.0(1.6)
17 °C Canberra						
7 h light	−0.9	0.5	−0.5	0.5	1.8	1.4(1.0)
	−1.0	0.5	−0.6	0.7	1.7	1.2(0.9)
	−1.0	0.8	−0.6	0.9	1.7	1.2(0.9)

[a]Values for the stoichiometry of O_2/CO_2 exchange in the light were calculated on the assumption that the observed rates of respiratory O_2 uptake and CO_2 evolution continued in the light. Numbers in parentheses are those calculated assuming a respiratory stoichiometry of O_2/CO_2 exchange of unity.

PEP from pyruvate via pyruvate, P_i dikinase, as follows:

$$PEP + CO_2 \longrightarrow OAA + P_i; OAA + NAD(P)H + H^+ \longrightarrow malate + NAD(P)^+ \quad (1.1)$$

$$malate + NAD(P)^+ \longrightarrow pyruvate + CO_2 + NAD(P)H + H^+ \quad (1.2)$$

$$pyruvate + P_i + 2\ ATP \longrightarrow PEP + 2\ P_i + 2\ AMP \quad (1.3)$$

These speculations presuppose that the covalent modification of CAM PEPC (Nimmo et al. 1990; see also Chap. 3), which sensitizes the enzyme to malate, is in itself insufficient to prevent futile cycling in this phase of CAM. Evidently, malate concentration late in Phase IV is insufficient to inhibit PEPC, as indicated by $^{14}CO_2$ incorporation into malate in the light (Kluge 1968; Osmond and Allaway 1974). Studies show that double carboxylation also occurs during Phase IV (Osmond and Allaway 1974; Ritz et al. 1986), and if this leads to futile cycling through malate, pyruvate and PEP, it would create an excess ATP demand. Thus, because the O_2/CO_2 stoichiometry in Phase IV is rather greater than unity, and because it is unlikely that these slow-growing plants engage in rapid nitrate assimilation, we propose that futile cycling through PEPC may be greater than previously thought.

1.4 Photosynthetic O_2/CO_2 Exchanges During Deacidification in Phase III

Phase III is the phase of CAM most analogous to the experimental conditions used in these experiments. During Phase III, malate serves as an internal CO_2 source in much the same way as the carbonate/bicarbonate buffer reservoir does in the leaf-disc electrode system originally devised by Delieu and Walker (1981). Previous studies have demonstrated that extraordinarily high CO_2 concentrations of over 4% (v/v) can build up behind closed stomata during deacidification in Phase III (Cockburn et al. 1979; Spalding et al. 1979). Because stomata close so tightly during Phase III, until now there has been no accurate way to assess the stoichiometry of photosynthetic O_2/CO_2 exchange during deacidification.

Such measurements are potentially able to distinguish the extent of malate decarboxylation during Phase III via two main pathways as follows:

$$malate + NAD(P)^+ \longrightarrow pyruvate + CO_2 + NAD(P)H + H^+ \quad (1.2)$$

$$pyruvate + P_i + 2\ ATP \longrightarrow PEP + 2\ P_i + 2\ AMP \quad (1.3)$$

$$PEP \longrightarrow PGA \longrightarrow PCR\ cycle \quad (1.4)$$

or

$$pyruvate \longrightarrow TCA\ cycle \longrightarrow CO_2 \quad (1.5)$$

We have previously excluded the occurrence of futile cycling [Eqs. (1.1–1.3)] in these CAM plants during Phase III on the basis of ^{13}C distribution in malate (Osmond et al. 1988). However, our recent studies, which show that the alternative oxidase is more actively engaged during Phase III (Robinson et al. 1992), demand that we attempt to evaluate the extent of the complete oxidation pathway (1.5) in these plants.

Qualitative representation of O_2/CO_2 exchange during deacidification (Fig. 1.1) shows that dark/light cycles lead to ever-increasing O_2 concentration in the experimental chamber as malate decarboxylation proceeds. The stoichiometry of this "O_2 pump" can be examined further to assess the pathways of malate decarboxylation. If malate is decarboxylated completely to CO_2, and the CO_2 is refixed via the PCR cycle, then we would expect a stoichiometry of O_2/CO_2 exchange of unity. If, however, malate is only partially decarboxylated to CO_2 and pyruvate, and if pyruvate is largely recovered by gluconeogenesis [Eqs. (1.2–1.4); Holtum and Osmond 1981], a stoichiometry of rather greater than unity would be expected. Recovery of pyruvate as 3-PGA via pyruvate, P_i dikinase and other enzymes requires 2 ATP per pyruvate [Eq. (1.3)], and a further 1.5 ATP to convert 3-PGA to the level of storage carbohydrate (see Chap. 2). The additional ATP required could be generated by cyclic electron transport processes, driven by further photo-oxidation of water and leading to increased O_2 evolution (see Chap. 6).

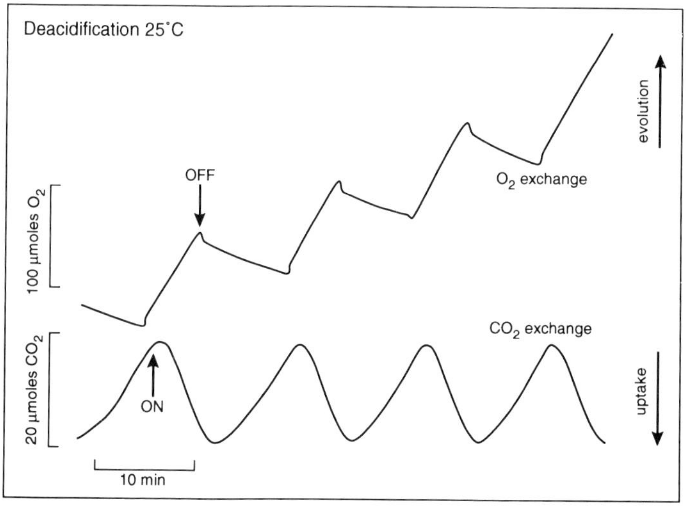

Fig. 1.1. Time course of O_2 and CO_2 exchange by leaf discs of *K. daigremontiana* in the dark and light during Phase III. The CO_2 electrode was calibrated between air and 1% (v/v) CO_2, and the O_2 concentration in the leaf disc chamber was reduced to 5% (v/v) with N_2. The CO_2 evolved in the dark intervals arises largely from deacidification of malate and was produced at rates far in excess of O_2 uptake or of CO_2 evolution in the dark in Phase IV (Beaufort experiment)

Table 1.2 shows photosynthetic stoichiometries obtained during deacidification. They have been calculated assuming that respiratory O_2 uptake continues in the light, and that CO_2 evolution from malate decarboxylation also continues in the light. The rates of CO_2 and O_2 exchange were higher in leaf discs from the plants used in Beaufort (kept in 27 °C/800 µmol photons m^{-2} s^{-1} for 10 h; 15 °C dark for 14 h) than from the plants used in Canberra (glasshouse-grown without temperature control). Nevertheless, it is obvious that the stoichiometry of O_2/CO_2 exchange during deacidification is considerably greater than unity (mean 1.8 ± 0.6). Highest stoichiometries were obtained in experiments at high temperatures and low CO_2 concentrations, and lowest stoichiometries were obtained in experiments with high CO_2 concentrations and low temperatures.

There may be both measurement and biological factors that contribute to these trends. However, we conclude from these experiments that it is unlikely that

Table 1.2. Stoichiometry of photosynthetic O_2 and CO_2 exchanges by leaf discs of *K. daigremontiana* in the light during deacidification in Phase III

Treatment	O_2 evolution (µmol m^{-2} s^{-1})	CO_2 uptake (µmol m^{-2} s^{-1})	O_2/CO_2
30 °C, 1.5 h light	20.4	8.0	2.6
(Beaufort, 0 to 1% CO_2)	21.4	9.1	2.4
3 h light	28.2	11.4	2.5
	29.2	11.0	2.7
4 h light	23.9	8.2	2.9
	24.7	8.5	2.9
30 °C, 2 h light	21.2	16.4	1.3
(Beaufort, 8% CO_2)	22.1	15.0	1.5
3.5 h light	20.2	15.1	1.4
	20.5	15.8	1.3
6 h light	15.8	12.6	1.3
	19.0	13.6	1.4
27 °C, 3 h light	9.4	6.2	1.5
(Canberra, 5% CO_2)	10.9	6.5	1.7
6 h light	9.5	5.0	1.9
	10.0	5.3	1.9
27 °C, 3 h light	7.1	4.4	1.6
	7.3	3.9	1.9
5 h light	8.1	4.6	1.8
	7.4	4.0	1.8
21 °C, 3 h light	22.3	8.5	2.6
(Beaufort, 0 to 1% CO_2)	21.7	9.2	2.4
18 °C, 4 h light	2.7	1.3	2.2
(Canberra, 5% CO_2)	2.2	1.4	1.5
8 h light	1.3	0.9	1.5
	1.4	1.3	1.1
18 °C, 3 h light	2.4	2.0	1.2
(Canberra, 5% CO_2)	2.7	2.5	1.1

malate is completely decarboxylated to CO_2 prior to photosynthesis. Rather our data show that carbon metabolism during deacidification creates a demand for increased photosynthetic O_2 evolution over that expected from photosynthetic CO_2 fixation alone (cf. Tables 1.1, 1.2). This demand may be associated with excess ATP requirements arising from gluconeogenic recovery of the 3-C portion of malate following decarboxylation.

1.5 Photosynthetic O_2/CO_2 Exchanges During Acidification in Phase I

Qualitative and quantitative representations of O_2 and CO_2 exchange during acidification in the early and late parts of the dark period are given in Fig. 1.2. The experimental treatments of successive 10-min dark and light cycles have marked effects on the course of CO_2 exchange but little effect on the exchange of O_2. Early in the dark period when the rate of dark CO_2 uptake is accelerating, successive light treatments which accelerate net CO_2 uptake eventually switch off dark CO_2 uptake altogether (Fig. 1.2a). At the end of the dark period, just before the natural photoperiod begins, successive light treatments have the same effect on dark CO_2 uptake, causing a switch to net CO_2 evolution in the dark, evidently as malate decarboxylation commences (Fig. 1.2b).

These changes in response during treatment, at times which correspond to Phase II and late Phase IV, were not observed throughout the rest of the dark period. Although illumination always elicited O_2 evolution, the induction period for maximum rate of O_2 evolution was much longer than during deacidification (data not shown, but see Chap. 6). It is evident that the activation state of ribulose 1,5-bisphosphate carboxylase/oxygenase (RUBISCO) in the dark in this CAM plant is not as low as in some others (Vu et al. 1984; cf. Servaites et al. 1986) and is not sufficient to retard the PCR cycle and electron transport when tissues are illuminated during Phase I.

The nature of accelerated CO_2 fixation in tissues exposed to illumination at intervals throughout the dark period is not clear. Table 1.3 shows that the extent of stimulation of CO_2 uptake by illumination is much greater at 30 °C or at 27 °C than at 17 °C. The stoichiometry of O_2/CO_2 exchange in the light in these treatments is unusual. If we assume that only the light-stimulated CO_2 uptake is associated with photosynthetic CO_2 fixation (that is, that PEPC-mediated dark CO_2 fixation continues in the 10-min period of illumination), and if we also assume that respiratory O_2 uptake continues in the light, then the O_2/CO_2 stoichiometry of photosynthesis averages 3.7 ± 0.9 (Table 1.4). If the assumption that CO_2 fixation via PEPC continues in the light is invalid, because the light-dependent covalent modification (Nimmo et al. 1990; see also Chap. 3) occurs rapidly and this enzyme is inactivated by malate levels in the cytoplasm, then the O_2/CO_2 stoichiometry of photosynthesis in the above experiments declines somewhat (Table 1.4; mean 2.8 ± 1.0), but remains well above that of C_3 photosynthesis (cf. Table 1.1).

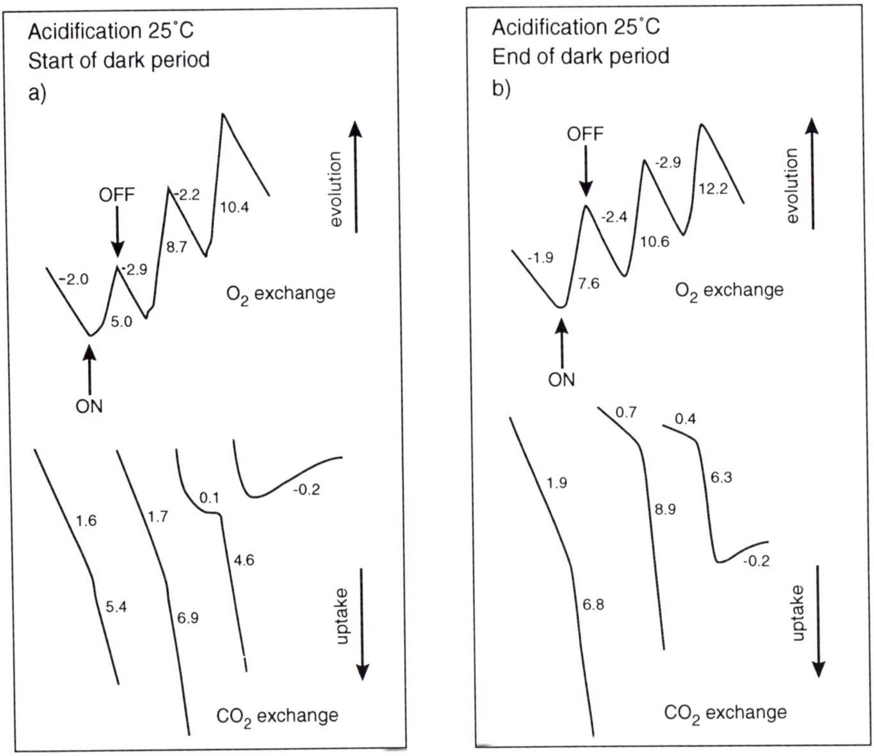

Fig. 1.2. **a** Time course of O_2 and CO_2 exchange by leaf discs of *K. daigremontiana* in the dark and light at the start of Phase I (5%(v/v) CO_2; 20%(v/v) O_2). The rate of CO_2 uptake again is stimulated by illumination, but the experimental treatments lead to an inhibition of CO_2 uptake in the dark, moving the tissue back into Phase IV (Canberra experiments). **b** Time course of O_2 and CO_2 exchange in *K. daigremontiana* in the dark and light at the end of Phase I [5%(v/v) CO_2; 20%(v/v) O_2]. The rate of dark CO_2 uptake declines with successive dark/light treatments as the experimental protocol moves the tissue forward into Phase II (Canberra experiments)

So far as we are aware, there are no labelling studies in which $^{14}CO_2$ or $^{13}CO_2$ have been applied in the light at intervals throughout the dark period which would help assess the relative activities of PEPC and RUBISCO at this time. We tentatively conclude that massive futile cycling through PEPC occurs, creating an extraordinary demand for ATP. Chlorophyll fluorescence quenching analyses (Winter et al. 1990; see also Chap. 6) during these interruptions to the dark period may prove instructive. Earlier studies suggest that fluorescence transients are significantly altered in Phase I (Osmond 1982; Everson et al. 1983; Chap. 6).

Table 1.3. Stimulation of CO_2 uptake when leaf discs of *K. daigremontiana* are illuminated at intervals throughout Phase I

Treatment	CO_2 fixation (μmol m^{-2} s^{-1})		
	Dark	Light	Light/Dark
30 °C, 3 h dark	2.4	5.8	2.4
(Beaufort, 0 to 1% CO_2)	2.1	6.1	2.9
5 h dark	2.1	8.4	4.0
	1.2	6.0	5.0
12 h dark	0.7	3.9	5.6
	0.8	5.7	7.1
27 °C, 4 h dark	3.2	5.1	1.6
(Canberra, 5% CO_2)	3.1	5.6	1.8
7 h dark	3.0	5.5	1.8
	2.9	6.9	2.4
9.5 h dark	2.9	5.9	2.0
	3.7	6.4	1.7
17 °C, 4 h dark	1.8	1.5	0.8
	2.4	2.5	1.0
6 h dark	2.1	2.0	0.9
	2.1	2.5	1.2

Table 1.4. Stoichiometry of photosynthetic O_2 and CO_2 exchanges following illumination of leaf discs of *K. daigremontiana* during acidification in Phase I

Treatment	O_2 evolution (μmol m^{-2} s^{-1})	CO_2 uptake (μmol m^{-2} s^{-1})	O_2/CO_2 [a]	O_2/CO_2 [b]
21 °C, 3 h dark	12.4	3.5	3.5	2.1
(Beaufort, 0 to 1% CO_2)	13.8	4.1	3.4	2.3
5 h dark	17.7	5.2	3.4	2.1
	15.6	4.8	3.3	2.6
12 h dark	12.5	3.2	3.9	3.2
	14.6	5.0	2.9	2.5
13.5 h dark	13.7	5.6	2.4	2.3
	15.5	3.4	4.7	4.2
27 °C, 4 h dark	10.8	1.9	5.7	2.1
(Canberra, 5% CO_2)	11.8	2.5	4.7	2.1
7 h dark	10.6	2.5	4.3	1.9
	14.5	4.0	3.6	2.1
9.5 h dark	7.7	3.0	2.6	1.3
	10.2	2.7	3.8	1.6

[a] Light-dependent CO_2 uptake.
[b] Calculated assuming total CO_2 uptake in the light (data not shown) is driven by photosynthetic electron transport.

1.6 Conclusions

It seems obvious that we need greater precision than can be afforded by net O_2- and CO_2-exchange measurements if we are to interpret the stoichiometry of O_2/CO_2 exchange in photosynthesis of CAM plants. Although natural-abundance isotope methods are useful for distinguishing the participation of photosynthetic carboxylases (Griffiths et al. 1990) and respiratory oxidases (Robinson et al. 1992), it is not clear whether these methods can contribute much more to our understanding of stoichiometric problems. On the other hand, gas-exchange studies using atmospheres enriched in ^{18}O and ^{13}C may have much to offer. These methods have been applied to CAM plants by Thomas and André (1987) and Thomas et al. (1987) with limited success. If we are to resolve processes in leaves or cladodes, at different phases of the 24-h cycle, in response to short-term changes in conditions, then the large-chamber, whole-plant studies undertaken by these authors are unsuitable. Rather, a combination of the O_2/CO_2 electrode system used here and the Kok inlet system used previously (Canvin et al. 1980) seems to hold promise.

It would be inexcusable to conclude without mention of *Clusia*. The unusually flexible CAM in these tropical stranglers (Ting et al. 1985; Popp et al. 1987; Franco et al. 1990; Chaps. 19 and 20) adds the fascinating dimension, still unexplained, of massive nocturnal citric-acid synthesis. Whether this acid arises by an unexplored carboxylation, such as the reversal of the isocitrate dehydrogenase-aconitase reaction sequence, or is simply withdrawn from the TCA cycle after condensation of OAA, derived from CO_2 fixation via PEPC, and acetyl CoA (Lüttge 1988), should be readily distinguishable by examination of ^{13}C or ^{14}C distribution in citrate after fixation of labelled CO_2. In the latter, less novel case, carboxyls 1C and 5C of the symmetrical molecule will be labelled. The carboxyl attached to 3C will label depending on the extent of 4C↔1C randomization in OAA and malate. In the former case, CO_2 fixation into 2-oxoglutarate will label only the carboxyl attached to 3C of the C_5 chain of isocitrate, aconitate and citrate. It remains to be seen whether these predictions can be evaluated using GCMS or other techniques, and whether the latter carboxylation reaction contributes yet another source of stoichiometric nightmares in these remarkable CAM plants.

Acknowledgements. This work was supported in part by an NSF grant DCB-90-06830 to CBO, and an ANU Visiting Fellowship and travel grant from the Deutsche Forschungsgemeinschaft to MP. One of the authors (CBO) is grateful to Dr. Joe Ramus, Director, Duke Marine Laboratory, for research hospitality during the early part of this work.

References

Bennet-Clark TA (1933) The role of organic acids in plant metabolism II. New Phytol 32: 128–161

Canvin DT, Berry JA, Badger MR, Fock H, Osmond CB (1980) Oxygen exchange in leaves in the light. Plant Physiol 66: 302–307

Cockburn W, Ting IP, Sternberg L (1979) Relationships between stomatal behavior and internal carbon dioxide concentration in crassulacean acid metabolism plants. Plant Physiol 63: 1029–1032
De la Torre A, Delgardo B, Lara C (1991) Nitrate dependent O_2 evolution in intact leaves. Plant Physiol 96: 898–901
De Saussure T (1804) Recherches chimiques sur la végétation. Chez la V.[e] Nyon, Paris
Delieu TD, Walker DA (1981) Polarographic measurement of photosynthetic O_2 evolution by leaf discs. New Phytol 89: 165–175
Everson G, Chen S-S, Black CC (1983) Diurnal variations in leaf fluorescence induction kinetics: variable fluorescence in crassulacean acid metabolism plants. Plant Physiol 72: 455–460
Franco AC, Ball E, Lüttge U (1990) Patterns of gas exchange and organic acid oscillations in tropical trees of the genus *Clusia*. Oecologia 85: 108–114
Gregory FG, Spear I, Thimann KV (1954) The interrelation between CO_2 metabolism and photoperiodism in *Kalanchoë*. Plant Physiol 29: 220–229
Griffiths H, Broadmeadow MST, Borland AM, Hetherington CS (1990) Short term changes in carbon-isotope discrimination identify transitions between C_3 and C_4 carboxylation during crassulacean acid metabolism. Planta 181: 604–610
Holtum JAM, Osmond CB (1981) The gluconeogenic metabolism of pyruvate during deacidification in plants with crassulacean acid metabolism. Aust J Plant Physiol 8: 31–44
Kluge M (1968) Veränderliche Markierungsmuster bei $^{14}CO_2$-Fütterung von *Bryophyllum tubiflorum* zu verschiedenen Zeitpunkten der Hell/Dunkelperiode. 1. Die $^{14}CO_2$-Fixierung unter Belichtung. Planta 88: 113–129
Lüttge U (1988) Day-night changes in citric acid levels in crassulacean acid metabolism: phenomenon and ecophysiological significance. Plant Cell Environ 11: 445–451
Nimmo HG, Carter PJ, Fewson CA, McNaughton GAL, Nimmo GA, Wilkins MB (1990) Regulation of phosphoenolpyruvate carboxylase: an example of a signal transduction via protein phosphorylation in higher plants. Adv Enzyme Regul 30: 121–131
Osmond CB (1976) CO_2 assimilation and dissimilation in the light and dark in CAM plants. In: Burris RH, Black CC (eds) CO_2 Metabolism and plant productivity. University Park Press, Baltimore, pp 217–233
Osmond CB (1982) Carbon cycling and the stability of the photosynthetic apparatus in CAM. In: Ting IP, Gibbs M (eds) Crassulacean acid metabolism. American Society of Plant Physiologists, Rockville, pp 112–127
Osmond CB, Allaway WG (1974) Patterns of CO_2 fixation in the CAM plant *Kalanchoë daigremontiana* I. Patterns of $^{14}CO_2$ fixation in the light. Aust J Plant Physiol 1: 503–511
Osmond CB, Holtum JAM, O'Leary MH, Roeske C, Wong OC, Summons RE, Avadhani PN (1988) Regulation of malic acid metabolism in crassulacean acid metabolism plants in the dark and light: in vivo evidence from ^{13}C labelling patterns after $^{13}CO_2$ fixation. Planta 175: 184–192
Popp M, Kramer D, Lee H, Diaz M, Ziegler H, Lüttge U (1987) Crassulacean acid metabolism in the tropical dicotyledonous trees of the genus *Clusia*. Trees 1: 238–247
Ritz D, Kluge M, Veith HJ (1986) Mass-spectrometric evidence for the double carboxylation pathway of malate synthesis by crassulacean acid metabolism plants in the light. Planta 167: 284–291
Robinson SA, Yakir D, Ribas-Carbo M, Giles L, Osmond CB, Siedow JN, Berry JA (1992) Measurements of the engagement of cyanide resistant respiration in the crassulacean acid metabolism plant *Kalanchoë daigremontiana* with the use of on-line oxygen-isotope discrimination. Plant Physiol 100: 1087–1091
Servaites JC, Parry MAJ, Gutteridge S, Keys AJ (1986) Species variation in the predawn inhibition of ribulose-1, 5-bisphosphate carboxylase/oxygenase. Plant Physiol 82: 1161–1163
Spalding MH, Stumpf DK, Ku MSB, Burris RH, Edwards GE (1979) Crassulacean acid metabolism and diurnal variations of internal CO_2 and O_2 concentrations in *Sedum praealtum* DC. Aust J Plant Physiol 6: 557–567
Thomas DA, Andre M (1987) Oxygen and carbon dioxide exchanges in crassulacean acid metabolism plants I. Effects of water stress on hourly and daily patterns. Plant Physiol Biochem 25: 85–93

Thomas DA, André M, Ganzin A-M (1987) Oxygen and carbon dioxide exchanges in crassulacean acid metabolism plants: II. Effects of CO_2 concentration and irradiance. Plant Physiol Biochem 25: 95–103

Thomas MD (1949) Physiological studies on acid metabolism in green plants. I. CO_2 fixation and CO_2 liberation in crassulacean acid metabolism. New Phytol 48: 390–420

Thomas MD, Beevers H (1949) Physiological studies on acid metabolism in green plants. II. Evidence of CO_2 fixation in *Bryophyllum* and the study of diurnal variation of acidity in this genus. New Phytol 48: 421–447

Ting IP, Sternberg LSL, De Niro MJ (1985) Crassulacean acid metabolism in the strangler *Clusia rosea* Jacq. Science 229: 969–971

Vu JCV, Allen LH, Bowes G (1984) Dark/light modulation of ribulose bisphosphate carboxylase activity in plants from different photosynthetic categories. Plant Physiol 76: 843–845

Walker DA (1988) The use of the oxygen electrode and fluorescence probes in simple measurements of photosynthesis. Oxygraphics, Sheffield

Winter K, Lesch M, Diaz M (1990) Changes in xanthophyll cycle components and in fluorescence yield in leaves of a crassulacean acid metabolism plant, *Clusia rosea* Jacq., throughout a 12 h photoperiod of constant irradiance. Planta 182: 181–185

Wolf J (1939) Beiträge zur Kenntnis des Säurestoffwechsels succulenter Crassulaceen V. Mikrorespiratorische Untersuchungen an Blattgeweben von *Bryophyllum calycinum*. Planta 29: 450–467

2 Alternative Carbohydrate Reserves Used in the Daily Cycle of Crassulacean Acid Metabolism

C.C. Black, J.-Q. Chen, R.L. Doong, M.N. Angelov, and S.J.S. Sung

2.1 Introduction

Each day a massive interlocked biochemical cycle occurs in the green tissues of crassulacean acid metabolism plants. The function of this interlocked cycle, in its simplest context, is to furnish most of the CO_2 for CAM plant photosynthesis. In addition, this diel (24 h) cycle produces the primary identifying marks of a CAM tissue through two ancillary cycles. One cycle involves a nocturnal acidification and its loss the next day, while the second concerns the depletion of a carbohydrate reserve at night and its replenishment the next day. Formally Benjamin Heyne (1815) is credited with writing, nearly two centuries ago, about the "acid as sorrel" taste of a succulent green plant at dawn and the "bland taste" caused by acidity loss later in the day. In fact, the exact origins of these observations are lost in antiquity, but certainly are referred to in Roman and Biblical writings. The circumstantial cause of the acidity was postulated to be an organic acid about a century ago and the bland taste later was associated with starch; but these ideas were not plainly coupled together in theory nor quantitatively studied until the late 1940s. Then, with the discovery of major portions of intermediary metabolism and the advent of additional quantitative biochemical procedures, the nature of the daily reciprocal relation between the acid and the bland taste was recognized and measured quantitatively. The acid taste is caused principally by malic acid, while the bland taste is caused by deacidification plus the reciprocal synthesis of a bland tasting carbohydrate, e.g. a polysaccharide such as starch.

Other daily ancillary cycles, e.g. CO_2 and O_2 exchange, stomatal functions, an internal pool of CO_2, etc. also exist as integral parts of CAM (Kluge and Ting 1978; Ting and Gibbs 1982; Winter 1985). The focus of this work, however, is on the type of carbohydrate and how each is metabolized in certain CAM plants when it functions as the daily carbon reservoir to provide the phospho*enol*-pyruvate (PEP) for nocturnal CO_2 fixation and organic acid synthesis. Today alternative pathways of intermediary carbohydrate metabolism in plants are known (Sung et al. 1988) and alternative carbohydrate reserves are recognized in specific CAM species (Black et al. 1982) which can be either a polysaccharide or a

Department of Biochemistry, Life Sciences Building, The University of Georgia, Athens, GA 30602, USA

Ecological Studies, Vol. 114
Winter/Smith (eds) Crassulacean Acid Metabolism

neutral hexose sugar. Unfortunately, these disparate bodies of knowledge have not been fully integrated with CAM. Therefore, in this unified presentation our aims are (1) to divide CAM plants into two metabolic groups, (2) to document the use of different carbohydrate reserves by each group, (3) to integrate the unique biochemical reactions within each group into characteristic sets of metabolic pathways, thereby depicting two metabolic sequences of carbon cycling in CAM, and (4) to compare the bioenergetics and other features of these two metabolic groups.

2.2 The Division of CAM Plants into Two Metabolic Groups

The division of CAM plants into distinct metabolic groups was proposed when two types of C_4-acid decarboxylases were found in CAM plants. Malic enzyme was first assayed in certain CAM plants in the mid 1950s, but it was not widely studied nor considered to act as a decarboxylase until the late 1960s, only several years after the 1965 discovery of C_4 photosynthesis (Black 1973). Thus in 1973 when a second decarboxylase, PEP carboxykinase, was found to be very active in specific CAM species, it was clear that two large groups of CAM plants existed (Dittrich et al. 1973). The details of these separate metabolic conversions were not clear, but, over the intervening two decades, new information about carbon metabolism has been found in higher plants and distinct metabolic variations have been discovered in other CAM plants.

Table 2.1 is a condensed presentation of earlier work on the taxonomic distribution of some carbon metabolism enzymes among CAM plants. In the initial work classifying CAM plants into two groups, the activities of PEP carboxykinase and pyruvate, P_i dikinase were used as the basis of division (Dittrich et al. 1973). Pyruvate, P_i dikinase shows a taxonomic pattern complementary to PEP carboxykinase (Table 2.1). But some enzymes were active in all CAM plants, however, with elevated activity in one group. For example, both types of malic enzyme tended to be expressed more in plants without PEP carboxykinase, whereas the pyrophosphate-dependent phosphofructokinase (PFK) tended to be more active in plants containing a highly active PEP carboxykinase. Even with these limitations, data as in Table 2.1 gave a reasonable basis for the division.

2.3 The Use of Soluble Sugars Versus Polysaccharides as a Carbohydrate Reserve

Starch was identified in early CAM work as the likely source of carbon for the nocturnal synthesis of organic acids (Bennet-Clark 1933; Wolf 1937). For several decades, research on CAM tissues was dominated by efforts to understand the night CO_2 fixation, O_2 metabolism, the respiratory quotient, and other daily gas exchange traits (Kluge and Ting 1978; Edwards and Walker 1983) (see also Chap. 1). Some clever ideas were published to "explain" the unusual gas exchange

Table 2.1. Taxonomic distribution of enzymes amongst CAM plants separated into two groups based on PEP carboxykinase activity

Family (number of species)	PEPCK[a]	NAD-ME[a]	NADP-ME[a]	PPDK[a]	PP_i-PFK[a]
		μmol mg^{-1} Chl h^{-1}			
Active PEP carboxykinase					
Asclepiadaceae(3)	173–440	10–90	2–44	nil[b]	–
Bromeliaceae(12)	193–999	16–140	6–91	nil	25–107[c]
Euphorbiaceae(2)	625–830	167–232	–	nil	–
Asphodelaceae(2)	178–480	29–62	–	nil	–
Vitaceae(1)	597	106	–	–	–
Little PEP carboxykinase					
Agavaceae(3)	ND[b]	43–785	144	40–50	–
Aizoaceae(2)	ND	195	–	–	–
Asteraceae(2)	ND	50	–	80	–
Cactaceae(3)	ND	47–920	131	30–200	2[c]
Crassulaceae(5)	ND	140–385	46–212	90–240	2–8[c]
Orchidaceae(3)	ND	60–190	151–217	30–200	–
Dracaenaceae(2)	ND	50–127	120	70	–

[a]PEPCK: PEP carboxykinase; NAD- or NADP-ME: malic enzyme; PPDK: Pyruvate, phosphate dikinase; PP_i-PFK: pyrophosphate-dependent phosphofructokinase. The level of detection via the assays employed was $< 5\,\mu$mol mg^{-1} Chl h^{-1}. Data collected about 1983 from Kluge and Osmond (1971); Dittrich et al. (1973); Sugiyama and Laetsch (1975); Black (1976); Dittrich (1976); Holtum and Osmond (1981); Black et al. (1982); and Carnal and Black (1983).
[b]Not detectable. A dash indicates no assay reported. Nil indicates a detectable activity but usually less than 1 μmol mg^{-1} Chl h^{-1}.
[c]Activity in μmol mg^{-1} protein h^{-1}.

traits of whole CAM plant tissues. Nevertheless, there was little meaningful understanding of how CAM functioned biochemically.

Hence, while the general phenomenon of an acid taste in succulent plants was recognized for centuries, no integrated model existed until the late 1940s. Then in a remarkable effort to unify the current knowledge about CAM gas exchange, M. Thomas presented a scheme [initially in the third edition of his textbook (Thomas 1947) and subsequently in a research paper (Thomas 1949)] for the interconversion of the carbohydrate and "vegetable acids" as given below:

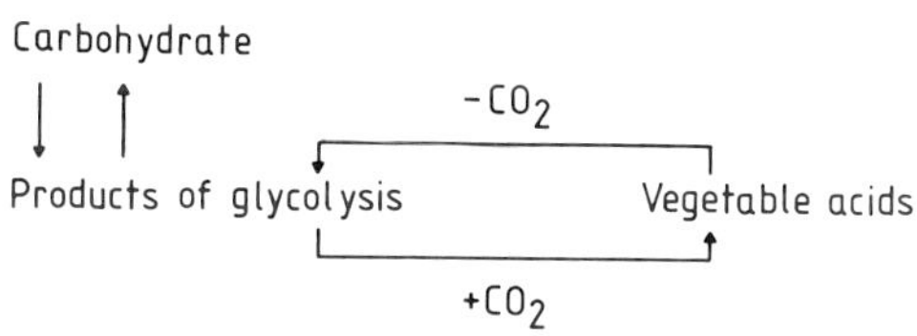

This simple loop-like scheme was given without further comment. Even so, in distilled essence, it is the model that guides the research on how CAM functions even today! These relationships were strongly supported in the series of quantitative studies on total acids and carbohydrates by H.B. Vickery and coworkers (Pucher et al. 1949; Vickery 1954) who demonstrated that these diel synthesis and degradation processes were the reverse of each other, occurring in a reciprocal fashion each day.

These quantitative relationships between starch and malic acid quickly allowed them to fit their data into a similar scheme. For example with *Bryophyllum calycinum* (= *Kalanchoë pinnata*), starch loss and acid accumulation balanced (Pucher et al. 1949). Indeed, in *K. pinnata,* starch loss at night was 50% higher than required for acid synthesis (Sutton 1975a, b). Even though putatively the carbohydrate was starch, in some cases, starch could not account for the amount of acids synthesized. For example, in *K. tubiflora* and *K. daigremontiana*, starch accounted for only two-thirds of the carbon (Sutton 1975a, b), and in *Opuntia aurantiaca* less than 40% of the carbon was derived from starch (Whiting et al. 1979). In each of these plants however, the glucan (low-molecular-weight polymers of glucose) pool change was sufficient to account for the remaining carbon and these authors concluded that soluble sugar pools did not contribute carbon for PEP synthesis. But in other studies, on the effects of changing environments, Vickery (1954) could not totally account for the malate carbon from starch even in *B. calycinum* (*K. pinnata*), which underscores the strong responses of CAM to environmental conditions (Kluge and Ting 1978).

It is somewhat ironic that, simultaneously with this excellent work done in Connecticut on balancing acidity with a polysaccharide, the first work with soluble sugars as a potential carbon reserve appeared in the literature with pineapple from Hawaii (Sideris et al. 1948). The exact pathway of carbohydrate metabolism in pineapple leaves had not been understood because substantial amounts of sugars were depleted each night concurrent with malate synthesis. Indeed, about 7 to 8% of a pineapple leaf total dry weight was lost as soluble sugars at night. The irony is that these workers failed to appreciate that pineapple leaves conduct CAM! In other work soluble sugars also were reported to decrease concurrent with organic acid formation in *Nopalea cochinellifera,* with this decrease supplying about 15% of the carbon needed for CAM (Master 1959).

One additional aspect of CAM came from measuring the daily carbohydrate turnover. From that work it was possible to calculate that between 8 and 20% of the total dry matter in green CAM leaves was committed to the daily cycle (Vickery 1954; Black 1976; Black et al. 1982). This is a massive investment by a green tissue!

We have reviewed the literature for measurements of the stoichiometries between acidity and various types of carbohydrate reserves. These data can be quantitatively compared over a night period or a day period since they must at least balance; in fact, the carbohydrate reserve should be in excess to account for normal respiration, translocation, etc. Some of the older work was just cited and Table 2.2 presents some more recent data with CAM plants. It is fundamental to note that all green plant tissues contain both polysaccharides and hexoses. However, for CAM to function, the massive acid accumulation each night must be balanced by a substantial carbohydrate supply. Also note that, energetically,

Table. 2.2. Changes in malate and carbohydrates from dusk to dawn for some CAM plants

Species[a]	Malate	Starch and glucans	Soluble sugars[b]	References
Ananas comosus	+84	−19	−19G, −20F, −3S	Carnal and Black (1989)
	+64, high PPFD	15% of CH_2O	−120	Borland and Griffiths (1989)
	+90, low PPFD	15% of CH_2O	−54	
Clusia minor[c]	+240[d]	−31	−174G, −165F, +8S	Popp et al. (1987)
	+124[d]	–	−92G, −79F, +5S	
Clusia alata	+63[d]	−22	−45G, −38F, +10S	
	+83[d]	–	−93G, −94F, −13S	
Kalanchoë daigremontiana	+225	−135	−5	Kenyon et al. (1981)
	+45	−60	–	Brunnhöfer et al. (1968)
Sedum telephium	+45, high[e] PPFD	−37	−25	Borland and Griffiths (1992)
	+31, high[e] PPFD	−40	−25	
	+10, high[e] PPFD	−19	−21	
	+32, low[e] PPFD	−6	−4	
	+11, low[e] PPFD	−3	−11	
	+4, low[e] PPFD	−2	−12	

[a]Data on other species are also available: Griffiths et al. (1989) on *Pyrrosia piloselloides*; Medina et al. (1986) on *Bromelia humilis*; Ball et al. (1991) on *Clusia rosea*; and Paul et al. (1993) on *Mesembryanthemum crystallinum*

[b]Soluble sugars are totals, or G = glucose, F = fructose, S = sucrose. All values are in µmol equivalents of hexoses or µ equivalents assuming $2H^+$ per malate per gramme fr. wt. of tissue.

[c]This species was originally described as *C. rosea* (Popp et al. 1987), but was subsequently shown to be *C. minor* (Franco et al. 1991).

[d]Collected at different seasons.

[e]Collected with tissues at different relative water contents.

carbohydrates cost much less to turn over than would a protein or a lipid, hence the latter are not considered likely carbon sources (see also Chap. 24). Vickery and others considered this idea earlier in CAM research and reached this conclusion also.

Again, from the data in Table 2.2, CAM plants can be divided into two groups based on balancing acidity either with a water-insoluble polysaccharide, e.g. starch or glucans, or with a water-soluble hexose. In some CAM plants the acid can be completely balanced by polysaccharides, e.g. *Kalanchoë daigremontiana*; but in *Ananas* and *Clusia* acidity is balanced by hexoses. In most measurements one or the other dominates (Table 2.2).

As a more detailed example, we have measured the stoichiometries of malate to various forms of carbohydrates in *Ananas* (pineapple) leaves (Black et al. 1982; Carnal and Black 1989, Table 2.2) that can form 84 μmol of malate at night per gramme fresh weight. Approximately 19 μmol (or 38 μmol malate equivalents) of glucose, as starch plus glucans, were used at night. Thus, glucans are not sufficient as a carbohydrate source to furnish the carbon to form PEP for malate synthesis. However, the soluble sugar pool in pineapple leaves is large and about 42 μmol of hexose equivalents were depleted overnight. The soluble sugar pools involved are glucose and fructose which, within pineapple leaf cells, are stored in the vacuole (Kenyon et al. 1985). Therefore, in these pineapple leaves counting polysaccharides and hexoses, a total of approximately 61 μmol of glucose equivalents were metabolized at night. Since 42 μmol of hexose equivalents would be needed to form 84 μmol of malate, the leaf has 19 glucose equivalents remaining for other types of night carbohydrate metabolism, translocation, etc. Therefore, with pineapple leaves in CAM, the soluble sugar pool furnishes, as a minimum, half of the carbon for PEP synthesis at night and maximally could account for all of the PEP. The leaf sucrose shows little diurnal change in its pool size; but note that pool size *is not* a measure of sucrose turnover. Presumably, sucrose must turn over rapidly to support the rapid growth of pineapple.

Such detailed work has not been performed with other hexose-utilizing CAM tissues. Nevertheless, from the data in Table 2.2 we can classify *Ananas* and *Clusia* as hexose-utilizing CAM plants. Also, a short report appeared on *Aloë arborescens* showing that mannose and galactose can serve as carbohydrates to balance the daily acidity (Verbücheln and Steup 1984). Unfortunately, there has been no further work on this species. We conclude that glucose and fructose are hexoses that CAM plants can commonly use for PEP synthesis each night and that other hexoses such as mannose and galactose can be used if the *Aloë* results are confirmed and the required hexose metabolism is elucidated.

The results in Table 2.2 also underscore how much the daily amplitude of CAM can vary with seasons of the year and other changes in environment. The *Clusia* data were collected at two seasons; and light intensity and tissue water content show strong influences on pineapple and *Sedum* (Table 2.2). This environmental responsiveness of CAM plants has been known for decades. In controlled-environment studies on photoperiod and thermoperiod, it was shown that extremes such as continuous light or darkness, or constant temperature, can depress the daily amplitude of CAM to zero in about 5 to 7 days. Conversely, CAM will recover in about the same time when the plants are given environmental parameters that have more normal daily fluctuations (Crews et al. 1976; Chang et al. 1981).

2.4 Sequences of Biochemical Reactions in the Daily Use of Hexoses Versus Starch in CAM

Accepting that a polysaccharide or a hexose can serve as the carbohydrate reserve in the diurnal cycle of CAM also leads to the understanding that the sequences of biochemical reactions will be different and the subcellular sites of carbohydrate storage will vary. Almost certainly, starch and glucans are synthesized and stored in the chloroplast. In contrast, free hexoses are most likely to be synthesized in the cytosol and stored in the vacuole. Therefore, a different sequence of biochemical reactions is required in each group of CAM plants. Collectively from these carbohydrate reserve studies (Table 2.2) we can realize that hexose-utilizing species also have a high activity of PEP carboxykinase, while the polysaccharide-utilizing species have malic enzymes as decarboxylases (Table 2.1).

In addition, new information has accrued in plants about intermediary carbohydrate metabolism that can be integrated into the pathways of carbon flow between, for example, pyruvate and sucrose. That information includes the use of PP_i as an energy source, the presence of several alternative routes to interconvert F 6-P and F 1,6-P_2, and the widespread use of various nucleoside triphosphates (NTPs) by kinases within intermediary metabolism (Sung et al. 1988).

Other unique features of plant metabolism also must be factored into their pathways of metabolism. In animal cellular metabolism, it is common to dephosphorylate G 6-P in order to translocate glucose in the bloodstream. But there is little convincing evidence that plant cells can dephosphorylate either glucose or fructose monophosphates. Hence it is not common for plants to synthesize free glucose or fructose. Rather, in plants free sugars are hydrolyzed from polymers, e.g. starch, glucans, fructans, or surcrose and its derivatives. Indeed, in plants the most direct biochemical route to form both glucose and fructose is first to synthesize sucrose and then to hydrolyze it with invertase. Therefore, this new information and these pecularities of plant metabolism must be integral portions of each daily CAM cycle.

Figures 2.1 and 2.2 are complete daily metabolic cycles for the CAM plants whose traits were illustrated in Tables 2.1 and 2.2. Figure 2.1 is for a polysaccharide (i.e. starch/glucan) cycle plant. This is the type of CAM cycle formulated essentially in parallel with similar discoveries about C_4 and C_3 phytosynthesis in the 1960s and 1970s. For example, it includes: a C_4-acid decarboxylation by either an NAD- or NADP-dependent malic enzyme; the conversion of pyruvate to PEP by pyruvate, P_i dikinase; the utilization of NADH by the cytosolic glyceraldehyde 3-P dehydrogenase; and the transport of triose phosphates across the chloroplast envelope. Also the two accumulating metabolites, starch and

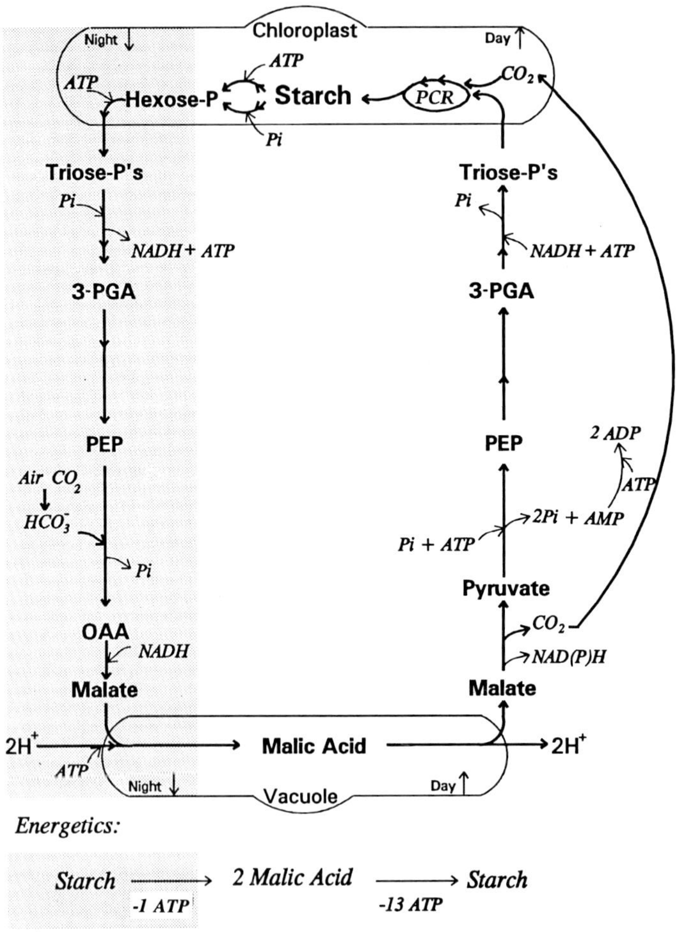

Fig. 2.1. Biochemical cycle for the daily flow of carbon in polysaccharide-cycling CAM plants. The major accumulating compounds, starch and malic acid, are temporally stored, day versus night, in the chloroplast and the vacuole. Starch breakdown in the chloroplast could be either amylolytic or phosphorolytic, but the overall energy budget for starch synthesis has been calculated assuming the latter. *PCR* Photosynthetic carbon reduction cycle

malic acid, are stored in separate organelles: the NAD-malic enzyme likely is in the mitochondrion, and most evidence shows the pyruvate, P_i dikinase is in the chloroplast. Hence, within these green CAM cells, the entire daily cycle involves three organelles and the cytosol.

The cycle in Fig 2.2 is a unified presentation that integrates current information about CAM plants that decarboxylate C_4 acids via PEP carboxykinase. This daily hexose cycle is for plants such as pineapple that utilize glucose and fructose as their carbohydrate reserve. We presented portions of this cycle earlier (Dittrich et al. 1973; Black et al. 1982; Carnal and Black 1989); but several newer features

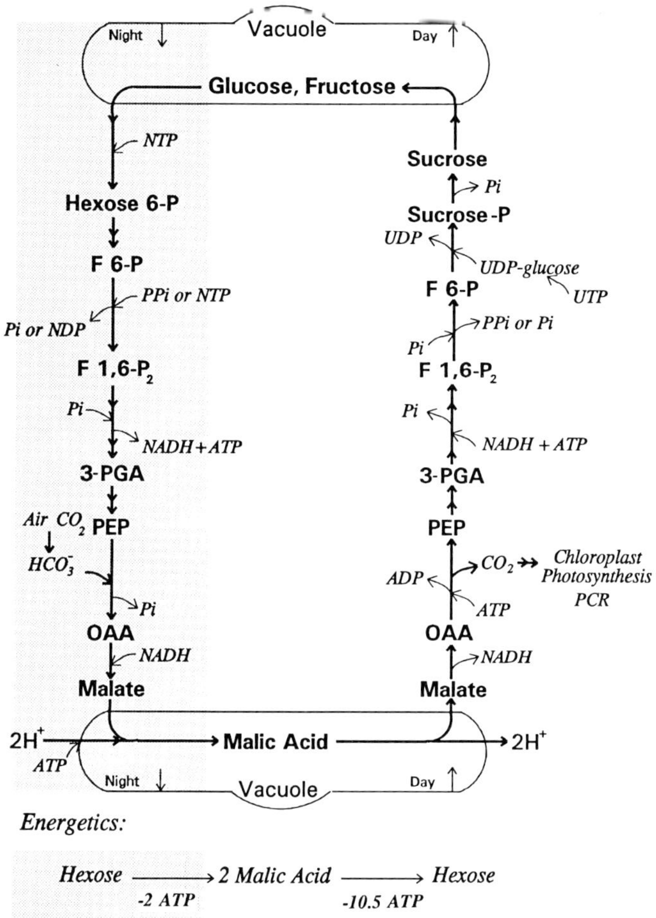

Fig. 2.2. Biochemical cycle for the daily flow of carbon in soluble-hexose-cycling CAM plants. The major accumulating compounds, hexoses and malic acid, are temporally stored, day versus night, in the same subcellular organelle, the vacuole. The overall energy budget has been calculated treating NTP and UTP as ATP equivalents

have been added. One is the use of alternative NTPs by hexokinases and phosphofructokinase at night and the other is the synthesis of sucrose to produce the storage hexoses, glucose and fructose, each day. As already stated, plants seemingly lack hexose monophosphate phosphatases; therefore they cannot produce free hexoses readily, particularly in the near equal quantities of glucose and fructose as found in pineapple leaves (Table 2.2). Rather the most direct biochemical route for an equal hexose synthesis is through sucrose synthesis followed by its cleavage with invertases. We postulate that the soluble acid invertase hydrolyses sucrose in the pineapple cell vacuole (Fig.2.2). The daily storage of hexoses in pineapple vacuoles has been demonstrated (Kenyon et al. 1985).

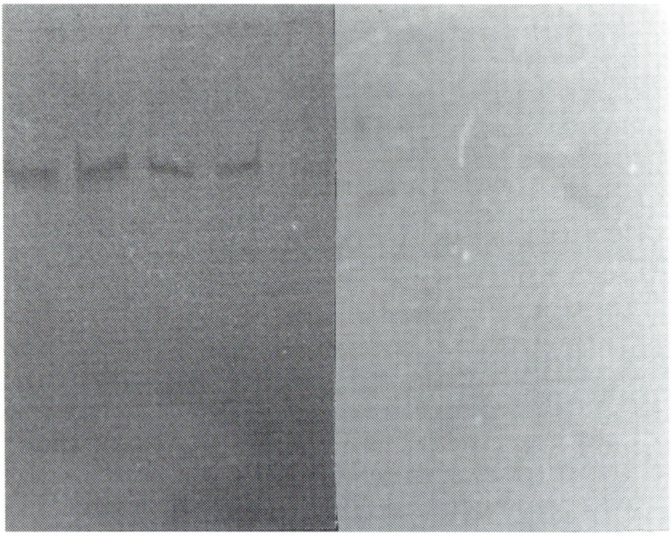

Fig. 2.3. Western immunoblotting for invertase in CAM plants. Equal amounts (20 µg) of soluble proteins from the following CAM plants were loaded in each lane of two SDS-PAGE gels. Both gels were run simultaneously, then they were transferred to nitrocellulose papers and immunoblotted with antibodies against soybean soluble acid invertase. *Left*: lane 1, *Billbergia amoena* (Bromeliaceae); lane 2, *Ananas comosus* (pineapple, Bromeliaceae); lane 3, *Hoya carnosa* (Asclepiadaceae); lane 4, *Aloë arborescens* (Asphodelaceae); lane 5, *Peperomia obtusifolia* (Piperaceae); *Right*: lane 1, *Kalanchoë daigremontiana* (Crassulaceae); lane 2, *Kalanchoë tubiflora* (Crassulaceae); lane 3, *Sedum telephium* (Crassulaceae); lane 4, *Pachyphytum oviferum* (Crassulaceae); lane 5, *Crassula lycopodioides* (Crassulaceae)

Therefore, in this group of CAM plants, the vacuole is the subcellular site for acid accumulation at night and for hexoses in the day, and the cytosol is the site of C_4-acid decarboxylation.

Recently, we also have studied some of the enzymatic requirements of both Figs. 2.1 and 2.2 in representative plants from these two CAM groups. Table 2.3 summarizes some of these results and shows that enzymes do tend to be expressed in accordance with both daily cycles in specific plants. While the enzyme data in Table 2.3 are compatible with Figs. 2.1 and 2.2, we think each of these species needs much more study to elucidate its unique traits. Also we have been studying the specific roles of invertases in plant tissue types (Chen 1992; Chen and Black 1992) and during this work we produced antibodies to soluble acid invertase from soybeans. Using that antibody, at near equal amounts of soluble leaf protein on each gel, we were unable to convincingly detect acid invertase in leaf extracts from polysaccharide-utilizing CAM plants. In contrast, the hexose-utilizing CAM plants stained positively in the same study (Fig. 2.3). Other interpretations, such as

Table 2.3. Taxonomic distribution of some enzymes of carbohydrate metabolism in CAM plants

Family, species	ADPG-PPase[a]	PP_i-PFK	ATP-PFK	ATP-FK	ATP-GK	AI	NI	SPS	PEPCK	NADP-ME	NAD-ME	PPDK
Asclepiadaceae, *Hoya carnosa*	30	624–872	2–26	6	7	27–45	6–9	32	610	207	52	12
Bromeliaceae, *Ananas comosus*	9	1772–2714	33–83	3	11	203–495	11–20	3	2160	142	18	6
Clusiaceae, *Clusia rosea*	107	20–29	53–127	174	ND[b]	42–262	16(3)	27	1260	ND	54	111
Asphodelaceae, 3[c]	4	15(5)	109(5)	10	6	17(5)	11(4)	40	1050	64	48	108
Crassulaceae, 5[d]	62(3)	31(6)	302(7)	4–13	2–35	97(5)	15(7)	31(3)	153(3)	178(3)	195(3)	124(3)

[a]ADPG-PPase: ADP glucose pyrophosphorylase
PP_i-PFK: pyrophosphate-dependent phosphofructokinase (also known as PFP, or pyrophosphate : fructose 6-phosphate 1-phosphotransferase)
ATP-PFK: ATP-dependent phosphofructokinase
ATP-FK: ATP-dependent fructokinase
ATP-GK: ATP-dependent glucokinase
All activities are in nmol min^{-1} mg^{-1} protein. All values followed by ( ) are means for the number of samples in parentheses
Assay temperature was 37 °C for SPS, 25 °C for others
AI or NI: acid or neutral invertase
SPS: sucrose 6-phosphate synthase
PEPCK: PEP carboxykinase
NADP-, NAD-ME: NAD(P)-dependent malic enzyme
PPDK: pyruvate, phosphate dikinase.

[b]Not detectable

[c]*Aloë arborescens, A. ciliaris, A. vera.*

[d]*Sedum telephium, Crassula rupestris, C. argentea, Kalanchoë fedtschenkoi, K. daigremontiana.*

poor antibody cross reactivity between plants, or the possible presence of competing protein inhibitors are possible (Chen 1992). We interpret these results as indicating that less acid invertase or a different antigenic protein is expressed in CAM plants that utilize polysaccharides. We emphasize that acid invertase activity is detectable in both PEPCK- and ME-CAM plants (Table 2.3), but the hexose-utilizing PEPCK-CAM plants show higher activities. This result supports the metabolic sequence for near equal hexose synthesis as depicted in Fig. 2.2 and measured in pineapple and *Clusia* (Table 2.2).

2.5 Bioenergetics in Different Groups of CAM Plants

To calculate the stoichiometries between acidity and malic acid we assume 2 H^+/malic acid and that each hexose is equivalent to 2 PEP molecules, hence 2 malic acid molecules. We have not calculated the energetics of all metabolites crossing membranes, e.g. a hexose-P or triose-P crossing the plastid envelopes; but we assume 1 ATP is used per 2 H^+ and 1 malic acid (Lüttge et al. 1981) transferred into the vacuole. Note that the exact biochemical route of starch breakdown in the chloroplast is not known. Thus, in CAM plants that use polysaccharides the net cost of nocturnal conversion and storage is 1 or 2 ATPs for 2 malic acids, depending on whether starch breakdown is phosphorolytic (Paul et al. 1993) or amylolytic, respectively (Fig. 2.1). The next day 13 ATPs are needed to convert 2 malic acids to a hexose equivalent stored as starch. The energy costs are similar in hexose-storing plants (Fig. 2.2), with 2 ATPs (or ATP equivalents) required per 2 malic acids at night and 10.5 ATPs (or ATP equivalents) per hexose synthesized the next day. These calculations include the PCR cycle.

These CAM cycles, as given in Figs. 2.1 and 2.2, are deceptively simple and straightforward. In fact, they have many unknown aspects; e.g. little information is available on the formation or potential utilization of PP_i as an energy source, either at the interconversion of F 6-P and F 1,6-P_2 or from the synthesis of sucrose or a polysaccharide. The movements of the substrates and energy sources into various cellular compartments is poorly understood, e.g. putatively the hexokinases needed in Fig. 2.2 are in mitochondria and control mechanisms for many of the enzymes are poorly studied.

The theory we propose for the overall daily carbon flow in both Fig. 2.1 and Fig. 2.2 is that there is a large carbon flow at night pulled, via a β-carboxylation of PEP, toward malic acid synthesis, by sequestering malic acid in the vacuole with a low overall energy cost (either 0.5 or 1 ATP, depending on the pathway of carbohydrate breakdown). Then, the next day, energy derived from photosynthesis would drive the reverse synthesis of carbohydrates. During the day the extra energy input, 2.25 to 3.5 ATP per malate, would mainly arise from light energy, whereas the lower energy requirement at night would be met by dark respiration (Figs. 2.1 and 2.2). The fixation of CO_2 in both instances would then have the additional energy requirements of 3 ATP : 2 NADPH : 1 CO_2.

2.6 Conclusions

In green CAM cells there is a massive biochemical commitment of between 8 and 20% of the total cell dry matter into a daily cycle. This diel cycle, in its simplest context, supplies much of the CO_2 for photosynthesis.

CAM plants can be divided into two large groups based upon the primary carbohydrate reservoir used in their daily cycle. Some CAM plants utilize polysaccharides as their carbohydrate reservoir, while others utilize soluble hexose sugars. These two groups of CAM plants mostly decarboxylate malate either with a malic enzyme or with PEP carboxykinase, respectively. Each group of CAM plants also expresses other characteristic sets of enzymes. Hence the sequence of metabolic reactions vary markedly between plants that store polysaccharides versus the storage of hexoses each day. The subcellular site of carbohydrate storage also varies, polysaccharides being stored in chloroplasts and soluble hexoses in the vacuole.

Acknowledgements. This work was partially supported by NSF and the U.S. Forest Service. Ms. Linda White skilfully constructed Figs 2.1 and 2.2.

References

Ball E, Hann J, Kluge M, Lee HSJ, Lüttge U, Orthen B, Popp M, Schmitt A, Ting IP (1991) Ecophysiological comportment of the tropical CAM-tree *Clusia* in the field. II. Modes of photosynthesis in trees and seedlings. New Phytol 117: 483–491

Bennet-Clark TA (1933) The role of organic acids in plant metabolism. Part I, II and III. New Phytol 32:37–71, 128–161, 197–230

Black CC (1973) Photosynthetic carbon fixation in relation to net CO_2 uptake. Annu Rev Plant Physiol 24: 253–286

Black CC (1976) Fractionation of stable carbon isotopes during crassulacean acid metabolism and the presentation of a unified concept of diurnal CO_2 metabolism in CAM plants. In: Benedict CR (ed) The fractionation of stable carbon isotopes by plants. Southern Section American Society of Plant Physiologists, Rockville, pp 51–73

Black CC, Carnal NW, Kenyon WH (1982) Compartmentation and the regulation of CAM. In: IP Ting, M Gibbs (eds) Crassulacean acid metabolism. American Society of Plant Physiologists, Rockville, pp 51–68

Borland AM, Griffiths H (1989) The regulation of citric acid accumulation and carbon recycling during CAM in *Ananas comosus*. J Exp Bot 210: 53–60

Borland AM, Griffiths H (1992) Properties of phosphoenolpyruvate carboxylase and carbohydrate accumulation in the C_3-CAM intermediate *Sedum telephium* L. grown under different light and watering regimes. J Exp Bot 43: 353–361

Brunnhöfer H, Schaub H, Egle K (1968) Die Beziehungen zwischen den Veränderungen der Malat- und Stärkekonzentrationen und dem CO_2- und O_2-Gaswechsel bei *Bryophyllum daigremontianum*. Z Pflanzenphysiol 60: 12–18

Carnal NW, Black CC (1983) Phosphofructokinase activities in photosynthetic organisms: the occurrence of pyrophosphate dependent 6-phosphofructokinase in plants and algae. Plant Physiol 71: 150–155

Carnal NW, Black CC (1989) Soluble sugars as the carbohydrate reserve for CAM in pineapple leaves. Implications for the role of pyrophosphate: 6-phosphofructokinase in glycolysis. Plant Physiol 90: 91–100

Chang NK, Vines HM, Black CC (1981) Nitrate assimilation and crassulacean acid metabolism in leaves of *Kalanchoë fedtschenkoi* variety *marginata*. Plant Physiol 68: 464–468

Chen JQ (1992) Biochemical properties and physiological roles of plant invertases. PhD Dissertation, University of Georgia, Athens, pp 258

Chen JQ, Black CC (1992) Biochemical and immunological properties of alkaline invertase isolated from sprouting soybean hypocotyls. Arch Biochem Biophys 295: 61–69

Crews CE, William SL, Vines HM, Black CC (1976) Changes in the metabolism and physiology of crassulacean acid metabolism plants grown in controlled environments. In: Burris RH, Black CC (eds) CO_2 Metabolism and plant productivity. University Park Press, Baltimore, pp 235–250

Dittrich P (1976) Nicotinamide adenine dinucleotide-specific "malic" enzyme in *Kalanchoë daigremontiana* and other plants exhibiting crassulacean acid metabolism. Plant Physiol 57: 310–314

Dittrich P, Campbell WH, Black CC (1973) Phosphoenolpyruvate carboxykinase in plants exhibiting crassulacean acid metabolism. Plant Physiol 52: 357–361

Edwards G, Walker D (1983) C_3, C_4: mechanisms, and cellular and environmental regulation, of photosynthesis. Blackwell, Oxford

Franco AC, Ball E, Lüttge U (1991) The influence of nitrogen, light and water stress on CO_2 exchange and organic acid accumulation in the tropical C_3-CAM tree, *Clusia minor*. J Exp Bot 42: 597–603

Griffiths H, Ong BL, Avadhani PN, Goh CJ (1989) Recycling of respiratory CO_2 during crassulacean acid metabolism: alleviation of photoinhibition in *Pyrrosia piloselloides*. Planta 179: 115–122

Haag-Kerwer A, Franco AC, Lüttge U (1992) The effect of temperature and light on gas exchange and acid accumulation in the C_3-CAM plant *Clusia minor* L. J Exp Bot 43: 345–352

Heyne B (1815) On the deoxidation of the leaves of *Cotyledon calycina*. Trans Linn Soc Lond 213: 815–816

Holtum JAM, Osmond CB (1981) The gluconeogenic metabolism of pyruvate during deacidification in plants with crassulacean acid metabolism. Aust J Plant Physiol 8: 31–44

Kenyon WH, Holaday AS, Black CC (1981) Diurnal changes in metabolite levels and crassulacean acid metabolism in *Kalanchoë daigremontiana* leaves. Plant Physiol 68: 1002–1007

Kenyon WH, Severson RF, Black CC (1985) Maintenance carbon cycle in crassulacean acid metabolism plant leaves. Plant Physiol 77: 183–189

Kluge M, Osmond CB (1971) Pyruvate, P_i dikinase in crassulacean acid metabolism. Naturwissenschaften 58: 414–415

Kluge M, Ting IP (1978) Crassulacean acid metabolism: analysis of an ecological adaptation. Springer, Berlin Heidelberg New York

Lüttge U, Smith JAC, Marigo G, Osmond CB (1981) Energetics of malate accumulation in the vacuoles of *Kalanchoë tubiflora* cells. FEBS Lett 126: 81–84

Master RWP (1959) Organic acid and carbohydrate metabolism in *Nopalea cochinellifera*. Experientia 15: 30–31

Medina E, Olivares E, Diaz M (1986) Water stress and light intensity effects on growth and nocturnal acid accumulation in a terrestrial CAM bromeliad (*Bromelia humilis* Jacq.) under natural conditions. Oecologia 70: 441–446

Paul MJ, Loos K, Stitt M, Ziegler P (1993) Starch-degrading enzymes during the induction of CAM in *Mesembryanthemum crystallinum*. Plant Cell Environ 16: 531–538

Popp M, Kramer D, Lee H, Diaz M, Ziegler H, Lüttge U (1987) Crassulacean acid metabolism in tropical dicotyledonous trees of the genus *Clusia*. Trees 1: 238–247

Pucher GW, Vickery HB, Abrahams MD, Leavenworth CS (1949) Studies on the metabolism of crassulacean plants: diurnal variation of organic acids and starch in excised leaves of *Bryophyllum calycinum*. Plant Physiol 25: 610–620

Sideris CP, Young HY, Chun HHQ (1948) Diurnal changes and growth rates as associated with ascorbic acid, titratable acidity, carbohydrate and nitrogenous fractions in the leaves of *Ananas comosus* (L.) Merr. Plant Physiol 23: 38–69

Sugiyama T, Laetsch WM (1975) Occurrence of pyruvate orthophosphate dikinase in the succulent plant, *Kalanchoë daigremontiana* Hamet et Perr. Plant Physiol 56: 605–607

Sung SJS, Xu DP, Galloway CM, Black CC (1988) A reassessment of glycolysis and gluconeogenesis in higher plants. Physiol Plant 72: 650–654

Sutton BG (1975a) The path of carbon in CAM plants at night. Aust J Plant Physiol 2: 377–387

Sutton BG (1975b) Glycolysis in CAM plants. Aust J Plant Physiol 2: 389–402

Thomas M (1947) Plant Physiology, 3rd edn. J & A Churchill, London

Thomas M (1949) Physiological studies on acid metabolism in green plants. I. CO_2 fixation and CO_2 liberation in crassulacean acid metabolism. New Phytol 48: 390–420

Ting IP, Gibbs M (eds) (1982) Crassulacean acid metabolism. American Society of Plant Physiologists, Rockville

Verbücheln O, Steup M (1984) Carbon metabolism and malate formation in the CAM plant *Aloë arborescens*. In: Sybesma C (ed) Advances in photosynthesis research, vol 3. Martinus Nijhoff, The Hague, pp 421–424

Vickery HB (1954) The effect of abnormally prolonged alternating periods of light and darkness upon the composition of *Bryophyllum calycinum* leaves. Plant Physiol 29: 520–526

Whiting BH, van de Venter HA, Small JGC (1979) Crassulacean acid metabolism in jointed cactus (*Opuntia aurantiaca*) Lindley. Agroplantae 11: 41–43

Winter K (1985) Crassulacean acid metabolism. In: Barber J, Baker NR (eds) Photosynthetic mechanisms and the environment. Elsevier, Amsterdam, pp 329–387

Wolf J (1937) Beiträge zur Kenntnis des Säurestoffwechsels sukkulenter Crassulaceen. II. Mitteilung. Untersuchungen über Beziehungen zwischen Sedoheptose und Äpfelsäure. Planta 26: 516–522

Wood WML (1952) Organic acid metabolism of *Sedum praealtum*. J Exp Bot 3: 336–355

3 Roles of Circadian Rhythms, Light and Temperature in the Regulation of Phospho*enol*pyruvate Carboxylase in Crassulacean Acid Metabolism

P.J. CARTER, C.A. FEWSON, G.A. NIMMO, H.G. NIMMO, and M.B. WILKINS

3.1 Introduction

Phospho*enol*pyruvate carboxylase (PEPC) plays a key role in the leaf tissue of CAM plants. It catalyses the nocturnal fixation of atmospheric CO_2 (as HCO_3^-) into oxaloacetate, which is subsequently reduced to malate and stored in the vacuole. During the day, malate released from the vacuole is decarboxylated and the resulting CO_2 is fixed via the Calvin cycle (e.g. Osmond and Holtum 1981). Consideration of the metabolic pathways involved in CAM suggests that mechanisms must exist to permit flux through PEPC at night and reduce or eliminate it during the day. In this chapter we describe the role of phosphorylation in the regulation of PEPC in the CAM plant *Bryophyllum (Kalanchoë) fedtschenkoi.*

Like the PEPCs from other types of higher plant (e.g. Andreo et al. 1987), CAM PEPC is an allosteric enzyme subject to activation by glucose 6-phosphate and inhibition by L-malate. It is widely thought that fluctuations in the cytosolic concentration of malate may play a role in regulating flux through PEPC. Some years ago, however, it became clear from the work of several groups that CAM PEPC must be controlled at an additional level, in that the enzyme is more sensitive to inhibition by malate (and therefore less active under physiological conditions) during the day than at night (Kluge et al. 1981; Winter 1982; Buchanan-Bollig and Smith 1984; Nimmo et al. 1984). Detailed investigation of the mechanism underlying this effect led to the discovery that *B. fedtschenkoi* PEPC is regulated by reversible phosphorylation (Nimmo et al. 1984, 1986), a finding since extended to other CAM plants.

B. fedtschenkoi is a "malic-enzyme type" CAM plant, in which the decarboxylation step is catalyzed by NAD- and/or NADP-dependent malic enzyme (see also Chap. 2); it exhibits CAM constitutively. It has been studied for many years, particularly in the context of circadian rhythms. Detached *Bryophyllum* leaves kept in constant environmental conditions exhibit persistent circadian rhythms of CO_2 metabolism (reviewed in Wilkins 1992). For example, in continuous darkness and CO_2-free air at 15 °C, leaves exhibit a rhythm of CO_2 output that is directly attributable to changes in flux through PEPC (Warren and Wilkins 1961); CO_2 that would otherwise be released is periodically refixed into malate. In continuous light and

Institute of Biomedical and Life Sciences, University of Glasgow, Glasgow G12 8QQ, UK

Ecological Studies, Vol. 114
Winter/Smith (eds) Crassulacean Acid Metabolism

normal air, leaves exhibit a rhythm of CO_2 assimilation that also appears to be due to periodic activity of PEPC (e.g. Wilkins 1992). These rhythms can be manipulated readily; the phase can be reset and the rhythms can be inhibited and then restarted at any specified time (e.g. Wilkins 1992). Since the effects of these treatments on the flux through PEPC are readily apparent, *Bryophyllum* affords an excellent experimental system for studies of the molecular mechanism that regulates PEPC and the physiological role of this mechanism.

3.2 Phosphorylation of PEPC in Intact Tissue

Diurnal changes in the activity of PEPC were studied systematically by examining the properties of the enzyme in freshly prepared and desalted extracts of leaf tissue from *Bryophyllum* plants adapted to a short day (8 h) photoperiod, and typical results are illustrated in Fig. 3.1. This work (Nimmo et al. 1984) revealed four important points. First, PEPC was some tenfold more sensitive to inhibition by malate during the light period (apparent K_i for malate 0.3 mM) than in the middle of the dark period (apparent K_i 3.0 mM). Second, the specific activity of PEPC measured under V_{max} conditions did not change significantly throughout the diurnal cycle. Third, the conversions of the "night" to the "day" form of PEPC

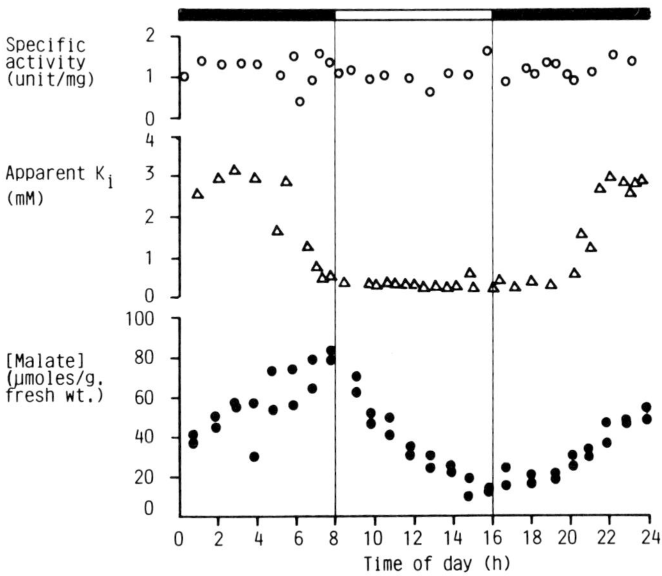

Fig. 3.1. Diurnal rhythm in the malate sensitivity of PEPC in *B. fedtschenkoi*. Dark period (■), light period (□), specific activity of PEPC (○), apparent K_i of PEPC for malate (Δ), leaf malate content (●). Taken from Nimmo et al. (1984) with permission

and vice versa coincided with the cessation and onset of malate accumulation, respectively, suggesting that these interconversions are very important in regulating the flux through PEPC. Finally, both conversions occurred during the dark period, about 1 to 2 h before the lights came on in the morning and about 4 to 6 h after the light went off at night. This information suggested that the conversions between the two kinetically distinct forms of PEPC represent some form of covalent modification of the enzyme and are controlled by a circadian rhythm rather than by light-dark transitions.

The molecular mechanism responsible for these conversions was then investigated, first by immunoprecipitation of PEPC from detached leaves prelabelled by $^{32}P_i$. This showed that the "night" form of the enzyme contained ^{32}P whereas the "day" form did not (Nimmo et al. 1984). On purification of the two forms of the enzyme it was found that the "night" form was phosphorylated on serine residues (Nimmo et al. 1986). Moreover, removal of the phosphate groups by treatment with alkaline phosphatase increased the malate sensitivity of the purified "night" form of PEPC to that of the "day" form. These data indicated that the conversion of the "day" to the "night" form of *Bryophyllum* PEPC was caused by protein phosphorylation. Similar conclusions have been reported for the PEPCs from other CAM species (e.g. Brulfert et al. 1986; Kluge et al. 1988; Baur et al. 1992).

These results raised the question of whether phosphorylation of PEPC could be responsible for the circadian changes in flux through the enzyme described above. Kinetic studies and immunoprecipitation from $^{32}P_i$-labelled leaves showed that, in continuous darkness and CO_2-free air at 15 °C, PEPC exhibited a persistent circadian rhythm of interconversions between a malate-sensitive, dephosphorylated form and a less sensitive, phosphorylated form. Importantly, the changes in the K_i of the PEPC for malate were exactly in phase with the rhythm of CO_2 output (Nimmo et al. 1987). Interconversions between the two forms of PEPC were also observed in leaves kept in continuous light. Overall, these results significantly strengthened the conclusions that phosphorylation of *Bryophyllum* PEPC (1) regulates flux through the enzyme, and (2) is controlled by an endogenous circadian rhythm rather than by changes in illumination. In order to understand the molecular basis of the circadian changes in the phosphorylation state of PEPC, the properties of the enzymes responsible for the phosphorylation and dephosphorylation of PEPC were investigated.

3.3 Properties and Regulation of PEPC Kinase and Phosphatase

Recently, a significant insight into plant protein phosphatases has been obtained by comparing them with their mammalian counterparts. Mammalian tissues contain four main types of protein-phosphatase catalytic subunit (termed 1, 2A, 2B and 2C) that dephosphorylate protein phosphoserine or phosphothreonine residues. These enzymes can be distinguished by their substrate specificities, sensitivities to certain inhibitors and requirements for activity (e.g. Cohen 1989). It is now clear from biochemical studies that higher plants contain activities that are very similar to the mammalian phosphatases 1, 2A and 2C (MacKintosh and Cohen 1989; Carter et al. 1990; MacKintosh et al. 1991).

Following up this work, Carter et al. (1990) showed that the phosphorylated form of *Bryophyllum* PEPC could be dephosphorylated by the purified catalytic subunit of protein phosphatase 2A from rabbit skeletal muscle, but not by the type 1 enzyme. Dephosphorylation of PEPC completely reversed the effects of phosphorylation on its sensitivity to malate. The type 2A catalytic subunit from *Bryophyllum* was partially purified, and it was found that it, too, could dephosphorylate PEPC (Carter et al. 1990). However, the activity of this phosphatase did not change significantly during the normal diurnal cycle in *Bryophyllum* (Carter et al. 1991), suggesting that the phosphorylation state of PEPC might be regulated largely through changes in the activity of PEPC kinase.

Carter et al. (1991) partially purified PEPC kinase from "night" *Bryophyllum* leaves and showed that it could phosphorylate the "day" form of PEPC to a stoichiometry approaching 1 per subunit. This caused essentially the same changes in the malate sensitivity of PEPC as are observed with intact tissue. The activity of PEPC kinase was not affected by Ca^{2+}, but the enzyme was inhibited by malate and glucose 6-phosphate (Carter et al. 1991). However, it is not yet clear if these effectors are physiologically significant.

Measurements of PEPC kinase activity in desalted leaf extracts prepared throughout the diel cycle in *Bryophyllum* proved very informative. Kinase activity appeared some 4 to 6 h after the lights went off at night and disappeared some 2 h before they went on in the morning. Leaf extracts that contained PEPC kinase activity also contained PEPC in the phosphorylated form with a high K_i for malate, whereas in those extracts lacking PEPC kinase activity, PEPC was in the dephosphorylated form (Carter et al. 1991). These results, allied to the observation that the activity of protein phosphatase 2A did not change significantly during the diurnal cycle, indicated that the major factor which determines the phosphorylation state of PEPC is the presence or absence of PEPC kinase activity.

The mechanism underlying the nocturnal appearance of PEPC kinase was investigated by allowing leaves to take up inhibitors of protein or RNA synthesis. Treatment of leaves with puromycin, cycloheximide, cordycepin, or actinomycin D all prevented the nocturnal appearance of PEPC kinase activity (Carter et al. 1991 and unpublished data). This indicates that de novo synthesis of a protein is required for the appearance of PEPC kinase. The simplest hypothesis is that the component synthesized is PEPC kinase itself. However, it could also be another component that is required to activate the kinase. Irrespective of the nature of the component, its synthesis must be followed by its destruction. *Bryophyllum* leaves kept in constant environmental conditions show a persistent rhythm of appearance and disappearance of PEPC kinase activity (Carter et al. 1991), indicating that the protein synthesis and destruction steps exhibit a circadian rhythm. This rhythm in PEPC kinase activity is presumed to play an important role in generating the rhythms of CO_2 metabolism (Wilkins 1992). A number of treatments that perturb the rhythms of CO_2 metabolism have been well defined (Wilkins 1992), and the effects of these treatments on PEPC kinase activity should provide much information about both the mechanism by which kinase activity is regulated and the metabolic role of the enzyme.

3.4 Effects of Light and Temperature on PEPC-Kinase Activity

The circadian rhythms of CO_2 metabolism in detached *Bryophyllum* leaves can be observed over the approximate temperature range 10 to 30 °C, but the circadian oscillator seems to be inhibited outside this range (Wilkins 1992). For the circadian rhythm in continuous darkness and 15 °C, short light and high temperature treatments cause a large phase-shift when applied between peaks of CO_2 output when CO_2 fixation is at a maximum, but no shift when applied at the crest of a peak when CO_2 fixation is zero. The converse is true for low temperature or high-CO_2 treatments. Measurement of leaf malate content suggests that low temperature and high-CO_2 favour carboxylation, and drive the rhythm to one fixed phase point, whereas high temperature and light favour malate removal and drive the rhythm to the opposite phase (Wilkins 1992). These observations prompted us to examine the effects of light and temperature on PEPC kinase activity.

As noted above, *Bryophyllum* leaves exhibit a rhythm of PEPC kinase activity, owing to protein synthesis and destruction, during the normal diel cycle or in constant environmental conditions. If leaves are subjected to a dark cycle at 5 °C instead of 15°C, PEPC kinase activity does not appear, indicating that the protein synthesis step is blocked. If leaves containing PEPC kinase activity (i.e. those in the middle of a normal dark period at 15 °C) are transferred to 5 °C in darkness, PEPC kinase activity remains stable. However, if such leaves are illuminated, PEPC kinase disappears within 2 h and PEPC becomes dephosphorylated. These results indicate that low temperature can block the protein destruction step but that this can be overcome by light. If leaves containing PEPC kinase activity are transferred to 30 °C in darkness, PEPC kinase again disappears and PEPC becomes dephosphorylated. These observations are summed up in Fig. 3.2. It is clear that treatments which cause a phase-shift when applied when CO_2 fixation is at a maximum (light and high temperature) are those that cause loss of PEPC kinase activity and conversion of PEPC to the less active, dephosphorylated form. Conversely, low temperature or high-CO_2 treatments, which cause a phase shift when applied when CO_2 fixation is zero, favour flux through PEPC (Wilkins 1992). Morever, the circadian rhythms of CO_2 metabolism in detached leaves are not observed at temperatures at which PEPC kinase activity is either stable (5 °C or below) or destroyed (above 30 °C). These observations emphasize the importance of changes in PEPC kinase activity and the phosphorylation state of PEPC in generating the rhythms of CO_2 metabolism.

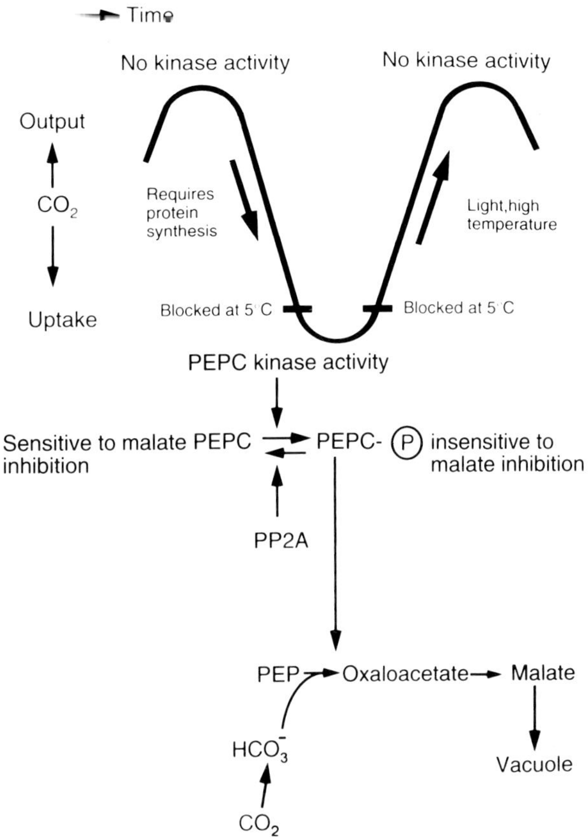

Fig. 3.2. Regulation of PEPC kinase activity. The *sinusoidal curve* represents the diel or circadian changes in kinase activity. *PP2A* represents protein-phosphatase 2A

3.5 Conclusions

Two interesting points emerge from this work. One is that changes in the activity of PEPC kinase reflect protein synthesis/degradation (possibly of the protein kinase itself). This is very unlike the typical animal paradigm, in which protein kinase activity is regulated by a second messenger, a metabolite or a protein-phosphorylation cascade without net synthesis/degradation of the kinase. Clearly, purification of PEPC kinase to homogeneity is an important objective that, once achieved, will allow a more detailed analysis of the mechanism regulating this enzyme to be made.

The second point is that PEPC kinase seems to play a major role in generating the circadian rhythms of CO_2 metabolism. The ultimate test of the importance of this enzyme will involve the generation of transgenic plants which express different levels of the kinase. In the meantime, however, it is clear that some environmental perturbations have significant effects on the activity of the kinase. These offer an excellent experimental system in which the regulation of PEPC kinase can be investigated.

Acknowledgement. Work from this laboratory was supported by the Agricultural and Food Research Council.

References

Andreo CS, Gonzalez DH, Iglesias AA (1987) Higher plant phosphoenolpyruvate carboxylase. FEBS Lett 213: 1–8

Baur B, Dietz KJ, Winter K (1992) Regulatory protein phosphorylation of phosphoenolpyruvate carboxylase in the facultative crassulacean-acid-metabolism plant *Mesembryanthemum crystallinum* L. Eur J Biochem 209: 95–101

Brulfert J, Vidal J, LeMarechal P, Gadal P, Queiroz O, Kluge M, Krüger I (1986) Phosphorylation-dephosphorylation process as a probable mechanism for the diurnal regulatory changes of phosphoenolpyruvate carboxylase in CAM plants. Biochem Biophys Res Commun 136: 151–159

Buchanan-Bollig IC, Smith JAC (1984) Circadian rhythms in crassulacean acid metabolism: phase relationships between gas exchange, leaf water relations and malate metabolism in *Kalanchoë daigremontiana*. Planta 161: 314–319

Carter PJ, Nimmo HG, Fewson CA, Wilkins MB (1990) *Bryophyllum fedtschenkoi* protein phosphatase 2A can dephosphorylate phosphoenolpyruvate carboxylase. FEBS Lett 263: 233–236

Carter PJ, Nimmo HG, Fewson CA, Wilkins MB (1991) Circadian rhythms in the activity of a plant protein kinase. EMBO J 10: 2063–2068

Cohen P (1989) The structure and regulation of protein phosphatases. Annu Rev Biochem 58: 453–508

Kluge M, Brulfert J, Queiroz O (1981) Diurnal changes in the regulation of PEP-carboxylase in crassulacean acid metabolism (CAM). Plant Cell Environ 4: 251–256

Kluge M, Maier P, Brulfert J, Faist K, Wollny E (1988) Regulation of phosphoenolpyruvate carboxylase in crassulacean acid metabolism: in vitro phosphorylation of the enzyme. J Plant Physiol 133: 252–256

MacKintosh C, Cohen P (1989) Identification of high levels of type 1 and type 2A protein phosphatases in higher plants. Biochem J 262: 335–339

MacKintosh C, Coggins J, Cohen P (1991) Plant protein phosphatases. Biochem J 273: 733–738

Nimmo GA, Nimmo HG, Fewson CA, Wilkins MB (1984) Diurnal changes in the properties of phosphoenolpyruvate carboxylase in *Bryophyllum* leaves: a possible covalent modification FEBS Lett 178: 199–203

Nimmo GA, Nimmo HG, Hamilton ID, Fewson CA, Wilkins MB (1986) Purification of the phosphorylated night form and dephosphorylated day form of phosphoenolpyruvate carboxylase from *Bryophyllum fedtschenkoi*. Biochem J 239: 213–220

Nimmo GA, Wilkins MB, Fewson CA, Nimmo HG (1987) Persistent circadian rhythms in the phosphorylation state of phosphoenolpyruvate carboxylase from *Bryophyllum fedtschenkoi* leaves and in its sensitivity to inhibition by malate. Planta 170: 408–415

Osmond CB, Holtum JAM (1981) Crassulacean acid metabolism. In: Hatch MD, Boardman NA (eds) The biochemistry of plants, vol 8, photosynthesis. Academic Press, New York, pp 283–328

Warren DM, Wilkins MB (1961) An endogenous rhythm in the rate of dark-fixation of carbon dioxide in leaves of *Bryophyllum fedtschenkoi*. Nature (Lond) 191: 686–688

Wilkins MB (1992) Circadian rhythms: their origin and control. New Phytol 121: 347–375

Winter K (1982) Properties of phospho*enol*pyruvate carboxylase in rapidly prepared, desalted leaf extracts of the crassulacean acid metabolism plant *Mesembryanthemum crystallinum* L. Planta 154: 298–308

4 Transport Across the Vacuolar Membrane in CAM Plants

J.A.C. Smith, J. Ingram, M.S. Tsiantis, B.J. Barkla, D.M. Bartholomew, M. Bettey, O. Pantoja, and A.J. Pennington

4.1 Introduction

Close metabolic parallels exist between the processes of CO_2 assimilation in C_4 plants and in CAM plants. In both types of plant, a C_4 cycle starts with the fixation of CO_2 (as HCO_3^-) by phospho*enol*pyruvate carboxylase (PEPC) and concludes with the release or CO_2 by decarboxylation of a C_4 dicarboxylate anion (malate or aspartate). This C_4 cycle is an ancillary pathway, in the sense that it does not mediate the net fixation of atmospheric CO_2. It simply passes on this CO_2, at greatly elevated concentration, to ribulose 1,5-bisphosphate carboxylase/oxygenase (RUBISCO) for assimilation through the standard C_3 photosynthetic carbon reduction cycle.

In stark contrast to C_4 photosynthesis, however, the entire pathway of photosynthetic carbon assimilation in CAM plants is contained within individual mesophyll cells, rather than being spatially subdivided between different tissues. This has profound implications for the compartmentation and regulation of metabolism in CAM plants. As is now well established, the large central vacuole of the mesophyll cells provides the main repository for nocturnally accumulated malic-acid, which is then released from the vacuole in the subsequent light period for decarboxylation in the cytoplasm. This temporal separation of malic-acid storage and remobilization raises important questions concerning both the mechanisms of malic-acid influx and efflux across the vacuolar membrane, as well as the way in which these fluxes are coordinated with carbon metabolism during the day–night cycle.

In the typical photosynthetic tissue of CAM plants, the subcellular structure of the mesophyll cells is spatially dominated by the large central vacuole. At the level of the transmission electron microscope, the cell cytoplasm is often barely discernible as a thin layer of electron-dense material appressed between the plasmalemma and tonoplast. In mature leaves of *Kalanchoë daigremontiana*, for example, the cytoplasm is estimated to constitute only about 1% of total cell volume, with the cell wall contributing a further 2% (Steudle et al. 1980). In CAM plants that possess a separate non-chlorophyll-containing water-storage tissue, these cells tend to be larger and even more highly

Department of Plant Sciences, University of Oxford, South Parks Road, Oxford OX1 3RB, UK

Ecological Studies, Vol. 114
Winter/Smith (eds) Crassulacean Acid Metabolism

vacuolate. Together, these tissues are responsible for the low surface area to volume ratio and low dry weight to fresh weight ratio that characterize the succulent leaves and stems of most CAM plants (Gibson 1982; Smith 1984).

In this chapter, we shall review the distinctive functional characteristics of the mesophyll cell vacuole in CAM plants. We consider first the role of the vacuole in the osmotic and ionic relations of CAM plants, before discussing how solute fluxes across the vacuolar membrane are regulated during the day–night cycle. Far from being merely an inert repository of cytotoxic ions and secondary metabolites, the vacuole plays a dynamic role in the regulation of carbon metabolism that is central to the process of photosynthetic CO_2 assimilation in CAM plants.

4.2 Osmotic and Ionic Relations of the Vacuole

The vacuole has an important role in plants in diverse cellular functions such as maintenance of charge (or acid–base) balance, ionic homeostasis, storage of metabolites, accumulation of potentially cytotoxic ions and secondary products, and sequestration of enzymes such as hydrolases and proteases (Boller and Wiemken 1986; Matile 1987; Raven 1987; Wink 1993). All these functions depend directly or indirectly on the tightly regulated transport of low-molecular-weight solutes or macromolecules across the vacuolar membrane. In CAM plants, the vacuole is so dominant volumetrically that the analysis of cell sap or bulk tissue extracts can be used to give reliable information on the vacuolar contents.

4.2.1 Osmotic Characteristics

One of the most distinctive features of the vacuolar sap of CAM plants as compared with other life-forms has been appreciated for a long time as a result of extensive measurements on desert succulents (Walter 1960). This is the low concentration of osmotically active solutes in the vacuole and the correspondingly low osmotic pressure of the cell sap, which is usually considerably less than 1 MPa even in plants suffering from prolonged water deficits. Together with the relatively high elasticity of the large, thin-walled mesophyll cells, these low osmotic pressures give the leaves or stems of CAM plants a large water-storage capacitance (Steudle et al. 1980; Smith 1984). Furthermore, storage capacitance is inversely related to turgor pressure, which tends to be low (commonly 0.2 MPa or less) in the mesophyll cells of CAM plants in which it has been quantified (Smith and Lüttge 1985; Rygol et al. 1987; Murphy and Smith 1994a). Consequently, the above-ground organs of CAM plants usually have a large capacity for storage and remobilization of water during periods of fluctuating water availability in the environment (Smith et al. 1987; Nobel 1988; Schulte et al. 1989).

4.2.2 Ionic Characteristics

The solute content of cell vacuoles in the photosynthetic tissue of CAM plants is characterized by a high concentration of carboxylic-acid anions such as malate, citrate and isocitrate (Smith 1987). Carboxylate accumulation is a distinctive feature of the above-ground parts of all terrestrial plants. Because of the limited apoplastic volume, shoot cells do not have direct access to a large extracellular medium for excretion of the excess negative charge (OH^- or HCO_3^-) resulting from nitrate assimilation (Osmond 1976; Raven and Smith 1976). Instead, the negative charge is balanced within the cell (and potential alkalinization counteracted) by synthesis of carboxylic acids from neutral carbohydrate precursors, the associated H^+ being removed from the cyto-plasm by active transport. This is the combined biochemical and biophysical 'pH stat' mechanism that is central to cytoplasmic ion homeostasis (Smith and Raven 1979).

In CAM plants, the pool of vacuolar carboxylates is made up of two components. One is involved in balancing the total cationic charge, made up largely by the inorganic cations K^+, Na^+, Ca^{2+} and Mg^{2+}. In mature leaves of *K. daigremontiana*, for example, the total negative charge contributed by the three carboxylates malate, citrate and isocitrate accounts for nearly 90% of the overall cationic charge equivalents (Phillips and Jennings 1976; Phillips 1980; A.-L.C. McWilliam, unpublished results). However, superimposed on top of this background pool of vacuolar carboxylates is another component, which is the malic acid (and sometimes citric acid) involved in the CAM cycle. These day–night changes in carboxylate content during the CAM cycle have been shown by extensive measurements of cell titratable acidity specifically to involve the free acid, i.e. 2 H^+ per malate and 3 H^+ per citrate (Lüttge et al. 1982; Franco et al. 1992; see Chap. 19). The magnitude of the pH changes observed depends on the buffering capacity of the vacuolar sap, which is high on account of the background pool of carboxylates, but pH values as low as 2.85 have been recorded towards the end of the nocturnal acidification phase in *Clusia* (Franco et al. 1990).

These large day–night changes in carboxylate content are very distinctive in being quantitatively charge balanced during the CAM cycle by H^+. This has important implications for possible mechanisms of carboxylate transport across the tonoplast, as will be discussed later. In the present context, it is interesting to compare carboxylate accumulation in the CAM cycle with that occurring in two other well-studied examples of vacuolar malate accumulation (Fig. 4.1). In the case of inorganic nitrate assimilation, synthesis of malic acid from neutral precursor serves both to neutralize the excess OH^- generated in metabolism and to allow transfer of negative charge from nitrate to malate: the net result is vacuolar accumulation of malate balanced by the one or more cations (frequently K^+) that accompanied nitrate uptake. Indeed, in C_3 plants the malate content of photosynthetic tissues typically *increases* during the day as a consequence of light-dependent nitrate assimilation (Martinoia and Rentsch 1994). In stomatal guard cells, however, where in most plants there is relatively little uptake of inorganic anion from the apoplast, the H^+ derived from malic-acid production is extruded from the cytosol by the plasmalemma H^+-ATPase, and this in turn can drive K^+ uptake. This has the interesting consequence that vacuolar pH *rises* in guard cells as malate accumulates (Willmer and Fricker 1995), because the cell is synthesizing a conjugate base that can buffer H^+. Whether vacuolar pH increases or decreases as a result of carboxylate accumulation is therefore determined quantitatively by the extent to which the carboxylate is charge balanced by inorganic cations. In principle, this change in pH could be predicted from the changes in ion content, since pH is a dependent and not an independent variable (Stewart 1983).

Another striking feature of the nocturnally accumulated malate in CAM plants is that it apparently behaves as an ideal osmoticum. Plots of the

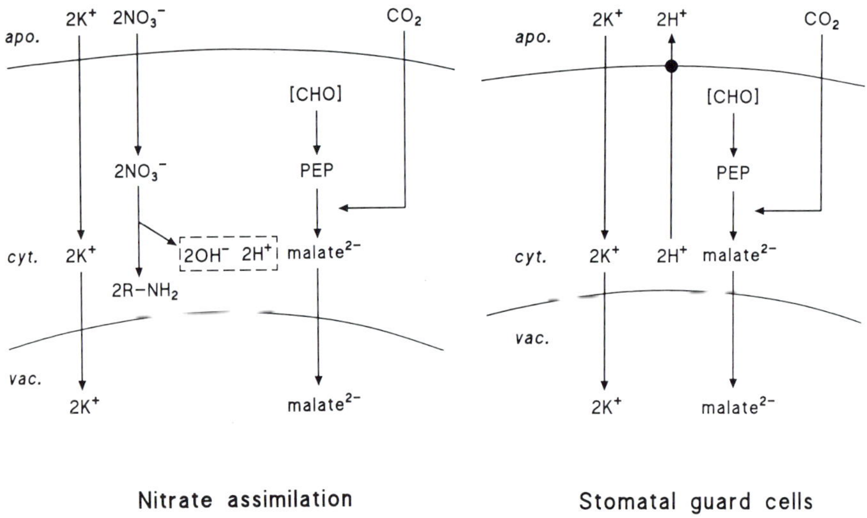

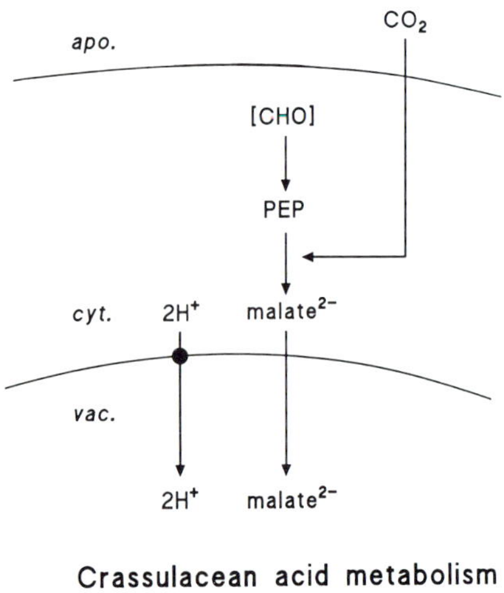

Fig. 4.1. Role of malic-acid synthesis and vacuolar malate accumulation in maintenance of intracellular charge balance. Fixation of CO_2 via PEPC in the cytosol results in the production of malic acid (as 2 H^+ plus malate^{2-}) when the substrate PEP is derived from neutral carbohydrate. In nitrate assimilation, the H^+ is balanced by excess OH^- resulting from NO_3^- reduction to the level of organic N (R-NH_2). In stomatal guard cells, the H^+ produced in malic-acid synthesis can be quantitatively removed from the cell by the plasmalemma H^+-ATPase and charge-balanced by K^+ uptake. In crassulacean acid metabolism, H^+ is removed into the cell vacuole by the tonoplast H^+-ATPase (or H^+-PP$_i$ase) and itself provides charge balance for the accumulated malate (*apo.*, apoplast; *cyt.*, cytosol; *vac.*, vacuole)

day–night changes in malate concentration against sap osmolality show a slope of approximately 1.0, indicating that the malate has an osmotic coefficient of unity (Lüttge and Nobel 1984; Smith and Lüttge 1985; Smith et al. 1986). Furthermore, by relating changes in sap malate concentration to changes in cell turgor pressure and water potential, it has been demonstrated that the reflection coefficient for malate is not significantly different from unity (Murphy and Smith 1994b).

Malate is thus a very effective osmoticum in CAM plants and plays an important role in cell water relations. As malate accumulation can drive the osmotic uptake of water, the CAM rhythm might be directly linked to 'water harvesting' from the environment (Smith and Lüttge 1985; Ruess and Eller 1985). However, quantitative analysis of predicted water flows in *Agave deserti* using an electrical-circuit analogue have shown that transpiration is by far the more important process in driving water uptake from the soil (Smith et al. 1987). Osmotically, the most important function for malate accumulation may be the maintenance of positive cell turgor pressure at times when transpiration rates are highest and cell water potentials are at their most negative.

If malate appears to approach ideality as a solute surprisingly closely, this conceals a more complex underlying chemistry. Many species of CAM plants, and most notably those in the Crassulaceae (Phillips and Jennings 1976; Kinzel 1989; Arata et al. 1992; A.-L.C. McWilliam, unpublished data), contain very high concentrations of soluble vacuolar calcium. This will be chelated by carboxylates to varying degrees depending on sap pH, as will the lower (but still significant) amounts of magnesium. (In contrast, the monovalent cations have low stability constants for chelation by carboxylates and will not form complexes to any significant degree.) As an example of the impact of the CAM cycle on cation-carboxylate chelation, Table 4.1 shows the principal forms of malate present in the vacuolar sap of *K. daigremontiana*, calculated using standard thermodynamic stability constants for all relevant complexes. In this example, a relatively high propertion of the total malate (48%) is chelated with calcium and magnesium at the beginning of the dark period, although the single most important species is malate^{2-}. As malic acid accumulates and sap pH decreases in the course of the dark period, the protonated species Hmalate^{1-} and H_2malate^0 start to predominate; more calcium and magnesium come to be complexed with malate, as the amount of chelating ligand increases, but at the end of the dark period these complexes only account for 16% of the total malate. Similar results were obtained by Arata et al. (1992) in a study of *K. daigremontiana*, except that their model did not incorporate values for vacuolar magnesium or citrate.

Knowledge of the precise chemical forms of vacuolar malate is needed to delineate the possible mechanisms of malate transport across the tonoplast. For many CAM plants, however, the occurrence of substantial day–night changes in citric-acid content introduces a further degree of complexity (Lüttge 1988). As yet, the implications of citric-acid accumulation for the ionic relations of the vacuole are not fully understood. Similarly, the cell vacuole is presumed (Smith and Bryce 1992) to be the main storage site for soluble sugars

Table 4.1. Chemical speciation of vacuolar malate at dusk and dawn during the CAM cycle in *Kalanchoë daigremontiana*

	Concentration (mM)	
Malate species	Dusk	Dawn
Mal^{2-}	4.66	3.43
$Hmal^{-}$	0.60	70.4
H_2mal^0	<0.01	50.3
$Ca.mal^0$	3.83	6.05
$Ca.Hmal^+$	0.06	15.6
$Mg.mal^0$	0.99	1.64
$Mg.Hmal^+$	<0.01	0.19

Values were calculated from measurements of the complete solute content of leaf-cell sap, sap pH, and the standard thermodynamic stability constants for all relevant complexes by a mass-action approach using the computer program SOLCON (White et al. 1990). The following were used as input parameters (dusk is beginning of the dark period, dawn is beginning of the light period): total $malate^{dusk} = 10.7$ mM, total $malate^{dawn} = 148$ mM; total $citrate^{dusk} = 11.0$ mM, total $citrate^{dawn} = 16.1$ mM; $pH^{dusk} = 5.72$, $pH^{dawn} = 3.52$. The other input parameters, which were constant over the day–night cycle, were: Ca = 49.2 mM; Mg = 14.7 mM; total isocitrate = 50.5 mM. (Unpublished data from J.A.C. Smith and N.J.A. Cook).

involved in the CAM cycle in phospho*enol*pyruvate carboxykinase (PEPCK)-type CAM plants, such as *Clusia* and at least some members of the Bromeliaceae (Chaps. 2, 19 and 25). As sugar accumulation in such plants occurs at the same time as malate decarboxylation, the net effect is that the day–night changes in sap osmolality are considerably reduced (Franco et al. 1992). More work, however, is needed on the osmotic and ionic relations of the vacuole in CAM plants that specialize in the accumulation of citric acid and soluble sugars.

4.3 Malic-Acid Accumulation in the Vacuole

Although the vacuole clearly plays a central role in the temporal co-ordination of the CAM cycle, the way in which solute transport across the tonoplast is regulated is still not completely understood. This requires consideration of both thermodynamic and kinetic factors, since it is important to identify not only the mechanisms of solute transport but also the controls under which these processes operate. Significant progress has been made in identifying some of the transport proteins involved at the molecular level, but our understanding of the way in which their activity is modulated in vivo is still fragmentary.

The transport model proposed by Lüttge's group (Lüttge and Ball 1979; Rona et al. 1980; Lüttge et al. 1982) to explain nocturnal accumulation of

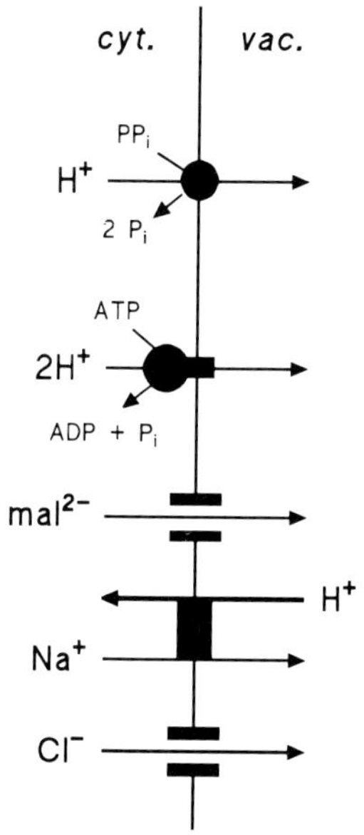

Fig. 4.2. Principal transport processes at the tonoplast of CAM plants responsible for solute influx into the vacuole. Solute accumulation is energized by the two primary H^+ pumps, the H^+-ATPase and H^+-PP_iase, although the relative importance of these two pumps may differ depending on the species concerned, cell type, developmental state, or environmental conditions. Malate is transported as the divalent ion (mal^{2-}) through a substrate-selective ion channel. In addition in halophytic CAM plants such as the Aizoaceae, secondary active transport of Na^+ into the vacuole occurs via an Na^+/H^+ antiporter, with Cl^- influx probably mediated through a separate ion channel

malic acid in the vacuole still appears to be consistent with all the evidence. Analysis of the electrochemical gradients across the tonoplast indicates that H^+ must be actively transported from the cytosol into the vacuole and that the malate^{2-} anion is close to equilibrium across the tonoplast and could probably be transported passively into the vacuole. These transport processes can be viewed as an essential response to the production of malic acid by glycolysis at the start of the dark period. Removal of H^+ into the vacuole is vital for the control of cytoplasmic pH (Smith and Raven 1979), and removal of malate is necessary to prevent feedback inhibition of PEPC, even with the enzyme in its phosphorylated, relatively malate-insensitive form (Chap. 2).

The large number of experimental observations over the last 10 years that have supported this general transport model have been reviewed previously (Lüttge 1987; Smith and Bryce 1992). Figure 4.2 summarizes the principal types of transport process responsible for solute influx into the vacuole that have been identified at the tonoplast of CAM plants. Here we shall consider recent progress in characterizing these transport systems at the molecular level, as well as the areas in which major questions still remain to be resolved.

4.3.1 Primary Active H^+ Transport

Plant vacuoles possess two pumps in the tonoplast capable of active transport of H^+ into the vacuolar lumen: an ATPase and an inorganic pyrophosphatase (PP_iase). Why the tonoplast should contain two enzymes apparently functioning in parallel is still not clear (Rea and Sanders 1987; White et al. 1990). Both enzymes are able to generate an inside-positive electrical potential difference across the tonoplast (Jochem and Lüttge 1987; Marquardt and Lüttge 1987; White and Smith 1989), which in turn is thought to provide the driving force for malate^{2-} influx via a separate transport protein. Values of about +25 mV (vacuole inside-positive) for the transtonoplast potential difference have been measured in situ with intracellular microelectrodes for *K. daigremontiana* (Rona et al. 1980). However, values as high as +85 mV have recently been recorded for this species using patch-clamp electrodes with isolated vacuoles in the current-clamp mode in the presence of 3 mM adenosine triphosphate (ATP) and 10 mM malate (O. Pantoja and J.A.C. Smith, unpublished results).

The vacuolar (or V-type) H^+-ATPase is now thought to be ubiquitous in eukaryotes, being found in the bounding membrane of a wide variety of acidic intracellular organelles such as lysosomes, endocytic vesicles, clathrin-coated vesicles and chromaffin granules (Forgac 1989; Nelson 1995). The enzyme has a complex multimeric structure, consisting of at least nine different types of subunit, with a total molecular mass in the range of 400 to 650 kDa. For the enzyme from leaves of *K. daigremontiana*, subunits with an approximate molecular mass of 72, 57, 48, 42, 39, 33 and 16 kDa have been identified (Warren et al. 1992). In a manner similar to the F-type ATP synthase of mitochondria, chloroplasts and eubacteria, the V-type ATPase is composed of a peripheral sector and a portion integral to the membrane. This results in a 'ball and stalk' structure that has been directly visualized in some high-quality images of freeze-fracture faces of the tonoplast of *Mesembryanthemum crystallinum* (Klink et al. 1990; Klink and Lüttge 1992). The high area densities of these particles provides strong corroborative evidence for the results from biochemical analyses, which suggest that the V-type ATPase can account for over 30% of the tonoplast protein in CAM plants (Bremberger et al. 1988; Klink et al. 1990; Bremberger and Lüttge 1992).

From studies so far of the plant enzyme, a consensus has emerged that the peripheral sector of the V-ATPase is composed of five (or possibly six) types of subunit (A–E, in order of decreasing size), with a likely stoichiometry of A_3B_3CDE (Sze et al. 1992; Ward and Sze 1992). Two of these subunits have been characterized at the molecular level. The A subunit (67–70 kDa) is responsible for ATP hydrolysis, and its sequence has now been deduced from cDNA clones from five species (*Daucus carota*, Zimniak et al. 1988; *Equisetum arvense* and *Psilotum nudum*, Starke et al. 1991; *Nicotiana tabacum*, Narasimhan et al. 1991; *Gossypium hirsutum*, T.A. Wilkins 1993). The B subunit (54–57 kDa) is a nucleotide-binding subunit that may have a regulatory function, and cDNAs have been isolated for this subunit from four species (*Arabidopsis thaliana*, Manolson et al. 1988; *Daucus carota*, Struve et al. 1990; *Gossypium hirsutum*, Wan and Wilkins 1994; *Hordeum vulgare*, Berkelman et al. 1994). As yet, neither the sequence nor the function of the smaller subunits in the peripheral sector has been established.

Of the three (or possibly four) subunits in the membrane-integral sector of the V-ATPase, only the 16-kDa subunit has so far been studied in any detail. This subunit forms the H^+-conducting pore through the membrane and is present in six copies per V-ATPase molecule (Sze et al. 1992). If, as seems probable, H^+ flow occurs through each copy of the 16-kDa polypeptide (Cross and Taiz 1990; Holzenburg et al. 1993), this ratio of 16-kDa to A subunits in the complete enzyme could also explain the observed reaction stoichiometry of 2 H^+ transported per ATP hydrolyzed (Smith et al. 1982; Bennett et al. 1984). The first full-length sequence for the 16-kDa subunit from plants was obtained for *Avena sativa* by Lai et al. (1991). Deduced sequences for this subunit have also been obtained from cDNA clones isolated recently from *K. daigremontiana* (Bartholomew et al. 1995) and *M. crystallinum* (Tsiantis et al. 1995). These show an unusually high level of amino-acid sequence identity to the *A. sativa* subunit (97.5% in both cases), given that these species cross the monocot–dicot divide. Indeed, this polypeptide appears to be phylogenetically the most highly conserved membrane protein so far known (Finbow et al. 1995). Interestingly, the 16-kDa subunit of the V-ATPase seems to represent a tandem repeat of the 8-kDa H^+-conducting subunit of the F-type ATP synthase, suggesting that the 16-kDa subunit originated from the duplication of an ancient gene ancestral to both the present-day V- and F-type enzymes (Mandel et al. 1988; Gogarten et al. 1992).

Although the molecular structure of the V-ATPase is now becoming clearer, little is known about the way in which its activity is regulated in vivo. Southern-blot analysis indicates that the 16-kDa subunit of the plant enzyme is encoded by a small multigene family (Lai et al. 1991; Bartholomew et al. 1995; Tsiantis et al. 1995). This implies a degree of specificity in the expression of different isoforms of this subunit in response to developmental or environmental signals. The most notable effects so far on V-ATPase expression have been observed in the halophyte *M. crystallinum*. In response to salt treatment there is a considerable increase in V-ATPase activity (Struve et al. 1985; Bremberger and Lüttge 1992; Barkla et al. 1995), although a component of this may be related to normal leaf ontogeny (Ratajczak et al. 1994b; see also Chap. 25). Remarkably, this response is associated with a significant increase in the diameter of the V-ATPase intramembranous particles seen in freeze-fracture faces of the tonoplast (Klink and Lüttge 1992; Rockel et al. 1994). Protein gels suggest that this reflects an increase in the amount of 16-kDa subunits (Rockel et al. 1994), and a substantial increase in transcript levels is also caused by salt (Tsiantis et al. 1995). Further types of subunit may also be added to the V-ATPase in response to salt (Bremberger and Lüttge 1992; Ratajczak et al. 1994b). The *M. crystallinum* enzyme is clearly one that can provide an excellent model for studies of the control of V-ATPase activity at the molecular level.

The significance of the second type of H^+ pump found at the tonoplast, the PP_iase, remains somewhat enigmatic. This enzyme is found throughout the plant kingdom (Rea and Poole 1993; Maeshima et al. 1994), and in vacuoles of

K. daigremontiana its catalytic activity is comparable with that of the V-ATPase (Marquardt and Lüttge 1987; White and Smith 1989). The significance of the enzyme is doubtless linked to the widespread importance of PP_i metabolism in plants, and thermodynamic considerations show that the PP_iase will act in the direction of substrate hydrolysis (Davies et al. 1993). Electrophysiological studies on isolated vacuoles have shown that the PP_iase can cotransport H^+ and K^+ into the vacuole with a 1:1 stoichiometry, suggesting that the enzyme may play a primary role, in at least some cell types, in active K^+ transport into the vacuole (Davies et al. 1992). Another function for the PP_iase might be as an ancillary enzyme to maintain the proton-motive force across the tonoplast at times when ATPase activity is restricted by substrate availability. This possibility is suggested by the observation that the PP_iase is allosterically activated by free Mg^{2+}, the concentration of which should increase considerably as ATP levels decline and Mg^{2+} is released from the $MgATP^{2-}$ complex (White et al. 1990).

One of the few experimental findings to help identify factors that control the expression of the PP_iase has also been made with *M. crystallinum*. This is the observation by Bremberger et al. (1988) that the activity of the PP_iase declines as CAM is induced in response to salt stress. This is not unexpected, given that the K^+-dependent activity of the PP_iase is competitively inhibited by Na^+ (White et al. 1990). However, this finding leads to the physiologically important conclusion that the V-ATPase in *M. crystallinum* will be the more important enzyme in energizing H^+ across the tonoplast in leaves performing CAM.

4.3.2 Malate Transport into the Vacuole

The simplest model for malate transport into the vacuole at night in CAM plants (Fig. 4.2) envisages malate^{2-} being transported passively down its electrochemical gradient to provide charge balance for the primary H^+ pump(s). Even though transport of the malate^{2-} species is thermodynamically downhill, once inside the malate^{2-} is protonated to give an equilibrium mixture with the Hmalate$^-$ and H_2malate^0 forms (and some becomes complexed with Ca^{2+} and Mg^{2+}, as shown in Table 4.1). Overall, this results in malate 'accumulation' in the vacuole, with a concentration of total malate many times higher than in the cytosol. However, it is not strictly accurate to regard this as an example of 'secondary active transport', since the transported species (malate^{2-}) is believed to move passively down its electrochemical gradient.

Thermodynamic considerations, however, do not provide any information about transport mechanisms at the molecular level. For a long time, malate influx into the vacuole was presumed to occur via some form of carrier or permease (Buser-Suter et al. 1982; Lüttge 1987; Nishida and Tominaga 1987; Martinoia and Rentsch 1994). A convenient method of exploring the characteristics of the influx mechanism was provided by studying the anion dependence of ATP- or PP_i-driven H^+ transport using fluorescent dyes. From these experiments, it could be inferred that the malate influx system had a relatively high (millimolar) K_m and exhibited a distinctive selectivity for four-carbon dicarboxylates with a *trans* carboxyl configuration (Jochem and Lüttge 1987; White and Smith 1989; Marquardt-Jarczyk and Lüttge 1990). Other characteristics, such as the inhibitor sensitivity of this transport system, were studied by monitoring uptake of radiolabelled substrate into non-energized vacuoles of tonoplast vesicles (Buser-Suter et al. 1982; Nishida and Tominaga 1987; Bettey and Smith 1993; Ratajczak et al. 1994a).

It was not until the patch-clamp technique was applied to isolated vacuoles from CAM plants that it became clear that most, if not all, of these transport characteristics could be ascribed to a malate-selective ion channel in the tonoplast. One of the most distinctive properties of this channel is its pronounced rectification, tending only to allow malate flow from the cytosolic side into the vacuole (Iwasaki et al. 1992; Smith and Bryce 1992). This is illustrated in Fig. 4.3 for a range of different anions, with the vacuoles incubated in 'symmetrical' solutions. Under these conditions, current only passes when the inside of the vacuole is clamped at positive voltages relative to the cytosolic surface. Remarkably, however, the relative magnitude of the currents passed matches exactly the anion permeabilities predicted from the earlier, more indirect, experiments with tonoplast vesicles (White and Smith 1989; Bettey

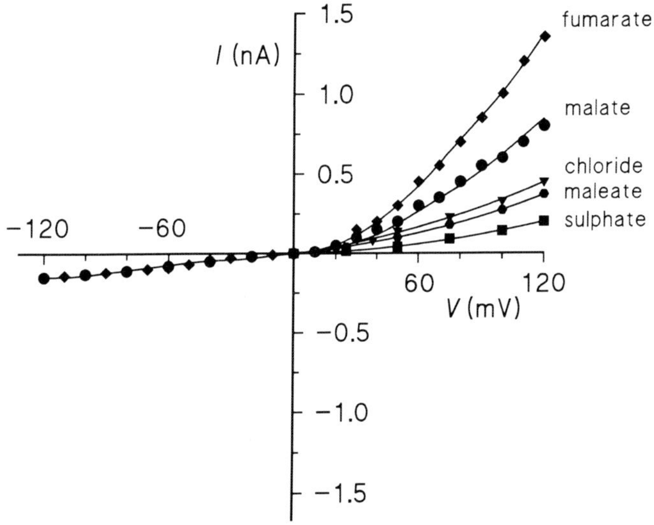

Fig. 4.3. Whole-vacuole current-voltage (*I/V*) plot derived from recordings of steady-state electrical currents in the presence of different anions (100 mM) for an isolated vacuole of *Kalanchoë daigremontiana*. Measurements were made using the whole-vacuole configuration of the patch-clamp technique with symmetrical bath and pipette solutions, containing 100 mM anion (balanced by *N*-methyl-D-glucamine) at pH 7.5 in the absence of added Ca^{2+}, as described by Smith and Bryce (1992)

and Smith 1993). Quantitative analysis also shows that the currents passed per unit area of membrane are sufficient to account for the observed rates of malate^{2-} transport into the vacuole at night (Smith and Bryce 1992).

These characteristics of the tonoplast malate channel suggest an obvious mechanism for vacuolar acidification at night in CAM plants. Activation of the tonoplast H^+ pump (or pumps) at the start of the dark period – presumably in response to the onset of cytoplasmic acidification when malic-acid synthesis is initiated – will render the transtonoplast potential more inside-positive. This in turn will increase the open probability of the malate channel, thereby allowing malate influx from the cytosol to proceed in parallel with H^+ transport. Towards the end of the night, the rate of malic-acid accumulation often slows down considerably: at this stage, it seems that the H^+ electrochemical potential difference against which the H^+-ATPase has to work approaches equilibrium with the free energy available from ATP hydrolysis (Lüttge et al. 1981; Smith et al. 1982). However, net malate efflux from the vacuole does not begin until the light period, and it is possible that the H^+ pump (or pumps) may then be down-regulated to prevent futile cycling or wasteful consumption of ATP. As yet, no information is available on the way in which pump activity is modulated in vivo, but further application of the fluorescence-indicator techniques described in Chap. 6 to CAM plants could provide a sensitive method for monitoring pump activity in real time.

Although the vacuolar malate currents observed in CAM plants are large, there is evidence that a qualitatively similar channel may be found in other plants such as *Beta vulgaris* (Pantoja et al. 1992). Indeed, there is some evidence from studies with tonoplast vesicles that the magnitude of the malate fluxes correlates with the degree of CAM expression (Struve and Lüttge 1987; White and Smith 1992); this has also been observed (P. J. White, unpublished results) with the C_3 to CAM shift that occurs during leaf development in *K. daigremontiana* (Ingram and Smith 1995). More direct measurements with the patch-clamp technique, however, indicate that the area density of the malate channel in the tonoplast does not change under such conditions (O. Pantoja and J. A. C. Smith, unpublished results), suggesting that the effect observed might be attributable to an increase in the amount of tonoplast membrane per unit leaf area or leaf dry mass.

To date, there has been no work on the mechanism of citrate transport at the tonoplast in CAM plants. Evidence from other species suggests that citrate may be transported into the vacuole by a similar type of transport process to malate (Marin et al. 1981; Rentsch and Martinoia 1991). Competition between malate and citrate is consistent with the two anions being transported by the same pathway (Rentsch and Martinoia 1991; Ratajczak et al. 1994a), but competition for binding sites can occur between ions that do not share the same translocation reaction. Further studies of transport at the tonoplast from CAM plants that are known citrate accumulators would be very valuable to clarify the mechanism of citrate transport.

4.3.3 Sodium Chloride Accumulation

In one family of CAM plants, the Aizoaceae, two other ions tend to accumulate in the vacuole to high concentrations, namely Na^+ and Cl^-. Halotolerant species such as *M. crystallinum* can contain over 1 M NaCl in the large epidermal bladder cells (Adams et al. 1992; Lüttge 1993). Consideration of the likely electrochemical gradients across the tonoplast indicates that active transport of Na^+ into the vacuole is clearly required. However, Cl^- might be passively distributed across this membrane, provided the cytosol can tolerate Cl^- concentrations of the order of 100 mM.

Energization of vacuolar Na^+ accumulation in *M. crystallinum* has recently been shown to be achieved by a substrate-specific tonoplast Na^+/H^+ antiporter (Barkla et al. 1995). This antiporter has a high K_m for Na^+ (48 mM), is selective for Na^+ over K^+, and catalyses the electroneutral exchange of H^+ for Na^+. When salt-treated plants are compared with controls of the same age, both Na^+/H^+ antiport activity and H^+-ATPase activity in the tonoplast fraction are increased to about the same degree, providing further evidence that the H^+-ATPase is primarily responsible for tonoplast energization in the salt-loaded (and CAM-induced) tissues (Barkla et al. 1995; cf. Ratajczak et al. 1994b). Cl^- influx is likely to be mediated by a specific anion channel, although this has not yet been investigated by patch-clamp techniques. At any rate, these Na^+ and Cl^- transport systems warrant inclusion in the suite of important ion fluxes that bring about solute accumulation in the vacuole of CAM plants (Fig. 4.2).

4.4 Malic-Acid Remobilization from the Vacuole

With the start of the light period, the nocturnally accumulated malic acid begins to be remobilized from the vacuole and is decarboxylated in the cytoplasm. As with the influx processes, there are important questions concerning both the nature of the ion fluxes across the tonoplast and the way in which they are regulated. However, compared with progress in studies of the tonoplast H^+ pumps and malate-influx mechanism, our understanding of the efflux processes is still very rudimentary.

The transport mechanisms that could conceivably be responsible for malate efflux out of the vacuole are shown in Fig. 4.4. With the vacuole in its acidified state at the beginning of the day, a large proportion of the total malate (e.g. over 80%, see Table 4.1) exists in protonated form ($Hmalate^-$ plus $H_2malate^0$). Because the concentration of these species in the cytosol is negligible, large passive driving forces exist for their downhill movement out of the vacuole. The undissociated acid, $H_2malate^0$, is only present in substantial quantities (10 mM or more) at the very lowest vacuolar pH values towards dawn. This species, however, is relatively lipophilic and could leak out of the vacuole by an uncatalysed passive-diffusion (or 'lipid-solution') mechanism.

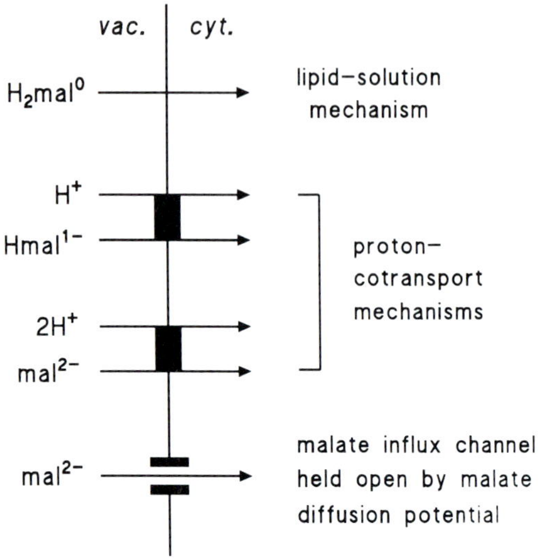

Fig. 4.4. Transport processes that might be responsible for malate efflux from the vacuole during the light period in CAM plants

Based on calculations of the predicted solute permeability coefficient for H_2malate^0, Lüttge and Smith (1984) argued that this flux could be significant at the lowest pH values. Being uncatalysed, it would also be essentially uncontrollable, so this backflux might set an upper limit to the maximum concentrations of malic acid attainable towards the end of the dark period.

For the major part of light period, it seems clear that malate efflux will occur principally in the form of Hmalate$^-$ or malate^{2-} (or both). One of the critical requirements of the efflux processes is that they should be able to mediate the stoichiometric efflux of 2 H^+ for every malate. For this reason, a H^+/malate cotransport mechanism (either H^+/Hmalate$^-$ or 2 H^+/malate^{2-}) would be very attractive, but experimental evidence for this type of transporter is not available. It is also possible that, if the malate^{2-} permeability of the tonoplast is high, the malate influx channel could be held open by an inside-positive malate diffusion potential. This, however, depends on the biophysical origins of the strong rectification of the malate-influx channel, which Arata et al. (1992) and Iwasaki et al. (1992) believe is gated in such a way as only to permit malate^{2-} transport *into* the vacuole. Instead, Iwasaki et al. (1992) have suggested that the slow vacuolar (SV) channel described by Hedrich and Neher (1987) might be responsible for malate^{2-} efflux. A serious difficulty with both of these channel models is that they suggest no plausible pathway for efflux of H^+ in the stoichiometric amounts needed to explain vacuolar deacidification.

In a manner complementary to the influx process, it seems likely that the efflux system is also subject to some form of kinetic control. Temperature is known to have an important influence on rates of malate efflux and might even poise the system between net influx and net efflux (see Chap. 5). However, the CAM cycle can proceed effectively under constant day–night temperature conditions, so control must presumably operate to prevent malate efflux from the vacuole at night. As M. B. Wilkins (1992) has recognized, components of these control processes at the membrane level might be responsible for the phase-shifting of circadian rhythms in *Kalanchoë*. First, however, a better understanding of the transport mechanisms at the tonoplast will be needed before it is possible to resolve how malate influx and efflux are co-ordinated with the day–night cycle of carbon metabolism in CAM plants.

Acknowledgements. This work has been supported by research grants from the Agricultural and Food Research Council, the Science and Engineering Research Council (now both part of the Biotechnology and Biological Sciences Research Council), the European Community, and by the Gatsby Charitable Foundation.

References

Adams P, Thomas JC, Vernon DM, Bohnert HJ, Jensen RG (1992) Distinct cellular and organismic responses to salt stress. Plant Cell Physiol 33: 1215–1223

Arata H, Iwasaki I, Kusumi K, Nishimura M (1992) Thermodynamics of malate transport across the tonoplast of leaf cells of CAM plants. Plant Cell Physiol 33: 873–880

Barkla BJ, Zingarelli L, Blumwald E, Smith JAC (1995) Tonoplast Na^+/H^+ antiport activity and its energization by the vacuolar H^+-ATPase in the halophytic plant *Mesembryanthemum crystallinum*. Plant Physiol (in press.)

Bartholomew DM, Rees DJG, Rambaut A, Smith JAC (1995) Isolation and sequence analysis of a cDNA encoding the 16 kDa subunit of a vacuolar-type H^+-ATPase from the CAM plant *Kalanchoë daigremontiana* (submitted for publication.)

Bennett AB, Spanswick RM (1984) H^+-ATPase activity from storage tissue of *Beta vulgaris*. II. H^+/ATP stoichiometry of an anion-sensitive ATPase. Plant Physiol 74: 545–548

Berkelman T, Houtchens KA, DuPont FM (1994) Two cDNA clones encoding isoforms of the B subunit of the vacuolar ATPase from barley roots. Plant Physiol 104: 287–288

Bettey M, Smith JAC (1993) Dicarboxylate transport at the vacuolar membrane of the CAM plant *Kalanchoë daigremontiana*: sensitivity to protein-modifying and sulphydryl reagents. Biochim Biophys Acta 1152: 270–279

Boller T, Wiemken A (1986) Dynamics of vacuolar compartmentation. Annu Rev Plant Physiol 37: 137–164

Bremberger C, Lüttge U (1992) Dynamics of tonoplast proton pumps and other tonoplast proteins of *Mesembryanthemum crystallinum* L. during the induction of crassulacean acid metabolism. Planta 188: 575–580

Bremberger C, Haschke H-P, Lüttge U (1988) Separation and purification of the tonoplast ATPase and pyrophosphatase from plants with constitutive and inducible crassulacean acid metabolism. Planta 175: 465–470

Buser-Suter C, Wiemken A, Matile P (1982) A malic acid permease in isolated vacuoles of a crassulacean acid metabolism plant. Plant Physiol 69: 456–459

Cross RL, Taiz L (1990) Gene duplication as a means for altering H^+/ATP ratios during the evolution of F_oF_1 ATPases and synthases. FEBS Lett 259: 227–229

Davies JM, Poole RJ, Rea PA, Sanders D (1992) Potassium transport into plant vacuoles energized directly by a proton-pumping inorganic pyrophosphatase. Proc Natl Acad Sci USA 89: 11701–11705

Davies JM, Poole RJ, Sanders D (1993) The computed free energy change of hydrolysis of inorganic pyrophosphate and ATP: apparent significance for inorganic pyrophosphate-driven reactions of intermediary metabolism. Biochim Biophys Acta 1141: 29–36

Finbow ME, Harrison M, Jones P (1995) Ductin – a proton pump component, a gap junction channel and a neurotransmitter release channel. BioEssays 17: 247–255
Forgac M (1989) Structure and function of vacuolar class of ATP-driven proton pumps. Physiol Rev 69: 765–796
Franco AC, Ball E, Lüttge U (1990) Patterns of gas exchange and organic acid oscillations in tropical trees of the genus *Clusia*. Oecologia 85: 108–114
Franco AC, Ball E, Lüttge U (1992) Differential effects of drought and light levels on accumulation of citric and malic acids during CAM in *Clusia*. Plant Cell Environ 15: 821–829
Gibson AC (1982) The anatomy of succulence. In: Ting IP, Gibbs M (eds) Crassulacean acid metabolism. American Society of Plant Physiologists, Rockville, pp 1–17
Gogarten JP, Starke T, Kibak H, Fishmann J, Taiz L (1992) Evolution and isoforms of V-ATPase subunits. J Exp Biol 172: 137–147
Hedrich R, Neher E (1987) Cytoplasmic calcium regulates voltage-dependent ion channels in plant vacuoles. Nature 329: 833–836
Holzenburg A, Jones PC, Franklin T, Pali T, Heimburg T, Marsh D, Findlay JBC, Finbow ME (1993) Evidence for a common structure for a class of membrane channels. Eur J Biochem 213: 21–30
Ingram J, Smith JAC (1995) Developmental regulation of NAD- and NADP-malic enzyme activities in leaves of the CAM plant *Kalanchoë daigremontiana* (submitted for publication.)
Iwasaki I, Arata H, Kijima H, Nishimura M (1992) Two types of channels involved in the malate ion transport across the tonoplast of a crassulacean acid metabolism plant. Plant Physiol 98: 1494–1497
Jochem P, Lüttge U (1987) Proton transporting enzymes at the tonoplast of leaf cells of the CAM plant *Kalanchoë daigremontiana*. I. The ATPase. J Plant Physiol 129: 251–268
Kinzel H (1989) Calcium in the vacuoles and cell walls of plant tissue. Forms of deposition and their physiological and ecological significance. Flora 182: 99–125
Klink R, Lüttge U (1992) Quantification of visible structural changes of the leaf tonoplast of *Mesembryanthemum crystallinum* by freeze fracture replicas prepared during the C_3-photosynthesis to CAM-transition. Bot Acta 105: 414–420
Klink R, Haschke H-P, Kramer D, Lüttge U (1990) Membrane particles, proteins and ATPase activity of tonoplast vesicles of *Mesembryanthemum crystallinum* in the C_3 and CAM state. Bot Acta 103: 24–31
Lai S, Watson JC, Hansen JN, Sze H (1991) Molecular cloning and sequencing of cDNAs encoding the proteolipid subunit of the vacuolar H^+-ATPase from a higher plant. J Biol Chem 266: 16078–16084
Lüttge U (1987) Carbon dioxide and water demand: crassulacean acid metabolism (CAM), a versatile ecological adaptation exemplifying the need for integration in ecophysiological work. New Phytol 106: 593–629
Lüttge U (1988) Day–night changes of citric acid levels in CAM: phenomenon and ecophysiological significance. Plant Cell Environ 11: 445–451
Lüttge U (1993) The role of crassulacean acid metabolism (CAM) in the adaptation of plants to salinity. New Phytol 125: 59–71
Lüttge U, Ball E (1979) Electrochemical investigation of active malic acid transport at the tonoplast into the vacuoles of the CAM plant *Kalanchoë daigremontiana*. J Membr Biol 47: 401–422
Lüttge U, Nobel PS (1984) Day–night variations in malate concentration, osmotic pressure, and hydrostatic pressure in *Cereus validus*. Plant Physiol 75: 804–807
Lüttge U, Smith JAC, (1984) Mechanism of passive malic-acid efflux from vacuoles of the CAM plant *Kalanchoë daigremontiana*. J Membr Biol 81: 149–158
Lüttge U, Smith JAC, Osmond CB, Marigo G (1981) Energetics of malate accumulation in the vacuoles of *Kalanchoë tubiflora*. FEBS Lett 126: 81–84
Lüttge U, Smith JAC, Marigo G (1982) Membrane transport, osmoregulation, and the control of CAM. In: Ting IP, Gibbs M (eds) Crassulacean acid metabolism. American Society of Plant Physiologists, Rockville, pp 69–91
Maeshima M, Mimura T, Sato T (1994) Distribution of vacuolar H^+-pyrophosphatase and a membrane integral protein in a variety of green plants. Plant Cell Physiol 35: 323–328

Mandel MY, Moriyama Y, Hulmes JD, Pan Y-CE, Nelson H, Nelson N (1988) cDNA sequence encoding the 16 kDa proteolipid subunit of chromaffin granules implies gene duplication in the evolution of H^+-ATPases. Proc Natl Acad Sci USA 85: 5521–5524

Manolson MF, Ouelette BFF, Filion M, Poole RJ (1988) cDNA sequence and homologies of the "57 kDa" nucleotide-binding subunit of the vacuolar ATPase from *Arabidopsis*. J Biol Chem 263: 17987–17994

Marin B, Smith JAC, Lüttge U (1981) The electrochemical proton gradient and its influence on citrate uptake in tonoplast vesicles of *Hevea brasiliensis*. Planta 153: 486–493

Marquardt G, Lüttge U (1987) Proton transporting enzymes at the tonoplast of leaf cells of the CAM plant *Kalanchoë daigremontiana*. II. The pyrophosphatase. J Plant Physiol 129: 269–286

Marquardt-Jarczyk G, Lüttge U (1990) Anion transport at the tonoplast of mesophyll cells of the CAM plant *Kalanchoë daigremontiana*. J Plant Physiol 136: 129–136

Martinoia E, Rentsch D (1994) Malate compartmentation – responses to a complex metabolism. Annu Rev Plant Physiol Plant Mol Biol 45: 447–467

Matile P (1987) The sap of plants. New Phytol 105: 1–26

Murphy R, Smith JAC (1994a) A critical comparison of the pressure-chamber and pressure-probe techniques for estimating cell turgor pressure in leaves of *Kalanchoë daigremontiana*. Plant Cell Environ 17: 15–29

Murphy R, Smith JAC (1994b) Derivation of a weighted-average reflection coefficient for mesophyll cell membranes of *Kalanchoë daigremontiana*. Planta 193: 145–147

Narasimhan ML, Binzel ML, Perez-Prat E, Chen Z, Nelson DE, Singh NK, Bressan RA, Hasegawa PM (1991) NaCl regulation of tonoplast ATPase 70-kilodalton subunit mRNA in tobacco cells. Plant Physiol 97: 562–568

Nelson N (ed) (1995) Organellar proton-ATPases. Springer, Berlin Heidelberg New York

Nishida K, Tominga O (1987) Energy-dependent uptake of malate into vacuoles isolated from CAM plant, *Kalanchoë daigremontiana*. J Plant Physiol 127: 385–393

Nobel PS (1988) Environmental biology of agaves and cacti. Cambridge University Press, Cambridge

Osmond CB (1976) Ion absorption and carbon metabolism in cells of higher plants. In: Lüttge U, Pitman MG (eds) Encyclopedia of plant physiology, new series, vol 2. Transport in plants II, part A, Cells. Springer, Berlin Heidelberg New York, pp 347–372

Pantoja O, Gelli A, Blumwald E (1992) Characterization of vacuolar malate and K^+ channels under physiological conditions. Plant Physiol 100: 1137–1141

Phillips RD (1980) Deacidification in a plant with crassulacean acid metabolism associated with anion-cation balance. Nature 287: 727–728

Phillips RD, Jennings DH (1976) Succulence, cations and organic acids in leaves of *Kalanchoë daigremontiana* grown in long and short days in soil and water culture. New Phytol 77: 599–611

Ratajczak R, Kemna I, Lüttge U (1994a) Characteristics, partial purification and reconstitution of the vacuolar malate transporter of the CAM plant *Kalanchoë daigremontiana* Hamet et Perrier de la Bâthie. Planta 195: 226–236

Ratajczak R, Richter J, Lüttge U (1994b) Adaptation of the tonoplast V-type H^+-ATPase of *Mesembryanthemum crystallinum* to salt stress, C_3-CAM transition and plant age. Plant Cell Environ 17: 1101–1112

Raven JA (1987) The role of vacuoles. New Phytol 106: 357–422

Raven JA, Smith FA (1976) Nitrogen assimilation and transport in vascular plants in relation to intracellular pH regulation. New Phytol 76: 415–431

Rea PA, Poole RJ (1993) Vacuolar H^+-translocating pyrophosphatase. Annu Rev Plant Physiol Plant Mol Biol 44: 157–180

Rea PA, Sanders D (1987) Tonoplast energization: two H^+ pumps, one membrane. Physiol Plant 71: 131–141

Rentsch D, Martinoia E (1991) Citrate transport into barley mesophyll vacuoles – comparison with malate-uptake activity. Planta 184: 532–537

Rockel B, Ratajczak R, Becker A, Lüttge U (1994) Changed densities and diameters of intra-membrane tonoplast particles of *Mesembryanthemum crystallinum* in correlation with NaCl-induced CAM. J Plant Physiol 143: 318–324

Rona J-P, Pitman MG, Lüttge U, Ball E (1980) Electrochemical data on compartmentation into cell wall, cytoplasm, and vacuole of leaf cells in the genus *Kalanchoë*. J Membr Biol 57: 25–35

Ruess BR, Eller BM (1985) The correlation between crassulacean acid metabolism and water uptake in *Senecio medley-woodii*. Planta 166: 57–66

Rygol J, Winter K, Zimmermann U (1987) The relationship between turgor pressure and titratable acidity in mesophyll cells of intact leaves of a crassulacean-acid-metabolism plant, *Kalanchoë daigremontiana*. Planta 172: 487–493

Schulte PJ, Smith JAC, Nobel PS (1989) Water storage and osmotic pressure influences on the water relations of a dicotyledonous desert succulent. Plant Cell Environ 12: 637–648

Smith FA, Raven JA (1979) Intracellular pH and its regulation. Annu Rev Plant Physiol 30: 289–311

Smith JAC (1984) Water relations in CAM plants. In: Medina E (ed) Physiological ecology of CAM plants. CIET (Unesco-IVIC), Caracas, pp 30–51

Smith JAC (1987) Vacuolar accumulation of organic acids and their anions in CAM plants. In: Marin B (ed) Plant vacuoles: their importance in solute compartmentation and their applications in plant biotechnology. Plenum, New York, pp 79–87

Smith JAC, Bryce JH (1992) Metabolite compartmentation and transport in CAM plants. In: Tobin AK (ed) Plant organelles. Cambridge University Press, Cambridge, pp 141–167

Smith JAC, Lüttge U (1985) Day–night changes in leaf water relations associated with the rhythm of crassulacean acid metabolism in *Kalanchoë daigremontiana*. Planta 163: 272–282

Smith JAC, Marigo G, Lüttge U, Ball E (1982) Adenine-nucleotide levels during crassulacean acid metabolism and the energetics of malate accumulation in *Kalanchoë tubiflora*. Plant Sci Lett 26: 13–21

Smith JAC, Griffiths H, Lüttge U, Crook CE, Griffiths NM, Stimmel K-H (1986) Comparative ecophysiology of CAM and C_3 bromeliads. IV. Plant water relations. Plant Cell Environ 9: 395–410

Smith JAC, Schulte PJ, Nobel PS (1987) Water flow and water storage in *Agave deserti*: osmotic implications of crassulacean acid metabolism. Plant Cell Environ 10: 639–648

Starke T, Linkila TP, Gogarten JP (1991) Two separate genes encode the catalytic 70 kDa V-ATPase subunit in *Psilotum* and *Equisetum*. Z Naturforsch 46c: 613–620

Steudle E, Smith JAC, Lüttge U (1980) Water-relation parameters of individual mesophyll cells of the crassulacean acid metabolism plant *Kalanchoë daigremontiana*. Plant Physiol 66: 1155–1163

Stewart PA (1983) Modern quantitative acid-base chemistry. Can J Physiol Pharmacol 61: 1444–1464

Struve I, Lüttge U (1987) Characteristics of $MgATP^{2-}$-dependent electrogenic proton transport in tonoplast vesicles of the facultative crassulacean-acid-metabolism plant *Mesembryanthemum crystallinum* L. Planta 170: 111–120

Struve I, Weber A, Lüttge U, Ball E, Smith JAC (1985) Increased vacuolar ATPase activity correlated with CAM induction in *Mesembryanthemum crystallinum* and *Kalanchoë blossfeldiana* cv. Tom Thumb. J Plant Physiol 117: 451–468

Struve I, Rausch T, Bernasconi P, Taiz L (1990) Structure and function of the promoter of the carrot V-type H^+-ATPase catalytic subunit gene. J Biol Chem 265: 7927–7932

Sze H, Ward JM, Lai S, Perera I (1992) Vacuolar-type H^+-translocating ATPases in plant endomembranes: subunit organization and multigene families. J Exp Biol 172: 123–135

Tsiantis MS, Bartholomew DM, Smith JAC (1995) Salt regulation of transcript levels for the 16 kDa subunit of a leaf vacuolar H^+-ATPase in the halophyte *Mesembryanthemum crystallinum* (submitted for publication.)

Walter H (1960) Einführung in die Phytologie, vol 3: Grundlagen der Pflanzenverbreitung. 1. Standortslehre (analytisch-ökologische Geobotanik), 2nd edn. Ulmer, Stuttgart

Wan CY, Wilkins TA (1994) Isolation of multiple cDNAs encoding the vacuolar H^+-ATPase subunit B from developing cotton (*Gossypium hirsutum* L.) ovules. Plant Physiol 106: 393–394
Ward JM, Sze H (1992) Subunit composition and organization of the vacuolar H^+-ATPase from oat roots. Plant Physiol 99: 170–179
Warren M, Smith JAC, Apps DK (1992) Rapid purification and reconstitution of a plant vacuolar ATPase using Triton X-114 fractionation: composition and substrate kinetics of the H^+-ATPase from the tonoplast of *Kalanchoë daigremontiana*. Biochim Biophys Acta 1106: 117–125
White PJ, Smith JAC (1989) Proton and anion transport at the tonoplast in crassulacean-acid-metabolism plants: specificity of the malate-influx system in *Kalanchoë daigremontiana*. Planta 179: 265–274
White PJ, Smith JAC (1992) Malate-dependent proton transport in tonoplast vesicles isolated from orchid leaves correlates with the expression of crassulacean acid metabolism. J Plant Physiol 139: 533–538
White PJ, Marshall J, Smith JAC (1990) Substrate kinetics of the tonoplast H^+-translocating inorganic pyrophosphatase and its activation by free Mg^{2+}. Plant Physiol 93: 1063–1070
Wilkins MB (1992) Circadian rhythms: their origin and control. New Phytol 121: 347–375
Wilkins TA (1993) Vacuolar H^+-ATPase 69-kilodalton catalytic subunit cDNA from developing cotton (*Gossypium hirsutum*) ovules. Plant Physiol 102: 679–680
Willmer CM, Fricker MD (1995) Stomata, 2nd edn. Chapman and Hall, London
Wink M (1993) The plant vacuole: a multifunctional compartment. J Exp Bot 44 [Suppl]: 231–246
Zimniak L, Dittrich P, Gogarten JP, Kibak H, Taiz L (1988) The cDNA sequence of the 69-kDa subunit of the carrot vacuolar H^+-ATPase: homology to the beta-chain of F_oF_1-ATPase. J Biol Chem 263: 9102–9112

5 The Tonoplast as a Target of Temperature Effects in Crassulacean Acid Metabolism

M. KLUGE and M. SCHOMBURG

5.1 Introduction

Temperature is known to be one of the most important factors bringing about phenotypic modulation of the diel and seasonal course of CAM (e.g. Kluge and Ting 1978; Medina and Osmond 1981; Winter 1985; Lüttge 1987). Temperature may affect CAM at several different levels. There are indirect effects, for instance modulation of the diel course of gas exchange via the leaf/air water-vapour pressure difference (Lüttge 1987; von Willert et al. 1992). Temperature may also act directly through the Q_{10} of the various processes involved in the metabolic pathways.

Kluge and Ting (1978) were the first to stress the view that a further temperature effect on CAM might be related to certain temperature-dependent properties of the tonoplast which controls the reversible storage of malic acid (see Chap. 4). In spite of support from physiological experiments (Wilkins 1983; Friemert et al. 1988; Kluge et al. 1991a,b), further evidence is necessary to verify the hypothesis that the properties of the vacuolar membrane are directly connected with temperature-mediated modulations of CAM. We have started to seek such evidence in biophysical studies of the temperature-dependent phase behaviour of the tonoplast lipids. This type of approach has become possible because of the development of reliable techniques for obtaining tonoplast material from CAM plants (Smith et al. 1984 a,b). In this chapter we will briefly consider the theoretical background to these studies and report on recent progress in these investigations.

5.2 Possible Implications of the Temperature-Dependent Phase Behaviour of Tonoplast Lipids for CAM

Storage of malic acid in the vacuole at night and its release during the following day belong to the key processes of CAM. Malic-acid efflux from the vacuole is thermodynamically passive (Lüttge and Smith 1984; Smith and Bryce 1992). At high malic-acid concentrations in the vacuole, with vacuolar pH being low, the

Institut für Botanik, Technische Hochschule Darmstadt, Schnittspahnstr. 10, D-64287 Darmstadt, Germany

Ecological Studies, Vol. 114
Winter/Smith (eds) Crassulacean Acid Metabolism

malic acid efflux from the vacuole may proceed largely by diffusion of undissociated malic-acid molecules through the lipid phase of the lipid-protein-mosaic constituting the membrane. During much of the day and generally under conditions when malic acid acumulation of CAM operates under suboptimal conditions, efflux must proceed as $Hmal^-$ and/or mal^{2-}, presumably via a carrier-catalyzed process (Smith and Bryce, 1992; see Chap. 4). Since efflux of H_2mal by a lipid-solution mechanism represents essentially an uncontrollable diffusion process, the accumulation of undissociated malic-acid molecules sets an upper limit to the maximum attainable vacuolar concentrations of malic acid (Lüttge and Smith 1984; Smith and Bryce 1992). Altogether, any process facilitating uncontrollable diffusion must lower the upper accumulation limit for malic-acid in the vacuole. We propose that enhancement of uncontrollable efflux is an important factor in explaining how elevation of temperatures may affect vacuolar storage of malic acid.

This proposal is consistent with the hypothesis of Wilkins (1983), who postulated that high temperatures enhance the release of malic acid from the vacuole by opening "gates" in the tonoplast, without giving a mechanistic explanation in terms of membrane structure. A reasonable such explanation can be derived from the observations of Papahadjopoulos et al. (1973) that phase transitions in a biomembrane occur with an increase in temperature, i.e. the formation of a mosaic of rigid and fluid lipid domains. According to Papahadjopoulos et al. (1973) this leads to the formation of "pores" which come into existence at the boundary surfaces between rigid and fluid lipid domains, and presumably also at the boundaries between the lipid domains and the membrane proteins of a membrane undergoing a phase transition. It is possible that the "gates" postulated by Wilkins (1983) in the tonoplast are identical with pores deriving from the phase transition of tonoplast lipids. Transport studies (Papahadjopoulos et al. 1973; Schreier-Mucillo et al. 1976) revealed that in the temperature range where phase transition apparently occurs, the efflux of ions and electrically neutral substances from preloaded membrane vesicles increased dramatically. This shows that pores between the lipid phases can lower the diffusion resistance and thereby enhance passive fluxes. Thus, the phase transition in the membrane lipids may also be highly relevant for the rate of passive malic-acid efflux from the vacuole in CAM.

Effective accumulation of malic acid in the vacuole at night would require a high diffusion resistance of the tonoplast to keep the rate of passive efflux low relative to the rate of active transport of malic acid into the vacuole. Thus, the temperature-dependent properties of the tonoplast of a CAM plant should be conditioned in such a way that a phase transition (and with it uncontrollable efflux) does not occur at normal night temperatures. Field studies on CAM succulents in the Namib desert (von Willert et al. 1992) have shown that plants began to deacidify instantaneously at night if a sudden hot wind (Föhn) increases the air temperature drastically. In the light period of CAM, however, the formation of pores by phase transition could be helpful to facilitate passive efflux of malic acid from the vacuole.

The phase behaviour of tonoplast lipids could also be relevant if malic-acid efflux does not proceed by diffusion through the lipid matrix itself but rather, as proposed by Arata et al. (1992), via specific protein channels (see also Chap. 4). It has been shown by Wolf (1978) that temperature-induced changes in the fluidity of the bulk membrane lipids can influence the conformation of membrane proteins and thereby the kinetics and regulatory properties of carriers and channels. This must have consequences for the rates of channel- or carrier-mediated transport, including malic-acid export from the vacuole.

Summing up, we believe that the temperature-dependent properties of the tonoplast, and particularly the phase behaviour of the membrane lipids, are highly relevant for the causal explanation of temperature effects on CAM and that these deserve more detailed study. In the following, we report on attempts made in this direction.

5.3 Experimental Approaches

Electron paramagnetic resonance (EPR) spectroscopy of spin-labelled lipids is a powerful tool for monitoring the temperature dependence of membrane fluidity (for theory see e.g. Marsh 1981; Galla 1988). We have applied this technique to the study of the temperature-dependent properties of the tonoplast in the obligate CAM plant *Kalanchoë daigremontiana* and the facultative CAM plant *Mesembryanthemum crystallinum* (Kluge et al. 1991a; Kliemchen et al. 1993). We found that, generally, as is to be expected, an increase in temperature leads to an increase in the fluidity of the tonoplast. However, a more detailed analysis (Kluge et al. 1991a,b) with tonoplast material from *K. daigremontiana* grown under our standard temperature regime (17 °C night/ 26 °C day) revealed that this fluidization does not proceed steadily over the whole range of tested temperatures (4 to 51 °C). Rather, at temperatures lower than 10 °C the membrane remains unchanged in a highly rigid state, and only above that temperature does the membrane begin to melt. The data showed clearly that between 10 and 14 °C the membrane is partially fluidized, leading to the mosaic-like structure of fluidized domains coexisting with rigid areas of quasi-crystalline lipids described above. Between 15 and 18 °C the rigidity of the membrane then remains unchanged. At temperatures higher than 18.5 °C, the membrane fluidity again begins abruptly to increase, indicating the onset of a second melting process. Thereafter, up to 51 °C (the highest temperature tested so far) the melting of the bulk membrane proceeds steadily.

The temperature-dependent behaviour of the tonoplast of *K. daigremontiana* fits well with the physiological characteristics of this plant. At temperatures lower than 10 °C, where we found the tonoplast to be in a highly rigid state, release of malic acid from the vacuole practically does not occur, i.e. in the light the plants do not deacidify (Kluge 1968; Kluge et al. 1991a). At 18 °C, where we observed in the tonoplast the sudden onset of the second melting process, there was a rather sudden increase in the mobilization of malic acid from the vacuole (Kluge et al.

1991a). It is thus tempting to speculate (but remains to be demonstrated) that both phenomena are causally linked.

As shown by Kliemchen et al. (1993), the fluidity of the tonoplast of CAM plants may undergo changes in response to temperature and salinity stress. In *K. daigremontiana*, acclimation of the plants to higher temperatures for 30 days, i.e. from previously 17 °C (night)/ 25 °C (day) to 25 °C (night)/ 34 °C (day), led to a marked increase in the rigidity of the membrane. In the acclimated plants, temperatures about 20 °C higher were required than for the standard-temperature plants to reach a given degree of tonoplast membrane fluidity. In other words, in the acclimated plants higher temperatures are required to elicit the phase transition in the membrane lipids. Moreover, malate efflux from the vacuole at a given temperature was significantly slower in the plants acclimated to high temperatures compared with the non-acclimated controls (Kliemchen et al. 1993). Berry and Raison (1981) also observed in *Nerium oleander* an increase in the rigidity of chloroplast membranes upon acclimation to high temperatures, although to a much lesser extent compared with the tonoplast of *K. daigremontiana*. Thus, the results we have obtained with the tonoplast of CAM plants might also apply to other membranes and represent a general feature of temperature acclimation.

As shown by Kliemchen et al. (1993) in *Mesembryanthemum crystallinum*, an increase in tonoplast rigidity also develops during salinity-mediated CAM induction. The tonoplast from plants in the CAM state required temperatures 3.3 °C higher to reach the same degree of fluidity as membranes from plants in the C_3 state. The shift from C_3 to CAM may increase the demands for storage of malic acid in the vacuole, and the observed shift towards higher rigidity of the tonoplast improves the ability for such storage.

The precise mechanisms behind the observed phenotypic changes in the fluidity of the tonoplast membrane are not yet clear. We have compared the fluidity of the complete tonoplast and of its lipid fraction deprived of membrane proteins. There was no difference in the lipid fraction of the tonoplast from acclimated and non-acclimated plants, whereas the complete membranes showed the large difference in fluidity as described above (unpubl.). This suggests that the observed fluidity changes are related more to the membrane proteins rather than the lipid matrix. We have also found that free Ca^{2+} decreased the fluidity of the membrane. It remains to be clarified whether or not this phenomenon is relevant for the control of the reversible malic-acid storage in the vacuole.

5.4 Outlook

Our investigations on the temperature-dependent phase behaviour of the tonoplast membrane of CAM plants are still at the onset, but the results obtained thus far suggest that temperature can influence CAM through its effect on the vacuole membrane. Up to now, the evidence that the observed phase behaviour is relevant for the passive malic-acid fluxes across the tonoplast derives exclusively from physiological experiments with living plant material (e.g. Friemert et al. 1988; Kluge et al. 1991b; Kliemchen et al. 1993) and is therefore rather indirect. Studies on the temperature profile of passive efflux with isolated tonoplast vesicles from CAM plants are now required parallel to the EPR measurements. Such studies could give more direct answers on the question whether the temperature-dependent phase behaviour of the tonoplast lipids is causally linked to the temperature responses of malic-acid accumulation and mobilization in CAM. The same holds true for likely interactions between the phase behaviour of the tonoplast lipids and the properties of the membrane proteins.

Finally, if temperature can modulate CAM activity via an effect on the tonoplast, there should exist not only phenotypic but also genotypic differences in the temperature-dependent phase behaviour of this membrane. The optimal temperature for storage of malic acid in the vacuole differs in different species of CAM plants and depends on the temperature regime in their natural habitats (for review see e.g. Winter 1985; Lüttge 1987). Thus, one might expect that in CAM plants derived from different climates (for instance, semi-deserts with cool nights and large day/night temperature differences, or tropical rain forests with high temperatures throughout the diel cycle), the phase behaviour of the tonoplast would be genotypically quite different. Further work will be needed to determine whether or not this expectation holds true.

References

Arata H, Iwasaki I, Kusumi K, Nishimura M (1992) Thermodynamics of malate transport across the tonoplast of leaf cells of CAM plants. Plant Cell Physiol 33: 873–880

Berry JA, Raison JK (1981) Responses of macrophytes to temperature. In: Lange OL, Nobel PS, Osmond CB, Ziegler H (eds) Encyclopedia of plant physiology New Series, vol 12 A. Physiological plant ecology I. Springer, Berlin Heidelberg New York, pp 277–338

Friemert V, Heininger D, Kluge M, Ziegler H (1988) Temperature effects on malic-acid efflux from the vacuoles and on the carboxylation pathways in crassulacean acid metabolism plants. Planta 174: 453–461

Galla H-J (1988) Spektroskopische Methoden in der Biochemie. Thieme, Stuttgart

Kaplan A, Gale J, Poljakoff-Mayber A (1976) Resolution of net fixation of carbon dioxide into its respiration and gross fixation components in *Bryophyllum daigremontianum*. J Exp Bot 27: 220–230

Kliemchen A, Schomburg M, Galla H-J, Lüttge U, Kluge M (1993) Phenotypic changes in the fluidity of the tonoplast membrane of crassulacean-acid-metabolism plants in response to temperature and salinity stress. Planta 189: 403–409

Kluge M (1968) Untersuchungen über den Gaswechsel von *Bryophyllum* während der Lichtperiode. II. Beziehungen zwischen dem Malatgehalt des Blattgewebes und der CO_2-Aufnahme. Planta 80: 359–377

Kluge M, Ting IP (1978) Crassulacean acid metabolism. Analysis of an ecological adaptation. Springer, Berlin Heidelberg New York

Kluge M, Wolf H, Fischer A (1991a) Crassulacean acid metabolism: temperature effects on the lag-phase in the photosynthetic oxygen evolution occurring at the onset of the light period. Plant Physiol Biochem 29: 83–90

Kluge M, Kliemchen A, Galla H -J (1991b) Temperature effects on crassulacean acid metabolism: EPR spectroscopy studies on the thermotropic behaviour of the tonoplast membranes of *Kalanchoë daigremontiana*. Bot Acta 104: 355–360

Lüttge U (1987) Carbon dioxide and water demand: crassulacean acid metabolism (CAM), a versatile ecological adaptation exemplifying the need for integration in ecophysiological work. New Phytol 106: 593–629
Lüttge U, Smith JAC (1984) Mechanism of passive malic-acid efflux from vacuoles of the CAM plant *Kalanchoë daigremontiana*. J Membr Biol 81: 149–158
Marsh D (1981) Electron spin resonance: In: Grell E (ed) Membrane spectroscopy. Springer, Berlin Heidelberg New York, pp 51–142
Medina E, Osmond CB (1981) Temperature dependence of dark CO_2 fixation and acid accumulation in *Kalanchoë daigremontiana*. Aust J Plant Physiol 8: 641–649
Osmond CB (1978) Crassulacean acid metabolism: a curiosity in context. Annu Rev Plant Physiol 29: 379–414
Papahadjopoulos D, Jacobsen K, Nir S, Isac T (1973) Phase transition in phospholipid vesicles. Fluorescence polarization and permeability measurements concerning the effect of temperature and cholesterol. Biochim Biophys Acta 311: 330–348
Schreier-Mucillo S, Marsh D, Smith ICP (1976) Monitoring the permeability profile of lipid membranes with spin probes. Biochemistry 172: 1–11
Smith JAC, Bryce JH (1992) Metabolite compartmentation and transport in CAM plants. In: Tobin AK (ed) Plant organelles. Cambridge University Press, Cambridge, pp 141–167
Smith JAC, Uribe EG, Ball E, Lüttge U (1984a) ATPase activity associated with isolated vacuoles of the crassulacean acid metabolism plant *Kalanchoë daigremontiana*. Planta 162: 299–304
Smith JAC, Uribe EG, Ball U, Heuer S, Lüttge U (1984b) Characterization of the vacuolar ATPase activity of the crassulacean-acid-metabolism plant *Kalanchoë daigremontiana*. Eur J Biochem 141: 415–420
von Willert D, Eller BM, Werger MJA, Brinckmann E, Ihlenfeldt H-D (1992) Life strategies of succulents in the desert. Cambridge University Press, Cambridge
Wilkins MB (1983) The circadian rhythm of carbon dioxide metabolism in *Bryophyllum*: the mechanism of phase-shift induction by thermal stimuli. Planta 157: 471–480
Winter K (1985) Crassulacean acid metabolism. In: Barber J, Baker NR (eds) Photosynthetic mechanisms and the environment. Elsevier, Amsterdam, pp 329–387
Wolfe J (1978) Chilling injury in plants - the role of membrane fluidity. Plant Cell Environ 1: 241–247

6 Regulation of Crassulacean Acid Metabolism in *Kalanchoë pinnata* as Studied by Gas Exchange and Measurements of Chlorophyll Fluorescence

U. HEBER, S. NEIMANIS, and W.M. KAISER

6.1 Introduction

As is well known, CAM plants open stomata at night and close them during the day to save water in regions characterized by periodic drought, hot days and cooler nights (see reviews by Winter 1985; Lüttge 1987; Griffiths 1988; Leegood and Osmond 1990; Smith and Bryce 1992). At night, they permit entry of atmospheric CO_2, which is transiently stored in the vacuole in the form of a carboxylic group of malic acid. In the light, malate is released from the vacuole and decarboxylated in the cytosol, usually behind closed stomata. The Calvin cycle of the chloroplasts and associated reactions then take care of the decarboxylation products CO_2 and pyruvate, which are converted into carbohydrates.

Little is known about the regulation of this complex specialized form of photosynthesis (Smith and Bryce 1992). The biochemical basis of circadian rhythms in the activity of phospho*enol*pyruvate carboxylase (PEPC) is now fairly well understood (Winter 1982; Carter et al. 1991; Wilkins 1992; see also Chap. 3), although the workings of the biological clock remain a mystery. A tonoplast ATPase and a pyrophosphatase are thought to be instrumental in energizing nocturnal transport of malic acid, which is formed by reduction of the primary carboxylation product oxaloacetic acid, from the cytosol into the vacuole. However, it is unknown how the tonoplast pumps are regulated. It is also unknown how malic acid manages to escape during daylight hours from the vacuole, and how its decarboxylation is geared to the consumption of the released CO_2 in photosynthesis. Once the store of malate is consumed, there is the question how the electron transport chain is protected against photoinactivation which would be the result of over-reduction, a fate not easy to escape when the photosynthetic substrate, CO_2, has been consumed and stomata remain closed under water stress (see Chap. 7).

The following experiments have been designed to address some of these problems. Leaves of *Kalanchoë pinnata* were cut from plants (or leaf discs were punched out) at different times of day and used for measurements of gas exchange and fluorescence, either immediately or after prolonged feeding of leaves (up to

Julius-von-Sachs-Institut für Biowissenschaften, Universität Würzburg, Mittlerer Dallenbergweg 64, D-97082 Würzburg, Germany

Ecological Studies, Vol. 114
Winter/Smith (eds) Crassulacean Acid Metabolism

12 h) with the fluorescent pH-indicating dyes pyranine (pK 7.3) or 5- (and 6-)carboxy-2′,7′-dichlorofluorescein (CDCF, pK 4.8) through the petiole. Dye concentrations were 2 mM (pyranine) or 0.5 mM (CDCF) in 2 mM 2-(N-morpholino)-ethanesulphonic acid (MES) at pH 6.8. Light scattering was measured according to Heber (1969) and chlorophyll fluorescence according to Schreiber et al. (1986). Transpiration and CO_2 exchange of leaves were measured by the IRGA technique with the Binos apparatus from Heraeus (Hanau, Germany) and O_2 exchange of leaf discs with the oxygen electrode of Hansatech (Bachofer, Reutlingen, Germany). Measurements were performed at 20 °C or, when PPFDs were high, at somewhat higher leaf temperatures. Electron transport rates are expressed as μmol m^{-2} s^{-1}, but can be recalculated on a chlorophyll basis, as 1 m^2 leaf surface contained 360 mg chlorophyll. The plants were grown in a glasshouse with additional illumination by fluorescent lamps (PPFD 80 μmol m^{-2} s^{-1}) under short-day conditions with a photoperiod of 10 h. Growth was slow and stomata usually remained closed during the daytime. Even at night, stomatal diffusion resistance was high. Titrations of leaf homogenates to a final pH of 6.5 established that the leaves contained about 200 mM H^+ at the end of the nocturnal carboxylation phase and essentially no titratable acid at 16.00 h in the afternoon.

6.2 Control of Photosystem II and of Linear Electron Transport

Figure 6.1 shows a simultaneous recording of photosynthetic O_2 evolution and of modulated chlorophyll fluorescence of a leaf disc punched out in the late morning from a leaf of *Kalanchoë pinnata*. The disc was sealed at the cut surface with Terostat IX (Teroson GmbH, Heidelberg, Germany) to prevent loss or entry of CO_2 through the wound and flashed with 1-s pulses of saturating light at a frequency of 0.01 Hz. The ratio of the flash-induced transient fluorescence rise $F_v = (F_m - F_o)$ to the maximum fluorescence F_m was about 0.8 in the dark. The ratio $(F'_m - F_s)/F'_m$ first decreased dramatically in the presence of actinic illumination (PPFD 500 μmol m^{-2} s^{-1}) and then increased until it stabilized at a low level, indicating first a decrease in the quantum efficiency of electron flow through PSII and then some increase (Genty et al. 1990). Even in the absence of external CO_2, O_2 was evolved after a lag phase which lasted about 6 min (upper part of Fig 6.1). In the presence of 5% (v/v) CO_2 (lower part of Fig. 6.1), the lag phase was extended to almost 15 min and the final rate of O_2 evolution was lower than in the absence of CO_2.

Obviously, net O_2 evolution in the absence of external CO_2 depends on the internal liberation of CO_2 from an internal store such as malate. Evolved O_2 is derived from water which is oxidized by photosystem II (PSII) of the chloroplast electron transport chain. Electron transport through PSII can be calculated according to Genty et al. (1990) from fluorescence data. Rates are proportional to $[(F'_m - F_s)/F'_m] \times$ PPFD. If CO_2 is the only electron acceptor (which should be expected if high CO_2 concentrations suppress photorespiration), a plot of the rate of O_2 evolution against $[(F'_m - F_s)/F'_m] \times$ PPFD should yield a straight line. The

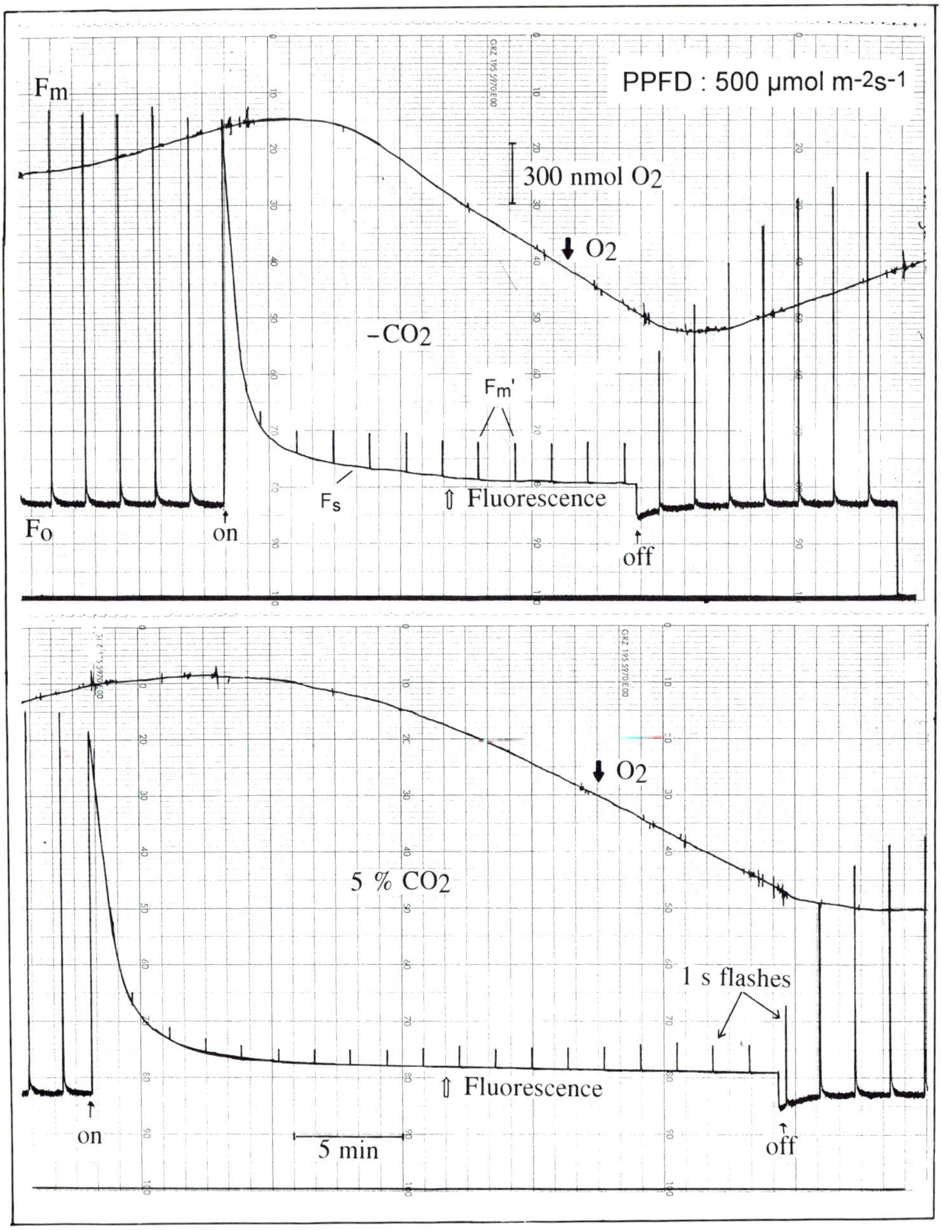

Fig. 6.1. Simultaneous recording of light-dependent O_2 evolution and modulated chlorophyll fluorescence of leaf discs of *Kalanchoë pinnata* in the morning after the nocturnal carboxylation phase. Actinic illumination with PPFD = 500 µmol m^{-2} s^{-1}. Saturating 1-s light pulses were given every 100 s. CO_2 was either absent (*upper part*) or present at a concentration of 5% (v/v) in air (*lower part*). F_o is the fluorescence level with fully oxidized quinone acceptor Q_A in the reaction centre of photosystem II (PSII), F_m (*darkened leaf*) or F'_m (*illuminated leaf*) the fluorescence level with fully reduced Q_A. F_s is steady-state fluorescence in the light. The zero level of fluorescence is indicated by a *baseline*

line depicted in Fig. 6.2 was used to calibrate electron flow through PSII against photosynthetic O_2 evolution. It can be seen that only experiments performed in the presence of high CO_2 levels fit the line. The majority of experiments performed during morning and afternoon hours yielded data on the right-hand side of the line. Obviously, in these cases (which also included some experiments with high CO_2) CO_2 was not the only electron acceptor. It must be emphasized that in all experiments stomata were largely or fully closed, offering variable, and sometimes apparently very high, diffusion resistance. This explains why high external CO_2 did not always suppress photorespiration. Despite these high diffusion resistances, release of internally generated O_2 could be detected.

Figure 6.3 shows simultaneous recordings of changes in apparent absorbance at 535 nm (A_{535}) and of chlorophyll fluorescence of two opposite leaves of a *Kalanchoë* plant in air (0.035% v/v CO_2) shortly before noon and after 22.00 h at night. Transpiration and CO_2 exchange were also measured, but are not shown because stomata were largely closed even after 22.00 h. The increased light scattering shown by the slow increases in A_{535} on illumination is interpreted to

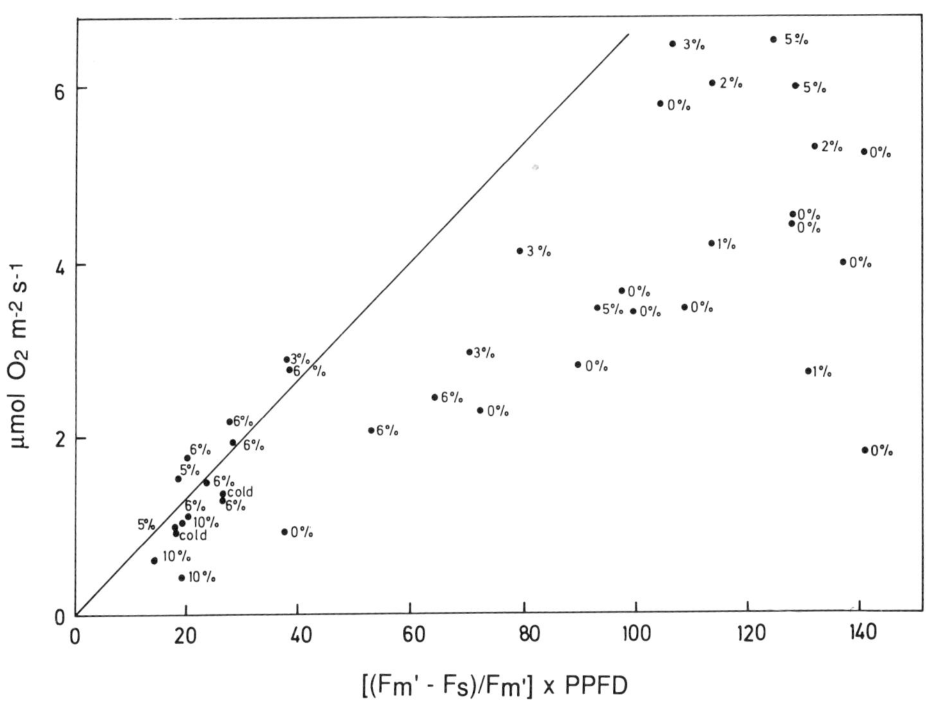

Fig. 6.2. Light-dependent O_2 evolution by leaf discs versus electron flow through photosystem II as calculated from the fluorescence parameters F'_m and F_s (see Fig. 6.1). Actinic illumination with PPFD = 500 µmol $m^{-2}s^{-1}$. CO_2 was either absent (0%) or present at the concentrations indicated in % (v/v). "Cold" refers to experiments where leaf discs were kept at 4 °C prior to the measurements to test possible effects of temperature on the subsequent CO_2 release in the light

indicate a protonation reaction at the thylakoid interior, which downregulates PSII activity when quantum fluxes cannot fully be used for photosynthetic purposes (Horton 1992; Heber et al. 1993). Increased scattering is not observed under steady-state conditions at low light intensities, which can still support a transthylakoid proton gradient fully sufficient for appreciable carbon reduction. Flash-induced fast positive spikes (frequency 0.01 Hz) and the fast dark/light transients shown in A_{535} have a difference spectrum which peaks at 522 nm. It is similar to the difference spectrum of the well-known electrochromic shift (P_{515}: Junge 1977; Katona et al. 1992) that indicates a transthylakoid membrane potential.

In the morning, light scattering was maximal only at the highest PPFD (950 μmol m^{-2} s^{-1}) and, after a transient increase at the beginning of illumination,

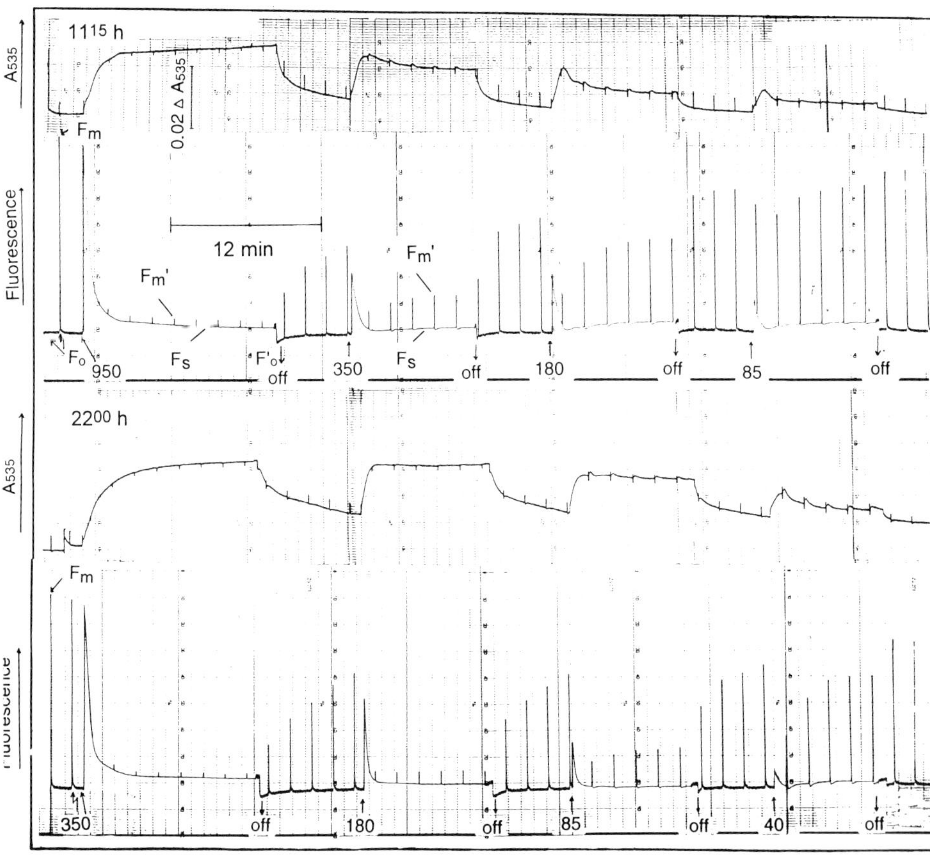

Fig. 6.3. Light-dependent changes in apparent absorbance at 535 nm (A_{535}) and of modulated chlorophyll fluorescence of two leaves of the same age of a *K. pinnata* plant in the morning and at night. *Numbers* denote PPFD in μmol m^{-2} s^{-1}. Saturating 1-s light pulses were given every 100s. $(F'_m - F_s)/F'_m$ is a measure of the relative quantum efficiency of PSII. The zero level of fluorescence is indicated by a *baseline*

was practically absent at the lowest PPFD (85 μmol m^{-2} s^{-1}). The initial and transient increase in scattering seen at the lower PPFDs is the result of the deactivation of Calvin cycle enzymes during the 6-min dark periods between the illumination cycles. Transiently increased scattering reflects the time needed for enzyme reactivation in the light. Dark inactivation of the Calvin cycle was faster and reactivation was slower in *Kalanchoë* compared to the C_3 plant spinach (data not shown).

After 22.00 h, scattering was maximal at a PPFD of 180 μmol m^{-2} s^{-1}, much below the PPFD of 950 μmol m^{-2} s^{-1} that produced maximum scattering after 11.00 h. It was still appreciable at a PPFD of 40 μmol m^{-2} s^{-1}. After 22.00 h, the quantum efficiency of electron transfer through PSII as expressed in fluorescence by $(F'_m - F_s)/F'_m$ was always below the quantum efficiency observed during the morning hours at comparable PPFDs.

Figure 6.4 shows electron transfer through PSII as calculated from fluorescence data according the line in Fig 6.2 at different times of day. Simultaneous measurements of transpiration and CO_2 exchange ascertained that stomata remained closed during the day. They opened only slightly at night. Electron flow was saturated at relatively low PPFDs. It is not known why calculated electron flow decreased somewhat at high PPFDs. Actually, it is possible that at high PPFDs electron flow through PSII is no longer strictly proportional to $[(F'_m - F_s)/F'_m] \times$ PPFD. Light-saturated electron flow was maximal during the morning hours. It declined in the afternoon and was lowest at 22.00 h. Since photosynthetic O_2 evolution was observed in the morning even in the absence of external CO_2 (Fig. 6.1), but was largely absent during the night (Fig. 6.5, upper part), there was the question which electron acceptors permitted slow electron transport through PSII when leaves were illuminated at night (see also fluorescence data of Fig. 6.3). Obvious candidates are O_2 in the Mehler reaction and in photorespiration (or malate, see below). Fig. 6.5 (lower part) shows that photosynthetic O_2 evolution is appreciable after a long lag phase even at night when CO_2 is made available and illumination is provided. Obviously, the Calvin cycle, which is deactivated during darkening, can be fully activated by light even during the night when PEPC has become active and competes with RUBISCO for CO_2.

When leaf discs were illuminated during the nocturnal carboxylation phase in the absence of external CO_2, net O_2 evolution was absent only initially (Fig. 6.5, upper part). The quantum efficiency of electron flow through PSII, which was very low initially as indicated by very low values of $(F'_m - F_s)/F'_m$ (see flash-induced fluoresence spikes in Fig. 6.5), increased steadily with illumination time, and some net O_2 evolution became discernible after an illumination time of about 30 min. Apparently, after a long lag phase, some malate which had previously been stored in the vacuole was released and became available for decarboxylation. Released CO_2 was then used not for carboxylation of PEP but rather of RuBP. Net O_2 evolution then indicated the reduction of the resulting 3-phosphoglycerate.

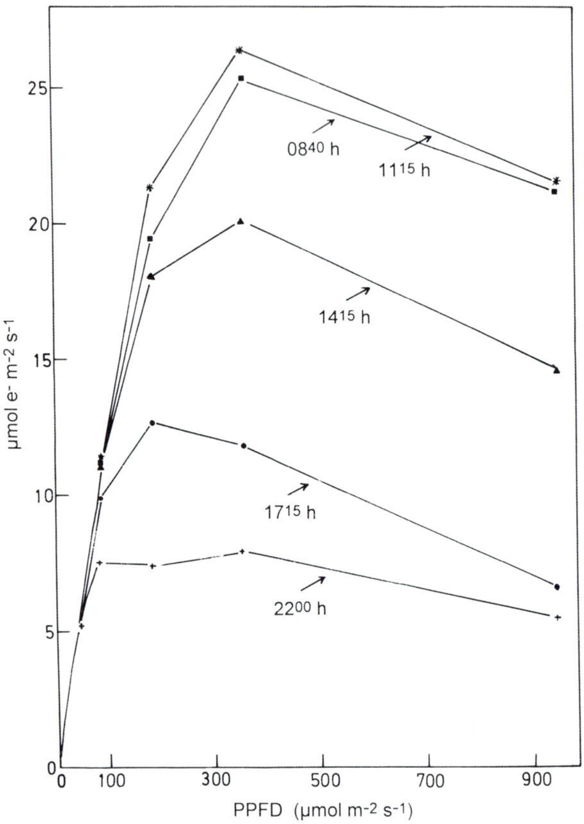

Fig. 6.4. Electron flow rates through PSII in air (calculated from $[(F'_m - F_s)/F'_m] \times$ PPFD, see Fig. 6.2) as a function of PPFD at different times of day

6.3 Malate Decarboxylation

Figure 6.6 shows that induction of photosynthesis as indicated by calculated electron flow is slow after prolonged darkening at night. Before illumination, the rate of respiratory CO_2 release was very low. It was increased after an illumination period (data not shown). When, after more than 30 min in the light, the O_2 concentration of the gas phase was decreased from 21 to 1% (v/v), electron flow decreased. This decrease was reversible. A similar decrease in electron flow was observed in spinach (data not shown) when the O_2 concentration of air was decreased, while CO_2 assimilation increased owing to suppression of photorespiration.

Figure 6.7 shows simultaneous measurements of changes in A_{535}, chlorophyll fluorescence, transpiration and CO_2 exchange in the morning, after a previous illumination had activated CO_2 release behind closed stomata, and in

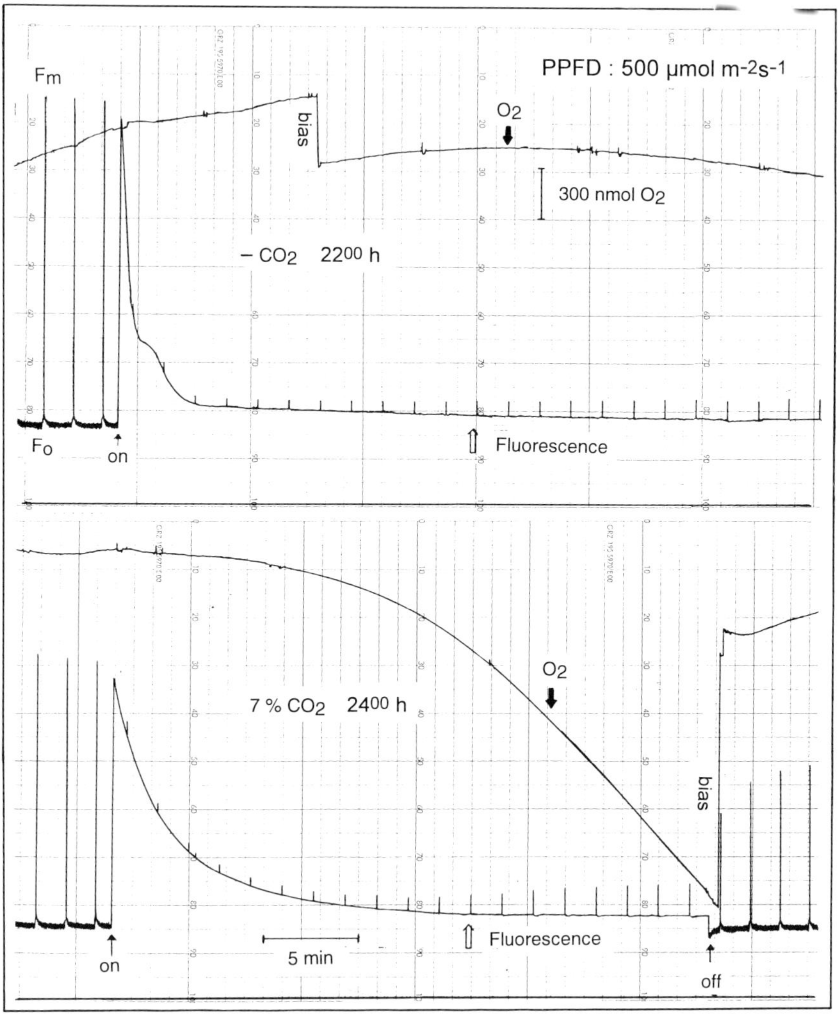

Fig. 6.5. Simultaneous recording of light-dependent O_2 evolution and modulated chlorophyll fluorescence of two leaf discs of *K. pinnata*, one at 22.00 h without added CO_2 and the other one at 24.00 h with 7% (v/v) CO_2 in the gas phase. Actinic illumination with PPFD = 500 µmol m^{-2} s^{-1}. Saturating 1-s light pulses were given every 100 s. The zero level of fluorescence is indicated by a *baseline*. *Bias*: readjustment of recording trace without change in sensitivity

the afternoon. Before illumination in the morning, CO_2 release was faster than CO_2 release in the afternoon. Illumination increased light scattering in the morning, but less so than in the afternoon, indicating that PSII photochemistry was more strongly downregulated in the afternoon. Increased fluorescence quenching in the afternoon indicated increased non-radiative dissipation of

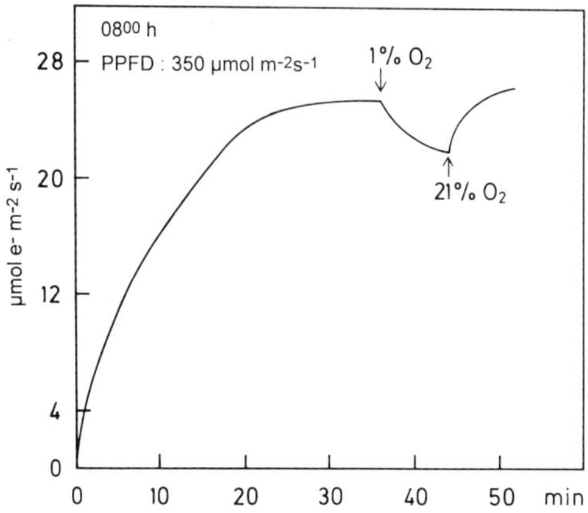

Fig. 6.6. Induction of photosynthesis after an 8-h dark period as expressed by the rate of electron flow through PSII in a leaf of *K. pinnata*. Rates are calculated from fluorescence by $[(F'_m - F_s)/F'_m] \times$ PPFD according to the calibration line shown in Fig. 6.2

excess excitation energy. When the light was turned off, fluorescence decreased transiently below the so-called F_o level to F'_o (see also Fig. 6.3). The ratio $(F_s - F'_o)/(F'_m - F'_o)$ is a measure of the state of reduction of the primary quinone acceptor Q_A in the reaction centre of PSII (Schreiber et al. 1986). The recordings show that Q_A was more reduced in the afternoon than in the morning. Stomata remained closed except for some slight opening at the end of the afternoon illumination period. Light-induced increase in leaf temperature accounts for most of the increased transpiration in the light. Very little net CO_2 uptake occurred after prolonged illumination, and only in the afternoon. The important point to be made about the experiment of Fig. 6.7 is that the calculated light-dependent electron flow was 34 μmol m^{-2} s^{-1} in the morning, whereas the reduction of the CO_2 released in the dark would only permit an electron flow of less than 4 μmol m^{-2} s^{-1}. The large discrepancy strongly suggests that most of the malate decarboxylation which accounts for the CO_2 release behind closed stomata in the morning hours is light-dependent. In fact, leaves kept darkened during the day maintained their level of acidity. Compensation of the dark CO_2 release in the afternoon by reduction in the light would require an electron flow of about 1.5 μmol m^{-2} s^{-1}, whereas the calculated electron flow was actually 14 μmol m^{-2} s^{-1}.

6.4 Photorespiration

Figure 6.2 indicated that much of the electron flow going through PSII is not used for the reduction of CO_2. Figure 6.6 showed that, immediately after the nocturnal carboxylation phase and the subsequent induction of photosynthesis, electron flow can be reduced by 1% (v/v) O_2, which suppresses photorespiration. However, Fig. 6.8 shows an experiment in which, somewhat later in the decarboxylation phase, a change from 21 to 1% (v/v) O_2 had little effect on the

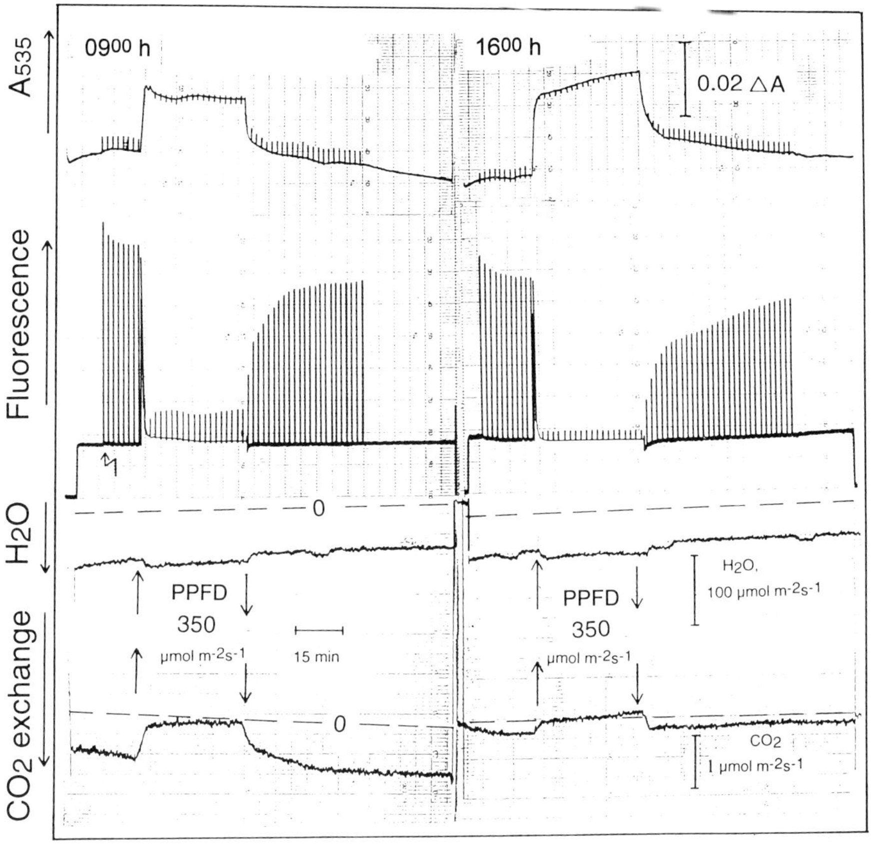

Fig. 6.7. Simultaneous recording of apparent absorbance at 535 nm (A_{535}), modulated chlorophyll fluorescence, transpiration (loss of H_2O) and CO_2 exchange of *K. pinnata* leaves after induction of photosynthesis in the morning and after deacidification in the afternoon. *Zero lines* denote absence of net gas exchange. Zero fluorescence is shown briefly at the beginning of the experiments, several min before the first 1-s saturating light pulses were given. Gas phase was air with 0.035% (v/v) CO_2

quantum efficiency of electron flow through PS II as shown by $(F'_m - F_s)/F'_m$. In the afternoon, the quantum efficiency of electron flow was decreased by a change from 21 to 1% (v/v) O_2. Light scattering, an indicator both of thylakoid energization and of downregulation of PSII (which increases when photorespiration is suppressed), was increased somewhat in the morning but more in the afternoon. In the latter case, stomata opened slightly when the internal CO_2 concentration decreased in the presence of 1% (v/v) O_2 owing to the inhibition of photorespiratory CO_2 release. They closed again as 21% (v/v) O_2 increased the partial pressure of CO_2 in the intercellular space. It is unclear to what extent the extra electron flow indicated in Fig. 6.2 on the right-hand side of the straight line,

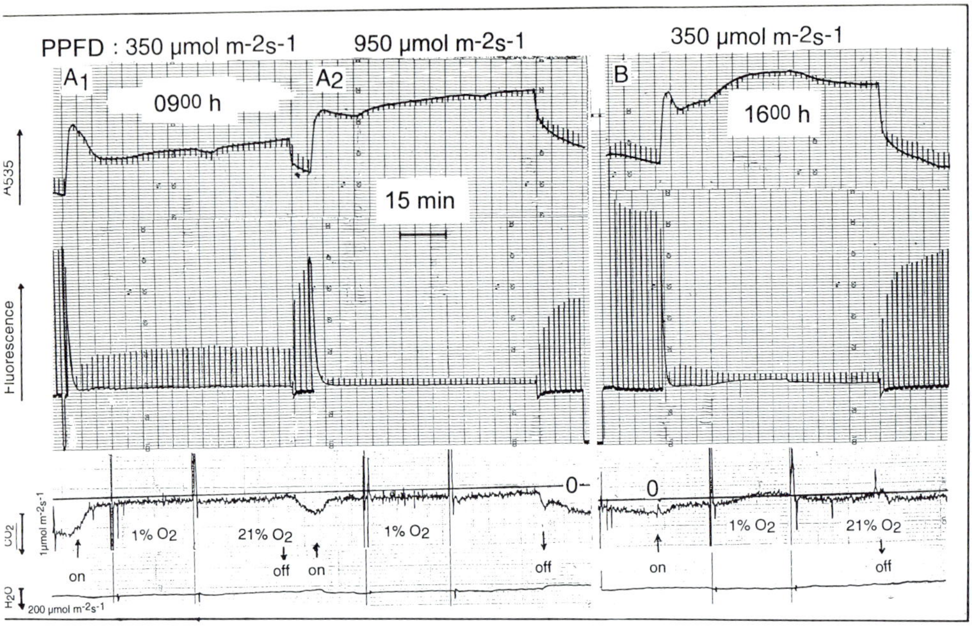

Fig. 6.8. Simultaneous recordings of apparent absorbance at 535 nm (A_{535}), modulated chlorophyll fluorescence, transpiration (loss of H_2O) and CO_2 exchange of *K. pinnata* leaves after induction of photosynthesis in the morning (two different PPFDs) and after deacidification in the afternoon. The gas phase was changed from air with 0.035% (v/v) CO_2 to 1% (v/v) O_2 as indicated

which is thought to reflect electron flow to CO_2 only, is electron flow to O_2 in the Mehler reaction or to O_2 during photorespiratory electron flow. However, since the apparent Michaelis constant for O_2 in the Mehler reaction is assumed to be low (about 60 µM in solution or a little more than 2% (v/v) in the gas phase; Radmer et al. 1978; Steiger and Beck 1981; Furbank et al. 1982; Schreiber and Neubauer 1990), it appears that much of the extra electron flow is caused by photorespiration.

6.5 pH-Sensitivity of Photosynthesis

CO_2 is not only a substrate of RUBISCO and, indirectly, of PEPC, it is also a potential acid, as protons are formed when its hydration product dissociates. Photosynthesis is sensitive to acidification of the chloroplast stroma (Enser and Heber 1980). In Figs. 6.1 and 6.5, it was shown that the lag phase of photosynthetic O_2 evolution is extended in the presence of high CO_2 concentrations. A similar extension of the lag phase is observed when isolated chloroplasts are illuminated in a medium with a pH appreciably below the optimum for photosynthesis, which is pH

7.6 (data not shown). Whereas isolated chloroplasts are not very effective in regulating their stromal pH (Wagner et al. 1990), very effective pH regulation may be observed in leaves of C_3 plants. Figure 6.9 compares the CO_2-dependence of steady-state photosynthetic O_2 evolution of the C_3 plant barley with that of *Kalanchoë pinnata*, after the acid-induced lag phase of photosynthesis had been overcome. In barley, maximum CO_2-dependent O_2 evolution was observed up to a CO_2 concentration of 20% (v/v) in air. Beyond this concentration, photosynthesis declined. In *Kalanchoë* leaf discs with unsealed cuts, CO_2-dependent O_2 evolution increased up to a CO_2 concentration of 3% (v/v), apparently because diffusion resistances still limited access of external CO_2. Internally generated CO_2 did not saturate photosynthesis. Beyond 3% (v/v) CO_2, O_2 evolution decreased, and it was very slow in the presence of 10% CO_2 (see also Fig. 6.2). The different response of photosynthesis of the C_3 and the CAM plant cannot be explained by drastically different buffering capacities of the chloroplast stroma. At physiological pH, phosphate and phosphate esters are the main buffers. Combined, their concentration is close to 15 mM (Lilley et al. 1977; Heldt et al. 1980). In C_3 plants such as barley, a main mechanism of cytoplasmic pH regulation appears to be proton/cation exchange at the level of the tonoplast (Heber et al. 1994). It is proposed that cytoplasmic acidification triggers the tonoplast proton pumps. Proton transport into the vacuole, which is compensated by a counter-transport of cations, restores the cytoplasmic pH in C_3 plants, permitting under appropriate conditions photosynthesis to proceed at its maximum rate over a wide range of CO_2 concentrations. Apparently, *Kalanchoë pinnata* cannot regulate the cytoplasmic pH as efficiently as barley by proton/cation exchange. This is indicated by long lag phases of O_2 evolution after acidification by high CO_2 as shown in Figs. 6.1 and 6.5, and by the steady-state O_2 evolution as a function of CO_2 concentration as shown in Fig. 6.9.

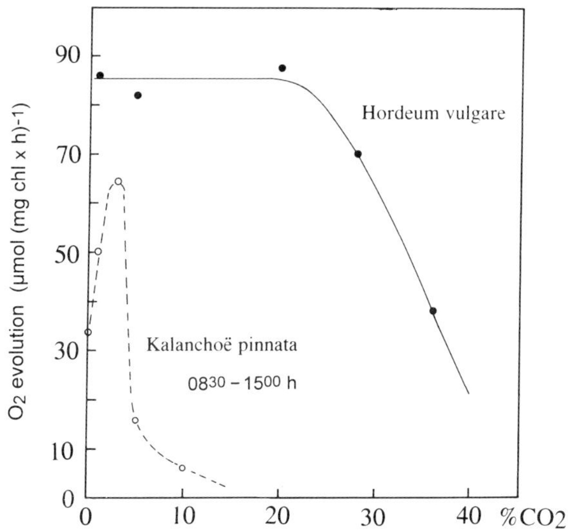

Fig. 6.9. Light-dependent O_2 evolution of leaves of barley (*Hordeum vulgare*) and of (unsealed) leaf discs of *K. pinnata*, at steady state after the initial CO_2-dependent lag phase, as a function of CO_2 concentration (% v/v) in air. Illumination was saturating for *Kalanchoë*, but not for barley

6.6 Proton Transport Across the Tonoplast

5- (and 6-)carboxy-2′,7′-dichlorofluorescein (CDCF) is a fluorescent dye with a p*K* of 4.8. Accordingly, it is a particularly sensitive probe for recording pH changes close to pH 4.8. A pH decrease is indicated by decreased fluorescence. Figure 6.10 shows that a *Kalanchoë* leaf that had been fed CDCF via the petiole for about 12 h exhibited light-dependent decreases in dye fluorescence at night, but not (or very little) during daylight hours. For comparison, chlorophyll fluorescence was recorded simultaneously. Interestingly, the pH decrease indicated by decreased CDCF fluorescence when the leaf was illuminated at night was not reversed on darkening, as it was in C_3 plants (Yin et al. 1990 a,b). Also in contrast to observations made with C_3 and C_4 plants (Raghavendra et al. 1993), it appeared to be light-saturated at very low PPFDs (Fig. 6.10A). In C_3 and C_4 plants, the light-dependent decrease of CDCF fluorescence was explained by increased energization of the tonoplast ATPase, which pumps protons into the vacuole. It is tempting to extend this interpretation to CAM plants. The absence of reversibility of the light-induced fluorescence change on darkening could then be explained as irreversible proton

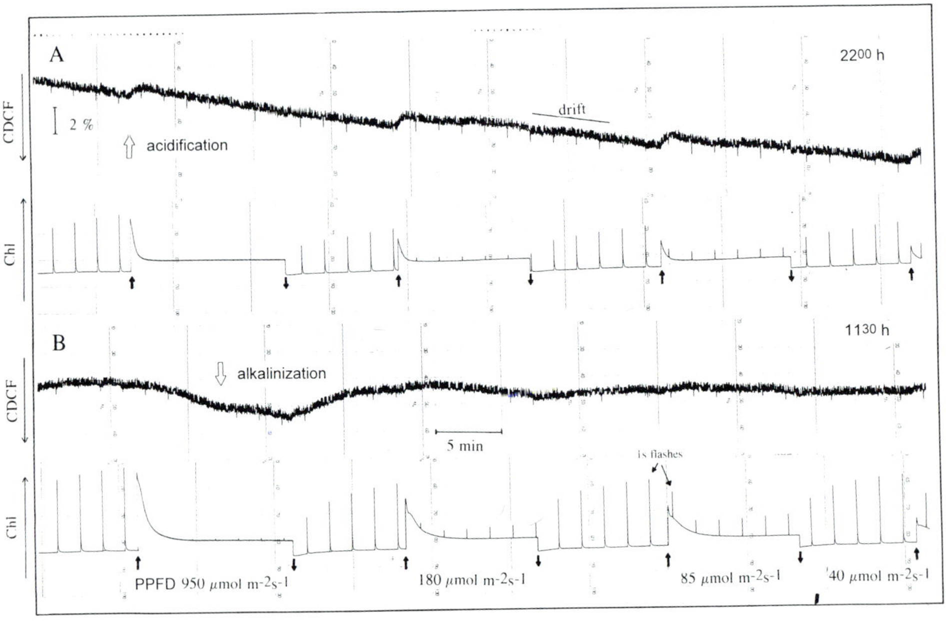

Fig. 6.10. Simultaneous recordings of fluorescence of the pH-indicating dye 5- (and 6-)-carboxy-2′, 7′-dichlorofluorescein (p*K* 4.8) and of chlorophyll of a leaf of *K. pinnata* that had been fed the dye through the petiole at night during the carboxylation phase (22.00 h) and in the morning after decarboxylation had been initiated (11.30 h). Illumination periods and PPFD are as indicated. Saturating 1-s light pulses were given every 100 s. *Spikes* in the fluorescence trace of the dye are caused by light leaks. Note that upward deflections of the CDCF recording represent decreases in pH

transport into the vacuoles under the conditions prevailing at night. It is obvious that such proton transport must be accompanied by movement of a charge-balancing ion, such as an anion into the vacuole.

No fluorescence change comparable to that seen at night was observed when the *Kalanchoë* leaf was illuminated during daylight hours (Fig. 6.10B). Rather, there was a slow light-dependent fluorescence increase. Its dependence on light intensity was different from the response observed at night. The slow alkalinization indicated by the fluorescence increase could be explained as efflux of malic acid from the vacuole during the day, but the reversibility of the fluorescence change seen at the highest PPFD of 950 μmol m^{-2} s^{-1} does not easily agree with such an explanation.

However, the differences in the fluorescence responses seen at night and during the daytime would be consistent with a deactivation of proton pumps of the tonoplast during the day, if the nocturnal acidification response really reflects increased transtonoplast proton transfer, when light increases the cytoplasmic energy level.

6.7 Light-Dependent Cytosolic Alkalinization

The pH-indicating dye pyranine has a p*K* of 7.3. Even though it is a trivalent sulphonate anion, it is capable of entering leaf mesophyll cells. It is believed to record mainly cytosolic pH changes (Yin et al. 1990a). Uptake of the dye through the petiole into *Kalanchoë* leaves was faster than uptake of CDCF, but much slower than uptake into leaves of C_3 plants. Like CDCF, pyranine increases its fluorescence upon deprotonation. Figure 6.11 shows a simultaneous recording of the fluorescence of chlorophyll and of pyranine of a leaf of *Kalanchoë pinnata*. At a PPDF of 350 μmol m^{-2} s^{-1}, pyranine fluorescence increased with biphasic kinetics. The first, small phase was much less pronounced than in C_3 plants (Yin et al. 1990a). The second slower and larger phase appeared to be kinetically related to increasing electron flow through PSII. At low PPFDs, the slow alkaline shift expressed by increasing pyranine fluorescence was smaller than at higher PPFDs. In contrast to CDCF fluorescence, pronounced differences in fluorescence kinetics could not be observed between morning hours and the nighttime, although signals observed at night tended to be smaller.

6.8 Metabolic Regulation of CAM

In C_3 plants, photosynthetic carbon reduction may be simplified as

$$6\,CO_2 + 6H_2O \rightarrow C_6H_{12}O_6 + 6\,O_2.$$

Light-dependent cytoplasmic alkalinization can then largely be accounted for by proton transport into the intrathylakoid space and into the vacuole. In CAM

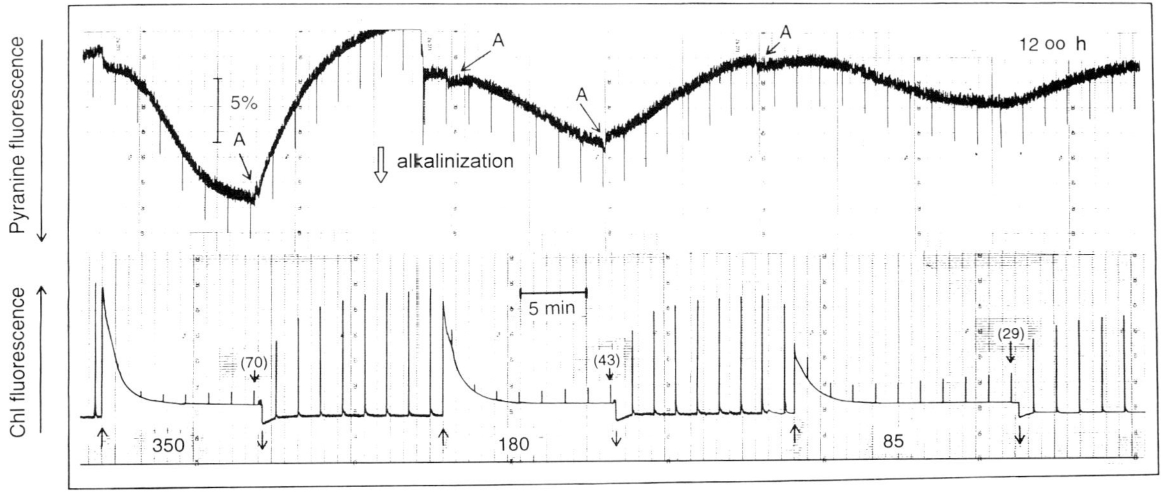

Fig. 6.11. Simultaneous recordings of fluorescence of the pH-indicating dye pyranine (p*K* 7.3) and of chlorophyll of a leaf of *K. pinnata* that had been fed the dye through the petiole. *Open numbers* indicate PPFDs in μmol m^{-2} s^{-1}, and *numbers in parentheses* the relative rates of electron flow through PSII. Changes in dye fluorescence denoted by A and *spikes* shown in the fluorescence trace of the dye are optical artefacts. Saturating 1-s light pulses were given every 100 s

plants, alkalinization could result from

$$3\,\text{malate}^{2-} + 9H_2O \rightarrow 2C_6H_{12}O_6 + 6OH^- + 3O_2,$$

if proton export from the vacuole lags behind malate export. The data of Fig. 6.11 suggest such a lag.

If this interpretation of the observations is correct, it would have considerable implications for the metabolic regulation of CAM. In C_3 plants, cytosolic acidification triggers a tonoplast proton pump, presumably the tonoplast ATP-ase (Ellebracht et al. 1994). Cytosolic alkalinization should then be expected to down-regulate the pump. Down-regulation would permit efflux of malate through open malate channels. Oxidation of the malate by malic enzyme creates a sink which would lead to continued efflux from the vacuole as long as malate channels are kept open. Proton efflux might lag behind malate efflux. This would create a diffusion potential, which could keep open the malate channels used for nocturnal malate influx (for an alternative view, see Chap. 4). Of primary importance for the regulation of CAM is obviously the regulation of malic enzyme by light. The enzyme is apparently inactive at night. This is shown by the slow rates of CO_2 release at the end of the carboxylation phase in the dark. This release is probably due to respiration. The initiation of CO_2 release in the light by decarboxylation of malate is slow, as can be seen from the long induction period of photosynthesis after an 8-h dark period (Fig. 6.6). However, once the enzyme has become active, it continues to decarboxylate malate even in the dark after a light period, as can be seen by considerable CO_2 release (Fig. 6.7). Nevertheless, decarboxylation is much slowed down in the dark compared to decarboxylation in the light (see discussion of Fig. 6.7), presumably owing to the accumulation of reaction products which cannot further be metabolized in the dark. Thus, regulation of CAM appears to be primarily regulation of PEPC, of malic enzyme and of proton pumps at the tonoplast. Regulation of the H^+ pumps appears to be governed by pH changes and is therefore secondary to the first two regulation levels.

Another important aspect of regulation concerns the chloroplast electron transport system of *Kalanchoë*. The electron-pumping function of PSII is essentially irreversible. PSII will oxidize water and pump electrons into the electron transport chain until the chain is reduced in the absence of electron acceptors. Full reduction of electron carriers in excess light is known to lead to rapid photo-inactivation of isolated chloroplasts. Inactivation of the photosynthetic apparatus of *Kalanchoë* would be unavoidable in the absence of a sensitive control mechanism which decreases the activity of PSII when CO_2 release from malate behind closed stomata ceases. Control of PSII is even needed during active CO_2 release from malate during Phase III of CAM (Smith and Bryce 1992), when PPFDs are very high, because the capacity of the Calvin cycle for carbon reduction is limited. In fact, effective control of PSII at high PPFDs is indicated by increased light scattering (Fig. 6.3), which indicates a proton gradient in excess of that sufficient for mere ATP synthesis. Such a proton gradient is needed for control of PSII, but cannot be established by carbon reduction or by photorespiratory reactions alone, because protons are released from the intrathylakoid space during the synthesis of the ATP which is consumed in these reactions. In contrast to assimilatory and photorespiratory reactions, reduction of oxygen in the Mehler reaction is unaccompanied by ATP consumption. The Mehler reaction should be able to support a proton gradient sufficient for the protonation reaction that appears to control PSII activity (Schreiber and Neubauer 1990). However, in C_3 plants such as spinach, O_2 reduction in the Mehler reaction appears to be too slow to prevent the excessive reduction of the electron transport chain (Wu et al. 1991). In C_3 plants under drought stress, the additional electron transport made possible by the cycling of CO_2 between photorespiratory release and assimilatory refixation behind closed stomata is necessary to avoid excessive reduction and to establish the redox poising needed for cyclic electron transport. It appears that a main role of cyclic electron transport is not only to provide ATP that cannot be produced sufficiently during linear electron flow but, no less important, to control PSII at excessive PPFDs (Heber and Walker 1992).

6.9 Conclusions

The experiments performed with *Kalanchoë pinnata* lead to the following conclusions:

1. Simultaneous measurements of O_2 evolution and chlorophyll fluorescence show that CO_2 is not the only electron acceptor in the light. Extra electron flow through PSII, which is often very considerable at the beginning and at the end of the light phase, appears to reduce oxygen in the sequence of photorespiratory reactions and/or in the Mehler reaction.
2. The rate of CO_2 release from malate is light-dependent.
3. Leaves of *Kalanchoë* are much less effective than leaves of C_3 plants in regulating cytoplasmic pH by proton/cation exchange across the tonoplast, as shown by high sensitivity of photosynthetic O_2 evolution to acidification.
4. Experiments with the pH-indicating dye 5- (and 6-)carboxy-2′,7′-dichlorofluorescein (p*K* 4.8) suggest that tonoplast proton pumps respond differently to illumination at night than during the day. A light-dependent decrease in vacuolar pH, which is not reversed on darkening, suggests increased and irreversible proton pumping into the vacuole at night on illumination.
5. Experiments with the pH-indicating dye pyranine (p*K* 7.3) reveal two light-dependent cytosolic alkalinization phases. The slower and larger one is related to electron flow. It is thought to show the alkalinization expected on illumination when malate is decarboxylated.
6. It is suggested that cytosolic acidification resulting from malic acid formation at night triggers the tonoplast proton pumps. Malate anions follow the protons using a tonoplast malate channel, transport being energized by the tonoplast membrane potential which is generated during energy-dependent proton flux.

Onset of cytoplasmic malate oxidation during the day causes cytosolic alkalinization, decreasing the activity of the pH-dependent tonoplast proton pumps. Passive channel-mediated efflux of malate from the vacuoles permits continued liberation of CO_2 in the cytoplasm and its subsequent refixation in the chloroplasts. Controlled efflux of protons from the vacuoles maintains charge balance.

Acknowledgements. This work could not have been done without the enthusiastic assistance of a student of chemistry, Mr. S. Wolf, and of Mrs. Hüttinger and Mrs. Noak, who were responsible for editorial procedures. This work was supported by the Sonderforschungsbereiche 176 and 251 of the University of Würzburg.

References

Carter PJ, Nimmo HG, Fewson CA, Wilkins MB (1991) Circadian rhythms in the activity of a plant protein kinase. EMBO J 10: 2063–2068

Ellebracht A, Sonnewald U, Heber U (1994) Light-stimulated proton transport into the vacuoles of leaf mesophyll cells does not require energization by the tonoplast pyrophosphatase. Planta 193: 203–204

Enser U, Heber U (1980) Metabolic regulation by pH gradients. Inhibition of photosynthesis by indirect proton transfer across the chloroplast envelope. Biochim Biophys Acta 592: 577–591

Furbank RT, Badger MR, Osmond CB (1982) Photosynthetic oxygen exchange in isolated cells and chloroplasts of C_3 plants. Plant Physiol 70: 927–931

Genty B, Harbinson J, Briantais JM, Baker NR (1990) The relationship between non-photochemical quenching of chlorophyll fluorescence and the rate of photosystem II chemistry in leaves. Photosynth Res 25: 249–257

Griffiths H (1988) Crassulacean acid metabolism: a re-appraisal of physiological plasticity in form and function. Adv Bot Res 15: 43–92

Heber U (1969) Conformational changes in chloroplasts induced by illumination of leaves in vivo. Biochem Biophys Acta 180: 302–319

Heber U, Walker DA (1992) Concerning a dual function of coupled cyclic electron transport in leaves. Plant Physiol 100: 1621–1626

Heber U, Katona E, Schönknecht G, Asada K (1993) Photosystem I can control photosystem II in leaves. In: Shima A, Ichihashi M, Fujiwara Y, Takebe H (eds) Frontiers of photobiology, International Congress Series 1021. Excerpta Medica, Amsterdam, pp 13–19

Heber U, Wagner U, Neimanis S, Bailey K, Walker D (1994) Fast cytoplasmic pH regulation in acid-stressed leaves. Plant Cell Physiol 35: 479–488

Heldt HW, Portis AR, Lilley RMcC, Mosbach A, Chon CJ (1980) Assay of nucleotides and other phosphate containing compounds in isolated chloroplasts by ion exchange chromatography. Anal Biochem 101: 278–287

Horton P (1992) Regulation of energy dissipation by chloroplast membranes. Photosynth Res 34: 106

Junge W (1977) Membrane potentials in photosynthesis. Annu Rev Plant Physiol 28: 503–536

Katona E, Neimanis S, Schönknecht G, Heber U (1992) Photosystem I-dependent electron transport is important in controlling photosystem II activity in leaves under conditions of water stress. Photosynth Res 34: 449–464

Leegood RC, Osmond CB (1990) The flux of metabolites in C_4 and CAM plants. In: Dennis DT, Turpin DH (eds) Plant physiology, biochemistry and molecular biology. Longman, Harlow, pp 274–298

Lilley R, Chon CJ, Mosbach A, Heldt HW (1977) The distribution of metabolites between spinach chloroplasts and medium during photosynthesis in vitro. Biochim Biophys Acta 460: 259–272

Lüttge U (1987) Carbon dioxide and water demand: crassulacean acid metabolism (CAM), a versatile ecological adaptation exemplifying the need for integration in ecological work. New Phytol 106: 593–627

Radmer R, Kok B, Ollinger O (1978) Kinetics and apparent K_m of the oxygen cycle under conditions of limiting carbon dioxide fixation. Plant Physiol 61: 915–917

Raghavendra A, Yin Z-H, Heber U (1993) Light-dependent pH changes in leaves of C_4 plants. II. pH response to CO_2 and oxygen in comparison to the response of C_3 plants. Planta 189: 278–287

Schreiber U, Neubauer C (1990) O_2-dependent electron flow, membrane energization and the mechanism of non-photochemical quenching of chlorophyll fluorescence. Photosynth Res 25: 279–293

Schreiber U, Schliwa U, Bilger W (1986) Continuous recording of photochemical and non-photochemical chlorophyll fluorescence quenching with a new type of modulation fluorometer. Photosynth Res 10: 51–62

Smith JAC, Bryce JH (1992) Metabolite compartmentation and transport in CAM plants. In: Tobin AK (ed) Plant organelles. Cambridge University Press, Cambridge, pp 141–167

Steiger H-M, Beck E (1981) Formation of hydrogen peroxide and oxygen dependence of photosynthetic CO_2 assimilation by intact chloroplasts. Plant Cell Physiol 22: 561–567

Wagner U, Kolbowski J, Oja V, Laisk A, Heber U (1990) pH homeostasis of the chloroplast stroma can protect photosynthesis of leaves during the influx of potentially acidic gases. Biochim Biophys Acta 1016: 115–120

Wilkins MB (1992) Circadian rhythms: their origin and control. New Phytol 121: 347–375

Winter K (1982) Properties of phosphoenolpyruvate carboxylase in rapidly prepared, desalted leaf extracts of the crassulacean acid metabolism plant *Mesembryanthemum crystallinum* L. Planta 154: 298–308

Winter K (1985) Crassulacean acid metabolism. In: Barber J, Baker NR (eds) Photosynthetic mechanisms and the environment. Elsevier, Amsterdam, pp 329–387

Wu J, Neimanis S, Heber U (1991) Photorespiration is more effective than the Mehler reaction to protect the photosynthetic apparatus against photoinhibition. Bot Acta 104: 289–291

Yin Z-H, Neimanis S, Wagner U, Heber U (1990a) Light-dependent pH changes in leaves of C_3 plants. I. Recording pH changes in different cellular compartments by fluorescent probes. Planta 182: 244–252

Yin Z-H, Neimanis S, Heber U (1990b) Light-dependent pH changes in leaves of C_3 plants. II. Effect of CO_2 and O_2 on the cytosolic and the vacuolar pH. Planta 182: 253–261

7 Energy Dissipation and the Xanthophyll Cycle in CAM Plants

W.W. ADAMS III and B. DEMMIG-ADAMS

7.1 Introduction

The response of photosynthetic organisms to high or excessive photon flux densities (PPFD) is a topic which interested several investigators in photosynthesis research during the first half of the twentieth century (e.g. Emerson 1936; Myers and Burr 1940; Kok 1956). Until recently, the phenomenon of photoinhibition, the inhibition of photosynthesis by excessive light, was equated with damage to the photosynthetic apparatus (Powles 1984; see also Demmig-Adams and Adams 1992a). In the majority of studies on photoinhibition, it is, in fact, likely that some form of damage was responsible for the majority of the decrease in photosynthetic competence observed, as most studies have involved the exposure of leaves, organisms, or even isolated chloroplasts or thylakoids to PPFD which was many times greater than that experienced during development.

During the latter part of the 1980s several investigators suggested that some of the characteristic changes observed in response to excessive light (e.g. reductions in photosynthesis and certain changes in chlorophyll fluorescence emission) might actually be associated with a photoprotective process involving the controlled thermal dissipation of energy in the pigment bed (Björkman 1987 a, b; Demmig and Björkman 1987; Demmig et al. 1987). Thus, although photoprotective in nature, this dissipation process could result in photoinhibition of photosynthesis, since such dissipation of energy is also a competitor with photosynthesis.

Photoinhibition was found to occur in several CAM species in a variety of different habitats (Adams 1988; Adams et al. 1987, 1988) at approximately the same time as the above studies were being carried out. It now seems likely that a large proportion, if not all, of the inhibiton of photosynthesis and photosystem II (PSII) photochemical efficiency observed in the naturally sun-exposed cladodes and leaves in these studies was due to photoprotective thermal energy dissipation rather than to damage of the photosynthetic apparatus. In the following we will review the evidence in support of a role of the xanthophyll cycle in energy dissipation, and present some of our findings on energy dissipation and the response of the xanthophyll cycle to light in CAM plants.

Department of Environmental, Population, and Organismic Biology, University of Colorado, Boulder, CO 80309-0334, USA

Ecological Studies, Vol. 114
Winter/Smith (eds) Crassulacean Acid Metabolism

7.2 Energy Dissipation and the Xanthophyll Cycle

7.2.1 Relationship Between Zeaxanthin Accumulation and Energy Dissipation

In a number of studies a positive and linear correlation between the leaf content of zeaxanthin and the level of energy dissipation has been observed for different species under a variety of conditions (for a review, see Demmig-Adams and Adams 1994). The xanthophyll cycle involves three carotenoids, violaxanthin, an intermediate antheraxanthin, and zeaxanthin, and two enzymes, violaxanthin de-epoxidase and zeaxanthin epoxidase (Fig. 7.1). When light is limiting (i.e., all of the absorbed light can be used in photosynthesis), zeaxanthin epoxidation is favored and violaxanthin is the predominant component of the xanthophyll cycle present in the leaf. As light levels become more and more excessive, violaxanthin de-epoxidation is favored, resulting in the rapid accumulation of greater and greater levels of zeaxanthin. This accumulation of zeaxanthin is accompanied by an increase in energy dissipation levels, as determined from measurements of chlorophyll fluorescence emission.

Given that fluorescence emanates from the singlet excited state of chlorophyll, energy dissipation must involve de-excitation of this species. Interaction between zeaxanthin and this excited state of chlorophyll could be either direct or indirect (Demmig-Adams 1990). Horton and his coworkers (Horton et al. 1991; Horton and Ruban 1992; Ruban and Horton 1992; Ruban et al. 1992) have proposed that zeaxanthin may amplify a de-excitation of the excited singlet state of chlorophyll suggested to occur in response to protonation and an aggregation of the light-harvesting chlorophyll complexes. On the other hand, Owens et al. (1992) have presented evidence that excess energy could be transferred directly from the

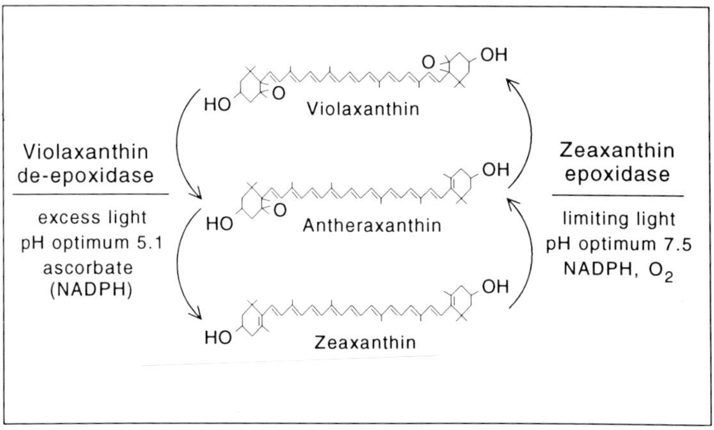

Fig. 7.1. Simple scheme showing the relationships between the three carotenoids of the xanthophyll cycle and some of the factors involved in their conversions

excited singlet state of cholorophyll to zeaxanthin, subsequently being dissipated as heat.

7.2.2 Evidence in Support of Zeaxanthin's Role in Energy Dissipation

7.2.2.1 Dithiothreitol, an Inhibitor of Violaxanthin De-Epoxidase

The reductant dithiothreitol (DTT) was found to inhibit the de-expoxidase reaction in vitro, thereby preventing the formation of zeaxanthin (Yamamoto and Kamite 1972). By placing the petiole of a leaf in water containing a low concentration of DTT, it was also found that DTT can inhibit the formation of zeaxanthin in vivo. In a number of studies the inhibition of zeaxanthin formation in intact leaves has been shown to be accompanied by an inhibition of energy dissipation activity (Bilger et al. 1989; Adams et al. 1990; Bilger and Björkman 1990, 1991; Demmig-Adams et al. 1990c). Although one might expect the inhibition of a photoprotective energy dissipation process to result in photodamage to the photosynthetic apparatus, further reductions in photosynthesis or photochemical efficiency resulting from such inhibition have been relatively small (Winter and Königer 1989; Adams et al. 1990; Bilger and Björkman 1990). Under such circumstances it is possible that high rates of repair, at least in high-light acclimated plants, are capable of minimizing any sustained damage (Adams and Demmig-Adams 1993; Demmig-Adams and Adams 1993a).

In some of the above studies it was found that a low level of energy dissipation activity could develop in leaves even in the absence of zeaxanthin, and at the time this was therefore thought to be unrelated to the xanthophyll cycle. A recent study by Gilmore and Yammoto (1993) has shown that DTT does not inhibit the activity of violaxanthin de-epoxidase completely, in that some antheraxanthin may accumulate in the presence of DTT. They have therefore suggested that the energy dissipation activity that develops in the presence of DTT can be accounted for by antheraxanthin (i.e., that antheraxanthin has the same ability to facilitate the energy dissipation process as does zeaxanthin), and this may also be the case in the above studies.

7.2.2.2 Energy Dissipation in Lichens

Blue-green algae, or cyanobacteria, offer an interesting system in which to study energy dissipation as they do not possess the xanthophyll cycle. Thus, we have compared energy dissipation activity and photoprotection in green algal lichens versus blue-green algal lichens (Demmig-Adams et al. 1990a,b). Green algal lichens, which do possess the xanthophyll cycle, exhibited changes in carotenoid content and in energy dissipation activity similar to those observed in higher plants (and both were furthermore inhibited by DTT). On the other hand, blue-green algal lichens that did not contain, nor form, any zeaxanthin exhibited very little development of energy dissipation activity upon exposure to excessive light. Furthermore, such blue-green algal lichens, which had developed in low light and therefore also presumably had very low rates of repair, did exhibit very sustained reductions in photosynthesis and photochemical efficiency, following the exposure to high light.

7.2.2.3 The Reduction State of Photosystem II

The degree to which electrons accumulate in PSII, or the reduction state of PSII, is a measure of the potentially dangerous accumulation of excitation energy. Comparisons between control leaves and leaves that have been treated with DTT, thereby preventing the formation of zeaxanthin and the development of energy dissipation, have shown that PSII becomes considerably more reduced under excessive light in the absence of thermal energy dissipation (Adams et al. 1990; Demmig-Adams and Adams 1990, 1994; Demmig-Adams et al. 1990c). Furthermore, theoretical estimations of the reduction state of PSII in the absence of thermal energy dissipation match those determined from actual measurements of leaves treated with DTT almost precisely (Demmig-Adams and Adams 1994). In addition, a comparison between a blue-green algal lichen that did not contain any zeaxanthin (and therefore did not exhibit thermal energy dissipation) and a green algal lichen, with similar rates of photosynthesis, revealed that the blue-green algal lichen experienced a considerably higher PSII reduction state under excessive light relative to the green algal lichen. All of these findings suggest that the energy dissipation process associated with zeaxanthin (and possibly antheraxanthin) does divert energy away from the reaction centers, thereby protecting the chlorophyll pigment bed against overexcitation.

7.2.2.4 Energy Dissipation in the Absence of Excess Energy

Gilmore and Yamamoto (1992) provided some very valuable evidence on the role of zeaxanthin in energy dissipation in some elegant experiments with isolated chloroplasts. They showed that ATP-dependent formation of a transthylakoid pH gradient in darkened chloroplasts resulted in a strong decrease in fluorescence yield (and thus an increase in thermal energy dissipation) in zeaxanthin-containing chloroplasts but not in zeaxanthin-free chloroplasts. Thus, prerequisites for the engagement of thermal energy dissipation include the presence of zeaxanthin as well as a pH gradient across the thylakoid membranes (or at least a low lumen pH), but not the presence of light per se.

In addition, sustained reductions in photochemical efficiency, even after a dark period of 12 h or more, have been observed in a number of species that had been exposed to various environmental stress factors in combination with high light. These depressions were suggested to have been due largely to sustained increases in energy dissipation activity associated with high levels of zeaxanthin. Such sustained changes in energy dissipation and/or zeaxanthin content have been found in plants experiencing water stress (Demmig et al. 1988; Demmig-Adams and Adams 1993b; Eickmeier et al. 1993), mangroves under saline conditions (Björkman et al. 1988), and in several evergreens during the winter (Adams and Demmig-Adams 1994, 1995). Such findings also suggest a strong link between the xanthophyll cycle and energy dissipation, even in the absence of excess light.

7.3 The Xanthophyll Cycle and the Light Environment

7.3.1 Diurnal Changes Under Natural Conditions

Full diurnal characterizations of the carotenoid compostion from the leaves of a number of plants have shown that the carotenoids of the xanthophyll cycle are the only ones to exhibit any diurnal changes during exposure to the normal growth light regime (Adams and Demmig-Adams 1992; Adams et al. 1992; Demmig-Adams and Adams 1992a). As incident PPFD rose and fell in the course of the day, violaxanthin levels fell and rose in an antiparallel manner to PPFD, whereas zeaxanthin (or zeaxanthin + antheraxanthin) levels mirrored those of PPFD. In all cases the maximum levels of zeaxanthin occurred during exposure to maximal daily PPFDs, regardless of the orientation of the leaf and the time of day during which peak incident PPFD occurred. In another study (Björkman and Demmig-Adams 1994), energy dissipation activity was observed to rise and fall concomitantly with the diurnal increase in zeaxanthin + antheraxanthin level in cotton leaves.

7.3.2 Acclimation to Different Light Environments

The rapid interconversions of the three components of the xanthophyll cycle in response to rapid changes in PPFD are not the only manner in which the cycle responds to the light environment. In comparing leaves of sun versus shade species, or leaves of the same species which have developed in high versus low light, the carotenoids of the xanthophyll cycle exhibit a greater positive change with increasing PPFD than any other carotenoid (Demmig-Adams et al. 1989; Thayer and Björkman 1990; Demmig-Adams and Adams 1992 a,b). The components of the xanthophyll cycle typically range from 10% of the total carotenoids in shade leaves to greater than 30% in sun leaves (which can represent a four- to fivefold difference in the total content of the xanthophyll cycle carotenoids). Furthermore, the zeaxanthin content of blue-green algae, blue-green algal lichens, and red algae (which also lack the xanthophyll cycle) has been found to increase with increasing growth light regime (Paerl et al. 1983; Cunningham et al. 1989; Millie et al. 1990; Adams et al. 1993). Such differences in the xanthophyll cycle or zeaxanthin content between light environments also suggest a role in acclimation to those environments, and a specific function in photoprotection seems likely.

7.4 Evidence from CAM Plants

7.4.1 Energy Dissipation in the Field

Comparisons of the changes in chlorophyll fluorescence emission from differently oriented photosynthetic surfaces in response to changes in natural sunlight have revealed some interesting phenomena associated with CAM (Fig. 7.2). For a southeast facing (SE) cladode surface which received light during the morning hours, thermal energy dissipation activity increased and decreased (Fig. 7.2C) as incident light increased and decreased (Fig. 7.2A). Photosynthesis rose more

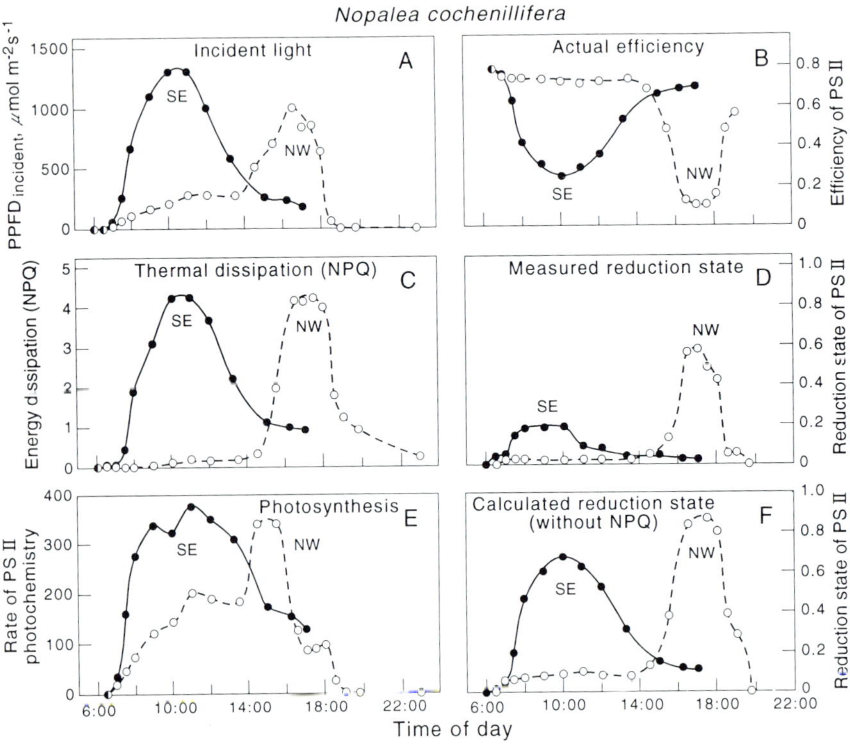

Fig. 7.2A–F. Changes in various parameters, calculated from measurements of chlorophyll fluorescence, in response to diurnal changes in light for two faces of a *Nopalea cochenillifera* cladode (*SE* southeast facing; *NW* northwest facing) on 30 March (*SE* face) and 31 March (*NW* face) 1988. *NPQ* nonphotochemical quenching = ratio of unquenched to quenched maximal fluorescence −1 (calculated from data in Adams et al. 1989. See Demmig-Adams and Adams 1993a, and Genty et al. 1989 for a description of the calculations.) The rates of electron transport are approximated from the product of the actual efficiency of PSII photochemistry and incident PPFD, and is therefore given in units of µmol photons $m^{-2} s^{-1}$. In order to convert this rate of photochemistry in µmol incident photons $m^{-2} s^{-1}$ to the actual rate of PSII electron transport, one would have to account for the absorptance of light by chlorophyll in the cladode and for the fact that 2 photons are required per electron in linear electron transport

rapidly and saturated, and remained somewhat more elevated as PPFD levels declined (Fig. 7.2E). The actual efficiency of PSII declined as PPFD levels rose (Fig. 7.2B), as one would expect given that more light was being absorbed than could be used, and thermal energy dissipation activity rose to high levels, competing with photochemistry for the absorbed energy. The reduction state of PSII increased to only approximately 20% during peak irradiance (Fig. 7.2D). However, had the cladode been incapable of thermally dissipating a considerable portion of the absorbed energy during peak irradiance, greater than 60% of the PSII centers would have been reduced (Fig. 7.2F).

In contrast, the northwest facing (NW) surface of the same cladode, which received direct illumination from the sun only in the late afternoon (Fig. 7.2A), exhibited markedly different responses from the southeast facing surface. The maximal level of energy dissipation activity was similar in the northwest face to those observed in the southeast face during peak irradiance (Fig. 7.2C), but occurred at PPFDs that were only 65% of the peak PPFDs experienced by the southeast facing surface. This was due to the fact that the rates of photosynthesis (PSII photochemistry), although transiently comparably high, declined dramatically in the late afternoon in the northwest facing surface while incident PPFD remained high (Fig. 7.2E). Presumably this dramatic decline in photosynthesis rate resulted from the exhaustion of the internal CO_2 supplied via the decarboxylation of malic acid, whereas the southeast facing surface continued to be supplied with sufficient CO_2 to sustain maximal rates of photosynthesis during the morning hours.

Given that the northwest facing surface received lower maximal PPFDs than the southeast facing surface, one might expect the actual efficiency of PSII and the reduction state of PSII to be higher and lower, respectively, in the northwest face relative to the southeast face during exposure to peak PPFDs. However, since the rate of electron transport declined so dramatically in the northwest face during direct illumination, the efficiency of PSII decreased considerably below that observed in the southeast face (Fig. 7.2B), and PSII became 60% reduced, compared with 20% in the southeast face during the morning (Fig. 7.2D). In the absence of thermal energy dissipation, however, the reduction state would have risen as high as 90% in the northwest facing cladode surface (Fig. 7.2F).

These marked differences in PSII efficiency and the reduction state of PSII between the two cladode surfaces therefore resulted primarily from the fact that the supply of CO_2 to sustain photosynthesis became depleted in the afternoon, as well as from the fact that the level of thermal energy dissipation did not increase in the northwest face above that observed in the southeast face. The dramatic nature of this decline in photosynthesis rate can be seen in Fig. 7.3. For the southeast face of the cladode the electron transport rates increased with increasing PPFD during the early morning and were even slightly higher as the PPFD declined during late morning (Fig. 7.3A). In the northwest face electron transport rates increased steeply with increasing PPFD as PPFD increased gradually during the morning. As incident PPFD increased from 700 to 1000 $\mu mol\, m^{-2} s^{-1}$ in the

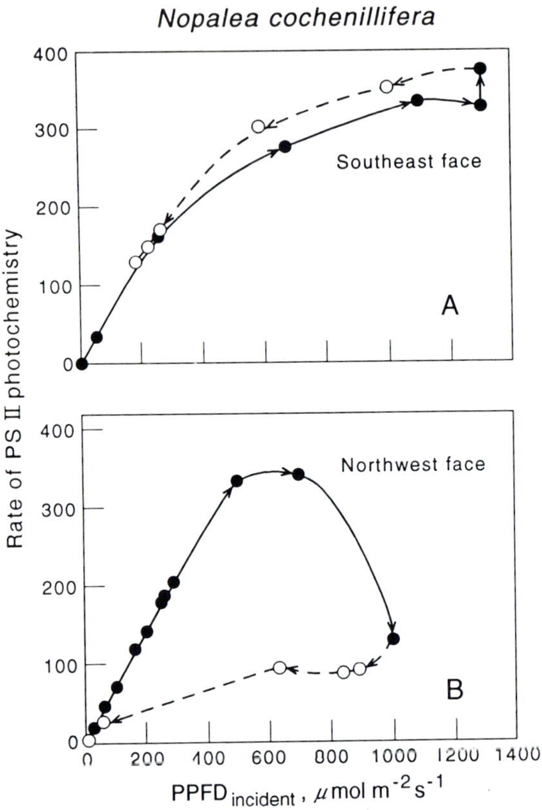

Fig. 7.3A,B. Changes in the rate of PSII photochemistry from two faces of a *Nopalea cochenillifera* cladode in response to diurnal changes in light. The *filled circles* and *solid lines* represent the data obtained as PPFDs were increasing, whereas the *open circles* and *dashed lines* represent the data obtained as PPFDs were decreasing. See the legend of Fig 7.2 for additional information

afternoon, however, electron transport declined sharply and continued to be much lower at a given PPFD than during the morning hours as PFD decreased (Fig. 7.3B).

Similar results were observed from the east and west faces of an *Opuntia ficus-indica* cladode, although in this case thermal energy dissipation activity did increase to a higher level in the west face that received light in the afternoon (data not shown; see Adams et al. 1989). In contrast to these observations that electron transport declined significantly in the exposed cladode surfaces during the afternoon as the internal supply of CO_2 became depleted, Winter and Lesch (1992) did not observe any decrease in electron transport rates at maximal PPFDs (approximately 1400 μmol m^{-2} s^{-1}) from the south face of an *O. ficus-indica* cladode, possibly due to the fact that the skies were partly overcast during the morning. Although Winter and Lesch (1992) did not determine the thermal energy dissipation activity in their *O. ficus-indica* cladode, they did observe an increase in the zeaxanthin content between dawn and early afternoon.

7.4.2 Acclimation

7.4.2.1 Low Light Versus High Light

Acclimation of the photosynthetic apparatus to different light environments involves an entire suite of changes, some of which contribute to changes in the capacity to utilize absorbed light energy and some of which contribute to changes in the capacity to protect the system against damage that might result from the absorption of excess light. Leaves of *Crassula argentea* that had acclimated to high light exhibited maximal rates of photosynthesis that were more than twice as great as those from leaves that had acclimated to very low light levels (Fig. 7.4A). The total pool of the xanthophyll cycle components was also more than twice as large in the high light-acclimated leaves than in the low light-acclimated leaves (Fig. 7.4B). This would suggest that the high light-acclimated leaves have a much greater capacity for thermal energy dissipation through the formation of higher levels of zeaxanthin (or zeaxanthin + antheraxanthin) than the low light-acclimated leaves. This is indeed the case, as the level of zeaxanthin (or zeaxanthin + antheraxanthin) upon exposure to light equivalent to full sunlight was more than three times as great in the high light-acclimated leaves relative to the low light-acclimated leaves (Figs. 7.4C, 4D). The capacity for thermal energy dissipation was more than two times higher in the high light-acclimated leaves relative to the low light-acclimated leaves (Fig. 7.4E).

The fact that the high light-acclimated leaves have both a greater capacity to utilize absorbed energy through photosynthesis (Fig. 7.4A), as well as a greater capacity to dissipate energy that is in excess (Fig. 7.4E), should result in a system that is less likely to experience damage when light is in excess. The reduction state of PSII in the low light-acclimated leaves was indeed more than twice as high as that determined from the high light-acclimated leaves during exposure to full sunlight (Fig. 7.4F), with almost 80% of the centers reduced. Under such conditions the likelihood of some chlorophyll molecules entering the triplet excited state, and potentially leading to the formation of destructive singlet oxygen, is considerably enhanced.

Another way to visualize the greater potential for damage to occur in the low light-acclimated leaves is to divide the absorbed energy into the fractions which are utilized in photosynthesis, dissipated thermally, or which remain in excess (Fig. 7.5). For the two sets of leaves illustrated in Fig. 7.4, illuminated with the equivalent of full sunlight, only 15% of the energy was utilized in photosynthesis, 31% dissipated thermally, and fully 54% was in excess and possibly resulting in damage to the photosynthetic system in the low light-acclimated leaves. In contrast, the high light-acclimated leaves were utilizing 33% of the energy through photosynthesis, dissipating 46% of the energy thermally, and only 21% remained in excess, the latter possibly being dealt with through other means (e.g. carotenoid quenching of triplet chlorophyll and singlet oxygen, other antioxidants, etc.).

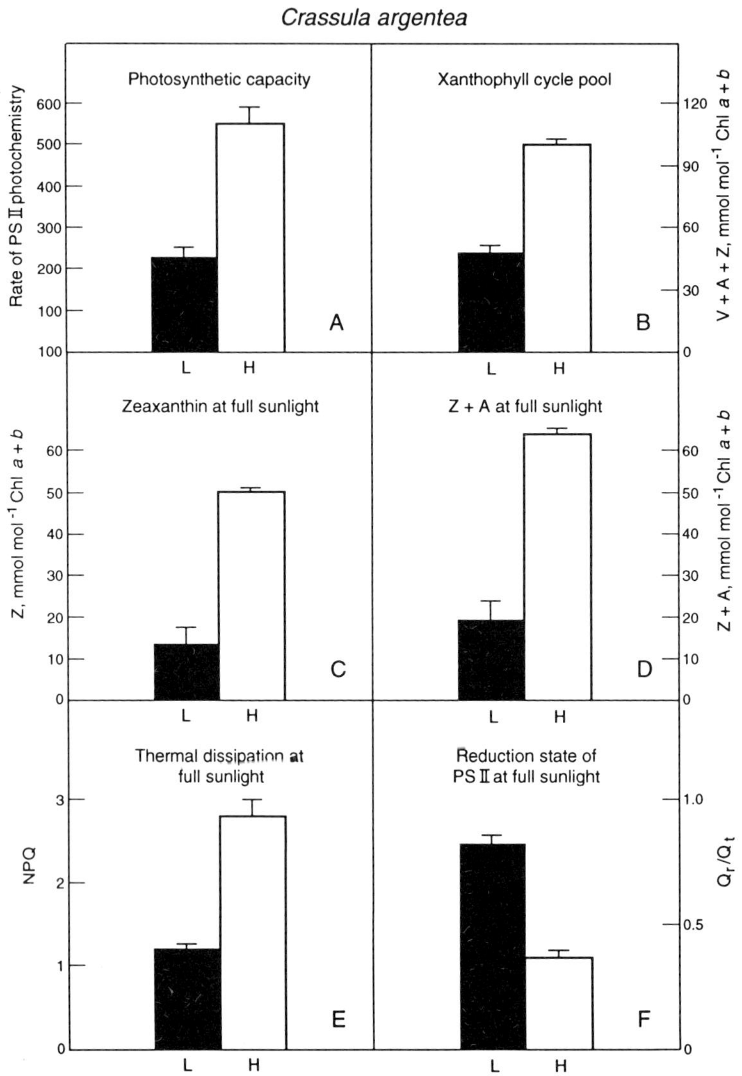

Fig. 7.4 A–F. Differences in photosynthetic capacity, several xanthophyll cycle parameters (*V* violaxanthin; *A* antheraxanthin; *Z* zeaxanthin), thermal energy dissipation activity, and the reduction state of PSII (Q_r/Q_t $Q_{reduced}/Q_{total}$) between leaves of *Crassula argentea* acclimated to either low light (*L* grown at 10 μmol photons m^{-2} s^{-1}) or high light (*H* peak PPFD of 1500 μmol m^{-2} s^{-1}) during exposure to 2050 μmol photons m^{-2} s^{-1} in air at 25 °C. Data are the mean ± standard deviation for three determinations. See the legend of Fig. 7.2 for additional information

7.4.2.2 Within a Leaf

Acclimation of the components of the xanthophyll cycle to different growth PPFDs can also be evaluated through expressing each carotenoid, and the

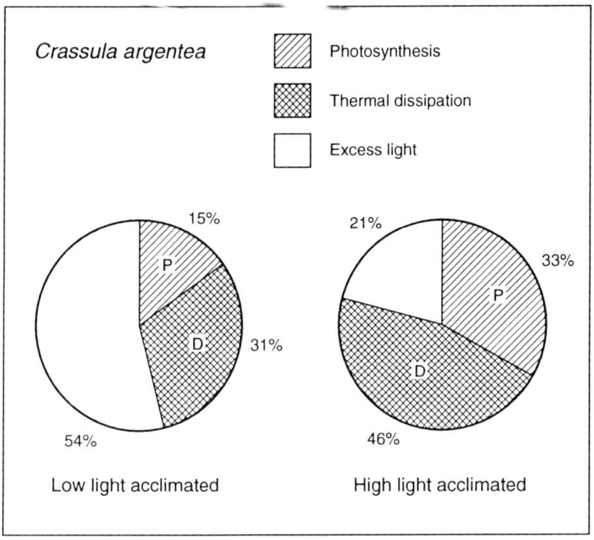

Fig. 7.5. The proportions of absorbed energy that are utilized through photosynthesis, dissipated thermally, or unaccounted for (light in excess; dealt with by antioxidant systems or potentially damaging the photosynthetic apparatus) for leaves of *Crassula argentea*, acclimated to either low light or high light, exposed to 2050 μmol photons m^{-2} s^{-1} (see the legend of Fig.7.3 for more details). These percentages were calculated from chlorophyll fluorescence parameters. Data are the mean of three determinations

xanthophyll cycle pool, as a fraction of the total carotenoids contained in a leaf. Such a comparison between the shade and sun leaves of *C. argentea* showed that the xanthophyll cycle components represented 12.5% of the carotenoids in the shade leaves, but 21.9% of the total carotenoids in the sun leaves (Fig. 7.6). Since the total carotenoid content was also higher in the high light-acclimated leaves, this actually represented an increase in the xanthophyll cycle pool of 107% on a chlorophyll basis between shade and sun.

Splitting of high light-acclimated leaves of *C. argentea* into two halves revealed a similar difference in the acclimation of the carotenoids between the lower, presumably more shade-acclimated, and upper halves (Fig. 7.6). Whereas the xanthophyll cycle pool represented a respectable 15.6% of the carotenoids in the lower half, it increased to 25.3% of the total carotenoids in the upper half. A finer subdivision of the leaves into more layers would no doubt reveal a gradient throughout the leaf, with the components of the xanthophyll cycle pool diminishing even further as one approached the bottom of the leaves (unless reflected light from below actually caused the pool to increase at the lower surface), and even higher quantities of the xanthophyll cycle pool in the uppermost layer of the leaves. Regardless, there is clearly an acclimation of the xanthophyll cycle components to the light gradient within a leaf.

Characterization of the xanthophyll cycle carotenoids from dawn to midday in the lower and upper halves of the high light-acclimated leaves in response to

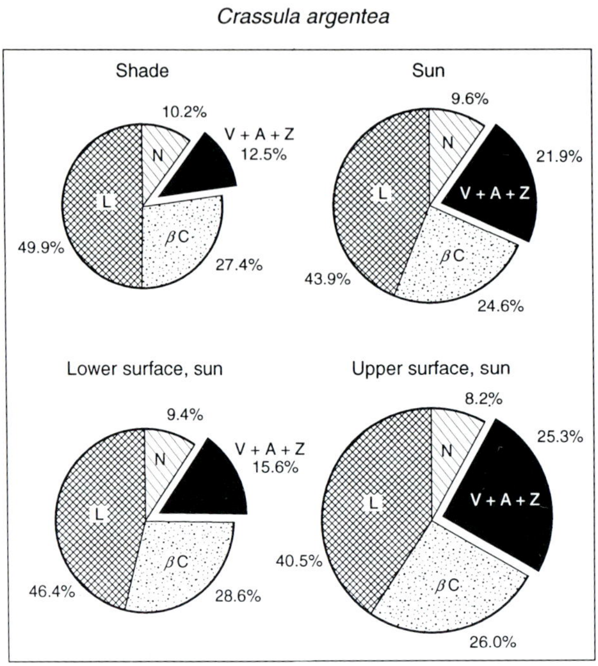

Fig. 7.6. Differences in the carotenoid composition between low light- and high light-acclimated leaves of *Crassula argentea* (n = 3), and between the lower and upper halves of high light- acclimated leaves (n = 6). *V* + *A* + *Z* Violaxanthin + antheraxanthin + zeaxanthin; *βC* β-carotene; *L* lutein; *N* neoxanthin

the normal dirunal changes in PPFD experienced by the plants during development shows that there were large diurnal changes in the composition of the xanthophyll cycle (Fig.7.7). Zeaxanthin, or zeaxanthin + antheraxanthin, increased significantly in both halves during midday (approximately a sixfold increase), consistent with a role in energy dissipation when light is in excess. Since the total pool of the xanthophyll cycle was greater in the upper halves than in the lower halves of the leaves, the total level of zeaxanthin, or zeaxanthin + antheraxanthin, was much greater in the upper halves on a chlorophyll basis (2 to 2.5 times greater at midday; Fig. 7.7, upper panels).

On the other hand, when expressed as a fraction of the total xanthophyll cycle pool (violaxanthin + antheraxanthin + zeaxanthin), the level of zeaxanthin (or zeaxanthin + antheraxanthin) was very similar in the lower and upper halves of the leaves (Fig. 7.7, lower panels). Thus the size of the entire pool of the xanthophyll cycle increases approximately in proportion to the increase in zeaxanthin (or zeaxanthin + antheraxanthin) at midday in the higher growth PPFD. Evidence such as this may provide a clue as to how the synthesis of the components of the xanthophyll cycle is regulated by light. It seems conceivable that the degree to which the xanthophyll cycle becomes de-epoxidized during peak irradiance is in some manner utilized as a signal to indicate whether or not

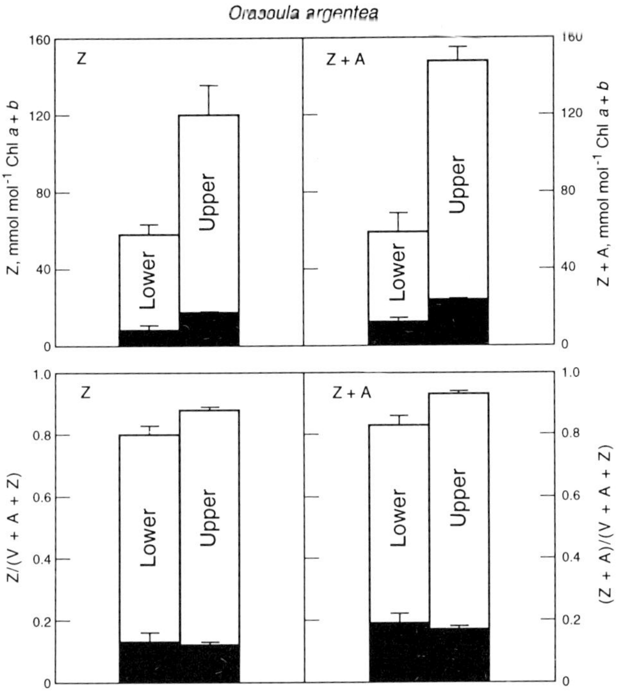

Fig. 7.7. Predawn (*darkened portions* of the columns) and midday (during exposure to 1500 μmol photons m^{-2} s^{-1} sunlight) determinations of several parameters related to the xanthophyll cycle (*V* violaxanthin; *A* antheraxanthin; *Z* zeaxanthin) from lower and upper halves of *Crassula argentea* leaves acclimated to high light. Data are the mean ± standard deviation of three determinations

more zeaxanthin should be synthesized from β-carotene and earlier precursors. In this particular species it would appear that the size of the pool might be regulated such that at peak irradiance, between 80 and 95% of violaxanthin becomes de-epoxidized, regardless of the absolute light level (compare lower and upper leaf halves, Fig. 7.7). Further de-epoxidation beyond this level might induce increased synthesis of zeaxanthin. In other species de-epoxidation of as little as 30% of the violaxanthin might be sufficient to result in a further increase in the xanthophyll cycle pool (Adams et al. 1992).

7.4.3 Photoinhibition

At the outset it was stated that decreases in the photochemical efficiency of PSII, as well as of the photon efficiency of photosynthesis, had been observed in various CAM species in the field (Adams et al. 1987, 1988; Adams 1988). At the time that this phenomenon of photoinhibition was reported, only a minimal interpretation as to the cause of these decreases could be offered. Utilizing some more recent approaches to the use of chlorophyll fluorescence, it is now possible to offer greater support to the idea that the photoinhibition observed in the field was due to increased and sustained thermal energy dissipation activity.

Some of the lowest predawn values of PSII photochemical efficiency were found in cladodes of *Opuntia basilaris* growing in Death Valley, California (Adams et al. 1987). In the spring of 1985, shortly after a rainfall event, tissue water potentials were relatively high and PSII photochemical efficiency was also relatively high in cladodes with various exposures to the sun (Table 7.1). Following a long period of reduced water availability, tissue water potentials decreased, and sustained depressions of PSII photochemical efficiency were observed from cladode surfaces that were exposed to the sun (south and west faces on 5 October; Table 7.1). The north-facing cladode surface, that did not experience exposure to direct sunlight, did not exhibit any significant reduction in PSII efficiency. Using the PSII efficiency of 0.75 observed in the north face on 21 May as the control value, which is conservative given that values as great as 0.87 have been found in some plants, estimates of the thermal energy dissipation activity responsible for the decreased levels of PSII photochemical efficiency, assuming that this was the sole factor responsible for such decreases, were generated (Table 7.1). The two largest values, of 2.28 and 3.17 in the exposed surfaces of the cladodes experiencing water stress, corresponding to PSII photochemical efficiencies of 0.48 and 0.42, are consistent with the relationship between the values of NPQ (approximately 4) and the intrinsic efficiency of PSII (approximately 0.3) observed in *N. cochenillifera* during exposure to peak PPFD (Fig. 7.2).

Table 7.1. Water potentials[a], photochemical efficiency of photosystem II[b], and calculated levels of energy dissipation activity (NPQ[c]) determined from different faces of *Opuntia basilaris* cladodes in Death Vallley, California, prior to sunrise

Date, 1985	Cladode face	Water potential, MPa	PSII efficiency F_v/F_m	NPQ
21 May[d]	North	−1.1	0.75	0
	South	−1.1	0.73	0.12
	East	−1.4	0.66	0.55
	West	−1.4	0.70	0.29
5 October[e]	North	−1.9	0.70	0.29
	South	−1.9	0.48	2.28
	West	−2.3	0.42	3.17

[a] Determined from a single tissue sample from the cladode for both surfaces. Data from Adams et al. (1987).

[b] Calculated from measurements of chlorophyll fluorescence from tissues frozen in liquid nitrogen. Data from Adams et al. (1987).

[c] The differences in measured PSII photochemical efficiency were used to calculate changes in the rate constant for energy dissipation in the pigment bed (Kitajima and Butler 1975), and this in turn was used to calculate values of F_m from which NPQ could be estimated (see Bilger and Björkman 1990; Demmig-Adams and Adams 1993a).

[d] Data obtained 11 days after a spring rainfall of 6.5 mm.

[e] Data obtained at the end of a hot and dry summer (13 mm of rainfall during the previous 4 months).

Thus a large part of the NPQ possible during exposure to direct sunlight may be maintained under water stress, even overnight in the absence of excess energy in the photochemical system. Furthermore, sustained decreases in PSII photochemical efficiency in *Nerium oleander* under a combination of high light and water stress were associated with increased levels of thermal energy dissipation and zeaxanthin (Demmig et al. 1988; Demmig-Adams and Adams 1993b). However, it still remains to be shown that sustained reductions in PSII photochemical efficiency in CAM plants in the field are accompanied by increased levels of zeaxanthin.

7.5 Conclusions

Photoprotection of the photosynthetic apparatus through a thermal energy dissipation process associated with the de-epoxidation of the xanthophyll cycle appears to be a ubiquitous phenomenon in higher plants, and CAM plants are no exception. In fact, the levels of energy dissipation observed in cacti in the field are higher than those found in any other plants to date. Given that obligate CAM plants such as cacti rely solely on the internal supply of CO_2 for photosynthesis during the day, perhaps the high capacity for energy dissipation activity should come as no surprise. Depletion of this internal supply of CO_2 is probably a common occurrence in CAM plants on sunny days, thus resulting in decreased rates of electron transport (utilization of the absorbed light energy; see Figs. 7.2 and 7.3). Under such circumstances employment of a photoprotective process that diverts and harmlessly dissipates excess energy within the chlorophyll pigment bed might be essential to prevent damage from occurring to the photosynthetic apparatus.

We have also shown that acclimation of the xanthophyll cycle pool and the resulting differences in the capacity to form zeaxanthin (and antheraxanthin) are accompanied by differences in the capacity for thermal energy dissipation in the chlorophyll pigment bed (Figs. 7.4, 7.5, and 7.6). In *C. argentea*, and several other species with CAM (unpubl.), high light-acclimated leaves with larger xanthophyll cycle pools exhibited much greater levels of energy dissipation activity accompanied by higher quantities of zeaxanthin (and antheraxanthin) during exposure to full sunlight than low light-acclimated leaves.

Given that there is a light gradient within leaves, one might expect to find acclimation of the level of xanthophyll cycle components within a leaf, and perhaps especially in the succulent photosynthetic organs of CAM plants. A simple subdivision of the leaves of *C. argentea* into two parts, lower and upper halves, revealed that there was significant acclimation of the xanthophyll cycle pool between the two halves, with a much larger pool present in the upper, more exposed half (Fig. 7.6). This resulted in a much greater capacity of the upper leaf halves to form the de-epoxidized products of the xanthophyll cycle during midday exposure to the peak growth PPFD (approximately 75% of full sunlight). However, in contrast, the ratio of zeaxanthin + antheraxanthin levels to the total xanthophyll cycle pool were similar between the lower and upper halves of the leaves (Fig 7.7). These latter findings may indicate that the stimulus for the synthesis of the xanthophyll cycle carotenoids by growth PPFD is through the degree to which the cycle becomes de-epoxidized.

Acknowledgements. We thank Susan Kiley for assistance with the determinations of electron transport rates, PSII reduction states, and energy dissipation activity in *C. argentea*, and David Barker for assistance with the predawn and midday collections of *C. argentea*. This work was supported by NSF Grant No. IBN-9207653 and a fellowship from the David and Lucile Packard Foundation.

References

Adams WW III (1988) Photosynthetic acclimation and photoinhibition of terrestrial and epiphytic CAM tissues growing in full sunlight and deep shade. Aust J Plant Physiol 15: 123–134

Adams WW III, Demmig-Adams B (1992) Operation of the xanthophyll cycle in higher plants in response to diurnal changes in incident sunlight. Planta 186: 390–398

Adams WW III, Demmig-Adams B (1993) Energy dissipation and photoprotection in leaves of higher plants. In: Yamamoto HY, Smith CM (eds) Photosynthetic responses to the environment. American Society of Plant Physiologists, Rockville, pp 27–36

Adams WW III, Demmig-Adams B (1994) Carotenoid composition and down regulation of photosystem II in three conifer species during the winter. Physiol Plant 92: 451–458

Adams WW III, Demmig-Adams B (1995) The xanthophyll cycle and sustained thermal energy dissipation activity in *Vinca minor* and *Euonymus kiautschovicus* in winter. Plant Cell Environ 18: 117–127

Adams WW III, Smith SD, Osmond CB (1987) Photoinhibition of the CAM succulent *Opuntia basilaris* growing in Death Valley: evidence from 77 K fluorescence and quantum yield. Oecologia 71: 221–228

Adams WW III, Terashima I, Brugnoli E, Demmig B (1988) Comparisons of photosynthesis and photoinhibition in the CAM vine *Hoya australis* and several C_3 vines growing on the coast of eastern Australia. Plant Cell Environ 11: 173–181

Adams WW III, Díaz M, Winter K (1989) Diurnal changes in photochemical efficiency, the reduction state of Q, radiationless energy dissipation, and non-photochemical fluorescence quenching in cacti exposed to natural sunlight in northern venezuela. Oecologia 80: 553–561

Adams WW III, Demmig-Adams B, Winter K (1990) Relative contributions of zeaxanthin-related and zeaxanthin-unrelated types of "high-energy-state" quenching of chlorophyll fluorescence in spinach leaves exposed to various environmental conditions. Plant Physiol 92: 302–309

Adams WW III, Volk M, Hoehn A, Demmig-Adams B (1992) Leaf orientation and the response of the xanthophyll cycle to incident light. Oecologia 90: 404–410

Adams WW III, Demmig-Adams B, Lange OL (1993) Carotenoid composition and metabolism in green and blue-green algal lichens in the field. Oecologia 94: 576–584

Bilger W, Björkman O (1990) Role of the xanthophyll cycle in photoprotection elucidated by measurements of light-induced absorbance changes, fluorescence and photosynthesis in leaves of *Hedera canariensis*. Photosynth Res 25: 173–185

Bilger W, Björkman O (1991) Temperature dependence of violaxanthin de-epoxidation and non-photochemical fluorescence quenching in intact leaves of *Gossypium hirsutum* L. and *Malva parviflora* L. Planta 184: 226–234

Bilger W, Björkman O, Thayer SS (1989) Light-induced spectral absorbance changes in relation to photosynthesis and the epoxidation state of the xanthophyll cycle components in cotton leaves. Plant Physiol 91: 542–551

Björkman O (1987a) High-irradiance stress in higher plants and interactions with other stress factors. In: Biggins J (ed) Progress in photosynthesis research, vol IV. Martinus Nijhoff, Dordrecht, pp 11–18

Björkman O (1987b) Low-temperature chlorophyll fluorescence in leaves and its relationship to photon yield of photosynthesis in photoinhibition. In: Kyle DJ, Osmond CB, Arntzen CJ (eds) Photoinhibition. Elsevier, Amsterdam, pp 123–144

Björkman O, Demmig-Adams B (1994) Regulation of photosynthetic light energy capture, conversion, and dissipation in leaves of higher plants. In: Schulze E-D, Caldwell MM (eds) Ecophysiology of photosynthesis. Springer, Berlin Heidelberg New York, pp 17–44

Björkman O, Demmig B, Andrews TJ (1988) Mangrove photosynthesis: response to high-irradiance stress. Aust J Plant Physiol 15: 43–61

Cunningham FX Jr, Dennenberg RJ, Mustardy L, Jursinic PA, Gantt E (1989) Stoichiometry of photosystem I, photosystem II, and phycobilisomes in the red alga *Porphyridium cruentum* as a function of growth irradiance. Plant Physiol 91: 1179–1187

Demmig B, Björkman O (1987) Comparison of the effect of excessive light on chlorophyll fluorescence (77 K) and photon yield of O_2 evolution in leaves of higher plants. Planta 171: 171–184

Demmig B, Winter K, Krüger A, Czygan F-C (1987) Photoinhibition and zeaxanthin formation in intact leaves. A possible role of the xanthophyll cycle in the dissipation of excess light energy. Plant Physiol 84: 218–224

Demmig B, Winter K, Krüger A, Czygan F-C (1988) Zeaxanthin and the heat dissipation of excess light energy in *Nerium oleander* exposed to a combination of high light and water stress. Plant Physiol 87: 17–24

Demmig-Adams B (1990) Carotenoids and photoprotection in plants. A role for the xanthophyll zeaxanthin. Biochim Biophys Acta 1020: 1–24

Demmig-Adams B, Adams WW III (1990) The carotenoid zeaxanthin and "high-energy-state quenching" of chlorophyll fluorescence. Photosynth Res 25: 187–197

Demmig-Adams B, Adams WW III (1992a) Photoprotection and other responses of plants to high light stress. Annu Rev Plant Physiol Plant Mol Biol 43: 599–626

Demmig-Adams B, Adams WW III (1992b) Carotenoid composition in sun and shade leaves of plants with different life forms. Plant Cell Environ 15: 411–419

Demmig-Adams B, Adams WW III (1993a) The xanthophyll cycle, protein turnover, and the high-light tolerance of sun-acclimated leaves. Plant Physiol 103: 1413–1420

Demmig-Adams B, Adams WW III (1993b) The xanthophyll cycle. In: Alscher R, Hess J (eds) Antioxidants in higher plants. CRC Press, Boca Raton, pp 91–110

Demmig-Adams B, Adams WW III (1994) Light stress and photoprotection related to the xanthophyll cycle. In: Foyer C, Mullineaux P (eds) Photooxidative stresses in plants: Causes and amelioration. CRC Press, Boca Raton, pp 105–126

Demmig-Adams B, Winter K, Winkelmann E, Krüger A, Czygan F-C (1989) Photosynthetic characteristics and the ratios of chlorophyll, β-carotene, and the components of the xanthophyll cycle upon a sudden increase in growth light regime in several plant species. Bot Acta 102: 319–325

Demmig-Adams B, Adams WW III, Czygan F-C, Schreiber U, Lange OL (1990a) Differences in the capacity for radiationless energy dissipation in green and blue-green algal lichens associated with differences in carotenoid composition. Planta 180: 582–589

Demmig-Adams B, Adams WW III, Green TGA, Czygan F-C, Lange OL (1990b) Differences in the susceptibility to light stress in two lichens forming a phycosymbiodeme, one partner possessing and one lacking the xanthophyll cycle. Oecologia 84: 451–456

Demmig-Adams B, Adams WW III, Heber U, Neimanis S, Winter K, Krüger A, Czygan F-C, Bilger W, Björkman O (1990c) Inhibition of zeaxanthin formation and of rapid changes in radiationless energy dissipation by dithiothreitol in spinach leaves and chloroplasts. Plant Physiol 92: 293–301

Eickmeier WG, Casper C, Osmond CB (1993) Chlorophyll fluorescence in the resurrection plant *Selaginella lepidophylla* (Hook. & Grev.) Spring during high-light and desiccation stress, and evidence for zeaxanthin-associated photoprotection. Planta 189: 30–38

Emerson R (1936) The effect of intense light on the assimilatory mechanism of green plants, and its bearing on the carbon dioxide factor. Cold Spring Harbor Symp Quant Biol 3: 128–137

Genty B, Briantais J-M, Baker NR (1989) The relationship between the quantum yield of photosynthetic electron transport and quenching of chlorophyll fluorescence. Biochim Biophys Acta 990: 87–92

Gilmore AM, Yamamoto HY (1992) Dark induction of zeaxanthin-dependent non-photochemical quenching mediated by ATP. Proc Natl Acad Sci USA 89: 1899–1903

Gilmore AM, Yamamoto HY (1993) Linear models relating xanthophylls and lumen acidity to non-photochemical fluorescence quenching. Evidence that antheraxanthin explains zeaxanthin-independent quenching. Photosynth Res 35: 67–78

Horton P, Ruban AV (1992) Regulation of photosystem II. Photosynth Res 34: 375–385

Horton P, Ruban AV, Rees D, Pascal AA, Noctor G, Young AJ (1991) Control of the light-harvesting function of chloroplast membranes by aggregation of the LHCII chlorophyll-protein complex. FEBS Lett 292: 1–4

Kitajima M, Butler WL (1975) Quenching of chlorophyll fluorescence and primary photochemistry in chloroplasts by dibromothymoquinone. Biochim Biophys Acta 376: 105–115

Kok B (1956) On the inhibition of photosynthesis by intense light. Biochim Biophys Acta 21: 234–244

Millie DF, Ingram DA, Dionigi CP (1990) Pigment and photosynthetic responses of *Oscillatoria agardhii* (Cyanophyta) to photon flux density and spectral quality. J Phycol 26: 660–666

Myers J, Burr GO (1940) Studies on photosynthesis: some effects of high intensity light on *Chlorella*. J Gen Physiol 24: 45–67

Owens TG, Shreve AP, Albrecht AC (1992) Dynamics and mechanism of singlet energy transfer between carotenoids and chlorophylls: light harvesting and non-photochemical fluorescence quenching. In: Murata N (ed) Research in photosynthesis, vol I. Kluwer, Dordrecht, pp 179–186

Paerl HW, Tucker J, Blant PT (1983) Carotenoid enhancement and its role in maintaining blue-green algal (*Microcystis aeruginosa*) surface blooms. Limnol Oceanogr 28: 847–857

Powles, SB (1984) Photoinhibition of photosynthesis induced by visible light. Annu Rev Plant Physiol 35: 15–44

Ruban AV, Horton P (1992) Mechanism of ΔpH-dependent dissipation of absorbed excitation energy by photosynthetic membranes I. Spectroscopic analysis of isolated light-harvesting complexes. Biochim Biophys Acta 1102: 30–38

Ruban AV, Rees D, Pascal AA, Horton P (1992) Mechanism of ΔpH-dependent dissipation of absorbed excitation energy by photosynthetic membranes II. The relationships between LHCII aggregation in vitro and qE in isolated thylakoids. Biochim Biophys Acta 1102: 39–44

Thayer SS, Björkman O (1990) Leaf xanthophyll content and composition in sun and shade determined by HPLC. Photosynth Res 23: 331–343

Winter K, Königer M (1989) Dithiothreitol, an inhibitor of violaxanthin de-epoxidation, increases the susceptibility of leaves of *Nerium oleander* L. to photoinhibition of photosynthesis. Planta 180: 24–31

Winter K, Lesch M (1992) Diurnal changes in chlorophyll *a* fluorescence and carotenoid composition in *Opuntia ficus-india*, a CAM plant, and in three C_3 species in Portugal during summer. Oecologia 91: 505–510

Yamamoto HY, Kamite L (1972) The effects of dithiothreitol on violaxanthin de-epoxidation and absorbance changes in the 500-nm region. Biochim Biophys Acta 267: 538–543

Part B
Environmental and Developmental Control of Crassulacean Acid Metabolism

Preface

This part of the book is largely devoted to the intraspecific versatility of carbon-assimilation patterns in CAM plants. Alterations in daytime versus nighttime carbon gain are generally related to (1) environmental or (2) developmental factors, or to both. Depending on ambient conditions, certain species are able to engage in either the C_3 or the CAM option of carbon acquisition (facultative CAM). In all CAM plants, however, an ontogenetic shift from the C_3 to the CAM pathway occurs during the early development of photosynthetic organs. Since environmental factors can influence developmental processes, it is sometimes difficult to attribute changes in CAM activity unequivocally to changes in the physical environment.

The first three chapters of this section (8,9,10) discuss recent findings on the physiology and molecular biology of a halophilic species that has become a widely studied model system in CAM research: *Mesembryanthemum crystallinum* (Aizoaceae). This species may exhibit all the features of a C_3 plant when well watered, but develops CAM in response to high soil salinity and drought stress (Winter and von Willert 1972). Chapter 8 presents a brief historical account of CAM research in *M. crystallinum* and considers the effects of photoperiod and externally applied growth regulators on certain physiological characteristics, such as changes in enzyme activity and gas-exchange patterns associated with the CAM pathway. Cushman and Bohnert (Chap. 9) review their detailed work since 1986 on identification and regulation of genes involved in the shift from the C_3 to the CAM pathway in *M. crystallinum*. Many of these genes seem to be up-regulated at the transcriptional level, although some of the responses appear to be associated with salt stress rather than the C_3 to CAM transition. Chapter 10 considers how, on a whole-plant basis, the effects of three plant growth regulators on CAM induction might be integrated and discusses the possible role of these substances in the transduction of environmental stress signals. All three chapters discuss the still controversial question of the relative importance of developmental and environmental factors in the induction of CAM in *M. crystallinum*.

We are now also beginning to understand the expression of CAM in other species with highly flexible gas-exchange characteristics besides *M. crystallinum*. Chapter 11 deals with the up- and down-regulation of CAM in a member of the Crassulaceae with a north-temperate distribution, *Sedum telephium*. In Chapter 12, the possible functional significance of recycling CO_2 via PEPC into malic acid is discussed, including species that only show net uptake of CO_2 during the daytime (CAM-cycling). Chapter 13 illustrates the importance of leaf ontogeny in the expression of CAM in the genus *Peperomia* (Piperaceae), a group with many species that exhibit intermediate gas-exchange characteristics. Versatility of carbon fixation via the C_3 and CAM pathways in different photosynthetic organs is considered using the example of *Cissus quadrangularis* (Vitaceae; Chap. 14), a species in which both leaves and stems engage in net carbon assimilation. In the last chapter of this section (Chap. 15), Borland and Griffiths describe the application of on-line carbon-isotope discrimination techniques for the continuous monitoring of carboxylation patterns in facultative CAM plants to evaluate the contributions to daily carbon gain of CO_2 uptake via RUBISCO and PEPC.

Reference

Winter K, von Willert DJ (1972) NaCl-induzierter Crassulaceensäurestoffwechsel bei *Mesembryanthemum crystallinum*. Z Pflanzenphysiol 67: 166–170

8 Factors Affecting the Induction of Crassulacean Acid Metabolism in *Mesembryanthemum crystallinum*

G.E. EDWARDS, Z. DAI, S.H. CHENG, and M.S.B. KU

8.1 Introduction

Some succulent plants initially perform C_3 photosynthesis but subsequently shift to CAM after developmental or environmental changes (e.g. photoperiod, water stress, day/night temperature). Species which have a strong dependency on the environment for expression of CAM are facultative, in contrast to obligate CAM plants which tend to function in the CAM mode under all conditions (Cockburn 1985; Winter 1985). Associated with this transition in facultative species is nocturnal opening of stomata with concomitant CO_2 uptake, and diurnal fluctuations of tissue acidity and malate content. Facultative CAM plants may maximize their growth by assimilating carbon via the C_3 pathway when environmental conditions are less stressful, and utilize the CAM mode when environmental conditions cause more potential for water stress.

The occurrence of CAM in a facultative species may be dependent on ontogeny, on environmental factor(s), or on both. If strictly dependent on ontogeny, then performance of CAM may occur during the life cycle without a dependence on the environmental conditions; if strictly induced by certain environmental cues, then its occurrence would not be inevitable. *Mesembryanthemum crystallinum* is a facultative CAM species which exhibits rapid induction of CAM under water stress. Since its discovery over 20 years ago as an inducible CAM species (Winter and von Willert 1972), it has proved to be a useful model system for studying molecular, biochemical, and environmental features of induction of CAM. In this chapter, factors which affect the induction of CAM in *M. crystallinum* are reviewed.

8.2 Discovery of Induction of CAM in *Mesembryanthemum crystallinum* by Water Stress in Controlled Environments

Mesembryanthemum crystallinum, the common ice plant, is an annual, halophytic species found in regions with Mediterranean-type climate such as the Pacific coast of the United States, the Mediterranean coastline, in the coastal desert of southern Africa, in the Canary Islands, and in Australia. Watering plants with

Department of Botany, Washington State University, Pullman, WA 99164-4238, USA

Ecological Studies, Vol. 114
Winter/Smith (eds) Crassulacean Acid Metabolism

high concentrations of salt (200 to 500 mM) have been found to induce CAM in *M. crystallinum* (Winter and von Willert 1972; Winter 1973a). This was not an effect of salt per se, since a number of stresses induce CAM in this species, including high salt, drought, high light combined with low humidity, and exposure of roots to low temperature or anoxia (Winter 1973b; Winter 1985). All of these conditions are thought to induce CAM by causing water stress, although the internal signals responsible for induction have yet to be discovered. This is not only induction of the performance of CAM, but induction of the biochemistry for the process. For example, it is accompanied by pronounced increases in the extractable activities of a number of enzymes associated with CAM, including phospho*enol*pyruvate carboxylase (PEPC) and malic enzyme (both NAD-ME and NADP-ME) (Holtum and Winter 1982). The performance of CAM is evident from measurements of nocturnal acidification and malate accumulation, and from gas-exchange patterns (Winter 1985; Winter and Gademann 1991) characteristic of the four phases of CAM (as described by Osmond 1978). As the degree of change in some of these parameters under different conditions will be discussed in this chapter, it is useful to note that following induction of CAM in *M. crystallinum* by water stress, PEPC activity can reach levels up to 1000 to 2000 $\mu mol\ mg^{-1}\ Chl\ h^{-1}$, with nocturnal acidification on a fresh weight (fr. wt.) basis up to 100 to 150 $\mu mol\ H^{+}\ g^{-1}$ (ca. 150 to 250 $mmol\ m^{-2}$).

8.3 Induction of CAM in a Natural Habitat

In the Mediterranean coastal area, *M. crystallinum* germinates during the rainy season in January and grows in the C_3 mode (Winter et al. 1978). The daily minimum temperature increases steadily from 8 to 10 °C in January–April to 20 °C in July, while the maximum daily temperature increases from about 20 °C in January–April to 28 °C in July. From February–April to May the maximum daily photosynthetic photon flux density (PPFD) increases by about 10% and the daylength increases from ca. 11 h to 14 h. The onset of CAM induction occurs about 3 months after germination: nocturnal accumulation of malate begins early April as the availability of soil water decreases. Reproductive growth is initiated at the end of May, and by July and August most of the plants have set seeds and died (Winter et al. 1978). Whether or not this apparent induction by water stress is ever dependent on, or enhanced by, other factors (increased daily PPFD, daylength, and day/night temperatures) in natural habitats remains to be determined (see recent views of Ostrem et al. 1987; Cheng and Edwards 1991; Piepenbrock and Schmitt 1991; Winter and Gademann 1991; Herppich et al. 1992).

8.4 Acceleration of Vegetative and Reproductive Growth Under Long Days

By growing *M. crystallinum* under long photoperiods under moderately high instantaneous PPFD (600 $\mu mol\ m^{-2}\ s^{-1}$) and high temperature (30 °C), and in a limited soil volume (60 × 60 mm peat pots), the life cycle is dramatically shortened

(Cheng and Edwards 1991). While daylength has little influence on the rate of formation of leaves from the main axis, the onset of secondary growth is earlier in plants growing in longer light periods. Axillary shoots begin to develop at ca. 3, 3.5 and 4 weeks, and reproductive growth begins at 5, 6 and 7 weeks after germination, when plants are grown under continuous light (24 h/0 h), 20 h/4 h and 16 h/8 h light/dark cycles, respectively. The total accumulative hours of illumination prior to flowering is similar in the three environments. Plants grown under a 12 h/12 h light/dark cycle begin flowering after 10 to 12 weeks. In a natural Mediterranean habitat, reproductive growth begins after about 5 months (Winter et al. 1978).

8.5 Effect of Growth Conditions on Induction of CAM by High Salinity

When *M. crystallinum* is grown under 12 h/12 h light/dark cycles, CAM can be induced by imposing water stress on plants which are 5 to 10 weeks old (see Winter 1985; Ostrem et al. 1987, and citations therein), and the nocturnal acidification can be accounted for by nocturnal synthesis of malate (Winter et al. 1981).

Ostrem et al. (1987) found PEPC can only be induced after 5 weeks of growth under a 12 h/12 h light/dark cycle (200 μmol photons m^{-2} s^{-1}), following the formation of five leaf pairs and the onset of secondary growth. This suggests the induction of CAM in *M. crystallinum* by water stress may be dependent on plant age or other environmental factors. Herppich et al. (1992) suggest induction of CAM in *M. crystallinum* is strictly under ontogenetic control and that the critical threshold is with the change from seedling to juvenile growth (rather than vegetative to reproductive growth at the onset of branching). Piepenbrock and Schmitt (1991) analyzed the effect of high salt application on gene transcripts and activity of PEPC with plants from 2 to 10 weeks of age (glasshouse-grown plants with supplemental lighting). Salt has a weak effect on 2-week-old plants, increasing thereafter with maximum induction at ca. 6 weeks. When plants are grown under longer light/dark cycles (16 h/8 h, 20 h/4 h, or 24 h/0 h at 600 μmol photons m^{-2} s^{-1} and 30 °C), large increases in activities of PEPC and NADP-ME are observed when treating plants at 2 to 3 weeks of age with 350 mM NaCl for 7 days (Cheng and Edwards 1991). With 6-week-old plants originally grown under 300 μmol photons m^{-2} s^{-1}, there is much greater induction of PEPC by salt under 850 than under 300 μmol photons m^{-2} s^{-1} (under a 16 h/8 h light/dark cycle), which suggests light intensity is an important factor in induction (McElwain et al. 1992). In general, these results indicate that induction of PEPC by salt is enhanced by growth at high PPFDs.

Conditions have been found which allow rapid induction of CAM in *M. crystallinum* when plants are only 2 weeks of age. This resulted partly from an interest in developing a screening system for isolating CAM mutants. When plants were grown initially under continuous light for 2 weeks, CAM was then induced in a few days by irrigating with 350 mM NaCl under a 16 h/8 h light/dark cycle. Induction occurred prior to secondary branching and resulted in large increases in PEPC activity, NADP-ME activity, and nocturnal acidification. Plants grown under continuous light have a high malate content and low tissue acidity, but after 7 days of induction under high salt, malate levels decrease in the light and there is nocturnal accumulation of malate (Fig. 8.1). Note that 1 mol of malate can generate from 0 to 2 mol of titratable acidity (H^+) depending on the counter ion in vivo. The degree of acidification during induction of CAM is greater than can be accounted for by nocturnal increases in malate. In fact, during the first 3 days of induction nocturnal acidification occurs without nocturnal accumulation of malate, which suggests some additional acid is formed in the dark (this is partly, but not totally, accounted for by nocturnal increase in citrate: J. Krall, S. Cheng, G. Edwards, unpubl.). In terms of developing conditions for rapid induction of CAM, experiments have not been performed which would distinguish between the relative importance of daylength, instantaneous PPFD, total daily PPFD, or diurnal changes in water status.

8.6 O_2 Evolution from Photosystem II and Net Rates of CO_2 Uptake Before and After Induction of CAM

In CAM plants, the reductant derived from photosynthetic O_2 evolution will be partitioned primarily between assimilation of atmospheric CO_2, photorespiration, and assimilation of CO_2 from decarboxylation of malate. The relative contribution of each at a given time during the day will be affected by the rate of carbon donation from malate and stomatal conductance.

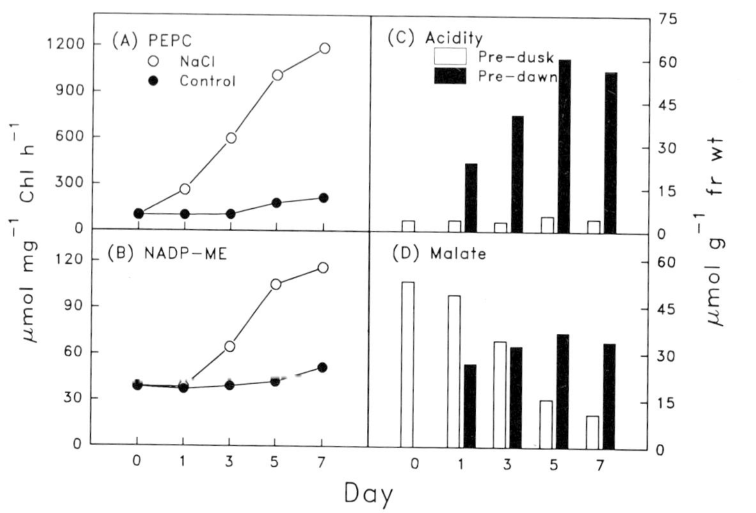

Fig. 8.1A–D. Induction of CAM in soil-grown *M. crystallinum* following watering with 350 mM NaCl under a 16 h/8 h light/dark cycle after growth of plants under continuous light for 2 weeks. Throughout growth temperature was 30 °C and PPFD was 600 μmol m^{-2} s^{-1}

During induction of CAM in *M. crystallinum* by high salinity there is a large decrease in stomatal conductance, resulting in a midday depression of the intercellular partial pressure of CO_2. This causes a midday decline in CO_2 uptake and the quantum yield of PSII, despite internal donation of carbon from malate (Winter and Gademann 1991). During photosynthesis in the C_3 mode under nonphotorespiring (NPR) conditions [800 μbar CO_2 (i.e. 80 Pa CO_2) and 2% (v/v) O_2], most of the PSII activity (Jo_2 estimated from fluorescence analysis) is linked to carbon assimilation in 5-week-old plants of *M. crystallinum*, since under this condition there is minimal utilization of reductant in photorespiration and the high rate of CO_2 assimilation constitutes the major sink for reductant (Fig. 8.2B). In the C_3 mode under photorespiratory (PR) conditions [330 μbar CO_2 (i.e. 33Pa CO_2), 21% (v/v) O_2], rates of photosynthesis are only about half that under NPR conditions due to the limited supply of CO_2 and higher levels of photorespiration (Fig. 8.2A). In this case, the rate of O_2 evolution (Jo_2) calculated from fluorescence analysis is about three times higher than the rate of net CO_2 uptake, which is presumably due to photorespiration becoming a major sink for electrons. Measurements of malate content showed there is no donation of carbon from malate in the C_3 mode (data not shown). High rates of O_2 evolution are maintained throughout the day during photosynthesis in the C_3 mode under both PR and NPR conditions: switching between PR and NPR conditions affects the partitioning of the utilization of reductant between photorespiration and CO_2 assimilation without affecting the rate of electron flow.

Following induction of CAM after salt stress for 7 days, there is a marked decline of CO_2 uptake during the day, but rates of O_2 evolution remain high (Fig. 8.2C). Part of the PSII activity is associated with assimilation of CO_2 donated from malate. However, in general the amount of malate accumulated in the dark (ca. 50 to 100 mmol m^{-2}) only provides enough inorganic carbon during malate decarboxylation to account for 5 to 10% of the O_2 evolved. In the above case, the malate content of the leaves decreases by ca. 40 mmol m^{-2} over the first 5 h of the day; averaged over that period the assimilation of CO_2 from malate would result in rates of Jo_2 of 2.3 μmol m^{-2} s^{-1}. This makes a significant contribution to carbon assimilation considering the low uptake of atmospheric CO_2, but it accounts for only ca. 10% of the PSII activity (assuming all the CO_2 generated is donated to the C_3 pathway). Thus, despite donation of carbon from malate, the intercellular partial pressure of CO_2 may be limiting (also see Winter and Gademann 1991), and much of the PSII activity may be linked to consumption of reductive power in photorespiration. This O_2-dependent PSII activity may have a role in dissipating excess energy and preventing photoinhibition when the supply of CO_2 is limited (Walker 1992). Under high CO_2 and low O_2 there is large increase in the net uptake of CO_2, particularly during phases II (early morning) and IV (late afternoon) of CAM (Fig. 8.2D). However, PSII activity is more limited under low O_2, which suggests uptake of atmospheric CO_2 and donation of carbon from malate are limiting PSII activity.

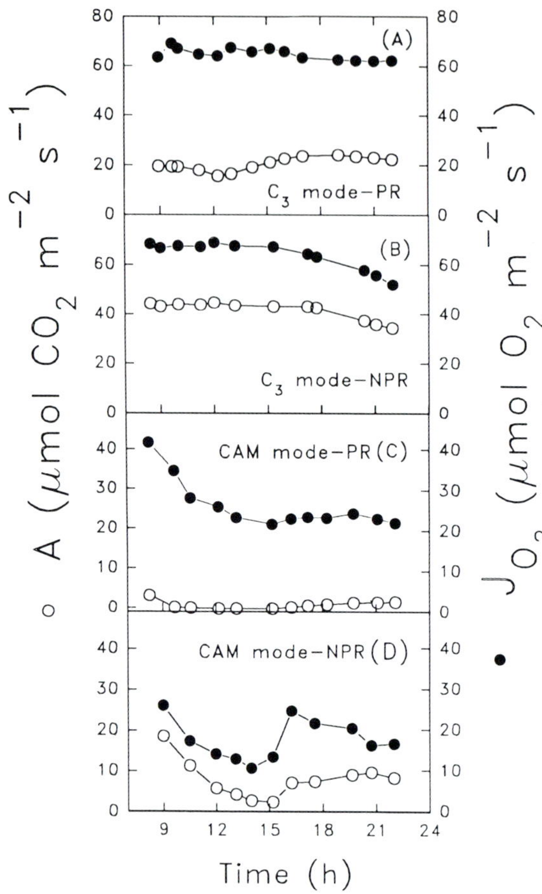

Fig. 8.2A–D. Rates of O_2 evolution from photosystem II (Jo_2) and net rates of CO_2 uptake (**A**) during measurements of photosynthesis on an individual leaf of *M. crystallinum* before and after induction of CAM by high salt (400 mM NaCl for 7 days). Plants were grown under 28 °C, 400 μmol photons m^{-2} s^{-1}, and 14h/10h light/dark cycle for 5 weeks. For each treatment the leaf was placed in the cuvette and illumination initiated at 08.00 h, which coincides with the beginning of the day. *PR* Photorespiratory conditions [21%(v/v) O_2, 340 μbar (34 Pa) CO_2]; *NPR* nonphotorespiratory conditions [2%(v/v) O_2, 800 μbar (80 Pa) CO_2]. The light absorbed by the leaf during the experiments was 1100 to 1200 μmol m^{-2} s^{-1}. Jo_2, the rate of O_2 evolution from photosystem II, was calculated from $Jo_2 = (\Phi_{PSII})$ (I_L) $(f)/4$, where Φ_{PSII} is the quantum yield of PSII, I_L is the light absorbed by the leaf, f is the fraction of I_L absorbed by PSII (taking $f = 0.5$ assuming equal absorption by PSI and PSII), divided by 4 (4 e^- transported/O_2 evolved). (See Krall and Edwards 1992)

8.7 Eventual Induction of CAM Under Well-Watered Conditions

Mature leaves of 10-week-old plants of *M. crystallinum* grown under a 12 h/12 h light/dark cycle exhibited a weak expression of CAM when grown under nonsaline conditions, but expression is enhanced under high light and low humidity (Winter 1973b). In well-watered plants grown in a glasshouse, the level of gene transcripts and activity of PEPC are low up to ca. 6 weeks of growth but then increase markedly from 7 to 10 weeks of growth, which coincides with a decreased growth rate (Piepenbrock and Schmitt 1991). Also, there is a progressive increase in PEPC and NADP-ME activities from 2 weeks up to 7 weeks in well-watered plants of *M. crystallinum* when grown under long days and under ca. 600 μmol photons m^{-2} s^{-1} (Fig. 8.3; Chu et al. 1990; Cheng and Edwards 1991). With growth under these conditions there is also induction of CAM, as shown by increased diurnal changes in malate and leaf acidity from 5 weeks of age, with a high level of CAM after 7 weeks (Fig. 8.3).

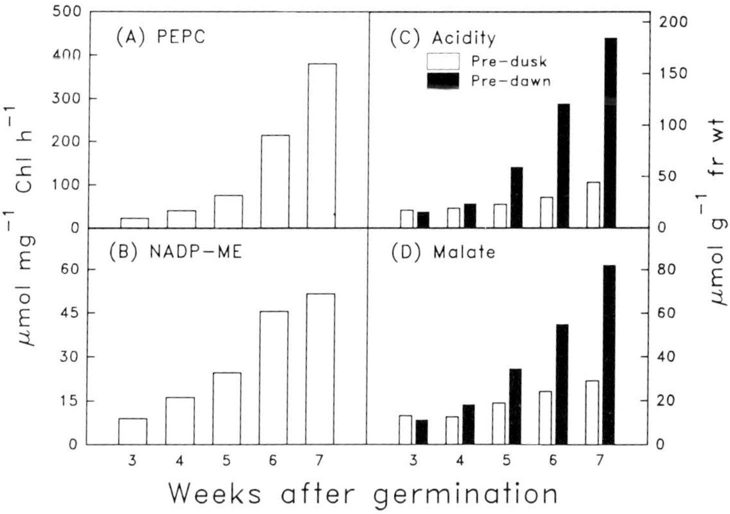

Fig. 8.3A–D. Eventual induction of CAM in *M. crystallinum* under well-watered conditions under 16 h/8 h light/dark cycles (28 °C/18 °C, 600 μmol photons m^{-2} s^{-1}). (Adapted with permission from Chu et al. 1990)

These results suggest that development of CAM in *M. crystallinum* under well-watered conditions may be enhanced by growth under long days with higher light intensities. Herppich et al. (1992) also report that after a threshold stage of development CAM is initiated in well-watered plants. However, it may be difficult to determine whether this is strictly dependent on development or on daily radiation, or whether it is linked to a degree of water stress. Leaves of well-watered *M. crystallinum* showed transient drops in photosynthesis, turgor pressure and water content during the day. Therefore, transient changes in water relations may be causally related to induction of CAM (Winter and Gademann 1991), particularly under long days and high PPFDs.

8.8 Conditions Resulting in Induction of Phospho*enol*pyruvate Carboxylase in the Absence of CAM

It is well documented that the induction of CAM in *M. crystallinum* by water stress in plants grown under a 12 h/12 h light/dark cycle is associated with an enormous rise in PEPC activity, and increases in a number of other enzymes of CAM (see Holtum and Winter 1982). However, there are situations where the levels of PEPC increase without performance of CAM. When plants are grown under a 20 h/4 h light/dark cycle under well-watered conditions there is ca. fivefold increase in the activities of PEPC between week 1 and week 4 of growth (up to ca. 300 μmol mg^{-1} Chl h^{-1}) without the occurrence of CAM (based on measurements of day/night changes in malate and titratable acidity). Also, under continuous light the activities of PEPC and NADP-ME have been shown to rise during growth, particularly under salt stress, but without performance of CAM (Cheng and Edwards 1991). Thus, the induction of PEPC is not obligately linked to growth under environmental conditions in which CAM will be functional. The mechanism for the uncoupling of CAM biochemistry and physiology in these cases remains unclear.

8.9 Conditions Resulting in Malate Synthesis in the Light in the Absence of CAM

In *M. crystallinum* growing in the field, prior to nocturnal accumulation of malate, a low level of malate accumulates in leaves in February and March (Winter et al. 1978). Thus, some synthesis of malate may be initiated prior to the full performance of CAM. In well-watered growth-chamber-grown plants under long photoperiods, high levels of malate accumulate in the leaves early in development (up to 3 to 4 weeks of growth) with little or no diurnal fluctuation in malate levels (Fig. 8.4). The levels of malate are similar to those which accumulate in the dark in plants performing CAM (e.g. 25 to 75 μmol g^{-1} fr. wt.). Apparently, the malate accumulated under long photoperiods is largely balanced by cations (e.g. Na_2-malate), since the titratable acidity is low (see Fig. 8.1C,D at day zero, and Cheng and Edwards 1991).

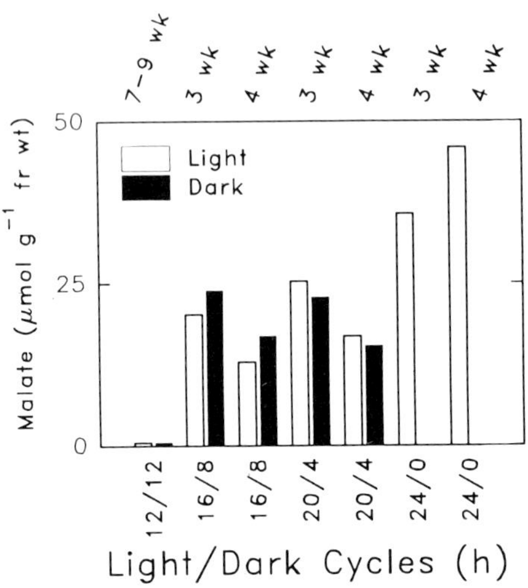

Fig. 8.4. Malate levels in well-watered *M. crystallinum* plants which were not performing CAM. [Adapted with permission from Winter 1973b (for 12 h/12 h) and from Cheng and Edwards 1991]. The weeks of growth are indicated at the *top* of the figure. Except for plants grown under continuous light, samples were taken at the end of the light and end of the dark period. There was little or no CAM under these conditions based on measurements of nocturnal acidification. (See Fig. 8.1 and Cheng and Edwards 1991)

Pulse-chase experiments ($^{14}CO_2/^{12}CO_2$) with well-watered plants of *M. crystallinum* grown under long photoperiods show a labelling pattern indicative of malate synthesis by a secondary carboxylation. This is observed with plants grown under 16 h/8 h, 20 h/4 h and 24 h/0 h light/dark cycles after 3 weeks (Cheng and Edwards 1991) and 4 weeks of growth (Fig. 8.5 and unpubl. results). A very low percentage of label appears in malate during the initial pulse, but following a 20-min chase ca. 20 to 30% of the label in soluble products appears in malate (Fig. 8.5 and results not shown). The label appearing in insolubles/starch after a 20-min chase is only 4 to 7% of the label in the soluble fraction. Although these plants were not performing CAM, the labelling pattern is like that observed during Phase IV (late-afternoon fixation of CO_2) of CAM in *Kalanchoë daigremontiana* (Osmond and Allaway 1974; Osmond 1978). During phase IV metabolism, there is evidence from mass spectrometer analysis of $^{13}CO_2$ fixation for synthesis of malate by a double carboxylation, in which case PEPC activity is dependent on RUBISCO to provide the precursor PEP from synthesis of PGA (Ritz et al. 1986; Osmond et al. 1988).

As noted above, very little malate is initially labelled in a pulse with $^{14}CO_2$, but malate constitutes about 25% of the soluble end-products of photosynthesis in the light in *M. crystallinum* grown under long photoperiods. If this occurs by utilization of PGA as a product of RUBISCO, the stoichiometry can be illustrated as follows. If 48 CO_2 react with RUBISCO, the net product would be 16 PGA, 4 of which could be utilized for synthesis of 4 malate, and 12 of which could be utilized for synthesis of 3 sucrose. This would result in 25% of the PGA being

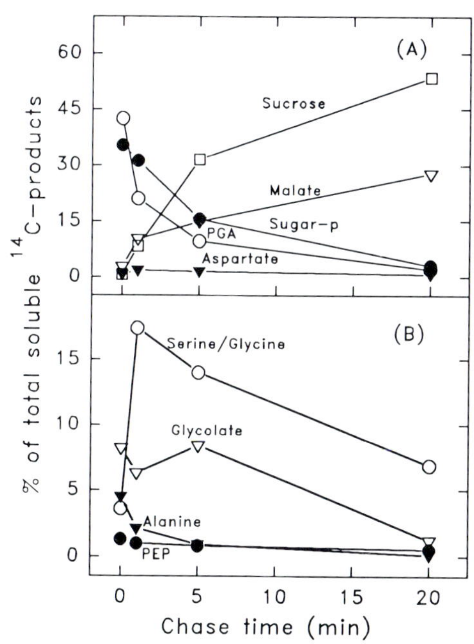

Fig. 8.5A,B. Appearance of ^{14}C in soluble organic products following a pulse-chase experiment with 4-week-old, well-watered plants of *M. crystallinum* grown under a 16 h/8 h light/dark cycle (30 °C, 600 µmol photons m^{-2} s^{-1}). Pulse with $^{14}CO_2$ for 12 s and chase with $^{12}CO_2$ up to 20 min (2 h prior to the end of the light period). The PPFD during the pulse chase was 1000 to 1200 µmol m^{-2} s^{-1}. *PGA* 3-Phosphoglycerate; *PEP* phospho*enol*pyruvate

utilized for synthesis of malate and 75% for synthesis of sucrose. For each 48 CO_2 reacting with RUBISCO, 4 CO_2 would react with PEP via PEPC, resulting in the synthesis of 4 malate. In this example, the PEPC activity is only 8% of the RUBISCO activity. The kinetics of malate labelling is consistent with the double carboxylation hypothesis (Fig. 8.5 and Cheng and Edwards 1991).

The results suggest that significant levels of malate are synthesized in the light in *M. crystallinum* under well-watered conditions with long photoperiods. There are several possible fates of the malate. It may be exported as an alternative product of photosynthesis to carbohydrates, accumulated as a means of balancing cations in the leaf (as occurs in *K. daigremontiana* under long days, resulting in a form of malate not available for utilization in CAM) (see Phillips 1980), accumulated as an osmoticum to protect against water stress (e.g., 25 μmol malate g^{-1} fr. wt. in the form of Na_2-malate has an osmotic pressure of ca. 0.2 MPa), or metabolized through a futile cycle via NADP-ME.

8.10 Induction of CAM by Growth Regulators

Since the importance of abscisic acid (ABA) in plants under water stress is well known, an obvious consideration is whether it has a role in induction of CAM in species like *M. crystallinum*, by its effect on either stomatal conductance or on synthesis of proteins required for CAM (Winter 1975; Ting 1981; Chu et al. 1990; McElwain et al. 1992). In earlier studies little or no induction of CAM was observed after spraying leaves of *M. crystallinum* with 100 μM ABA, but some side effects were noticed (ageing symptoms or wilting, see Winter 1975, 1985). Ting (1981) showed that application of high levels of ABA (1 mM) to leaves of the inducible CAM plant *Portulacaria afra* causes stomatal closure during the day, resulting in CAM-like activity. There was little effect of the ABA treatment on PEPC activity; it was suggested that the shift in carbon metabolism in inducible CAM plants may be associated with daytime closure of stomata. We find that application of low levels of ABA (1 to 10 μM) to 4- to 5-week-old plants of *M. crystallinum* (grown under a 16 h/8 h light/dark cycle, 600 μmol photons $m^{-2} s^{-1}$) can substitute for salinity or water stress in inducing CAM (Chu et al. 1990; Edwards et al. 1990; Fig. 8.6). This is observed when ABA is applied to leaves, to the roots in hydroponic culture, or by watering plants growing in soil with ABA solutions. Induction by low concentrations of ABA occurs without affecting the relative water content of tissue measured at the end of the day (Chu et al. 1990), but more detailed analyses of the daily change in water relations during induction by ABA are needed [e.g. according to Winter and Gademann (1991) during induction by high salt]. Application of ABA to the roots gives the most reproducible results with respect to magnitude of induction of CAM, probably because of better control of uptake into leaves. We have consistently observed induction by ABA in plants of *M. crystallinum* grown under these conditions; reports to the contrary are likely to be related to either concentrations of ABA used or growth of plants under insufficient light. Application of high concentrations of ABA should be avoided, as this can cause wilting and obvious damage to roots (e.g., this is observed in hydroponic cultures with concentrations as low as 10 μM ABA). It was recently shown that induction by ABA is dependent on light intensity, as high induction occurs under 850 μmol photons m^{-2} s^{-1}, while little induction occurs under 300 μmol photons m^{-2} s^{-1} with 16-h photoperiods (McElwain et al. 1992) or 12-h photoperiods (Thomas et al. 1992).

We have also found that farnesol can induce CAM in *M. crystallinum* when applied in micromolar concentrations (maximum induction with 10 μM solutions), equivalent to the induction by 350 mM NaCl (Dai et al. 1990). Farnesol is a naturally occurring isoprenoid alcohol (3, 7, 11-trimethyl-2, 6, 10-dodecatrien-1-ol), which like ABA is an antitranspirant. Farnesol causes an increase in ABA when applied to spinach leaves and it is a derivative of farnesyl pyrophosphate, which is a possible precursor to ABA synthesis (Ogunkanmi et al. 1974; Zeevaart and Creelman 1988). The levels of CAM induction by application of farnesol or ABA is influenced by growth conditions. When plants are grown under a shorter daylength (14 h) and a lower temperature regime (25 °C/18 °C), the magnitude of CAM induction is generally 30–40% lower than that under 16 h/8 h light/dark cycles and a temperature of 30 °C/18 °C (results not shown).

In testing other growth regulators, no induction of CAM was found in *M. crystallinum* by feeding micromolar concentrations of the synthetic auxin 2,4-dichlorophenoxyacetic acid (2,4-D), gibberellin, or methyl jasmonate in nutrient media. Surprisingly, however, application of cytokinins via the roots also induces CAM in *M. crystallinum* (Dai et al. 1990, Fig. 8.6), which is supported by evidence for cytokinin-induced expression of CAM-specific PEPC (Thomas et al. 1992).

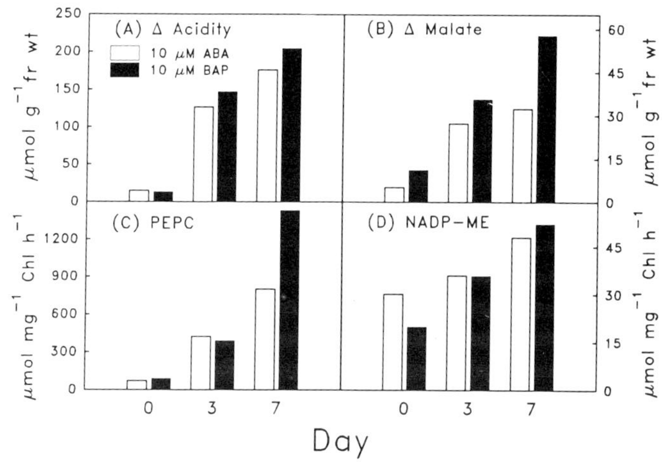

Fig. 8.6A–D. Induction of CAM in 4-week-old soil-grown plants of *M. crystallinum* by watering with 10 μM ABA or 10 μM BAP twice per day for 7 days. Plants were grown under a 16 h/8 h light/dark cycle, 28 °C/18 °C, ca. 600 μmol photons m^{-2} s^{-1}. Data on acidity and malate represent nocturnal accumulation. Day *zero* represents measurements prior to addition of hormones. Similar induction is obtained by watering with 350 mM NaCl (results not shown)

The time course of induction following applications of 10 μM solutions of the cytokinin benzylaminopurine (BAP) is similar in magnitude to that of ABA. Also, induction occurs when feeding either BAP or ABA in combination with NaCl to the roots (results not shown).

Besides measurements of induction based on activities of PEPC and NADP-ME, and nocturnal increases in titratable acidity and malate levels (Fig. 8.6), a further definitive test of induction can be made from measuring gas exchange patterns. Application of salt, ABAor BAP gave very similar patterns of induction of CAM in *M. crystallinum* based on diurnal changes in net rates of CO_2 uptake and stomatal conductance (Fig. 8.7). The patterns of gas exchange are also similar to those previously observed following induction by high salt (Winter and Gademann 1991). By day 7 in all treatments there is substantial nocturnal uptake of CO_2, and marked changes in the daily pattern of CO_2 uptake and in stomatal conductance. Salt-, ABA-, or BAP-treated plants showed the four phases typical of CAM in many species (nighttime CO_2 uptake in Phase I, transient daytime uptake in Phase II, midday depression of CO_2 uptake in Phase III, and

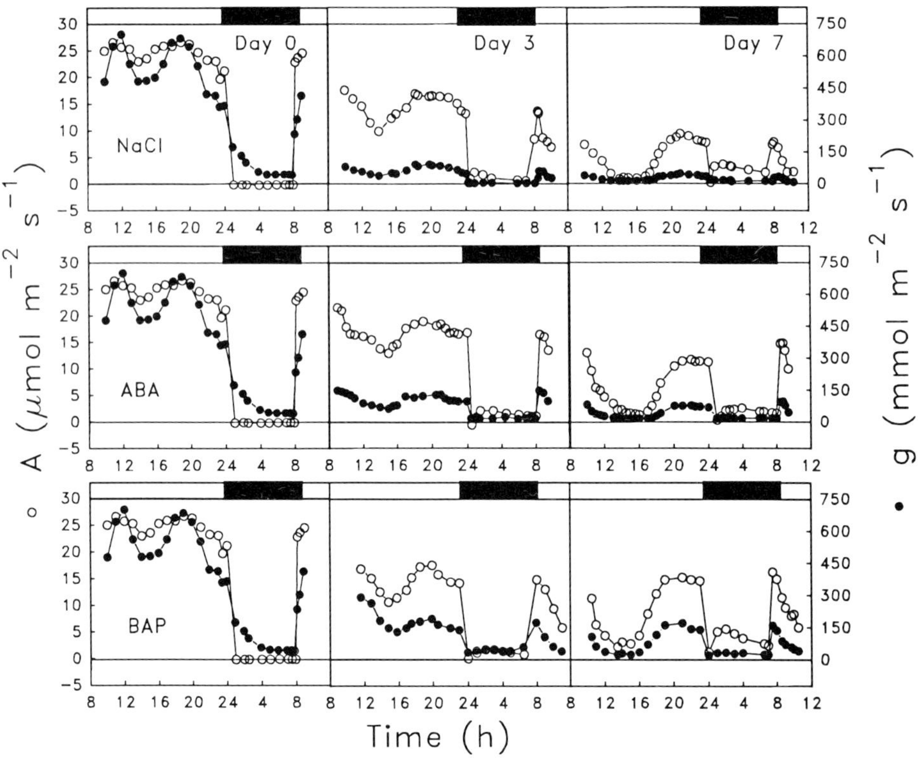

Fig. 8.7. Changes in net rates of CO_2 uptake (*A*) and stomatal conductance (*g*) measured on an individual leaf following induction by watering 5-week-old soil-grown plants of *M. crystallinum* with solutions of 350 mM NaCl, 1 μM ABA or 10 μM BAP. At *top* of figure the *open bar* represents the light period (1000 μmol photons m^{-2} s^{-1}) and the *closed bar* the dark period

late-afternoon uptake in phase IV).

Induction of PEPC transcripts has been observed during dehydration of excised leaves of *M. crystallinum* (Schmitt and Piepenbrock 1992). Feeding micromolar levels of BAP or ABA via the petiole to excised leaves also induces a large increase in PEPC activity equivalent to that induced by allowing excised leaves to dehydrate (Fig. 8.8), which supports our earlier observations on

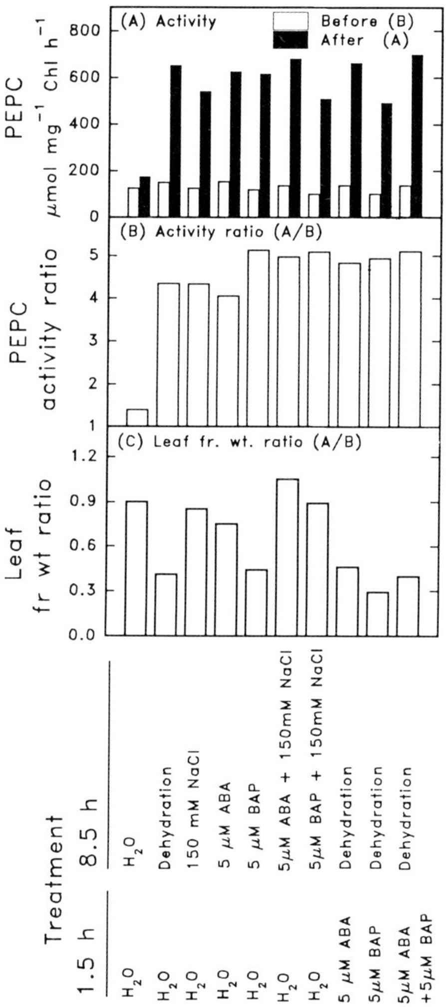

Fig. 8.8. Induction of PEPC activity in excised leaves of *M. crystallinum* by dehydration, or by feeding NaCl, ABA or BAP via the petiole. Mature leaves were excised from plants 5 to 6 weeks of age (16 h/8 h light/dark cycle, 28/18 °C, 400 μmol photons m^{-2} s^{-1}) and incubated in the growth room during the day. During the first 1.5 h, petioles were immersed in water or solutions of ABA and BAP as indicated; in the subsequent 8.5 h, leaves were dehydrated or fed solutions via the petiole as indicated

induction by feeding these hormones to roots. However, BAP when fed to cut leaves in higher concentrations inhibits the subsequent increase in PEPC activity by dehydration (not shown). This supports the observations of Schmitt and Piepenbrock (1992), who found that micromolar levels of cytokinins, when applied to leaves, inhibited the induction by salinity in intact plants, and inhibited the induction of PEPC transcripts by dehydration in excised leaves. Further work is required to elucidate the mechanism(s) of action of cytokinin under these conditions (see also Chap. 10).

As discussed above, ABA can induce CAM without addition of salt. Induction by salt may also be possible without a dependence on ABA, as suggested from results with Lovastatin, an inhibitor of isoprenoid synthesis (Thomas et al. 1992). However, ABA, unlike cytokinins, increases in leaves of *M. crystallinum* under salt stress in parallel with increases in PEPC mRNA (Thomas et al. 1992). In considering the actual stimuli involved in vivo, it would be of interest to follow the water status and hormone content of leaves during induction under natural conditions in the field.

In summary, there are still unresolved questions about the requirements for CAM induction in *M. crystallinum*. Is induction by water stress dependent on plants reaching a certain stage of development, or is there a growth condition under which water stress will induce CAM in very young plants? Is there an absolute requirement for water stress for induction of CAM in *M. crystallinum*, or can it be induced eventually during development under certain environments without needing to be triggered by water stress? Current evidence suggests a developmental requirement prior to induction which can be accelerated under certain growth conditions, a role for ABA in induction by water stress, and an effect of PPFD on the degree of induction by water stress. Development, water stress, and PPFD are all likely to contribute to the signalling mechanism(s) for induction in this species. In the future, with respect to dependence of induction on development, measurements should be made of the total carbon gain under different growth conditions versus the degree of inducibility of CAM. In evaluating the mechanism of induction and whether there is an absolute requirement for water stress during induction under various growth conditions, or by application of hormones, careful analyses need to be made of the water status of the plant.

Acknowledgements. Our own research on this subject has been supported by the USDA Competitive Grants Program including Grant 90-37280-5706. The authors appreciate all other support which has been provided for presentation and publication of this paper.

References

Cheng S-H, Edwards GE (1991) Influence of long photoperiods on plant development and expression of crassulacean acid metabolism in *Mesembryanthemum crystallinum*. Plant Cell Environ 14: 271–278

Chu C, Dai Z, Ku MSB, Edwards GE (1990) Induction of crassulacean acid metabolism in the facultative halophyte *Mesembryanthemum crystallinum* by abscisic acid. Plant Physiol 93: 1253–1260

Cockburn W (1985) Variations in photosynthetic acid metabolism in vascular plants: CAM and related phenomenon. New Phytol 101: 3–24

Dai Z, Ku MSB, Edwards GE (1990) Induction of crassulacean acid metabolism in *Mesembryanthemum crystallinum* by growth regulators. Plant Physiol Suppl 93: 124

Edwards GE, Cheng S-H, Chu C, Ku MSB (1990) Environmental and hormonal dependence of induction of crassulacean acid metabolism in *Mesembryanthemum crystallinum*. In: Baltscheffsky M (ed) Current research in photosynthesis, vol IV. Kluwer, Dordrecht, pp 393–396

Herppich W, Herppich M, von Willert DJ (1992) The irreversible C_3 to CAM shift in well-watered and salt-stressed plants of *Mesembryanthemum crystallinum* is under strict ontogenetic control. Bot Acta 105: 34–40

Holtum JAM, Winter K (1982) Activities of enzymes during induction of crassulacean acid metabolism in *Mesembryanthemum crystallinum* L. Planta 155: 8–16

Krall JP, Edwards GE (1992) Relationship between photosystem II activity and CO_2 fixation in leaves. Physiol Plant 86: 180–187

McElwain EF, Bohnert HJ, Thomas JC (1992) Light mediates the induction of phospho*enol*-pyruvate carboxylase by NaCl and abscisic acid in *Mesembryanthemum crystallinum*. Plant Physiol 99: 1261–1264

Ogunkanmi AB, Wellburn AR, Mansfield TA (1974) Detection and preliminary identification of endogenous antitranspirants in water stressed sorghum plants. Planta 117: 293–302

Osmond CB (1978) Crassulacean acid metabolism: a curiosity in context. Annu Rev Plant Physiol 29: 379–414

Osmond CB, Allaway WG (1974) Pathways of CO_2 fixation in the CAM plant *Kalanchoë daigremontiana*. I. Patterns of $^{14}CO_2$ fixation in the light. Aust J Plant Physiol 1: 503–511

Osmond CB, Holtum JAM, O'Leary MH, Roeske C, Wong OC, Summons RE, Avadhani PN (1988) Regulation of malic acid metabolism in crassulacean acid metabolism plants in the dark and light: in vivo evidence for ^{14}C-labelling patterns after $^{14}CO_2$ fixation. Planta 175: 184–192

Ostrem JA, Olsen SW, Schmitt JM, Bohnert HJ (1987) Salt stress increases the level of translatable mRNA for PEPC in *Mesembryanthemum crystallinum*. Plant Physiol 84: 1270–1275

Phillips RD (1980) Deacidification in a plant with crassulacean acid metabolism associated with anion-cation balance. Nature 287: 727–728

Piepenbrock M, Schmitt JM (1991) Environmental control of phospho*enol*pyruvate carboxylase induction in mature *Mesembryanthemum crystallinum* L. Plant Physiol 97: 998–1003

Ritz D, Kluge M, Veith HJ (1986) Mass-spectrometric evidence for the double-carboxylation pathway of malate synthesis by crassulacean acid metabolism plants in light. Planta 167: 284–291

Schmitt JM, Piepenbrock M (1992) Regulation of phospho*enol*pyruvate carboxylase and crassulacean acid metabolism induction in *Mesembryanthemum crystallinum* L. by cytokinin. Plant Physiol 99: 1664–1669

Thomas JC, McElwain EF, Bohnert HJ (1992) Convergent induction of osmotic stress responses. Abscisic acid, cytokinin and the effects of NaCl. Plant Physiol 100: 416–423

Ting IP (1981) Effects of abscisic acid on CAM in *Portulacaria afra*. Photosynth Res 2: 39–48

Walker D (1992) Excited leaves. New Phytol 121: 325–345

Winter K (1973a) Zum Problem der Ausbildung des Crassulaceensäurestoffwechsels bei *Mesembryanthemum crystallinum* unter NaCl-Einfluß. Planta 109: 135–145

Winter K (1973b) CO_2-Fixierungsreaktionen bei der Salzpflanze *Mesembryanthemum crystallinum* unter variierten Außenbedingungen. Planta 114: 75–85

Winter K (1975) Die Rolle des Crassulaceen-Säurestoffwechsels als biochemische Grundlage zur Anpassung von Halophyten an Standorte hoher Salinität. Doctoral Thesis, Technische Hochschule Darmstadt

Winter K (1985) Crassulacean acid metabolism. In: Barber J, NR Baker, (eds), Photosynthetic mechanisms and the environment. Elsevier, Amsterdam, pp 329–387

Winter K, Gademann R (1991) Daily changes in CO_2 and water vapor exchange, chlorophyll fluorescence, and leaf water relations in the halophyte *Mesembryanthemum crystallinum* during the induction of crassulacean acid metabolism in response to high NaCl salinity. Plant Physiol 95: 768–776

Winter K, von Willert DJ (1972) NaCl-induzierter Crassulaceensäurestoffwechsel bei *Mesembryanthemum crystallinum.* Z Pflanzenphysiol 67: 166–170

Winter K, Lüttge U, Winter E, Troughton JM (1978) Seasonal shift from C_3 photosynthesis to crassulacean acid metabolism in *Mesembryanthemum crystallinum* growing in its natural environment. Oecologia 34: 225–237

Winter K, Edwards GE, Holtum JM (1981) Nocturnal accumulation of malic acid occurs in mesophyll tissue without proton transport to epidermal tissue in the inducible crassulacean acid metabolism plant *Mesembryanthemum crystallinum.* Plant Physiol 68: 355–357

Zeevaart JAD, Creelman RA (1988) Metabolism and physiology of abscisic acid. Annu Rev Plant Physiol Plant Mol Biol 39: 439–473

9 Transcriptional Activation of CAM Genes During Development and Environmental Stress

J.C. Cushman[1] and H.J. Bohnert[2]

9.1 Introduction

Crassulacean acid metabolism (CAM) can be defined as the ability to perform significant CO_2 assimilation into C_4 acids (mainly malic acid) in the dark. Reutilization of the stored C_4 acids results in a characteristic diel (24 h) pattern of organic-acid fluctuation along with a reciprocal pattern of fluctuating levels of storage glucan. CAM is found primarily, although not exclusively, in succulent plants, particularly those having large chloroplast-containing parenchyma cells with the ability to store large amounts of malic acid and water. CAM plants are mainly tropical or subtropical in origin and typically inhabit arid environments with periodic water deficits such as semi-deserts, or regions with Mediterranean climates, or epiphytic habitats in tropical forests (Kluge and Ting 1978). Their ecological distribution and the observation that CAM plants open their stomata at night and close them during the day to avoid excessive evaporative water loss have led to the general consensus that CAM is a functional adaptation to dry environments and a means of water conservation. However, not all CAM plants occur in arid environments. *Isoetes* and related species which grow in aquatic environments display CAM (Keeley and Busch 1984). In these instances, CAM provides the plants with a source of carbon when levels of CO_2 in the aquatic environment become limiting during the day.

Crassulacean acid metabolism exhibits great plasticity in its manifestations and variations on the central theme of CAM, such as CAM-idling (Ting 1985), a response to severe drought conditions. Many plants display CAM on a strictly constitutive basis, whereas others, such as *Peperomia,* clearly exhibit CAM on a developmental basis (Sipes and Ting 1985; Holthe et al. 1987; see also Chap. 13). Various other environmental stimuli can cause CAM to become elicited in a facultative manner. In certain plants, CAM is regulated by daylength (Brulfert et al. 1982) or season (Guralnick et al. 1984). CAM can also be brought about by exposure to water stress in various species such as *Portulacaria afra* (Ting and

[1]Department of Biochemistry and Molecular Biology, Oklahoma State University, Stillwater, OK 74078-0454, USA
[2]Departments of Biochemistry, of Molecular and Cellular Biology, and of Plant Sciences, The University of Arizona, Tucson, AZ 85721, USA

Ecological Studies, Vol. 114
Winter/Smith (eds) Crassulacean Acid Metabolism

Hanscom 1977), *Sedum telephium* (Groenhof et al. 1986; see also Chap. 11), and *Mesembryanthemum crystallinum* (Winter and von Willert 1972).

Mesembryanthemum crystallinum is the best-studied example of CAM inducibility at both the biochemical and physiological levels (for a review, see Bohnert et al. 1992). A number of other attributes, such as an interesting developmental profile, amenability to tissue culture regeneration (Meiners et al. 1991), a short life-span, and small genome size (Meyer et al. 1990; DeRocher et al. 1990) make this model even more attractive and should aid future research. Additional avenues of investigation have been initiated by the characterization of other adaptive mechanisms to salinity and drought stress, such as osmoprotective solute biosynthesis (Vernon and Bohnert 1992a) in transgenic plants (Tarczynski et al. 1993; Vernon et al. 1993b).

9.2 CAM Evolution

CAM is currently recognized in 33 plant families (Chap. 26), its widespread taxonomic distribution arguing strongly for multiple, independent origins (Moore 1982). One hypothesis suggests that CAM biochemistry arose from the multiplication and alteration of genes encoding enzymes for malate formation and decarboxylation in guard cells (Teeri 1982; Cockburn 1983). However, many other cell types exhibit the ability to produce malate. All enzymes required for a functional CAM pathway are derived from existing genes essential for photosynthetic carbon metabolism (Cockburn 1983). These enzymes perform indispensible reactions in glycolysis, gluconeogenesis, mobilization of carbon compounds into the TCA cycle and photosynthetic carbon flow (Fig. 9.1). The genetic alterations needed to transform guard-cell-specific gene expression to other cell types and bring about functional CAM are likely to be few. The necessary changes were to include adjustments in the spatial and/or temporal patterns of gene expression and evolutionarily selected alterations in regulatory properties. A discussion of the origins of CAM has been presented (Monson 1989), and is discussed further in Chapters 24 and 25.

Aside from alterations in expression patterns, a second important adaptation might have involved changes in the enzymes themselves. Very little work has been conducted to address the question of enzyme adaptation to a specific mode of photosynthetic carbon fixation. One of the best-studied enzymes in CAM is phospho*enol*pyruvate carboxylase (PEPC), the enzyme responsible for the primary fixation of CO_2 (as HCO_3^-) into oxaloacetate (Fig. 9.1). Tracing the evolution of this enzyme and the evolution of the genes that encode PEPC might give us clues about the rate of CAM evolution and potential molecular changes which the enzyme underwent in its different forms in CAM, as well as in C_3 and C_4 metabolism. Analyses of this type have been conducted for C_3 and C_4 species of *Flaveria* (Hermans and Westhoff 1992) and for the C_4 species *Sorghum* (Lepiniec et al. 1993). As more genes for this enzyme have been characterized, a larger number of sequences, including the CAM-specific form of PEPC from

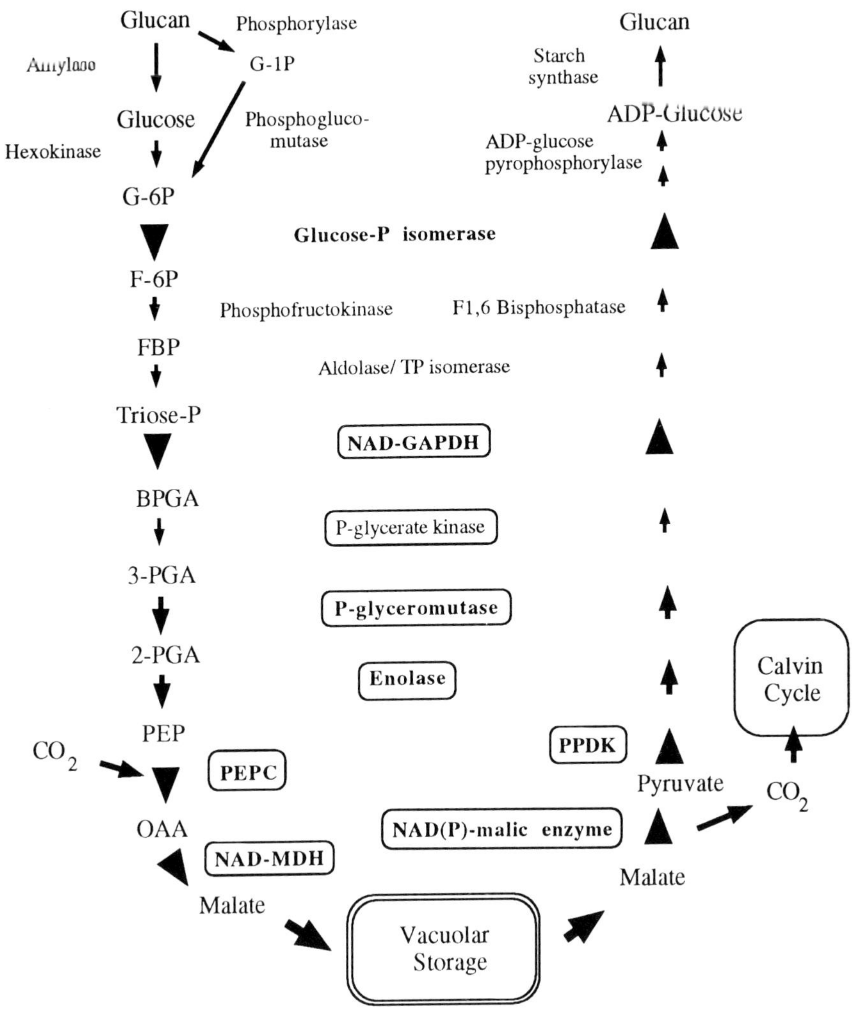

Fig. 9.1. Carbon flow, enzymes of CAM and their genes in *Mesembryanthemum crystallinum*. The schematic depiction of carbon flow in CAM omits compartmentation (only the vacuole is included) and instead focuses on the characteristics of induced enzymes and their genes as *Mesembryanthemum crystallinum* switches from C_3 to CAM during development and under stress. The large arrows, and enzyme names in boldface, indicate that the corresponding enzyme activities increase during CAM induction (Edwards et al. 1982; Foster et al. 1982; Holtum and Winter 1982). Enzyme names that are *framed* indicate that the genes or transcripts have been cloned, although not all genes have been characterized in detail. All the genes for enzymes that are highlighted by *frames* are transcriptionally induced

Mesembryanthemum crystallinum, can be analyzed in the context of other classes of plants. Based on the phylogenetic analysis of PEPC (Fig. 9.2), this CAM enzyme does not fall into a class of its own, but appears more closely related to the C_3 enzymes from sorghum and corn, as well as to the C_4 enzyme from sugarcane.

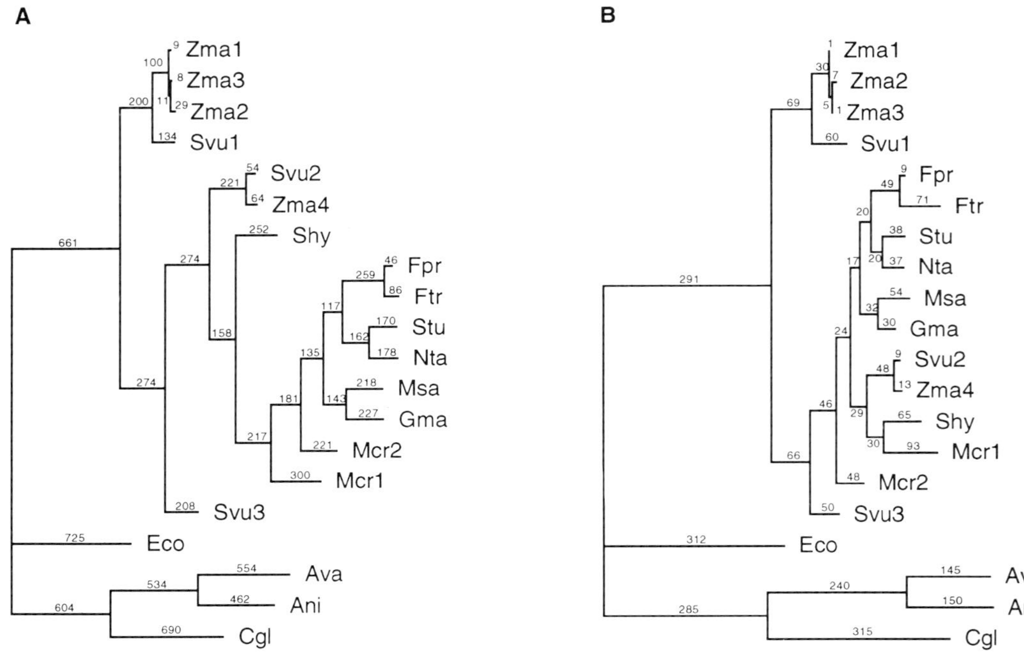

Fig. 9.2A,B. Evolutionary relation of PEPC enzymes and their genes. The two panels indicate results from parsimony analysis of the nucleotide sequence of the coding region (**A**) and of the encoded protein (**B**). The bacterial and cyanobacterial sequences were considered as an "outgroup" for the evolutionary analysis. *Eco E. coli; Mcr1 Mesembryanthemum crystallinum* CAM isogene; *Mcr2 M. crystallinum* C_3 form, *Zma1-4 Zea mays* (the C_4 forms are *Zma1*, *Zma2* and *Zma3*; *Zma4* is the C_3 form); *Svu1-3 Sorghum vulgare* (the C_4 form is *Svu1* and *Svu2* is C_3 form; *Svu3* is C_4 form); *Fpr Flaveria pringlei* (C_3); *Ftr Flaveria trinervia* (C_4) *Ava Anabaena variabilis*; *Ani Anacystis nidulans*; *Nta Nicotiana tabacum*; *Stu Solanum tuberosum*; *Msa Medicago sativa* (root nodule form); *Gma Glycine max*; *Shy Saccharum hybridus*; *Cgl Corynebacterium glutamicum*. References for the individual sequences have not been included here (see Lepiniec et al. 1993). *Numbers* alongside the branches indicate the number of inferred changes according to the parsimony reconstruction (Swofford 1993)

Table 9.1. Features that distinguish juvenile and adult growth phases of *Mesembryanthemum crystallinum*

Character	Juvenile	Adult
Leaf shape	Obovate with entire margins; net-veined; symmetrical; nearly sessile; only slightly succulent; may be large ($>10\,cm^2$) declining in size from leaf pair three on	Bilaterally symmetrical; smaller; more succulent; bearing anthocyanic mucro at apex when young, more pronounced with age; leaves becoming smaller with age
Leaf position	Along primary axis; primary leaves always in pairs; opposite five to seven pairs in total	Along secondary axes; oldest secondary leaves in pairs and opposite; becoming alternate later
Growth habit	Orthotropic; primary axis stunted	Plagiotropic; terminal flower at primary meristem
Branches	Equal to number of primary leaves or less	Unlimited side branch formation; terminal flowers
Anthocyanin	Absent[a]	Strong in flowers and young leaves; along secondary and tertiary axes
Flowers	Absent	Present
Bladder cells	Relatively small, appressed	Large, turgid, protruding; salt storage function
Ploidy categories	2 to 32 N; cell-specific	Increased ploidy in many, but not in all, cell types
DNA methylation[b]	*CAG-methylation high	*CAG-methylation lower within 2 days after stress started
Pinitol/Ononitol	absent or low concentration	High and increasing with distance from primary axis

[a]Anthocyanin may be induced by red light and/or long light periods (E. J. DeRocher, unpubl.).
[b](*CAG) DNA methylation was measured using differential 4′-and/or 5′-cytosine methylation status with the help of the restriction methylases *Eco*RII and *Bst*NI (Y. Buryanov, unpubl.).

The phylogenetic relationships are quite similar whether DNA or amino-acid sequence data are used (Fig. 9.2 A, B). The characterization of additional sequences from other CAM species should provide further insight into the independent evolutionary origins of PEPC and other CAM-specific enzymes. The conclusions that can be drawn from parsimony analysis (PAUP program; Swofford 1993) are, because of a relatively low number of enzyme sequences, at present rather limited.

9.3 Life Cycle of *Mesembryanthemum crystallinum*

To understand the genctic and molecular reactions that underlie CAM inducibility in the ice plant, it is important to understand the growth and devclopment of

M. crystallinum (Table 9.1). The plant shows clearly distinguishable juvenile and adult growth phases and a dramatic change in development as it switches to the adult phase. The plant is adapted to a Mediterranean climate where water is plentiful only during the initial part of the growing season. Conditions in the plant's natural habitat cannot usually be reproduced adequately in the laboratory, where low light intensities and small soil rooting volumes frequently limit plant growth. While plants grown under our standard conditions (see below) will typically reach a weight of 1 kg, plants grown in the field in Arizona have been found to weigh up to 20 kg (Adams et al. 1992).

The plants grow under our conditions (approx. 400 μmol photons $m^{-2} s^{-1}$ for 12 h, 25 °C/12 h dark, 17 °C; daily watering during the first 5 weeks of growth; initial stress by 500 mM NaCl in 0.5 Hoagland's repeated twice weekly) by developing one leaf pair/week along the primary axis (Table 9.1). Side shoots appear when the plants are 5 weeks old, the first of which are derived from the axillary bud above leaf pair two, later from the axillary buds of leaf pairs of higher order. Along these side shoots, small leaves develop that are morphologically different from the primary leaves and eventually flowers develop. We have never seen more than seven primary leaf pairs and the number five (and higher) leaves remain relatively small. Leaves of the pairs two and three are usually the largest primary leaves.

Several controversial observations with respect to CAM induction, which can be found in the literature, may be due to differences in growth conditions (Chu et al. 1990; Winter and Gademann 1991; Adams et al. 1992; McElwain et al. 1992; Schmitt and Piepenbrock 1992; Thomas et al. 1992a,b; see also Chap. 8). When the plants are 5 to 6 weeks old, coinciding with the emergence of side shoots, the gene expression changes that lead to CAM induction are readily inducible by salt stress treatment using 500 mM NaCl. At this age CAM is strongly induced by transcriptional activation (Cushman et al. 1990; Cushman and Bohnert 1992), while stress induction of transcription at other periods of development are lower in magnitude. Older plants left unstressed will show increases in transcription, mRNA and translation of CAM-gene transcripts in a slow and gradual manner. In contrast, stress of such older plants leads to a rapid induction of the pathway presumably as a result of the slow accumulation of transcripts being rapidly utilized once the induction process is initiated (Cushman et al. 1990; Vernon and Bohnert 1992b; Vernon et al. 1993a). Since chronological age does not always reflect physiological age when different growth conditions are used (see also Chap. 10.), differences have been reported in the time interval within which CAM genes are induced, ranging from a few hours to 2 days. In summary, the changes which the ice plant undergoes in its transition from juvenile to adult growth (Table 9.1) are similar or even more dramatic than classical examples for phase changes in plant development (Sussex 1989; Poethig 1990). Included in Table 9.1 are several observations, for example ploidy levels (DeRocher et al. 1990) and nuclear DNA methylation studies (Y. Burjanov and H.J. Bohnert, unpubl.), which indicate the mechanisms by which the transition to the adult growth phase may be accomplished. For example, changes in the methylation status of the maize C_4-PEPC promoter have been correlated with increased expression of the gene and enzyme accumulation (Langdale et al. 1991). The developmental phase-change coincides with a peak in CAM inducibility, which may indicate a correlation between the two processes. Much more attention needs to be given to the study of possible interrelationships between plant development and the degree and timing of stress responses that modulate metabolism.

9.4 Requisites for Environmental Stress Tolerance

Associated with CAM and halophilicity, *Mesembryanthemum crystallinum* has evolved a number of biochemical and morphological features, such as succulence, accumulation of solutes, and the presence of epidermal bladder cells, that contribute to the long-term survival of the plant. We believe these features are connected to a series of gene expression programmes designed for the maintenance of photosynthesis and integrity of the chloroplast, osmotic stabilization of the intracellular environment, and reutilization of cellular components. Collectively, these features help to create the environmental stress-tolerance phenotype that assures reproductive success (Winter and Ziegler 1992).

9.4.1 Maintaining a Functional Chloroplast

M. crystallinum demonstrates various strategies to protect its photosynthetic capacity during stress. When the plants are exposed to high levels of NaCl (400 to 500 mM) in the rooting medium, Na^+ and Cl^- concentrations within the chloroplast increase to a lesser extent than in the total cell volume (Demmig and Winter 1986). Furthermore, osmoprotective substances, such as pinitol that accumulates in the chloroplast and cytosol (see below; Paul and Cockburn 1989) and proline that accumulates preferentially in the cytosol (Demmig and Winter 1986; Treichel 1986; Thomas et al. 1992a), serve to maintain the integrity of the organelle. Preservation of the photosynthetic apparatus has been demonstrated in *M. crystallinum* by Köster and Anderson (1989) for the protein compositions of PSI and PSII, which did not change significantly under salt-stress conditions. Some alterations in the PSI/PSII ratios were interpreted to be functional in the prevention of photo-oxidative damage. Measurements of electron transport and fluorescence indicated that the photosynthetic machinery, at least during early periods of severe salt stress, remains largely unaffected by stress (Köster and Anderson 1989). It is perhaps not surprising that expression of nuclear genes for essential chloroplast functions, such as the expression of ferredoxin-NADP-reductase or phosphoribulokinase (Michalowski et al. 1989b, 1992), continues under stress. The regulation of expression of these genes is complex. There appears to be transcriptional and post-transcriptional control (DeRocher and Bohnert 1993), and there may also be post-translational control.

9.4.2 Osmotic Adjustment

Even before CAM becomes functional and contributes to water conservation, seawater-strength salinity does not severely damage the plants. If the NaCl level is raised from below 50 mM to 400 or 500 mM, plants will wilt for 2 days at most and will then recover, in contrast to the behavior of non halophilic plants, indicating that *Mesembryanthemum* shows osmotic adjustment and avoids toxic sodium effects.

To understand the molecular basis of salt tolerance, the role of several genes that are not involved in CAM per se, but accompany the induction of CAM, have been studied. Recently, a cDNA encoding the enzyme myo-inositol O-methyl transferase (Imt1) was characterized from *M. crystallinum* (Vernon and Bohnert 1992a). This gene encodes the first of two enzymes in a biosynthetic pathway that leads from the ubiquitous plant metabolite myo-inositol to pinitol (Paul and Cockburn 1989). Imt1 catalyzes the methylation of inositol to ononitol, which is slowly epimerized to pinitol. Expression of the *Imt1* gene and the Imt1 activity increase markedly during salt stress and low-temperature stress, but not under drought conditions (Vernon et al. 1993a). Interestingly, unlike the expression of certain CAM genes, *Imt1* expression is not dependent on plant development, but can be maximally elicited by stress at any age (Vernon and Bohnert 1992b). If polyol accumulation is viewed as a water-stress ameliorating factor, plants would be at an advantage if polyol biosynthesis were uncoupled from development, responding only to the perception and signalling of environmental stresses. The *Imt1* gene has been transferred into other plants to test directly its efficacy in providing protection against osmotic stress (Vernon et al. 1993b). The characterization and manipulation of genes involved in the polyol biosynthetic pathway represents a lucid demonstration of the value of studying a halophytic, non-crop species. Exploitation of this pathway for crop cultivation may have universal applicability, as well as serving to advance our basic understanding of stress physiology.

9.4.3 Magnitude of Stress-Induced Gene Expression

During the transition from C_3 to CAM the expression of several hundred genes is altered – more genes are down-regulated than up-regulated (Meyer et al. 1990). A rough estimation is that perhaps 100 genes are up-regulated in the above-ground tissues, indicating that we may have identified only 15 to 20% of these genes. Among these other salt-stress-induced genes expressed in leaves, a thiol protease (Table 9.2) (Ostrem et al. 1990; J.C. Cushman and J. A. Ostrem, unpubl.) has been isolated. We think this protease, *Sep1*, and several other transcripts for closely related functions which have not yet been characterized in depth, may be involved in general protein turnover that is significant as the plants switch to CAM. Such proteases may be specific for altering components of the transcription machinery, or responsible for the large-scale turnover of many of the down-regulated proteins to provide amino acids for production of the enzymatic machinery associated with CAM (Holtum and Winter 1982). The induction of synthesis of most of these transcripts, of *Imt1*, *Sep1*, or the induction of transcripts for CAM-related functions, is under transcriptional control.

9.5 Regulation of CAM Gene Expression

In response to salt stress, biochemical changes occur that are indicative of CAM. Malic acid levels begin to fluctuate diurnally and the activities of many enzymes associated with malate metabolism and glycolysis/gluconeogenesis increase (Edwards et al. 1982; Holtum and Winter 1982; Winter et al. 1982; Fig. 9.1). The increases in enzyme activities are directly linked to increases in enzyme protein levels (Winter et al. 1982; Michalowski et al. 1989a; Schmitt et al. 1989) that are a direct result of de novo protein synthesis in response to salt stress (Foster et al. 1982; Höfner et al. 1987). We have observed that many CAM genes display increased steady-state mRNA levels upon salt stress. For example, steady-state mRNA levels for PEPC, PPDK (Michalowski et al. 1989a; Schmitt et al. 1989) and NAD-glyceraldehyde 3-phosphate dehydrogenase, an enzyme of the glycolytic pathway which supplies the substrate for PEPC, increase following salt stress (Ostrem et al. 1990).

9.5.1 Transcript Amounts

The increased enzyme activities for many CAM enzymes are controlled by changes in gene expression that are mediated mainly by transcriptional activation. For example, one member of the PEPC gene family is encoded by a CAM-specific isogene (*Ppc1*) whose expression is transcriptionally induced by salt stress (Table 9.2; Cushman and Bohnert 1989a; Cushman et al. 1989). In contrast, another gene family member (i.e. *Ppc2*) encoding an alternate form of the enzyme displays an expression pattern not enhanced by stress (Table 9.2; Cushman and Bohnert 1989b; Cushman et al. 1989). Other examples of transcriptional activation of CAM-specific genes include *Gpd1* (Vernon et al. 1993a), *Ppdk1* (J.C. Cushman, unpubl.), a cytosolic NADP-malic enzyme (NADP-ME) (Cushman 1992), and a chloroplast-localized NADP-malate dehydrogenase (NADP-MDH) (Cushman 1993). Like *Ppc1*, all of these genes display a pattern of increased mRNA accumulation that is at least partially the result of increased transcription rate (Table 9.2).

For certain of these genes, the contribution of their gene products to CAM is not clearly resolved. Earlier work demonstrated that *M. crystallinum* contains an NAD-dependent form of malic enzyme localized to the mitochondria that also displayed increased activity during salt stress treatment (Holtum and Winter 1982). During the induction process, the activities of both an NADP- and NAD-dependent form of MDH increase (Holtum and Winter 1982; Winter et al. 1982). The reduction of OAA to malate is thought to be carried out primarily by

Table 9.2. Genes of different metabolic pathways and their response to salt stress in *Mesembryanthemum crystallinum*

Gene	mRNA	Enzyme	Location	Transcriptional induction	Developmental induction	Reference
(*Malate metabolism*)						
Ppc1	3.3 kb	Phospho*enol*pyruvate carboxylase (EC 4.1.1.31)	Cytosol	+ + +	+ +	Cushman et al. (1989)
Mdh1	1.75 kb	NADP-malate dehydrogenase (EC 1.1.1.82)	Chloroplast	+ +	+ +	Cushman (1993)
Mdh2	1.5 kb	NAD-malate dehydrogenase (EC 1.1.1.37)	Mitochondria/ cytosol	+ +?	?	J.C. Cushman (unpubl.)
Mod1	2.15 kb	NADP-malic enzyme (EC 1.1.1.40)	Cytosol	+ +	+ +	Cushman (1992)
Mod2	?	NAD-malic enzyme (EC 1.1.1.39)	Mitochondria	+ +?	?	–
Ppdk1	3.4 kb	Pyruvate orthophosphate dikinase (EC 2.7.9.1)	Chloroplast	+ + +	+ +	Fißlthaler et al. (1995)
(*Glycolysis/gluconeogenesis*)						
Eno1	1.7 kb	Enolase (EC 4.2.1.11)	Cytosol	+ +	?	–
Pgm1	1.55 kb	Phosphoglyceromutase (EC 2 7.5.3)	Cytosol	+ +	?	J. Simpson, C.B. Michalowski (unpubl.) D.M. Vernon (unpubl.)
Pgk1	1.55 kb	Phosphoglyceratekinase (EC. 2.7.2.3)	Cytosol	+ +	?	J. Simpson, C.B. Michalowski (unpubl.) D.M. Vernon (unpubl.)
Gpd1	2.1 kb	NAD-glyceraldehyde-3-phosphate dehydrogenase (EC 1.2 1.12)	Cytosol	+ + +	+ +	Ostrem et al. (1990)
Gpi1	?	Glucosephosphate isomerase (EC 5.3.1.9)	Cytosol	+	?	–

(Induced by salt stress)						
Imt1	1.6 kb	Myo-inositol *O*-methyl transferase	Cytosol	+ + + (Transiently: roots)	–	Vernon and Bohnert (1992a)
Sep1	1.3 kb	Sulphydryl endopeptidase	Chloroplast?	+ + +	?	J.C. Cushman and J.A. Ostrem (unpubl.)
(Unaffected/repressed by salt stress)						
Ppc2	3.3 kb	Phospho*enol*pyruvate carboxylase	Cytosol	+/–	–	Cushman and Bohnert (1989a)
RbcS1–6	0.75kb	Small subunit of RUBISCO	Chloroplast	–	–	DeRocher et al. (1993) DeRocher and Bohnert (1993)
RbcL	1.75 kb	Large subunit of RUBISCO	Chloroplast	–	–	C.B. Michalowski (unpubl.)
Prk1	1.5 kb	Phosphoribulokinase	Chloroplast	(Transiently down)	?	Michalowski et al. (1992)
PetH	1.5 kb	Ferredoxin-NADP-reductase	Chloroplast	–	?	Michalowski et al. (1989b)
Cab1–7	1.05 kb	Chlorophyll *a/b* binding protein	Chloroplast	–	?	C.B. Michalowski (unpubl.)
Act	?	Actin	Cytosol	+/–	?	–
Tub	?	Tubulin	Cytosol	+/–	?	–

an NAD-dependent form of MDH located either in the mitochondria or the cytosol. Using a probe for the gene that encodes the NADP-dependent form of this enzyme, we have detected the existence of alternate isogenes that likely encode the NAD-dependent form of MDH. Sorting out the contribution of these different isozymes to the CAM pathway is facilitated by having the cloned genes available for transformation studies involving overexpression and supression of gene expression using antisense techniques (see below).

The regulation of other salt-stress-induced genes expressed in leaves (Table 9.2; Ostrem et al. 1990; Vernon et al. 1993a) is not in all cases identical to the regulation of CAM genes. This observation indicates that several gene control programmes exist and that different environmental stimuli have different effects on the genes, which were all detected and isolated because of their differential expression during salt stress. For example, *Ppc1*, *Gpd1*, *Imt1* and *Sep1* (initially termed B5; Ostrem et al. 1990) all respond to salt stress by displaying approximately similar rates (2.5- to 5-fold) of transcriptional activation in in-vitro nuclear run-on experiments (Vernon et al. 1993a). However, only *Ppc1* showed mRNA accumulation under drought stress conditions. Exposure to cold temperatures (4 °C) resulted in the strong induction of *Imt1* transcript levels, no significant change in *Ppc1* transcript levels, and a down-regulation in *Sep1* (B5) transcript levels (Vernon et al. 1993a).

9.5.2 Transcription of CAM Genes

The coordinate increase in steady-state transcript levels for the set of genes involved in diverse functions such as glycolysis, gluconeogenesis, carboxylation and decarboxylation of malate, compatible solute biosynthesis and protein turnover suggests that there might be a common controlling mechanism responsible for transcriptional activation of these genes by salt stress. The transcriptional induction process is likely to involve binding of a transcriptional activator within the 5′-flanking regions upstream of the transcription start sites. Interaction of a factor or a combination of factors with the basic transcription apparatus is required to activate, enhance, or repress transcription rates.

To begin to explore the mechanism of this transcriptional activation process, we have characterized the 5′-flanking regions of two stress-induced genes, *Ppc1* (Cushman and Bohnert 1992) and *Gpd1* (J.C. Cushman, unpubl.; Fig. 9.3). For comparison, we have also sequenced the 5′-flanking region of the nonresponsive *Ppc2* gene (J.C. Cushman, unpubl.). In general terms, the regions share little similarity, with only a few regions of intermittent, widely scattered sequence homologies being obvious between, for example, the *Ppc1* and *Gpd1* promoters (Cushman et al. 1993a,b). A detailed analysis of the occurrence of potential DNA-binding sites was conducted by comparing these upstream sequences to binding-site consensus sequences of known eukaryotic transcription factors. We find that there are multiple potential binding-site motifs for conserved eukaryotic transcription activators (Fig. 9.3), such as ATF (Hai et al. 1989) and AP4 (Hu et al.

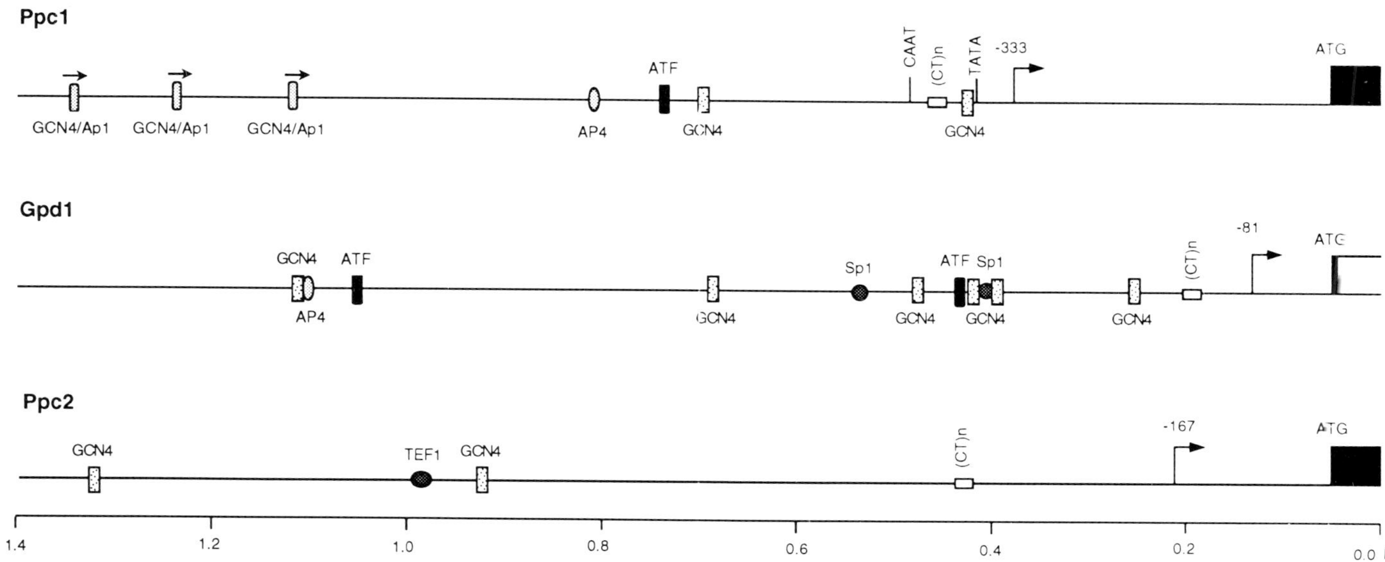

Fig. 9.3. Conserved eukaryotic sequence elements of *Mesembryanthemum crystallinum Ppc1*, *Gpd1* and *Ppc2* promoters. Comparisons of the nucleotide sequences of the 5′-flanking regions of the inducible CAM *Ppc1* and *Gpd1* genes and the constitutively expressed *Ppc2* gene reveal conserved *cis*-elements for the binding of several eukaryotic transcription factors. Transcription start sites are indicated by *bent arrows*. For each gene the initiation codons and first exon are indicated above the *solid boxes*. Potential transcription-factor binding sites that can be identified include those for GCN_4 (Arndt and Fink 1986), $(CT)_n$ or $[(GA)_n]$ (Gilmour et al. 1989), GCN4/AP1 (Lee et al. 1987), AP4 (Hu et al. 1990), ATF (Hai et al 1989), Sp-1 (Kadonaga et al. 1986) and TEF1 (Davidson et al. 1988)

1990), that are shared by *Ppc1* and *Gpd1*. Other sites are found in all three gene control regions, such as $(CT)_n$ [complementary: $(GA)_n$], which is a potential binding site for GAGA-binding factor (Gilmour et al. 1989). Other sites such as those for the positive transcription factor GCN4 (Arndt and Fink 1986) are common to all three promoters, but are more numerous in the two transcriptionally activated promoters. In general, it is obvious from this level of analysis that the inducible genes possess 5′-flanking regions that are rich in potential binding sites for evolutionarily conserved factors known to activate transcription. Aside from these sequences, few discrete sequence motifs are found that may serve as binding sites for either novel or previously observed transcription factors. In particular, it is impossible to identify by comparisons a potential "CAM induction" consensus sequence element in the upstream control region of the promoters of the *Ppc1* and *Gpd1* genes. Several different approaches are being taken to define such sequence elements that must play a role in the activation of CAM genes by water stress.

To characterize potential DNA-binding proteins present in the nucleus of the ice plant, nuclear extracts have been prepared and analyzed by gel-mobility-shift assays. Multiple binding interactions have been documented in the 5′-flanking region of *Ppc1* (Cushman and Bohnert 1992). Binding interactions are detected along the entire length of the promoter and are associated with regions that contain potential evolutionarily conserved transcriptional activation sequences. One particularly abundant factor, called PCAT-1, has been characterized by having an AT-rich binding site (although it probably is not an HMG (high-mobility group) protein, such as are known to bind to AT-rich sequences in an unspecific manner). More importantly, PCAT-1 is more abundant, or has a greater binding affinity, in nuclear extracts isolated from stressed plants. Determination of the exact role of this factor in the regulation of CAM-gene expression will require more detailed analysis of this *trans*-acting factor and ultimately isolating cDNA clones encoding the factor.

9.5.3 Analysis of Transcription Control

A common strategy used to characterize *cis*-element function in transcription is to link these elements to a reporter gene and introduce the chimeric genes back into the original host to assay expression patterns resulting from the stress response. Constructs fusing the promoter *Ppc1* with the coding region of the β-glucuronidase (GUS) gene exhibit high levels of expression in transgenic tobacco; however, this expression does not appear to be salt-stress-inducible in mature plants (Cushman et al. 1993a). In *M. crystallinum*, which is not readily transformable, an alternative approach to stable transformation is being pursued. We have succeeded in developing a transient assay system that uses microprojectile bombardment of intact *M. crystallinum* leaves that faithfully mimics the transcriptional activation in response to salt stress found in whole plants. Unstressed leaves show little GUS activity, whereas leaves from stressed plants show an approximately fivefold increase in GUS activity. A series of promoter deletions has been conducted and regions that effect both enhancement and silencing of reporter-gene expression have been uncovered (J.C. Cushman, unpubl.). Further refinement of this approach is being used to define more accurately those specific *cis*-acting sequence elements necessary for transcriptional activation.

Identification of specific *cis*-elements that mediate CAM-gene expression will permit the isolation of cognate *trans*-acting transcription factors involved in transcriptional activation. Once in hand, we can use these factors to begin to dissect the signalling mechanisms operative during responses to stress and during CAM induction. By defining the components of the systems and their regulation, we can begin to understand the true nature of the signals actually perceived by plants when CAM is induced, whether these signals be turgor, water potential, ion accumulation or ionic imbalances on opposite sides of membranes, growth regulators, or differences in osmotic pressure in different cell types.

9.5.4 Transcription and mRNA Stability

Other factors controlling the overall expression of a particular gene include alterations in the turnover rates of transcripts mediated by changes in mRNA stability. In vitro nuclear run-on experiments, which involve the labelling of nascent RNA chains by incorporation of radioactive nucleotides into isolated nuclei, allow the distinction between mRNA transcription rates and the accumulation of steady-state levels of transcripts. Levels of transcripts are governed by both the rate that transcripts are synthesized and readied for export to the cytosol and by mRNA degradation rates. For the CAM genes that have been analyzed, such as *Ppc1*, *Ppdk1*, *Mdh1*, *Gpd1* and *Mod1* (Table 9.2), CAM induction is preceded by increases in the rate of transcription of these genes (Cushman et al. 1989, 1993b; Cushman 1992, 1993). However, by comparing transcription rates with steady-state mRNA levels, we can infer that regulation of mRNA-turnover rates is likely to be also operative in the control of gene expression during *M. crystallinum* development and stress responses (Cushman et al. 1990). First, transcripts associated with the basic photosynthetic apparatus (e.g. *RbcS*) decline transiently, but then rebound after stress (DeRocher and Bohnert 1993), as had been observed before (Michalowski et al. 1989a). It is particularly interesting that transcription rates and mRNA amounts for various *RbcS* gene family members in *M. crystallinum* behave differently during stress (DeRocher and Bohnert 1993). Transcript levels decline, while transcription rates remain constant or even increase for some of these genes. The overall long-term decline in RUBISCO activity in leaves in the face of increasing *RbcS* transcription would indicate that post-transcriptional control, possibly involving selective destabilization of messages, is operative. Second, transcripts of CAM-related genes increase over the same time period, with the buildup being affected by possible increases in mRNA stabilization (Cushman et al. 1990). The possible role of differential regulation of transcript stability during CAM induction and plant development must, however, be directly addressed by measuring actual mRNA half-life and changes in the stability of specific transcripts.

To investigate possible post-transcriptional control mechanisms that regulate mRNA stability during stress, we have recently focused our attention on the role of RNA-binding proteins. We have isolated several proteins by ssDNA affinity chromatography that resemble RNA-binding proteins (RBP) (H. Breiteneder and H.J. Bohnert, unpubl.). A subset of these proteins, from which direct amino-terminal protein micro-sequencing data could be obtained, belong to a previously characterized group of chloroplast RNA-binding proteins (cRBP) (Li and Sugiura 1990). Based on this amino-acid sequence information, we used degenerate oligonucleotide primers and the polymerase chain reaction (PCR) to isolate and characterize clones for cRBP transcripts. A single gene is responsible for several transcripts with different size of their untranslated 3′-end. These transcript sizes tend to be replaced under stress by a very long transcript of the gene including a 407 bp 3′-end that contains four nucleotide-sequence domains conserved among other RBP transcripts (H. Breiteneder, unpubl.). The presence of such an extraordinarily long 3′-untranslated portion of this transcript, containing conserved sequence motifs, implicates their involvement in stabilizing cRBP transcripts. The function of cRBPs within chloroplasts is not entirely known, but they may be involved in altering chloroplast transcript stability during stress treatment. We think that the long 3′-end may be responsible for transcript stability of this chloroplast RNA-binding protein and that increased amounts of cRBP, after chloroplast import, may enhance chloroplast translation resulting in the protection of this compartment, as has been observed before by several investigators (see above).

9.6 Transduction Mechanisms of Environmental Stress

The mechanisms by which *Mesembryanthemum crystallinum* senses changes in its environment and transduces those signals into physiological reactions are likely to be complex. Based on the information about signalling acquired thus far, it becomes increasingly clear that a simple model of coordinate regulation in response to a single signal is inadequate to explain the intricate patterns of gene responses to environmental stresses. We already know that drought, salt, low temperature, light, and the growth regulators ABA and cytokinin share certain portions of signalling pathways, while each environmental stress and endogenous factor triggers in addition unique responses which are furthermore modulated by interacting stress factors and development (McElwain et al. 1992; Thomas et al. 1992b; Vernon et al. 1993a). Understanding the integration of these individual environmental stresses and various endogenous growth regulators in controlling gene expression will be a goal of future studies.

Among the growth regulators, ABA has been implicated in the induction of CAM by osmotic stress (Ting 1985). In *Peperomia*, ABA application resulted in stomatal closure and fluctuations in organic-acid levels could be elicited (Hanscom and Ting 1978). In well-watered *Portulacaria afra*, exogenously applied ABA has been shown to mimic water stress by inducing CAM characteristics (Ting 1985). The correlation between ABA and CAM induction is further supported by the observed increase in endogenous ABA following osmotic stress in *Peperomia camptotricha* (Sipes and Ting 1985; see also Chap. 13). In *M. crystallinum*, increases in enzyme activities associated with CAM (PEPC and NADP-ME) appear to be initiated by exogenous application of ABA (Chu et al. 1990; see also Chap. 8). Recent studies allowed us to quantitate endogenous ABA levels. After salt stress, endogenous ABA increased by a factor of 8 to 10 in the leaves and by a factor of 2 in the root system (Thomas et al. 1992b). Depending on experimental conditions, CAM transcripts may (McElwain et al. 1992; Cushman et al.1993a) or may not (McElwain et al. 1992; Thomas et al. 1992b) increase after ABA addition to the medium of hydroponically grown plants. Further studies indicated that the induction of *Ppc1* by ABA is dependent upon light and temperature conditions. The ability of ABA to elicit PEPC protein or transcript accumulation is observed under high light intensity and longer illumination periods (McElwain et al. 1992). Exogenous ABA, therefore, cannot be considered a simple mediator of drought- or salinity-induced CAM genes. Furthermore, despite the inhibition of ABA accumulation with Lovastatin during salt stress, proline and PEPC transcripts and protein levels were still induced (Thomas et al. 1992b), suggesting that other factors are likely to be involved in the induction process. In conclusion, although we do not rule out a role for ABA in partially mediating stress responses, it is clear that it does not play a pivotal role in controlling gene expression in *M. crystallinum*.

Another growth regulator, cytokinin, plays some role in regulating gene expression. Exogenous applications (addition to the hydroponic solutions) of different cytokinins resulted in increased PEPC expression (mRNA and protein) and in proline accumulation, while endogenous amounts of cytokinins did not change significantly under these conditions or under salt stress (Thomas et al. 1992b). Cytokinin has also been shown to increase the rate of transcription of the genes *Ppc1* (CAM-PEPC) and *Sep1* (thiol protease) (Vernon et al. 1993a). In a different experimental set-up using *M. crystallinum*, Schmitt and Piepenbrock (1992) obtained a different result, namely an inhibitory effect of applied cytokinin on PEPC expression (see also Chaps. 8 and 10). The reasons for these discrepancies are not yet fully understood.

Aside from growth regulators, others have suggested that age-related fluctuations in leaf turgor may be related to CAM enzyme induction (Winter and Gademann 1991). This hypothesis is supported by the fact that various environmental stress treatments (i.e. high salinity, drought, low temperature, low humidity, high light intensity) affect the water status of the plants. Various genes, however, such as *Ppc1*, *Imt1* and *Sep1*, respond differently to different types of water stress (Vernon et al. 1993a). This would argue against age-related wilting of mature leaves as the single, dominant stress factor mediating all gene responses involved in CAM induction. This underscores our notion that both developmental and environmental signals are integrated into the plant's response circuitry. In summary, it would appear unlikely that one stress effector can be singled out as a general response mediator. Rather, we expect a concomitant, synergistic and overlapping series of transduction processes to be involved in the CAM induction pathway.

9.7 Genetics and Transformation of *Mesembryanthemum crystallinum*

The work carried out by many groups over the last 20 years allows *Mesembryanthemum crystallinum* to stand out as one of the most advanced systems for the physiological, biochemical and molecular analysis of environmental stress responses. The plant, however, is not currently a model for genetic analysis and has not been amenable, up to now, to gene transfer experiments. Before we can fully exploit the large amount of data that has been accumulated for this halophyte, a genetic system and a transformation/regeneration system must be developed.

Establishing genetic studies with *Mesembryanthemum* will open new avenues of research on mechanisms of stress tolerance. Established models, arabidopsis for example, and plants of economic importance, such as tomato, corn, or rice, are of little use because these plants are less tolerant to drought and salinity. Analysis of restriction fragment length polymorphisms (RFLP) and mapping of quantitative trait loci (QTL) have been used in some crop species to identify regions on chromosomes that appear to be related to water-use efficiency (Martin et al. 1989) and thermotolerance (Ottaviani et al. 1991). These studies have not yet resulted in a better mechanistic understanding of stress-tolerance reactions. In other studies, salt-sensitive and salt-tolerant wheat cultivars and species have been compared and crossbred, resulting in the identification of specific differences in gene expression (Galvez 1992; Gulick and Dvorak 1992). Considering these findings, a strong case can be made for developing a biochemically well-characterized plant model, such as *M. crystallinum*, into a genetic model.

Toward this end, we have initiated work to establish a collection of mutants for genetic analysis. Seeds have been irradiated with fast neutrons (range 20 to 120 krad gamma-rays from a cobalt-60 source) at the reactor of the International Atomic Energy Agency (Vienna, Austria). The choice of fast neutrons over chemical mutagenesis was made because we wished to obtain large alterations, deletions and chromosome breaks, which should lead to easily observable phenotypic changes. A range around 50 krad proved to be most suitable for our purposes. At higher irradiation doses practically all seeds died shortly after germination, and at a lower dose of 20 or 30 krad practically no phenotypic changes were observed in the generation emerging from the seeds, although chromosome breaks might have occurred at low frequency. Various phenotypic alterations observed in seedlings from intermediate irradiation levels included mild to severe morphological changes of tissues or cells, alterations in pigmentation, altered morphology of leaves, stems and roots, alterations in growth habit and changes in the stress response (H.J. Bohnert, F. Lehle and R. Khan, unpubl.). We are now generating a large seed population from this population of mutated plants.

Through the accumulation and characterization of mutants generated by classical means, genetic studies are underway. Establishing procedures for the transformation and regeneration of *M. crystallinum* has, however, proved to be more time-consuming than anticipated. Regeneration of multiple shoots has been accomplished (Meiners et al. 1991) and recent improvements of this protocol have resulted in increased regeneration frequencies (A. Bornhouser and J.C. Thomas, unpubl.). The combination of *Agrobacterium tumefaciens* transformation and regeneration protocols has, in our hands, resulted in transformed callus tissue expressing reporter genes, but regeneration has, up to now, not been achieved. An alternative transformation strategy has been successful using *A. rhizogenes* (strain R1000 and wild-type strains) (Andolfatto et al. 1993). We are at the early stages of testing transformation/regeneration protocols based on *A. rhizogenes* vectors which hopefully will result in transgenic *M. crystallinum* plants (A. Bornhouser and J.C. Thomas, pers. comm.). Once regeneration of transformed plants is accomplished, a number of experimental approaches will become possible which are crucial to further our understanding of the plant's tolerance reactions. For example, expression of antisense constructs of genes such as *Ppc1* and *Imt1* will allow us to evaluate the functional significance of individual enzymes to the overall CAM and polyol biosynthesis pathways, respectively. Experiments with transgenic *M. crystallinum* will also provide information about the stress perception, signalling and gene-activation processes.

9.8 Perspectives

CAM has been recognized as a carbon-concentration mechanism and a carbon-assimilation mode for conserving water which thereby confers upon plants an advantage in their ability to colonize CO_2-poor and arid environments. CAM has become a model for the mechanistic understanding of the interrelationships between environmental cues and changes in gene expression in plants. The biochemical complexity of the pathway, including compartmental interactions and regulatory mechanisms for enzyme activation, are matched by complex gene-activation mechanisms as plants switch from C_3 to CAM. Equally important are the genetic mechanisms leading to morphological changes that help establish and maintain CAM.

Our studies of CAM-gene expression are aimed at understanding both the transcription and post-transcriptional mechanisms involved in responding to environmental stress in the context of plant development. We have begun to uncover only the most basic mechanisms of transcription control involving *cis*-elements and protein–DNA interactions. The complex structure of the two CAM-specific promoters characterized thus far points, perhaps not surprisingly, to many potential control mechanisms. Furthermore, our results suggest that post-transcriptional events are at work, adding yet another level of gene-expression control. For example, PEPC, with its pivotal role in controlling the flux of carbon into C4 acids, is tightly regulated by a variety of metabolic effectors and post-translational modifications of the enzyme (Jiao and Chollet 1991; see also Chap. 3). Such multiple control of important enzymatic steps is not unheard of. For example, 3-hydroxy-3-methylglutaryl Co A reductase, controlling the entry of metabolites into the mevalonate pathway, has been shown to be regulated at many levels: transcription, mRNA stability, translation, protein stability, and protein inactivation by phosphorylation (Goldstein and Brown 1990).

The interplay of CAM and other strategies of survival in water-limited environments appears equally complex. Our experiments (Cushman et al. 1990; Vernon and Bohnert 1992b; Vernon et al. 1993a) indicate that CAM, which is induced incompletely and inefficiently in very young plants, is unlikely to account solely for the water-stress tolerance of *M. crystallinum*. Other mechanisms, such as a developmentally programmed osmotic adjustment in the bladder cells of old plants (Adams et al. 1992) and inducible osmotic adjustment in young plants (Vernon and Bohnert 1992b; Vernon et al. 1993a), appear to be equally important. The gene encoding the first committed step in the pinitol biosynthetic pathway is equally inducible in young plants and in old plants (Vernon and Bohnert 1992a, b). Polyol accumulation appears to be an important component of osmotic adjustment and turgor maintenance.

Mesembryanthemum plants die from the soil up, losing roots, primary leaves and stems, secondary and tertiary leaves and stems consecutively until only flowers, fruits and seed pods remain photosynthetically active and alive. Amounts of several polyols increase during this process. The expression of CAM in the leaves along with the accumulation of high levels of osmoticum (mainly polyols and, under salt stress, also sodium chloride) in the seed pods (Steudle et al. 1975; Lüttge et al. 1978; Adams et al. 1992) provides *Mesembryanthemum crystallinum* with a means of living off its own substance to produce seeds in the absence of moisture. A better understanding of the complex interplay between these stress responses will be derived from future studies employing genetic, development, biochemical and physiological analyses.

Acknowledgements. Funding for H.J.B. and J.C.C by USDA-NRI-CRGP (Plant Responses to the Environment) and by the Arizona Agricultural Experiment Station is gratefully acknowledged. We wish to thank Jeanette Simpson, Chris Michalowski, Heimo Breiteneder, Dan Vernon, Angela Bornhouser, Jay DeRocher, Patricia Adams and John Thomas for communicating unpublished data and for their willingness to discuss practically any topic anytime. We thank Mark Fishbein (Dept. Ecology & Evolutionary Biology, Tucson, AZ) and Phil Reid (Smith College, Northampton, MA) for help with Table 9.1.

References

Adams P, Thomas JC, Vernon DM, Bohnert HJ, Jensen RG (1992) Distinct cellular and organismic responses to salt stress. Plant Cell Physiol 33: 1215–1223

Andolfatto P, Bornhouser A, Bohnert HJ, Thomas JA (1993) Transformed hairy roots of *Mesembryanthemum crystallinum*: gene expression patterns upon salt stress. Physiol Plant 90: 708–714

Arndt K, Fink GR (1986) GCN4 protein, a positive transcription factor in yeast, binds general control promoters at all 5′-TGACTC-3′ sequences. Proc Natl Acad Sci USA 83: 8516–8520

Bohnert HJ, Vernon DM, DeRocher EJ, Michalowski CB, Cushman JC (1992) Biochemistry and molecular biology of CAM. In: Wray JL (ed) Inducible plant proteins. Cambridge University Press, Cambridge, pp 113–137

Brulfert J, Müller M, Kluge M, Queiroz O (1982) Photoperiodism and crassulacean acid metabolism. Planta 154: 326–331

Chu C, Dai Z, Ku MSB, Edwards GE (1990) Induction of crassulacean acid metabolism in the facultative halophyte *Mesembryanthemum crystallinum* by abscisic acid. Plant Physiol 93: 1253–1260

Cockburn W (1983) Stomatal mechanism as the basis of the evolution of CAM and C_4 photosynthesis. Plant Cell Environ 6: 275–279

Cushman JC (1992) Characterization and expression of NADP-malic enzyme cDNA induced by salt stress from the facultative CAM plant, *Mesembryanthemum crystallinum*. Eur J Biochem 208: 259–266

Cushman JC (1993) Molecular cloning and expression of chloroplast NADP-malate dehydrogenase during crassulacean acid metabolism induction by salt stress. Photosynth Res 35: 15–27

Cushman JC, Bohnert HJ (1989a) Nucleotide sequence of the *Ppc2* gene encoding a housekeeping isoform of phosphoenolpyruvate carboxylase from *Mesembryanthemum crystallinum*. Nucleic Acids Res 17: 6743–6744

Cushman JC, Bohnert HJ (1989b) Nucleotide sequence of the gene encoding a CAM specific isoform of phosphoenolpyruvate carboxylase from *Mesembryanthemum crystallinum*. Nucleic Acids Res 17: 6745–6746

Cushman JC, Bohnert HJ (1992) Salt stress alters A/T-rich DNA-binding factor interactions within the phosphoenolpyruvate carboxylase promoter from *Mesembryanthemum crystallinum*. Plant Mol Biol 20: 411–424

Cushman JC, Meyer G, Michalowski CB, Schmitt JM, Bohnert HJ (1989) Salt stress leads to differential expression of two isogenes of phospho*enol*pyruvate carboxylase during crassulacean acid metabolism induction in the common ice plant. Plant Cell 1: 715–725

Cushman JC, Michalowski CB, Bohnert HJ (1990) Developmental control of crassulacean acid metabolism inducibility by salt stress in the common ice plant. Plant Physiol 94: 1137–1142

Cushman JC, Meiners MS, Bohnert HJ (1993a) Expression of a phosphoenolpyruvate carboxylase promoter from *Mesembryanthemum crystallinum* is not salt-inducible in mature transgenic tobacco. Plant Mol Biol 21: 561–566

Cushman JC, Vernon DM, Bohnert HJ (1993b) ABA and the transcriptional control of CAM induction during salt stress in the common ice plant. In: Verma DPS (ed) Control of plant gene expression. CRC Press, Boca Raton, pp 287–300

Davidson I, Xiao JH, Rosales R, Staub A, Chambon P (1988) The HeLa cell protein TEF-1 binds specifically and cooperatively to two SV40 enhancer motifs of unrelated sequence. Cell 54: 931–942

Demmig B, Winter K (1986) Sodium, potassium, chloride and proline concentrations of chloroplasts isolated from a halophyte, *Mesembryanthemum crystallinum* L. Planta 168: 421–426

DeRocher EJ, Bohnert HJ (1993) Expression of the *rbcS* gene family in *Mesembryanthemum crystallinum* during development and under environmental stress. Plant Cell 5: 1611–1625

DeRocher EJ, Harkins KR, Galbraith DW, Bohnert HJ (1990) Developmentally regulated systemic endopolyploidy in succulents with small genomes. Science 250: 99–101

DeRocher EJ, Quigley F, Mache R, Bohnert HJ (1993) The six genes of the Rubisco small subunit multigene family from *Mesembryanthemum crystallinum*, a facultative CAM plant. Mol Gen Genet 239: 450–462

Edwards GE, Foster JG, Winter K (1982) Activity and intracellular compartmentation of enzymes of carbon metabolism of CAM plants. In: Ting IP, Gibbs M (eds) Crassulacean acid metabolism. American Society of Plant Physiologists, Rockville, pp 92–111

Fißlthaler B, Meyer G, Bohnert HJ, Schmitt JM (1995) Age-dependent induction of pyruvate, orthophosphate dikinase in *Mesembryanthemum crystallinum* L. Planta 196: 492–500

Foster JG, Edwards GE, Winter K (1982) Changes in the levels of phosphoenolpyruvate carboxylase with induction of CAM in *M. crystallinum* L. Plant Cell Physiol 23: 585–594

Galvez AF (1992) Genetic responses to salinity stress in halophytic *Lophopyrum elongatum*, glycophytic wheat and the *L. elongatum* × wheat amphiploid. PhD Thesis, University of California, Davis

Gilmour DS, Thomas GH, Elgin SCR (1989) *Drosophila* nuclear proteins bind to regions of alternating C and T residues in gene promoters. Science 245: 1487–1490

Goldstein JL, Brown MS (1990) Regulation of the mevalonate pathway. Nature 343: 425–430

Groenhof AC, Bryant JA, Etherington JR (1986) Photosynthetic changes in the inducible CAM plant *Sedum telephium* L. following the imposition of water stress. Ann Bot 57: 689–695

Gulick PJ, Dvorak J (1992) Coordinate gene response to salt stress in salt tolerant *Lophopyrum elongatum* immediately after initiation of stress. Plant Physiol 100: 1384–1388

Guralnick LJ, Rorabaugh PA, Hanscom Z (1984) Seasonal shifts of photosynthesis in *Portulacaria afra* (L.) Jacq. Plant Physiol 76: 643–646

Hai TW, Liu F, Coukos WJ, Green MR (1989) Transcription factor ATF cDNA clones: an extensive family of leucine zipper proteins able to selectively form DNA-binding heterodimers. Genes Dev 3: 2083–2090

Hanscom Z, Ting IP (1978) Irrigation magnifies CAM-photosynthesis in *Opuntia basilaris* (Cactaceae). Oecologia 33: 1–15

Hermans J, Westhoff P (1992) Homologous gene for the C4 isoform of phosphoenolpyruvate carboxylase in a C_3 and C_4 *Flaveria* species. Mol Gen Genet 234: 275–284

Heun AM, Gorham J, Lüttge U, Wyn Jones RG (1981) Changes of water-relation characteristics and levels of organic cytoplasmic solutes during salinity induced transition of *M. crystallinum* from C_3-photosynthesis to crassulacean acid metabolism. Oecologia 50: 66–72

Höfner R, Vazquez-Moreno L, Winter K, Bohnert HJ, Schmitt JM (1987) Induction of crassulacean acid metabolism in *M. crystallinum*: mass increase and de-novo synthesis of PEP-carboxylase. Plant Physiol 83: 915–919

Holthe PA, Sternberg LW, Ting IP (1987) Developmental control of CAM in *Peperomia scandens*. Plant Physiol 84: 743–747

Holtum JAM, Winter K (1982) Activity of enzymes of carbon metabolism during the induction of crassulacean acid metabolism in *Mesembryanthemum crystallinum* L. Planta 155: 8–16

Hu YF, Lüscher B, Admon A, Mermod N, Tijan R (1990) Transcription factor AP-4 contains multiple dimerization domains that regulate dimer specificity. Genes Dev 4: 1741–1752

Jiao J, Chollet R (1991) Posttranslational regulation of phosphoenolpyruvate carboxylase in C_4 and crassulacean acid metabolism plants. Plant Physiol 95: 981–985

Kadonaga JT, Jones KA, Tijan R (1986) Promoter-specific activation of RNA polymerase II transcription by Sp1. Trends Biochem Sci 11: 20–33

Keeley JE, Busch G (1984) Carbon assimilation characteristics of the aquatic CAM plant *Isoetes howellii*. Plant Physiol 76: 525–530

Kluge M, Ting IP (1978) Crassulacean acid metabolism. Analysis of an ecological adaptation. Springer, Berlin Heidelberg New York

Köster S, Anderson JA (1989) The photosynthetic apparatus of C_3 and CAM-induced *Mesembryanthemum crystallinum* L. Photosynth Res 19: 251–264

Langdale JA, Taylor WC, Nelson T (1991) Cell-specific accumulation of maize phosphoenolpyruvate carboxylase is correlated with demethylation at a specific site > 3kb upstream of the gene. Mol Gen Genet 225: 49–55

Lee W, Mitchell P, Tijan R (1987) Purified transcription factor AP-1 interacts with TPA-inducible enhancer elements. Cell 49: 741–752
Lepiniec L, Keryer E, Philippe H, Gadal P, Cretin C (1993) *Sorghum* PEPC gene family: structure, function and molecular evolution. Plant Mol Biol 21: 487–502
Li Y, Sugiura M (1990) Three distinct ribonucleoproteins from tobacco chloroplasts: each contains a unique amino-terminal acidic domain and two ribonucleoprotein consensus motifs. EMBO J 9: 3059–3066
Lüttge U, Fischer E, Steudle E (1978) Membrane potential and salt distribution in epidermal bladders and photosynthetic tissue of *Mesembryanthemum crystallinum*. Plant Cell Environ 1: 121–129
Martin B, Nienhuis J, King G, Schaefer A (1989) Restriction fragment length polymorphisms associated with water use efficiency in tomato. Science 243: 1725–1728
McElwain EF, Bohnert HJ, Thomas JC (1992) Light moderates the induction of phospho*enol*pyruvate carboxylase by NaCl and abscisic acid in *Mesembryanthemum crystallinum*. Plant Physiol 99: 1261–1264
Meiners MS, Thomas JC, Bohnert HJ, Cushman JC (1991) Regeneration of multiple shoots and plants from *Mesembryanthemum crystallinum*. Plant Cell Rep 9: 563–566
Meyer G, Schmitt JM, Bohnert HJ (1990) Direct screening of a small genome: estimation of the magnitude of plant gene expression changes during adaptation to high salt. Mol Gen Genet 224: 347–356
Michalowski CB, Olson SW, Piepenbrock M, Schmitt JM, Bohnert HJ (1989a) Time course of mRNA induction elicited by salt stress in the common ice plant (*M. crystallinum*). Plant Physiol 89: 811–816
Michalowski CB, Schmitt JM, Bohnert HJ (1989b) Expression during salt stress and nucleotide sequence of cDNA for ferredoxin-NADP$^+$ reductase from *Mesembryanthemum crystallinum*. Plant Physiol 89: 817–823
Michalowski CB, DeRocher EJ, Bohnert HJ, Salvucci ME (1992) Phosphoribulokinase from ice plant: transcription, transcripts and protein expression during environmental stress. Photosynth Res 31: 127–138
Monson RK (1989) On the evolution of pathways resulting in C_4 photosynthesis and crassulacean acid metabolism (CAM). Adv Ecol Res 19: 57–110
Moore PD (1982) Evolution of photosynthetic pathways in flowering plants. Nature 295: 647–648
Osmond CB (1978) Crassulacean acid metabolism: a curiosity in context. Annu Rev Plant Physiol 29: 379–414
Ostrem JA, Vernon DM, Bohnert HJ (1990) Increased expression of a gene coding for NAD: glyceraldehyde-3-phosphate dehydrogenase during the transition from C_3 photosynthesis to crassulacean acid metabolism in *Mesembryanthemum crystallinum*. J Biol Chem 265: 3497–3502
Ottaviani E, Sari Gorla M, Pé E, Frova C (1991) Molecular markers (RFLPs and HSPs) for the genetic dissection of thermotolerance in maize. Theor Appl Genet 81: 713–719
Paul MJ, Cockburn W (1989) Pinitol, a compatible solute in *Mesembryanthemum crystallium* L. J Exp Bot 40: 1093–1098
Poethig RS (1990) Phase change and the regulation of shoot morphogenesis in plants. Science 250: 923–930
Schmitt JM, Piepenbrock M (1992) Regulation of phospho*enol*pyruvate carboxylase and crassulacean acid metabolism induction in *Mesembryanthemum crystallinum* L. by cytokinin. Plant Physiol 99: 1664–1669
Schmitt JM, Höfner R, Abou-Mandour AA, Vazquez-Moreno L, Bohnert HJ (1989) CAM-induction in *Mesembryanthemum crystallinum*: protein expression. In: Singhal GS (ed) Photosynthesis-molecular biology and bioenergetics. Springer, Berlin Heidelberg New York, pp 259–268
Sipes DL, Ting IP (1985) Crassulacean acid metabolism and crassulacean acid metabolism modifications in *Peperomia camptotricha*. Plant Physiol 77: 59–63
Steudle E, Lüttge U, Zimmermann U (1975) Water relations of the epidermal bladder cells of the halophytic species *Mesembryanthemum crystallinum*: direct measurements of hydrostatic pressure and hydraulic conductivity. Planta 126: 220–246

Sussex IM (1989) Developmental programing of the shoot meristem. Cell 56: 225–229

Swofford DL (1993) PAUP: Phylogenetic analysis using parsimony, version 3.1. Computer program distributed by the Illinois Natural History Survey, Champaign, Illinois

Tarczynski MC, Jensen RG, Bohnert HJ (1993) Stress protection in transgenic tobacco producing a putative osmoprotectant, mannitol. Science 259: 508–510

Teeri JA (1982) Photosynthetic variation in the Crassulaceae. In: Ting IP, Gibbs M (eds) Crassulacean acid metabolism. American Society of Plant Physiologists, Rockville, pp 244–259

Thomas JC, DeArmond RL, Bohnert HJ (1992a) The influence of NaCl on growth, proline and phosphoenolpyruvate carboxylase levels in *Mesembryanthemum crystallinum* suspension cultures. Plant Physiol 98: 626–631

Thomas JC, McElwain EF, Bohnert HJ (1992b) Convergent induction of osmotic stress-responses. Abscisic acid, cytokinin, and the effects of NaCl. Plant Physiol 100: 416–423

Ting IP (1985) Crassulacean acid metabolism. Annu Rev Plant Physiol 36: 595–622

Ting IP, Hanscom Z (1977) Induction of acid metabolism in *Portulacaria afra*. Plant Physiol 59: 511–514

Treichel S (1986) The influence of NaCl on pyrroline-5-carboxylate reductase in proline-accumulating cell suspension cultures of *Mesembryanthemum nodiflorum* and other halophytes. Physiol Plant 67: 173–181

Vernon DM, Bohnert HJ (1992a) A novel methyl transferase induced by osmotic stress in the facultative halophyte *Mesembryanthemum crystallinum*. EMBO J 11: 2077–2085

Vernon DM, Bohnert HJ (1992b) Increased expression of a myo-inositol methyl transferase in *Mesembryanthemum crystallinum* is part of a stress response distinct from crassulacean acid metabolism induction. Plant Physiol 99: 1695–1698

Vernon DM, Ostrem JA, Bohnert HJ (1993a) Stress perception and response in a facultative halophyte: the regulation of salinity-induced genes in *Mesembryanthemum crystallinum*. Plant Cell Environ 16: 437–444

Vernon DM, Tarczynski MC, Jensen RG, Bohnert HJ (1993b) Cyclitol production in transgenic tobacco: analysis of the biochemical pathway from myo-inositol to pinitol. Plant J 4: 199–205

Winter K (1985) Crassulacean acid metabolism. In: Barber J, Baker NR (eds) Photosynthetic mechanisms and the environment. Elsevier, Amsterdam, pp 329–387

Winter K, Gademann R (1991) Daily changes in CO_2 and water vapor exchange, chlorophyll fluorescence, and leaf water relations in the halophyte *Mesembryanthemum crystallinum* during the induction of crassulacean acid metabolism in response to high salinity. Plant Physiol 95: 768–776

Winter K, von Willert DJ (1972) NaCl-induzierter Crassulaceensäurestoffwechsel bei *Mesembryanthemum crystallinum*. Z Pflanzenphysiol 67: 166–170

Winter K, Ziegler H (1992) Induction of crassulacean acid metabolism in *Mesembryanthemum crystallinum* increases reproductive success under conditions of drought and salinity stress. Oecologia 92: 475–479

Winter K, Foster JG, Edwards GE, Holtum JAM (1982) Intracellular localization of enzymes of carbon metabolism in *Mesembryanthemum crystallinum* exhibiting C_3 photosynthetic characteristics or performing crassulacean acid metabolism. Plant Physiol 68: 300–307

10 Environmental Control of CAM Induction in *Mesembryanthemum crystallinum* – a Role for Cytokinin, Abscisic Acid and Jasmonate?

J.M. Schmitt, B. Fißlthaler, A. Sheriff, B. Lenz, M. Bäßler, and G. Meyer

10.1 Introduction

The common ice plant (*Mesembryanthemum crystallinum* L.) is a halophilic species with a growth optimum at 50 to 100 mM sodium chloride both for whole plants (Winter 1973a; Höfner 1986) and cell suspension cultures (Treichel et al. 1988). When plants are grown at high salt concentrations (200 to 500 mM), salt is accumulated in the leaves (Winter 1973a) and leaf osmotic pressure increases drastically (Heun et al. 1981). Turgor, however, only decreases transiently after the onset of stress, but recovers and eventually increases above pre-stress levels (Winter and Gademann 1991). In the presence of high salt, the mode of primary carbon fixation switches from C_3 to CAM (Winter and von Willert 1972).

This transition from C_3 to CAM is characterized by quantitative changes in the inventory of cellular enzymes (Table 10.1). In a few cases, among them PEPC (Höfner et al. 1987), PPDK and NADP-ME (Schmitt et al. 1989), it could be shown that the increases in apparent activity are correlated with enzyme protein as measured immunochemically.

C_3-type PEPC is different from CAM-type PEPC as indicated by the ratio of enzyme activity to precipitin area in the Mancini immunodiffusion test. From the reaction of identity in the Ouchterlony test it was concluded that the specific activity of the CAM enzyme is higher (Höfner et al. 1989). The C_3 and CAM enzymes are also distinguished by their apparent mobilities in SDS gels. The C_3 subunit with an apparent molecular mass of 110 kDa differs from a salt-induced subunit with a molecular mass of 100 kDa (Höfner et al. 1989). In gels of higher resolution, two more subunits may be seen whose role in metabolism is unclear (Thomas et al. 1992b). It is interesting to note that the actual function of CAM is not required for enzyme induction. PEPC and NADP-ME can be induced during growth of *M. crystallinum* in continuous light without a light/dark cycle, a condition under which CAM is not functioning (Cheng and Edwards 1991; see also Chap. 8)

Approximately 100 to 200 different species of mRNAs are increased in abundance under salt stress (Meyer et al. 1990). Only a few of these have been identified (Table 10.1). In the cases where transcript levels have been determined

---

Institut für Pflanzenphysiologie und Mikrobiologie, Freie Universität Berlin, Königin-Luise-Str. 12–16, D-14195 Berlin, Germany

Ecological Studies, Vol. 114
Winter/Smith (eds) Crassulacean Acid Metabolism

Table 10.1. Changes in enzyme, protein and mRNA levels in *Mesembryanthemum crystallinum* in response to salt stress

Protein	Gene	Copy number	Expression after salt treatment: Enzyme activity	Protein level	mRNA level	Reference
Alanine aminotransferase		n.d.	+	n.d.	n.d.	Holtum and Winter (1982)
Aspartate aminotransferase		n.d.	+	n.d.	n.d.	Holtum and Winter (1982)
Cytochrome *c* oxidase		n.d.	0	n.d.	n.d.	Winter et al. (1982)
Enolase		n.d.	+	n.d.	n.d.	Holtum and Winter (1982)
Ferredoxin NADP-oxidoreductase		1–3	n.d.	0	0	Michalowski et al. (1989)
Fructose 1,6-bisphosphatase		n.d.	0	n.d.	n.d.	Holtum and Winter (1982)
Germin-like protein		n.d.	n.d.	n.d.	+	Michalowski and Bohnert (1992)
Glucose 6-phosphate dehydrogenase		n.d.	0	n.d.	n.d.	Holtum and Winter (1982)
Glucosephosphate isomerase		n.d.	+	n.d.	n.d.	Holtum and Winter (1982)
Glutamate dehydrogenase		n.d.	+	n.d.	n.d.	Whitehouse et al. (1991)
Glutamine synthetase		n.d.	0	0, –[a]	n.d.	Whitehouse et al. (1991)
Glycolate oxidase		n.d.	0	n.d.	n.d.	Whitehouse et al. (1991)
Hexokinase		n.d.	0	n.d.	n.d.	Holtum and Winter (1982)
Hydroxypyruvate reductase		n.d.	0	n.d.	n.d.	Whitehouse et al. (1991)
Myo-inositol *O*-methyl transferase		n.d.	n.d.	+	+	Vernon and Bohnert (1992a) Vernon and Bohnert (1992b)
NAD-glyceraldehyde 3-phosphate dehydrogenase	*Gpd1*	1	+	n.d.	+	Holtum and Winter (1982) Ostrem et al. (1990)
NAD-malate dehydrogenase		n.d.	+	n.d.	n.d.	Holtum and Winter (1982)
NAD-malic enzyme		n.d.	+	n.d.	n.d.	Holtum and Winter (1982)
NADP-glyceraldehyde 3-phosphate dehydrogenase		n.d.	0	n.d.	n.d.	Holtum and Winter (1982)
NADP-malate dehydrogenase		n.d.	+	n.d.	n.d.	Holtum and Winter (1982)
NADP-malic enzyme	*Mod1*	>2	+	+	+	Cushman (1992) Schmitt et al. (1989) Holtum and Winter (1982)

PEPC						
	Ppc 1	1	+	+	+	Schmitt et al. (1988)
						Schmitt et al. (1989)
						Cushman et al. (1989)
	Ppc 2	1	n.d.	0	0	Cushman et al. (1989)
						Höfner et al. (1989)
Phosphofructokinase		n.d.	0	n.d.	n.d.	Holtum and Winter (1982)
Phosphoglycerate kinase		n.d.	+	n.d.	n.d.	Holtum and Winter (1982)
Phosphoglyceromutase		n.d.	+	n.d.	n.d.	Holtum and Winter (1982)
Phosphoribulokinase	*Prk*	n.d.		0	–	Michalowski et al. (1992)
PPDK	*Ppd 1*	1	n.d.	+	+	Schmitt et al. (1988)
						Fißlthaler et al. (1995)
Putative reverse transcriptase		1–2	n.d.	n.d.	–	G. Meyer (unpubl.)
Pyruvate kinase		n.d.	+	n.d.	n.d.	Holtum and Winter (1982)
Ribulose 1, 5-bisphosphate carboxylase/oxygenase	*RbcS1–6*	6[b]	0	0	0	Höfner et al. (1987)
						Saitou et al. (1991)
						DeRocher et al. (1987)
						DeRocher et al. (1991)
Tonoplast ATPase		n.d.	+	+[c]	n.d.	Bremberger and Lüttge (1992)
Tonoplast pyrophosphatase		n.d.	–	n.d.	n.d.	Bremberger and Lüttge (1992)

+/– indicates an at least twofold increase/decrease compared to control level; 0 indicates a less-than-twofold change relative to the control level; n.d. indicates not determined.

[a] Two isoforms of GS were detected, but the chloroplast isozyme was not detectable in CAM-type plants.

[b] Gene copy number for the nuclear-encoded small subunit.

[c] Two subunits of 27 and 31 kDa are more abundant in CAM plants.

by Northern-type hybridization, increased steady-state levels of mRNA correlated with increased enzyme activity and protein (Table 10.1). PEPC induction has been studied in some detail. During induction, the CAM-specific isoform (Höfner et al. 1989) is synthesized de novo (Höfner et al. 1987). Translation is directed by mRNA newly transcribed from a stress-specific PEPC isogene. Transcription of the C_3-specific isogene is essentially unchanged (Cushman et al. 1989). Distribution of the CAM and the C_3 function to two specific genes is, however, not a general pattern: for NAD-glyceraldehyde 3-phosphate dehydrogenase and PPDK, only one gene has been found per haploid genome (Table 10.1).

Under suitable conditions, enzyme induction in response to water stress can be a rapid process. PEPC mRNA is induced after only 2 h in excised, drought-stressed leaves (Schmitt 1990), and within 6 to 8 h in intact plants stressed by salt (Thomas et al. 1992b) or drought (Schmitt and Piepenbrock 1992b). The activities of PEPC and NADP-ME may increase within 24 h of salt treatment (Cheng and Edwards 1991). The rapid response to stress together with transcriptional regulation of CAM-specific enzymes suggests that the signal transduction chain may be short and amenable to molecular analysis. A nuclear protein designated PCAT-1 has been identified which is increased in abundance in leaf extracts from salt-stressed tissue. PCAT-1 binds to two distinct sites in the promoter region for the CAM-specific PEPC (Cushman and Bohnert 1992).

The role of gene expression in response to salinity has been reviewed recently (Bohnert et al. 1988, 1989a,b; Cushman et al. 1990a, 1993). In this chapter we focus on the two interrelated questions: (1) is CAM induction basically a developmentally programmed process which is only enhanced by salt stress, and (2) what is the possible role of plant growth regulators in the induction process?

10.2 The Concept of Stress

As can be inferred from the growth optimum of 100 mM NaCl for *M. crystallinum* (see Introduction), growth in the presence of salt is not necessarily stressful. The CAM pathway, which is thought to be a water-conserving adaptation to salt-induced water deficits, is minimal at 100 mM NaCl (Winter 1973b). If, by the criteria of optimal growth and CAM induction, plants grown at about 100 mM NaCl show a minimum reaction of strain (Levitt 1980), then both growth in the absence of salt and growth in higher concentrations of salt must be termed stress. There is little doubt that the no-salt stress and the high-salt stress lead to different types of strain. Both can, however, elicit the same effect – CAM induction. We therefore prefer the term well-watered rather than unstressed to indicate that plants were grown in the absence of added salt.

Another indicator of high-salt strain is the accumulation of compatible osmolytes like sugar alcohols (Paul and Cockburn 1989; Adams et al. 1992; Vernon and Bohnert 1992a) and proline (Heun et al. 1981; Demmig and Winter

1986; Thomas et al. 1992a). In excised, salt-treated leaves, proline accumulates rapidly, and proline levels are linearly correlated with chloride levels, both in salt-treated and in control leaves (Fig. 10.1).

10.3 Environmental or Developmental Control of CAM Induction?

10.3.1 CAM Induction in Well-Watered Plants

Induction of CAM-related enzymes and expression of CAM in well-watered plants have been observed in several laboratories (Winter 1973a,b; Rygol et al. 1986; Cushman et al. 1990b; Cheng and Edwards 1991; Piepenbrock and Schmitt 1991; Herppich et al. 1992). Under a given set of environmental conditions, induction occurs at a certain age. The age-related increase in CAM or in CAM-related enzyme activities has led to the hypothesis that this process is fully or partially under developmental control (Cushman et al. 1990b; Herppich et al. 1992).

A related but slightly different hypothesis states that CAM *inducibility* is under developmental control (Cushman et al. 1990b; Adams et al.1992; Herppich et al. 1992). It implies that the competence to switch to CAM is acquired at a certain developmental stage. When stressing plants of different age, it can indeed be observed that older plants induce PEPC more strongly and more rapidly (Cushman et al. 1990b; Piepenbrock and Schmitt 1991).

To be able to assess the possible contribution of developmental factors, it must be clear what is meant by the term development. Development is a programmed sequence of phenotypic changes, under temporal, spatial, and quantitative control

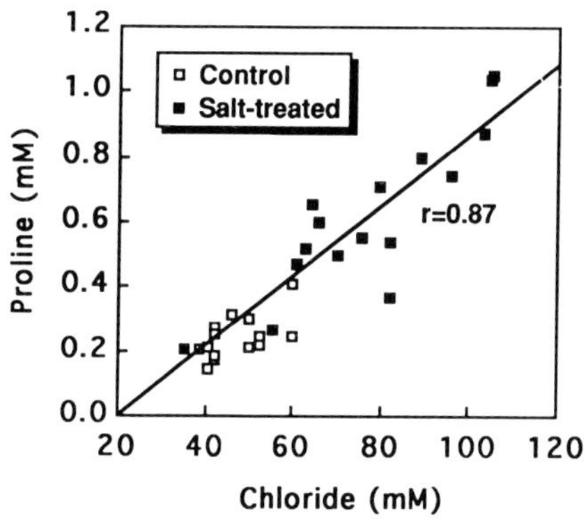

Fig. 10.1. Correlation of chloride and proline concentrations in the cell sap of salt-stressed leaves. Leaves from layer 3 (Piepenbrock and Schmitt 1991) of 7-week-old plants were excised in the morning and immersed with the petioles in 100, 200, 300, 400 and 500 mM NaCl, 1 mM $CaCl_2$ (closed symbols) as described previously (Schmitt 1990). Controls were the opposite leaves from the same plants left attached (open symbols). After 8 h, leaves were harvested by freezing in liquid nitrogen and proline was determined using a modification (Lenz 1992) of the method of Troll and Lindsley (1955). Chloride was measured with a chloride-sensitive electrode

which is irreversible, or at least difficult to reverse, under normal developmental conditions (Rieger et al. 1991). Increasing age is therefore not necessarily equivalent to development. In addition, other criteria must be met.

Four morphological stages of development are readily distinguished in *M. crystallinum* (see also Chap. 9). After germination, the main axis grows. Six or more leaf pairs are formed in sequence. The second phase is characterized by growth of axillary shoots on which smaller leaves develop. The leaves on the main axis lose turgor and show symptoms of senescence, beginning with the older and progressing to the younger leaves. Both processes change the weight ratio of stems to leaves. This ratio has been used as a quantitative indicator of developmental status (Winter et al. 1978). The next phase is characterized by anthesis and development of the seeds, and in the fourth and last phase, senescence and death occur. The life cycle of the plants under natural growth conditions, for example in Israel, is completed within about 7 months (Winter et al. 1978).

The plant age at which PEPC induction in well-watered plants is observed varies from 2 weeks under very long photoperiods (Cheng and Edwards 1991) to 8 to 9 weeks under more normal photoperiods (Cushman et al. 1990b). Plants which are 10 weeks of age or older may, however, show no signs of CAM. Ten-week-old plants grown in pots of 6.5 l have levels of PEPC mRNA typical of C_3 plants (Schmitt and Piepenbrock 1992a), and 11- to 12-week-old plants show no signs of CAM when grown in moderate light and high relative humidity (Winter 1973b).

Induction in well-watered plants is clearly related to plant age only under a given set of conditions: it is not age-related in the sense that induction occurs at a specific, constant time after germination. Genetic diversity is probably not responsible for these effects, or at least in some instances a genetic influence can be excluded (Winter 1973a; Winter 1973b; Schmitt and Piepenbrock 1992a). The wide variability in the absolute age at which induction occurs in the absence of salt stress strongly suggests that environmental conditions play a paramount role. In other words, the environment rather than chronological age determines the time of onset of CAM in well-watered plants.

One environmental variable – small soil volume – correlates well with early induction of PEPC mRNA in well-watered plants. Plants grown in small pots (0.16 l) are small in size and induce earlier than plants grown in larger pots up to 6.5 l in volume (Schmitt and Piepenbrock 1992a). PEPC mRNA induction is not correlated with the developmental status of the plants as measured by the weight ratio of leaves to stems, but rather with a decrease in leaf water content (Schmitt and Piepenbrock 1992a). While soil volume has a strong effect on growth and onset of induction, it has little effect on development of the plants. The light regime, however, is able to influence ontogeny. The onset of secondary growth and anthesis is earlier in plants grown in longer light periods. Flowering can be observed as soon as 5 weeks after germination when plants are grown in small pots under continuous light (Cheng and Edwards 1991; see also Chap. 8).

Could inducibility be developmentally controlled? We have tried to address this question by measuring induction of PEPC in very young plants. The youngest plants tested were 1 week old with a shoot fresh weight of 25 mg (median). They were able to induce PEPC transcripts after 8 h of wilting. Two-week-old seedlings with a shoot fresh weight of less then 1% of the mature plants (as grown in 0.7-l pots) were able to induce PEPC transcripts to 400% of the unstressed level in a 6-day stress period (Piepenbrock and Schmitt 1991). Freshly germinated seedling weighing 6.5 mg (median) have not been tested.

10.3.2 Relief from Stress

Developmentally regulated processes are controlled by a genetic programme. We perceive the results of the genetic programme as ontogeny. Ontogeny is directional. A seedling, for instance, will never shrink and dehydrate to become a seed again. If a given process is readily reversible as a consequence of the change in environmental conditions, developmental control over it is loose or absent.

Salt-treated *M. crystallinum* plants can tolerate a change back to high water potentials in the rooting medium (Winter 1974a; von Willert et al. 1976; Vernon et al. 1988). There is general agreement that PEPC transcript levels are decreased markedly within the first few hours after the change to low-salt conditions in intact plants (Vernon et al. 1988) or excised leaves (Schmitt 1990; Piepenbrock and Schmitt 1991; Schmitt and Piepenbrock 1992b). Likewise, PEPC activity (von Willert et al. 1976; Vernon et al. 1988) and protein (Vernon et al. 1988) are decreased. Whether nocturnal malate accumulation and other CAM features are maintained (Winter 1974a; Herppich et al. 1992) or reduced (Winter 1974b; A. Sheriff and J.M. Schmitt, unpubl.) seems to be dependent on experimental conditions.

We have compared the effect of different pretreatments on the down-regulation of PEPC mRNA upon rehydration. Five-week-old and 9-week-old plants were either well-watered or salt-treated. From these plants, leaves were excised and hydrated by immersing the petiole in 1 mM $CaCl_2$. After 8 h of rehydration, PEPC mRNA levels were very similar and comparable to C_3 levels, regardless of the pretreatment of the plants. The rapid and extensive reversibility of PEPC transcript accumulation by hydration of leaves – regardless of plant age – suggests that leaf water status rather than developmental status is the decisive regulatory factor.

We propose that well-watered old plants differ form young plants with respect to PEPC induction and inducibility mainly in the intensity and duration of previous water deficits. Mild water deficits, caused by diurnal transient decline in leaf water content (Heun et al. 1981; Winter and Gademann 1991), are sufficient to lead with time to induction in well-watered plants, and to intensify the response to salt stress in intact plants or wilting in detached leaves (Piepenbrock and Schmitt 1991). Diurnal leaf water deficits are dependent on environmental conditions like soil volume, light intensity and relative humidity. In the few cases where leaf water content of well-watered plants has been measured, water deficit was correlated with induction. The criterion to be met has been stated by Winter and Gademann: the age-related occurrence of CAM in *M. crystallinum* would be truly developmental if it can be demonstrated in the absence of any leaf water deficit (Winter and Gademann 1991).

10.3.3 Leaf Water Content

Wilting of leaves is a very effective way to increase PEPC mRNA abundance. Induction in detached, wilting leaves is triggered by minor water deficits. As little as 5% water loss leads to a near-maximal increase in PEPC transcripts (Piepenbrock and Schmitt 1991). It must be kept in mind that normal diurnal changes in leaf water content are often in the order of 15%, i.e. about three times the amount inducing PEPC mRNA in an excised leaf (Winter and Gademann 1991). This suggests that well-watered plants should undergo PEPC transcript induction during the day and transcript reduction due to rehydration at night (Vernon et al. 1988; Schmitt 1990; Winter and Gademann 1991). Diurnal changes in PEPC transcript abundance are, indeed, observed, but only in salt-stressed (Höfner et al. 1990) and not in well-watered plants (R. Höfner, pers. comm.).

The high sensitivity of excised leaves to even minor water deficits together with the absence of concomitant diurnal alterations in mRNA levels suggests that regulatory factors from other parts of the plant may exert a negative control on induction in the intact plant. These factors could originate from the rest of the shoot (e.g. the apex) or from the roots.

10.3.4 The Role of the Roots

Land plants have the problem of balancing carbon gain against water loss in a constantly changing environment. Stomatal movement and assimilate partitioning can only be regulated successfully if the water status of the plant and the availability of soil water can be sensed (Davies and Zhang 1991). It has been shown that stomatal closure may occur in maize plants growing in drying soil even before leaf water relations change appreciably (Blackman and Davies 1985). This suggests that non-hydraulic signals from the roots may regulate stomatal movement.

Little is known about the possible functions of roots in CAM induction. Impaired root function will lead to a restricted transpiration stream. When roots are cooled and shoots are maintained at normal temperature, leaves suffer water deficit and both PEPC and CAM are induced. Similar effects can be observed when the nutrient solution is not aerated (Winter 1974c). In such experiments, the transport of water and solutes is inhibited but apparently not abolished. Similarly, when *M. crystallinum* plants are grown in limited soil volume, growth is severely restricted, leaf water content decreases and CAM is induced (Cheng and Edwards 1991; Schmitt and Piepenbrock 1992a).

A direct comparison between excised leaves and whole plants suggests that other parts of the plant modulate induction in a given leaf. When whole plants are wilted, induction of PEPC transcripts is less pronounced than in excised leaves, although the water loss is slightly more severe in the whole plant (Schmitt and Piepenbrock 1992b). The stronger induction in excised leaves is not caused by wounding stress, since excised and therefore wounded leaves show strong decreases in PEPC transcripts when they are rehydrated (Schmitt 1990; Piepenbrock and

Schmitt 1991; Schmitt and Piepenbrock 1992b). These experiments show that the roots are not required for PEPC mRNA induction. Nonetheless, it is clear that the role of the roots in CAM induction requires more attention. In particular, we need more information on whether root-derived signals (other than hydraulic ones) are involved in the regulation of leaf metabolism.

10.4 Modulation of PEPC and CAM Induction by Growth Regulators

10.4.1 Abscisic Acid (ABA)

The CAM-related enzymes PEPC and NADP-ME, as well as CAM as indicated by nocturnal malate accumulation, are induced in well-watered plants by treatment with up to 10 μM ABA (Chu et al. 1990). ABA-induced expression of CAM does not require a decrease in leaf water content (Chu et al. 1990). This suggests that ABA may have a primary effect on gene expression, perhaps by stimulation of transcription, which is afforded by ABA-responsive elements (ABREs) in the promoter regions of plant genes (Katagiri and Chua 1992). Sequences similar to ABREs can be found in the promoters of the stress-specific PEPC (Cushman and Bohnert 1992) and PPDK genes (Fig. 10.2).

Induction by ABA is influenced by environmental conditions. Exogenously added ABA (10 μM to the hydroponic culture medium) fails to induce PEPC mRNA under low light (Thomas et al. 1992b), but stimulates PEPC mRNA and protein accumulation under high light (McElwain et al. 1992). Leaves from salt-stressed *M. crystallinum* plants accumulate ABA (Thomas et al. 1992b). When ABA accumulation is suppressed (although not totally inhibited) by preincubation

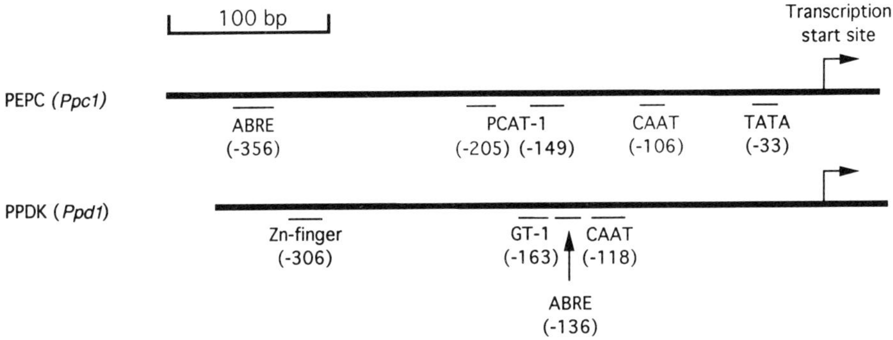

Fig. 10.2. Comparison of the promoter regions of the genes coding for the salt-induced isoform of PEPC and PPDK. The PEPC promoter has been redrawn from data in Cushman and Bohnert (1992). Putative regulatory elements (Katagiri and Chua 1992) are indicated. *ABRE* ABA-responsive element; *CAAT* and *TATA* sequences for initiation by RNA polymerase II; *GT-1* light-responsive elements; *PCAT-1* binding site for a nuclear protein (Cushman and Bohnert 1992); *Zn-finger* binding site for a zinc-finger protein

with the inhibitor Lovastatin and the plants are challenged with salt, both PEPC transcripts and proline are accumulated (Thomas et al. 1992b).

10.4.2 Cytokinin

Very little is known about the role of cytokinin in the regulation of gene expression in plants during normal vegetative growth. In general, exogenous growth regulators will only cause a response if the tissue under study is not saturated by endogenous phytohormones. Cytokinin-induced changes in mRNA levels have therefore been measured mainly in hormone-depleted systems like excised leaves (Lerbs et al. 1984; Sugiharto et al. 1992), cultured cells (Teyssendier de la Serve et al. 1985) or in cytokinin-dependent intact plants grown heterotrophically in the dark (Flores and Tobin 1986).

The bulk of cytokinin is synthesized in the roots (Goodwin et al. 1978) and transported to the shoots, and thus this hormone could be a candidate for a possible signal substance mediating root-shoot communication under water shortage. An early observation suggests that a decreased cytokinin flow to the shoot may indeed be a response to drought stress. Cytokinin activity in xylem exudates from water-stressed sunflower plants is reduced in comparison with the exudate from well-watered controls (Itai and Vaadia 1965). Cytokinin could thus fulfill the role of a root-derived non-hydraulic signal molecule, indicating water availability in the soil.

Two groups have measured the effect of exogenously added cytokinin on CAM induction. As described below, the results depend strongly on the organ – shoot or roots – to which the growth regulator is applied.

10.4.2.1 Cytokinin Treatment of Shoots

In salt-treated plants or wilting, excised leaves, exogenous cytokinin suppresses PEPC transcript accumulation when sprayed onto the leaves. Cytokinin (6-benzylaminopurine; BA) used in concentrations from 5 to 500 μM (1) inhibits the up-regulation of PEPC transcripts, enzyme activity and CAM induction in salt-stressed intact plants, (2) inhibits the accumulation of PEPC mRNA in leaves from well-watered plants, (3) down-regulates PEPC transcripts within 8 h in pre-stressed, intact plants after a single spraying of an individual leaf, (4) inhibits accumulation of PEPC transcripts in excised, wilting leaves, and (5) accelerates the net decrease of PEPC transcripts in excised leaves from pre-stressed plants under rehydration conditions. Water content in BA-treated leaves is decreased, clearly indicating that the down-regulation of PEPC and CAM is not a secondary stress effect (Schmitt and Piepenbrock 1992b).

These data are consistent with a negative regulatory function of root-derived cytokinin on CAM-specific PEPC gene expression in leaves. Under favourable soil water conditions, cytokinin would be transported to the leaves via the transpiration stream and suppress PEPC transcript accumulation by overriding daily transient conditions of leaf water deficits (Winter and Gademann 1991), which in an isolated leaf would be sufficient to cause induction (Piepenbrock and Schmitt 1991). Limited water in the soil would lead to a decreased transport of the putative regulator to the shoot and thereby trigger induction. Cytokinin can decrease PEPC transcripts within 8 h in the intact plant, even when water stress is not relieved (Schmitt and Piepenbrock 1992b). It is, however, not clear whether cytokinin is a necessary element in the signal chain leading to the down-regulation of PEPC. The finding that detached, rehydrating leaves can decrease PEPC mRNA in the absence of added cytokinin (Schmitt 1990; Piepenbrock and Schmitt 1991; Schmitt and Piepenbrock 1992b) would argue against such an exclusive function if one assumes that the bulk of cytokinin is synthesized in the roots (Goodwin et al. 1978).

10.4.2.2 Cytokinin Treatment of Roots

When applied to the hydroponic culture solution, BA induces PEPC protein and mRNA. Levels of endogenous zeatin and zeatin riboside were not significantly changed by salt treatment of 6-week-old plants (Thomas et al. 1992b). Other cytokinins were not measured. It is interesting to note that mRNA levels for PEPC in the C_4 plant maize are also increased by cytokinin when detached leaves derived from N-starved plants are fed with nitrate and cytokinin (Sugiharto et al. 1992).

There is the question why the same growth regulator can have opposing effects depending on the mode of application. The kinetics of the process may be of significance. BA applied to the leaves decreases PEPC transcripts within 4 h (Schmitt and Piepenbrock 1992b). A strong inducing effect of BA applied to the roots is seen after 7 days, whereas the first weak induction of PEPC is seen after 3 days (Thomas et al. 1992b). The 3- to 7-day period leaves sufficient time for secondary effects like development of leaf water deficits as a consequence of a cytokinin-induced inhibition of root growth (Su and Howell 1992). Furthermore, preliminary results indicate that 10 μM BA strongly stimulates microbial growth in – and decreases the pH of – the nutrient solution (Piepenbrock and Schmitt, unpubl.). Accumulation of proline in response to root-supplied cytokinin (Thomas et al. 1992b) suggests that water stress might be involved in the reaction. Whatever the cause of the differing effects, the need for further experimentation is obvious.

10.4.3 Jasmonate

Jasmonates are natural growth regulators which – when added externally – influence numerous processes in plants, for example the reactions to wounding, senescence and osmotic stress. Proteinase inhibitors, thionins and other proteins of unknown function are induced by jasmonate (see Sembdner and Parthier 1993 for a recent review). By comparing protein synthesis in situ and in an mRNA-primed wheat-germ system, it has been found that (±)methyl jasmonate

(JA-Me) apparently decreases the synthesis of many "housekeeping" proteins whose mRNA levels in the cell are essentially unchanged (Mueller-Uri et al. 1988).

Preliminary results from our laboratory indicate that jasmonate down-regulates PEPC and CAM. In salt-stressed plants, exogenous JA-Me sprayed daily at concentrations from 0.1 to 100 μM inhibited the up-regulation of PEPC transcripts, enzyme activity and CAM induction.

There is no evidence that the down-modulation of CAM and the preceding PEPC induction are caused by indirect effects, i.e. by preventing leaf water loss which is known to precede PEPC mRNA induction in well-watered (Schmitt and Piepenbrock 1992b) and CAM induction in salt-stressed plants (Winter and Gademann 1991). This is in accordance with the result that jasmonate is not a short-term regulator of stomata (Horton 1991). Our results indicate that for the period of 6 days water content of leaves is hardly influenced by concentrations of JA-Me that strongly modulate PEPC gene expression. If anything, a slight increase in leaf water content was seen in JA-Me treated stressed plants. This increase is too small to explain a down-modulation of CAM induction and its partial reactions (Winter and Gademann 1991; Schmitt and Piepenbrock 1992a).

JA-Me also accelerated the reduction in PEPC transcripts in excised leaves derived from pre-stressed plants under rehydration conditions. The levels of mRNA specific for the small subunit of ribulose 1,5-bisphosphate carboxylase (*RbcS*) used as a control were slightly increased in JA-Me-treated plants, but were strongly decreased in JA-Me-treated excised leaves undergoing rehydration.

Inhibition of the induction of PEPC mRNA, enzyme activity and CAM by JA-Me in salt-stressed intact plants is strongly reminiscent of the similar effects of shoot-supplied cytokinin on PEPC and CAM induction (Schmitt and Piepenbrock 1992b). This was surprising, since JA-Me is considered to be a cytokinin antagonist in several experimental systems (Sembdner and Parthier 1993). We therefore investigated whether BA given in addition to JA-Me would lead to a stronger effect. Moderate concentrations of each of the growth regulators sprayed alone to salt-stressed plants resulted, as expected, in a mild attenuation of induction at the levels of mRNA, PEPC activity and CAM. When both growth regulators were applied in combination, the effect was much stronger. Acid fluctuation and PEPC mRNA levels were comparable to the unstressed controls in growth-regulator-treated, salt-stressed plants, while the ratio fresh weight to dry weight was decreased to 20, a value typical of salt-stressed plants. Similar additive effects of BA and JA-Me were seen in drought-stressed leaves and in rehydrated leaves derived from pre-stressed plants.

10.4.4 Combinations of Growth Regulators

In *M. crystallinum*, the timing of the switch from C_3 to CAM must be carefully controlled to optimize growth. Too early a switch would burden the plant with a growth handicap, while too late a switch could mean death by desiccation.

Positive (ABA) and negative (cytokinin) control of PEPC by phytohormones whose levels may be controlled by water status (Morgan 1990) would enable regulation of PEPC enzyme levels in opposite directions. A finely regulated PEPC level by hormones responsive to water status would be consistent with the view that induction is tightly controlled and regulated predominantly by environmental factors.

We have provided evidence that – besides cytokinin and ABA – jasmonate as a third growth regulator may be implicated in the modulation of CAM induction. Other growth regulators may enhance or suppress the jasmonate effects. As a general rule, jasmonate and ABA act similarly, whereas jasmonate and cytokinin act antagonistically (Sembdner and Parthier 1993). ABA is known to induce both PEPC enzyme and CAM (Chu et al. 1990). Contrary to expectation, JA-Me was found to influence these processes in the opposite direction, acting as a down-modulator of PEPC mRNA, protein and enzyme accumulation and suppressing CAM induction. Moreover, the rule of antagonistic action between cytokinin and jasmonate is also violated in *M. crystallinum*. The two hormones were found to have similar and in certain situations even additive effects, showing that it is too early for generalizations.

The possible role of growth regulators in CAM induction has only been recognized relatively recently. We are still far from understanding their putative roles in the signal transduction chain. It will be interesting to determine whether internal growth regulators change in concentration when plants or leaves are stressed or rehydrated. An exciting possibility will be to increase the internal concentration of cytokinin in transgenic plants carrying a gene responsible for cytokinin synthesis under an inducible promoter (Schmülling et al. 1989; Gatz et al. 1991; Smart et al. 1991).

Acknowledgements. Our work was supported by the Deutsche Forschungsgemeinschaft. Some of the experiments of G.M. were done in the laboratories of Dr. Ulrich Heber (Würzburg) and Dr. Hans-Jürgen Bohnert (Tucson), whose support is gratefully acknowledged. B.F. and M.B. were recipients of stipends from the Studienstiftung des Deutschen Volkes. B.F. was supported by FNK der Freien Universität Berlin.

References

Adams P, Thomas JC, Vernon DM, Bohnert HJ, Jensen RG (1992) Distinct cellular and organismic responses to salt stress. Plant Cell Physiol 33: 1215–1223

Blackman PG, Davies WJ (1985) Root to shoot communication in maize plants and the effects of soil drying. J Exp Bot 36:39–48

Bohnert HJ, Ostrem JA, Cushman JC, Michalowski CB, Rickers J, Meyer G, DeRocher EJ, Vernon DM, Krueger M, Vazquez-Moreno L, Velten J, Höfner R, Schmitt JM (1988) *Mesembryanthemum crystallinum*: a higher plant model for the study of environmentally induced changes in gene expression. Plant Mol Biol Rep 6: 10–28

Bohnert HJ, Cushman JC, Meyer G, Ostrem JA (1989a) Changes in gene expression in response to salt stress. In: Tazawa M (ed) Plant water relations and growth under stress. Myu KK Press, Osaka, pp 143–150

Bohnert HJ, Ostrem JA, Schmitt JM (1989b) Changes in gene expression elicited by salt stress in *Mesembryanthemum crystallinum*. In: Cherry JH (ed) Environmental stress in plants – biochemical and physiological mechanisms. Springer, Berlin Heidelberg New York, pp 159–171

Bremberger C, Lüttge U (1992) Dynamics of tonoplast proton pumps and other tonoplast proteins of *Mesembryanthemum crystallinum* L. during the induction of crassulacean acid metabolism. Planta 188: 575–580

Cheng SH, Edwards GE (1991) Influence of long photoperiods on plant development and expression of crassulacean acid metabolism in *Mesembryanthemum crystallinum*. Plant Cell Environ 14: 271–278

Chu C, Dai Z, Ku MSB, Edwards GE (1990) Induction of crassulacean acid metabolism in the facultative halophyte *Mesembryanthemum crystallinum* by abscisic acid. Plant Physiol 93: 1253–1260

Cushman JC (1992) Characterization and expression of NADP-malic enzyme cDNA induced by salt stress from the facultative crassulacean acid metabolism plant, *Mesembryanthemum crystallinum*. Eur J Biochem 208: 259–266

Cushman JC, Bohnert HJ (1992) Salt stress alters A/T-rich DNA-binding factor interactions within the phosphoenolpyruvate carboxylase promoter from *Mesembryanthemum crystallinum*. Plant Mol Biol 20: 411–424

Cushman JC, Meyer G, Michalowski CB, Schmitt JM, Bohnert HJ (1989) Salt stress leads to differential expression of two isogenes of phospho*enol*pyruvate carboxylase during CAM induction in the common ice plant. Plant Cell 1: 715–725

Cushman JC, DeRocher EJ, Bohnert HJ (1990a) Gene expression during adaptation to salt stress. In: Kattermann FJ (ed) Environmental injury to plants. Academic Press, New York, pp 173–203

Cushman JC, Michalowski CB, Bohnert HJ (1990b) Developmental control of CAM inducibility by salt stress in the common ice plant. Plant Physiol 94: 1173–1142

Cushman JC, Vernon DM, Bohnert HJ (1993) ABA and the transcriptional control of CAM induction during salt stress in the common ice plant. In: Verma DPS (ed) Control of plant gene expression. CRC Press, Boca Raton, pp 287–300

Davies WJ, Zhang JH (1991) Root signals and the regulation of growth and development of plants in drying soil. Annu Rev Plant Physiol Plant Mol Biol 42: 55–76

Demmig B, Winter K (1986) Sodium, potassium, chloride and proline concentrations of chloroplasts isolated from a halophyte, *Mesembryanthemum crystallinum* L. Planta 168: 421–426

DeRocher EJ, Michalowski CB, Bohnert HJ (1991) cDNA sequences for transcripts of the ribulose-1,5-bisphosphate carboxylase/oxygenase small subunit gene family of *Mesembryanthemum crystallinum*. Plant Physiol 95: 976–978

DeRocher EJ, Ramage RT, Michalowski CB, Bohnert HJ (1987) Nucleotide sequence of a cDNA encoding rbcS from the desert plant *Mesembryanthemum crystallinum*. Nucleic Acids Res. 15: 6301

Fißlthaler B, Meyer G, Bohnert HJ, Schmitt JM (1995) Age-dependent induction of pyruvate, orthophosphate dikinase in *Mesembryanthemum crystallinum* L. Planta 196: 492–500

Flores S, Tobin EM (1986) Benzyladenine modulation of the expression of two genes for nuclear-encoded chloroplast proteins in *Lemna gibba*: apparent post-transcriptional regulation. Planta 168: 340–349

Gatz C, Kaiser A, Wendenburg R (1991) Regulation of a modified CaMV 35S promoter by the Tn10-encoded Tet repressor in transgenic tobacco. Mol Gen Genet 227:229–237

Goodwin PB, Gollnow Bl, Letham DS (1978) Phytohormones and growth correlations. In: Letham DS, Goodwin PB, Higgins TJV (eds) Phytohormones and related compounds–a comprehensive treatise, vol II. Elsevier, Amsterdam, pp 215–249

Herppich W, Herppich M, von Willert DJ (1992) The irreversible C_3 to CAM shift in well-watered and salt-stressed plants of *Mesembryanthemum crystallinum* is under strict ontogenetic control. Bot Acta 105: 34–40

Heun AM, Gorham J, Lüttge U, Wyn Jones RG (1981) Changes of water-relation characteristics and levels of organic cytoplasmic solutes during salinity induced transition of *Mesembryanthemum crystallinum* from C_3-photosynthesis to crassulacean acid metabolism. Oecologia 50: 66–72

Höfner R (1986) Phospho*enol*pyruvatcarboxylase als Produkt der in vivo und in vitro Proteinbiosynthese von *Mesembryanthemum crystallinum* L. bei Induktion des Crassulaceensäurestoffwechsels. Diploma Thesis, Universität Würzburg, Würzburg

Höfner R, Vazquez-Moreno L, Winter K, Bohnert HJ, Schmitt JM (1987) Induction of crassulacean acid metabolism in *Mesembryanthemum crystallinum* by high salinity: mass increase and de novo synthesis of PEP-carboxylase. Plant Physiol 83: 915–919

Höfner R, Vazquez-Moreno L, Abou-Mandour AA, Bohnert HJ, Schmitt JM (1989) Two isoforms of phospho*enol*pyruvate carboxylase in the facultative CAM plant *Mesembryanthemum crystallinum*. Plant Physiol Biochem 27: 803–810

Höfner R, Fißlthaler B, Piepenbrock M, Schmitt JM (1990) Circadian mRNA fluctuations in salt-stressed *Mesembryanthemum crystallinum* L. In: Hoffmann P, Rank B (eds) Photosynthese–die Spezifik des pflanzlichen Energiestoffwechsels. Humboldt-Universität, Berlin, pp 33–41

Holtum JAM, Winter K (1982) Activity of enzymes of carbon metabolism during the induction of crassulacean acid metabolism in *Mesembryanthemum crystallinum* L. Planta 155: 8–16

Horton RF (1991) Methyl jasmonate and transpiration in barley. Plant Physiol 96: 1376–1378

Itai C, Vaadia Y (1965) Kinetin-like activity in root exudate of water-stressed sunflower plants. Physiol Plant 18: 941–944

Katagiri F, Chua NH (1992) Plant transcription factors – present knowledge and future challenges. Trends Genet 8: 22–27

Lenz B (1992) Prolinakkumulation und Spiegel von CAM-spezifischen mRNAs bei *Mesembryanthemum crystallinum*. Diploma Thesis, Freie Universität, Berlin

Lerbs S, Lerbs W, Klyachko NL, Romanko EG, Kulaeva ON, Wollgiehn R, Parthier B (1984) Gene expression in cytokinin- and light-mediated plastogenesis of *Cucurbita* cotyledons: ribulose-1,5-bisphosphate carboxylase/oxygenase. Planta 162: 289–298

Levitt J (1980) Responses of plants to environmental stresses. Academic Press, Orlando

McElwain EF, Bohnert HJ, Thomas JC (1992) Light moderates the induction of phospho*enol*-pyruvate carboxylase by NaCl and abscisic acid in *Mesembryanthemum crystallinum*. Plant Physiol 99: 1261–1264

Meyer G, Schmitt JM, Bohnert HJ (1990) Direct screening of a small genome – estimation of the magnitude of plant gene expression changes during adaptation to high salt. Mol Gen Genet 224: 347–356

Michalowski CB, Bohnert HJ (1992) Nucleotide sequence of a root-specific transcript encoding a germin-like protein from the halophyte *Mesembryanthemum crystallinum*. Plant Physiol 100: 537–538

Michalowski CB, Schmitt JM, Bohnert HJ (1989) Expression during salt stress and nucleotide sequence of cDNA for ferredoxin-$NADP^+$ reductase from *Mesembryanthemum crystallinum*. Plant Physiol 89: 817–822

Michalowski CB, DeRocher EJ, Bohnert HJ, Salvucci ME (1992) Phosphoribulokinase from ice plant–transcription, transcripts and protein expression during environmental stress. Photosynth Res 31: 127–138

Morgan PW (1990) Effects of abiotic stresses on plant hormone systems. In: Alscher RG, Cumming JR (eds) Stress responses in plants: adaptation and acclimation mechanisms. Wiley-Liss, New York, pp 113–146

Mueller-Uri F, Parthier B, Nover L (1988) Jasmonate-induced alteration of gene expression in barley leaf segments analyzed by in-vivo and in-vitro protein synthesis. Planta 176: 241–247

Ostrem JA, Vernon DM, Bohnert HJ (1990) Increased expression of a gene coding for NAD-glyceraldehyde-3-phosphate dehydrogenase during the transition from C_3 photosynthesis to crassulacean acid metabolism in *Mesembryanthemum crystallinum*. J Biol Chem 265: 3497–3502

Paul MJ, Cockburn W (1989) Pinitol, a compatible solute in *Mesembryanthemum crystallinum* L.? J Exp Bot 40: 1093–1098

Piepenbrock M, Schmitt JM (1991) Environmental control of phospho*enol*pyruvate carboxylase induction in mature *Mesembryanthemum crystallinum* L. Plant Physiol 97: 998–1003

Rieger R, Michaelis A, Green MM (1991) Glossary of genetics. Springer, Berlin Heidelberg New York

Rygol J, Büchner K-H, Winter K, Zimmermann U (1986) Day/night variations in turgor pressure in individual cells of *Mesembryanthemum crystallinum* L. Oecologia 69: 171–175

Saitou K, Nakamura Y, Kawamitsu Y, Matsuoka M, Samejima M, Agata W (1991) Changes in activities and levels of pyruvate, orthophosphate dikinase with induction of crassulacean acid metabolism in *Mesembryanthemum crystallinum* L. Jap J Crop Sci 60: 146–152

Schmitt JM (1990) Rapid concentration changes of phospho*enol*pyruvate carboxylase mRNA in detached leaves of *Mesembryanthemum crystallinum* L. in response to wilting and rehydration. Plant Cell Environ 13: 845–850

Schmitt JM, Piepenbrock M (1992a) Induction of mRNA for phospho*enol*pyruvate carboxylase is correlated with a decrease in shoot water content in well-irrigated *Mesembryanthemum crystallinum*. Plant Physiol 99: 759–761

Schmitt JM, Piepenbrock M (1992b) Regulation of phospho*enol*pyruvate carboxylase and crassulacean acid metabolism induction in *Mesembryanthemum crystallinum* L. by cytokinin: modulation of leaf gene expression by roots? Plant Physiol 99: 1664–1669

Schmitt JM, Michalowski CB, Bohnert HJ (1988) Gene expression during CAM induction under salt stress in *Mesembryanthemum*: cDNA library and increased levels of mRNA for phospho*enol*pyruvate carboxylase and pyruvate orthophosphate dikinase. Photosynth Res 17: 159–171

Schmitt JM, Höfner R, Abou-Mandour AA, Vazquez-Moreno L, Bohnert HJ (1989) CAM induction in *Mesembryanthemum crystallinum*: protein expression. In: Singhal GS (ed) Photosynthesis – molecular biology and bioenergetics. Springer, Berlin Heidelberg New York, pp 259–268

Schmülling T, Beinsberger S, De Greef J, Schell J, Van Onckelen H, Spena A (1989) Construction of a heat-inducible chimaeric gene to increase the cytokinin content in transgenic plant tissue. FEBS Lett 249: 401–406

Sembdner G, Parthier B (1993) Biochemistry, physiological and molecular actions of jasmonates. Annu Rev Plant Physiol Plant Mol Biol 44: 569–589

Smart CM, Scofield SR, Bevan MW, Dyer TA (1991) Delayed leaf senescence in tobacco plants transformed with *tmr*, a gene for cytokinin production in *Agrobacterium*. Plant Cell 3: 647–656

Su WP, Howell SH (1992) A single genetic locus, ckr1, defines *Arabidopsis* mutants in which root growth is resistant to low concentrations of cytokinin. Plant Physiol 99: 1569–1574

Sugiharto B, Burnell JN, Sugiyama T (1992) Cytokinin is required to induce the nitrogen-dependent accumulation of mRNAs for phospho*enol*pyruvate carboxylase and carbonic anhydrase in detached maize leaves. Plant Physiol 100: 153–156

Teyssendier de la Serve B, Axelos M, Péaud-Lenoël C (1985) Cytokinins modulate the expression of genes encoding the protein of light-harvesting chlorophyll *a*/*b* complex. Plant Mol Biol 5: 155–163

Thomas JC, De Armond RL, Bohnert HJ (1992a) Influence of NaCl on growth, proline, and phospho*enol*pyruvate carboxylase levels in *Mesembryanthemum crystallinum* suspension cultures. Plant Physiol 98: 626–631

Thomas JC, McElwain EF, Bohnert HJ (1992b) Convergent induction of osmotic stress-responses. Abscisic acid, cytokinin, and the effects of NaCl. Plant Physiol 100: 416–423

Treichel S, Hettfleisch H, Eilhardt S, Faist K, Kluge M (1988) A possible induction of CAM by NaCl-stress in heterotrophic cell suspension cultures of *Mesembryanthemum crystallinum*. J Plant Physiol 133: 419–424

Troll W, Lindsley J (1955) A photometric method for the determination of proline. J Biol Chem 215: 655–660

Vernon DM, Bohnert HJ (1992a) Increased expression of a myo-inositol methyl transferase in *Mesembryanthemum crystallinum* is part of a stress response distinct from crassulacean acid metabolism induction. Plant Physiol 99: 1695–1698

Vernon DM, Bohnert HJ (1992b) A novel methyl transferase induced by osmotic stress in the facultative halophyte *Mesembryanthemum crystallinum*. EMBO J 11: 2077–2085

Vernon DM, Ostrem JA, Schmitt JM, Bohnert HJ (1988) PEPCase transcript levels in *Mesembryanthemum crystallinum* decline rapidly upon relief from salt stress. Plant Physiol 86: 1002–1004

von Willert DJ, Treichel S, Kirst GO, Curdts E (1976) Environmentally controlled changes of phospho*enol*pyruvate carboxylases in *Mesembryanthemum*. Phytochemistry 15: 1435–1436

Whitehouse DG, Rogers WJ, Tobin AK (1991) Photorespiratory enzyme activities in C_3 and CAM forms of the facultative CAM plant, *Mesembryanthemum crystallinum* L. J Exp Bot 42: 485–492

Winter K (1973a) Zum Problem der Ausbildung des Crassulaceensäurestoffwechsels bei *Mesembryanthemum crystallinum* unter NaCl-Einfluß. Planta 109: 135–145

Winter K (1973b) CO_2 Fixierungsreaktionen bei der Salzpflanze *Mesembryanthemum crystallinum* unter variierten Außenbedingungen. Planta 114: 75–85

Winter K (1974a) CO_2-Gaswechsel von an hohe Salinität adaptiertem *Mesembryanthemum crystallinum* bei Rückführung in glykisches Anzuchtmedium. Ber Dtsch Bot Ges 86: 467–476

Winter K (1974b) NaCl-induzierter Crassulaceen-Säurestoffwechsel bei der Salzpflanze *Mesembryanthemum crystallinum*. Abhängigkeit des CO_2-Gaswechsels von der Tag/Nacht-Temperatur und von der Wasserversorgung der Pflanzen. Oecologia 15: 383–392

Winter K (1974c) Einfluß von Wasserstreß auf die Aktivität der Phosphoenolpyruvat-Carboxylase bei *Mesembryanthemum crystallinum* (L.). Planta 121: 147–153

Winter K, Gademann R (1991) Daily changes in CO_2 and water vapor exchange, chlorophyll fluorescence, and leaf water relations in the halophyte *Mesembryanthemum crystallinum* during the induction of crassulacean acid metabolism in response to high NaCl salinity. Plant Physiol 95: 768–776

Winter K, von Willert DJ (1972) NaCl-induzierter Crassulaceensäurestoffwechsel bei *Mesembryanthemum crystallinum*. Z Pflanzenphysiol 67: 166–170

Winter K, Lüttge U, Winter E, Troughton JH (1978) Seasonal shift from C_3 photosynthesis to crassulacean acid metabolism in *Mesembryanthemum crystallinum* growing in its natural environment. Oecologia 34: 225–237

Winter K, Foster JG, Edwards GE, Holtum JAM (1982) Intracellular localization of enzymes of carbon metabolism in *Mesembryanthemum crystallinum* exhibiting C_3 photosynthetic characteristics or performing crassulacean acid metabolism. Plant Physiol 69: 300–307

11 Regulation of Crassulacean Acid Metabolism by Water Status in the C_3/CAM Intermediate *Sedum telephium*

N. Smirnoff

11.1 Introduction

Sedum telephium L. (Crassulaceae) is a member of a genus whose species show great variability in the expression of CAM. Some species carry out full CAM, others always have C_3 photosynthesis, while yet others show intermediate carbon-isotope composition along with fluctuations in malic acid, with or without some degree of nocturnal stomatal opening (Bender et al. 1973; Knopf and Kluge 1979; Teeri 1982; Pilon-Smits et al. 1991; see also Chap. 23). The last group is of interest ecologically and physiologically because there is a large degree of environmental control over the patterns of gas exchange and the extent of malate fluctuation which results, as in the case of *S. telephium*, in plants which exhibit CAM, CAM-cycling and CAM-idling and C_3 photosynthesis (Lee and Griffiths 1987). These plants can be termed C_3/CAM intermediates. Soil and plant water deficits are particularly important in causing a switch between C_3 and CAM in this group of species of which *S. telephium* has been most widely studied. As well as their great interest in terms of evolution and plant adaptation to the environment, C_3/CAM intermediates offer a model system that can be used to investigate metabolic regulation and gene expression in relation to environmental factors.

In this chapter the relationship between CAM and plant water status in *S. telephium* and other C_3/CAM intermediate *Sedum* species is described. It will be shown that this plant is "poised" to carry out CAM, and the capacity is not expressed when the plant is well-watered. The ability to switch between C_3 photosynthesis and CAM may be of particular importance to a temperate plant such as *S. telephium* which lives in shallow soil where water availability can fluctuate rapidly.

11.2 Characteristics of the C_3-CAM Switch in *Sedum telephium*

The changes in gas-exchange patterns and titratable-acidity fluctuations in *S. telephium* that occur in response to water deficit and irradiance have been

Department of Biological Sciences, Hatherly Laboratories, University of Exeter, Prince of Wales Road, Exeter EX4 4PS, UK

Ecological Studies, Vol. 114
Winter/Smith (eds) Crassulacean Acid Metabolism

described in some detail (Groenhof et al. 1986, 1990; Lee and Griffiths 1987; Borland and Griffiths 1990, 1992; see also Chap. 15). In summary, well-watered plants have C_3 photosynthesis and gas exchange, but as water deficit increases day–night fluctuations in malic acid appear along with an increase in phosphoenolpyruvate carboxylase (PEPC) activity. The gas-exchange patterns indicate a small net uptake of carbon dioxide at night as water deficit intensifies, but most of the malate is derived from recycling of respired carbon dioxide. All the acid fluctuation is accounted for by malic acid (Kenyon et al. 1985; Groenhof et al. 1986; Borland and Griffiths 1990). The time course of appearance of malate fluctuations and PEPC activity are shown in Fig. 11.1. Although these measurements were made during a time course of drying, and therefore confound water status with time, the transition between C_3 and CAM occurs over a decrease in leaf water potential of only 0.2 MPa (Lee and Griffiths 1987). Indeed, it has been our experience that it can be difficult to maintain the plants in their C_3 state during bright summer days in the glasshouse.

11.3 Regulation of Malate Accumulation by Water Status in *Sedum telephium*

11.3.1 Relationship Between Water Status and Malate Accumulation

The first observations of the inducibility of CAM in *S. telephium* showed that malate fluctuations increase with time after withholding water from soil (Fig. 11.1; Groenhof et al. 1986). Because drying of the soil takes time, it was not possible to determine either how rapidly malate fluctuations could be triggered or its exact relationship with soil or plant water status. To minimize this problem, two experimental approaches have been employed. Firstly, leaves have been detached from the plant and allowed to dehydrate when placed in enclosed petri

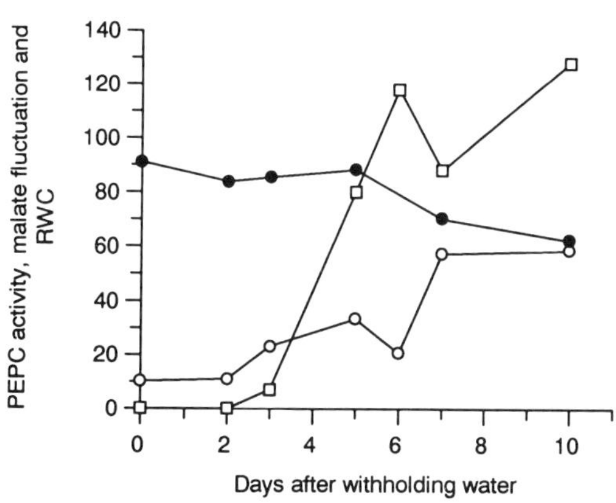

Fig. 11.1. Time course of changes in leaf relative water content (● RWC, %), dawn to dusk malate fluctuations (□ μmol H^+ g^{-1} fr. wt.) and phospho*enol*pyruvate carboxylase (PEPC) activity (○ nmol NADH min^{-1} mg^{-1} protein) in *Sedum telephium* leaves after withholding water from the soil. (Redrawn using data in Groenhof et al. 1990)

dishes. Control leaves are hydrated by floating them on water or keeping their cut bases immersed in water. When leaves from well-watered plants (RWC = 92%) are detached in the middle of a 12-h light period, an increase in titratable acidity and malate during the following dark period occurs in those deprived of water, while the controls do not accumulate malate (Groenhof et al. 1990; Fig. 11.2). This malate accumulation is triggered rapidly by a small decrease in leaf RWC from 92% to between 89 and 84% over the dark period. The malate is decarboxylated during the following light period. When leaves from previously droughted plants are treated in the same way, rehydration results in a smaller malate accumulation than in leaves without a water supply (Fig. 11.2). It is clear that malate accumulation is very closely related to the hydration of the leaves, irrespective of whether they have been previously well-watered or droughted.

The second approach has been to grow the plants in sand and to apply a range of different water potentials to the roots by watering with polyethylene glycol (PEG 8000) solutions. PEG toxicity has been reported (Lawlor 1970), but no toxicity to *S. telephium* was evident. When treatment is started in the middle of the light period, malate accumulation is triggered during the subsequent dark period (Conti and Smirnoff 1992; Fig. 11.3A). The extent of decarboxylation of malate during the following light period is also increased by decreasing the water potential of the rooting medium (Fig. 11.3A). A similar increase in decarboxylation has been noted in *S. spectabile* allowed to dry out rapidly in soil (Brulfert et al. 1988). This species is closely related and morphologically similar to

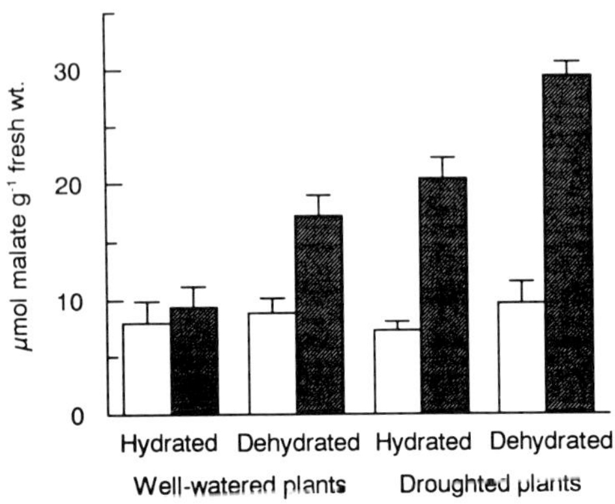

Fig. 11.2. The effect of leaf water status on malate accumulation in detached *Sedum telephium* leaves. Leaves were detached from well-watered (92% RWC) and droughted (82% RWC) plants in the middle of a 12-h light period. They were then hydrated by placing their petioles in water, or dehydrated by allowing them to lose water in a humid atmosphere. Leaf RWC at the beginning and end of the subsequent dark periods were 89 and 84%, respectively, for the well-watered plants and 81 and 80%, respectively, for the droughted plants. Rehydrated leaves reached 100% RWC by the end of the dark period. *White bars* Malate concentration at the beginning of the dark period; *shaded bars* malate concentration at the end of the dark period. The concentration is expressed on the basis of initial leaf weight. *Bars* are standard deviation (n = 6). (N. Smirnoff, unpubl.)

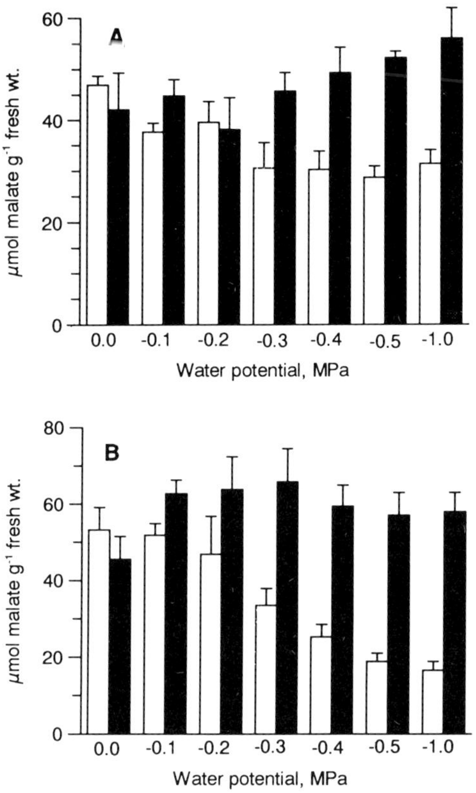

Fig. 11.3A,B. The effect of rooting medium water potential on malate fluctuations in *Sedum telephium* leaves. The plants were watered in the middle of the light period with polyethylene glycol (PEG 8000) dissolved in the nutrient medium to provide a range of water potentials. Malate was measured at the end of the following dark period and at the end of the subsequent day (**A**) and after 7 days (**B**). *White bars* Malate concentration at the beginning of the dark period; *shaded bars* malate concentration at the end of the dark period. *Bars* are standard errors (n = 5). (Conti and Smirnoff 1992)

S. telephium and also exhibits a rapid C_3/CAM switch. In *S. pulchellum*, water loss also increases acid fluctuations within 24 h of treatment; however, in this species acid fluctuation is already present in well-watered plants which are carrying out CAM-cycling (Martin et al. 1988). After 7 days of exposure to PEG, malate fluctuations are directly related to the water potential of the rooting medium, becoming greater as water potential decreases (Fig. 11.3B).

It can be concluded that malate accumulation is triggered rapidly by small losses of water (10% of initial weight) from the leaves (cf. Lee and Griffiths 1987). This initial accumulation and the subsequent light/dark fluctuations are closely related to soil and leaf water status. The close relationship between malate fluctuation and water status is also evident in the other C_3/CAM intermediate *Sedum* species noted above.

These relationships between water status and malate have been noted previously in the constitutive CAM plant *Kalanchoë daigremontiana* (Lüttge et al. 1975, 1977). Decreasing the water potential of the external medium around leaf slices with mannitol influences nocturnal malate accumulation in the same way as noted above in *S. telephium*. At high external water potential, malate accumulation is completely suppressed, while it progressively increases as water potential decreases to approximately −0.6 MPa (Lüttge et al. 1975). On the other hand, fully hydrating *K. tubiflora* leaves does not prevent malate accumulation, so there may be differences between species in response to hydration (Kluge et al. 1980).

Further experiments with leaf slices of *K. daigremontiana* at the beginning of the light period showed that malate efflux into the medium was almost completely stopped in the presence of mannitol at −0.5 MPa. Release of malate from the vacuole is therefore inhibited by low turgor (Lüttge et al. 1977). This response is different to that described above for intact *S. telephium* and *S. spectabile* plants, where reduced external water potential increases the extent of malate decarboxylation or metabolism during the day. The observations made by Lüttge and his colleagues were used in support of a model to explain the daily oscillations of malate in terms of control of its transport into and out of the vacuole by turgor-sensitive active influx and passive efflux mechanisms. The change from influx to efflux was suggested to occur when malate accumulation caused a sufficient increase in turgor, which then inhibits influx and triggers efflux. This model depends on the existence of turgor-sensitive transport processes at the tonoplast, despite the possibility of only very small pressure gradients across this membrane. A direct effect of turgor on malate metabolism was discounted.

Turgor-sensitive membrane transport is now well established. For example, activity of the plasma membrane H^+-ATPase is influenced by turgor, which affects potassium and sugar uptake. Such responses may be involved in regulating osmotic adjustment and the loading of sugars into developing seeds (Reinhold et al. 1984; Wyse et al. 1986; Wolswinkel 1990). Stretch-sensitive ion channels occur in the plasma membrane (Schroeder and Hedrich 1989; Ding and Pickard 1993) and in the tonoplast of red beet (Alexandre and Lassalles 1991). In the latter, a potassium channel is activated by a difference in osmotic pressure across the membrane. The authors concluded that this channel could act as an osmoreceptor that is not directly dependent on turgor, but on a small osmotic imbalance between cytoplasm and vacuole. There is therefore a clear possibility that water status-dependent transport of malate at the tonoplast could be mediated by mechanisms of this type. The question is whether such mechanisms can fully account for daily malate fluctuations and the responses of C_3/CAM intermediates to water deficit, or whether effects on metabolism are also involved. Indeed, it is obvious that changed malate accumulation must be allied with a changed rate of its synthesis, so the problem remains as to how this is regulated. The simplest explanation is that the well-known feedback inhibition of phospho*enol*pyruvate carboxylase (PEPC) by malate is involved.

11.3.2 Effect of Water Deficit on PEPC and Malic Enzyme Capacity

PEPC has been intensively investigated in CAM plants because it is considered to have a key role in controlling malate accumulation. Particular importance has been attached to day/night oscillations in its sensitivity to inhibition by malate (Andreo et al. 1987) There is strong evidence that lower sensitivity of the enzyme extracted at night is related to phosphorylation (Brulfert et al. 1986; Nimmo et al. 1986; Carter et al. 1991; see also Chap. 3). It has also been suggested that the regulatory properties of the enzyme, particularly in *Crassula argentea* and *Zea mays* are influenced by its degree of oligomerization, which may change from day to night or with assay conditions (Wu and Wedding 1985; Podesta and Andreo 1989). We have investigated both the total capacity of PEPC and its kinetic properties in relation to the triggering of malate accumulation by water deficit in *S. telephium*.

When water is withheld from intact *S. telephium* plants, the gradual increase in acidification is associated with an increase in the total capacity of PEPC when assayed under near-optimum conditions (i.e. high pH and saturating substrates). Consequently there is a close relationship between ΔH^+ and PEPC capacity when they are plotted against RWC (Fig. 11.1) (Groenhof et al. 1990). The increase in catalytic capacity is accompanied by an increase in PEPC protein detected by immunoblotting (Blanchard 1991). In comparison with the inducible CAM plant *Mesembryanthemum crystallinum* (Bohnert et al. 1992; see also Chaps. 9 and 10), it seems likely that such a large increase in PEPC would involve an increase in gene transcription. Very similar increases and time courses of PEPC induction occur in *S. spectabile* (Brulfert et al. 1988), *S. rupestre* (Pilon-Smits et al. 1990) and *Crassula sieberiana* (Brulfert et al. 1991). In the last species this is not associated with increased malate synthesis.

A more detailed investigation of PEPC capacity in *S. telephium* has been carried out by applying water deficit with PEG 8000. In this way the effect of time and water status can be separated as described above. One day after application of PEG to the root system there is little change in PEPC capacity. After 7 days, PEPC capacity is increased as the water potential of the rooting medium decreases (Fig. 11.4). The triggering of malate accumulation after 1 day (Fig. 11.3) is therefore not associated with an increase in PEPC capacity. A similar response after 1 day is found with detached leaves, where acidification occurs when water is withheld, but PEPC capacity is unaffected (Groenhof et al. 1990). The extent of malate breakdown during the day is also increased by PEG treatment (Fig. 11.3), but this is not associated with any increase in the capacity of NAD- or NADP-malic enzyme (N. Smirnoff and S. Conti, unpubl.).

11.3.3 Effect of Water Deficit on the Properties of PEPC

It is possible that the increased activity of PEPC could be caused by the increase of a CAM-specific isoform. Müller and Kluge (1983) proposed from immunological evidence that a CAM-specific PEPC isoform is present in the C_3/CAM intermediate species *S. telephium*, *S. spectabile*, *S. coeruleum*, *S. sieboldii* and *S. acre* when exhibiting CAM, but not in those with C_3 photosynthesis. They found PEPC subunits of 105 and 115 kDa. Only the latter increased in CAM-performing

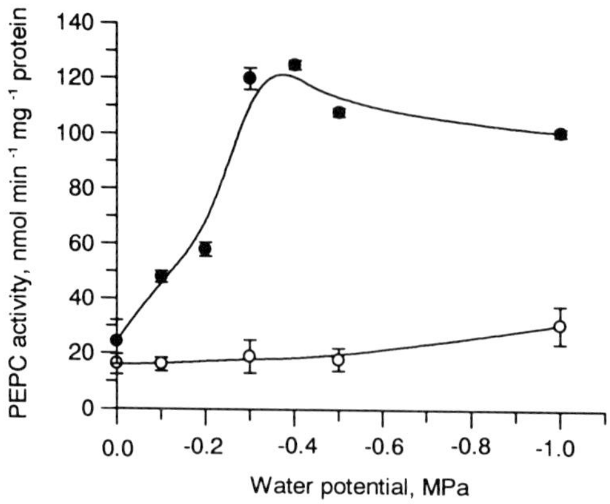

Fig. 11.4. The effect of rooting medium water potential on PEPC activity in *Sedum telephium* leaves. The treatments are described in Fig. 11.3. ○ 1 day after treatment; ● 7 days after treatment. (Conti and Smirnoff 1992)

plants. In comparison, immunoblots of *S. telephium* PEPC reveal two subunits of 101 and 106 kDa (Blanchard 1991). The subunit composition of the native enzyme is not known. On the other hand, only one subunit of 100 kDa has been detected in *S. rupestre* (Pilon-Smits et al. 1990). Non-denaturing electrophoresis and gel filtration suggest the presence of monomeric and dimeric forms of PEPC in extracts (Müller et al. 1982; Groenhof et al. 1988). We recently reported the separation of two PEPC isoforms by ion-exchange chromatography of *S. telephium* extracts and showed that only one of these forms increased during water deficit (Groenhof et al. 1990). Further experiments do not confirm these results. Despite this, the other reports suggest there is a strong possibility that isoforms of PEPC exist in CAM plants, even if they are not separable by ion-exchange chromatography (Höfner et al. 1989).

The properties of PEPC from *S. telephium* vary with extraction and assay conditions, as well as with time of day and growth conditions. Under the extraction conditions normally employed in assays near pH 7.0, the day and night forms of PEPC from well-watered plants are sensitive to inhibition by malate. After a period of water deficit has caused an increase in maximum enzyme capacity, the sensitivity to malate is much less in the night form while the day form retains a similar sensitivity to well-watered plants (Groenhof et al. 1985, 1990; Borland and Griffiths 1992). These results can be interpreted as evidence for the induction by water deficit of a form of PEPC which, being less sensitive to malate inhibition at night, is able to catalyze greater rates of malate synthesis. In all the measurements of PEPC kinetics described above, the enzyme exhibits Michaelis-Menten kinetics with respect to PEP at all times of day and also in the presence of malate. This is the case for most of the CAM species (Jones et al. 1978; Nott and Osmond 1982; Winter 1982).

In order to minimize alteration of PEPC during extraction, we have used a rapid extraction and desalting procedure in which protease and phosphatase inhibitors are added to the extraction buffer. Using this technique, PEPC from droughted plants tends to exhibit Michaelis-Menten kinetics with respect to PEP concentration from the middle to the end of the dark period (Fig. 11.5A). During the day, PEPC from droughted plants has allosteric kinetics (positive cooperativity) with Hill numbers approaching 2 (Fig. 11.5B). Addition of 2 mM malate causes PEPC to exhibit allosteric kinetics during light and dark periods, and inhibition is less in the night form (Fig. 11.5A,B). PEPC from well-watered plants tends to show greater positive cooperativity during the day and night compared with PEPC from droughted plants. Activity from well-watered plants was too low for accurate determination after malate addition.

Changes in the properties of PEPC over a 24-h cycle are shown in Fig. 11.6, where the ratio between activity at 0.1 and 2 mM PEP ($V_{0.1}/V_{2.0}$ ratio) is shown, along with the effect of malate on V_{max}. The ratio reflects variation both in positive cooperativity and $S_{0.5}$, larger values indicating higher affinity for PEP and a closer approach to Michaelis-Menten kinetics. The decrease in the $V_{0.1}/V_{2.0}$ ratio during the day occurs in both well-watered and droughted plants. The main difference noted so far is a higher affinity for PEP at the end of the dark period in PEPC from droughted leaves, which is discussed further below. Malate also decreases the ratio in droughted plants and decreases V_{max} more during the day. Reliable measurements have not yet been obtained for these parameters for PEPC from well-watered plants assayed with malate because activity is very low. Allosteric kinetics have not been noted in previous investigations of *S. telephium* PEPC, even in the presence of malate (Groenhof et al. 1985, 1990; Borland and Griffiths 1992), so it appears that rapid extraction and inclusion of protease and phosphatase inhibitors may influence the properties of PEPC. The malate-sensitive day form of PEPC can rapidly lose sensitivity to malate and show an increase in affinity for PEP after extraction (Winter 1982). However, the difference in malate sensitivity between the day and night forms in *Bryophyllum fedtschenkoi* is stabilized by the presence of protease inhibitors, which prevent cleavage of a small peptide fragment from the enzyme (Nimmo et al. 1986). It appears that inclusion of protease inhibitors has the same effect on *S. telephium* PEPC. The tendency to sigmoidal kinetics for PEPC in the absence of malate noted in the day and with well-watered forms has rarely been reported in CAM plants (Nott and Osmond 1982; Winter 1982; Wu and Wedding 1985; Wedding and Black 1986). An exception is *S. praealtum* where rapid extraction and assay at pH 7.0 also reveals the same pattern as described above for droughted *S. telephium* (Manetas 1982). Because these characteristics are so strongly dependent on extraction and assay conditions, it is difficult to extrapolate them to PEPC activity in vivo with certainty.

Since the PEPC induced by drought has different properties to PEPC from well-watered plants, it is possible that the modulation of malate accumulation caused by rapid changes in hydration of detached leaves (Fig 11.2) is associated with alterations to PEPC properties while the total capacity is unaffected. Preliminary measurements show that when PEPC is extracted towards the end of the dark period, dehydration causes a decrease in K_m (PEP) while rehydration increases the K_m (Fig. 11.7). Malate sensitivity is not altered by hydration state.

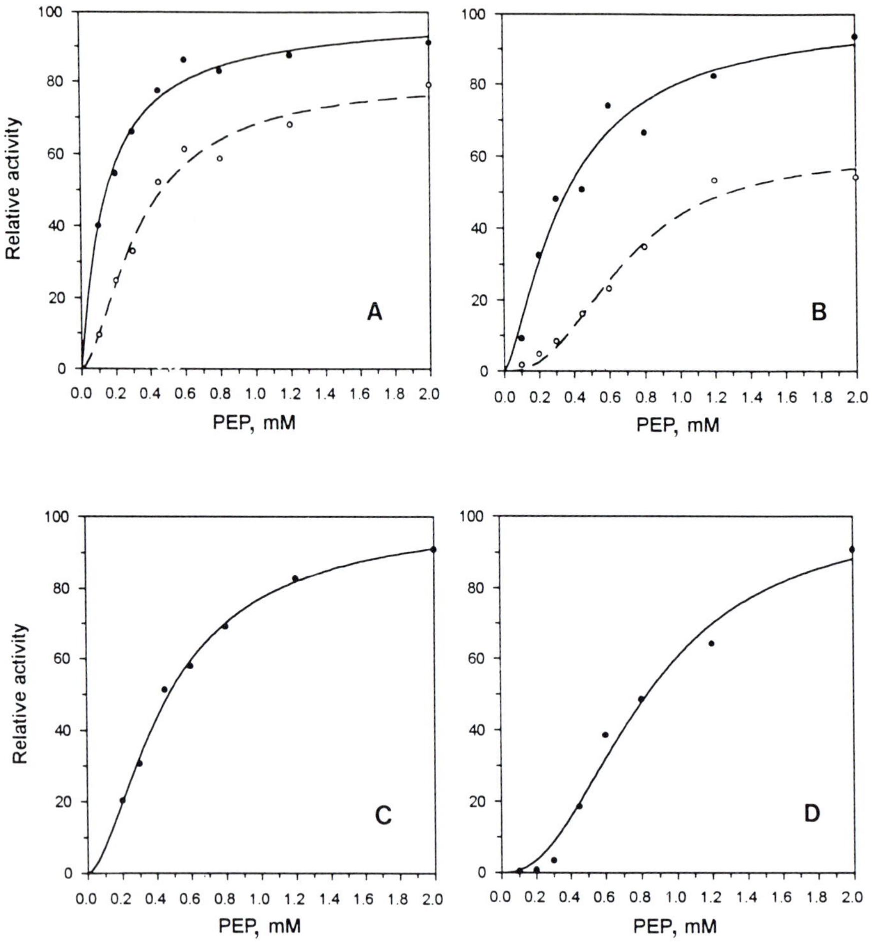

Fig. 11.5 A–D. The effect of long-term water deficit on activity (% of V_{max} at pH 7.0) versus PEP concentration of PEPC in desalted crude extracts of *Sedum telephium* leaves. The plants were droughted until PEPC capacity was 5 to 10 times greater than in well-watered plants. **A,B** droughted plants; **C,D** well-watered plants; **A,C** middle of dark period; **B,D** middle of light period. *Dotted lines* indicate assays in the presence of 2 mM malate. Rates with malate were too low to measure in well-watered plants. The $S_{0.5}$ values (mM PEP) are: night/droughted 0.15; day/droughted 0.24; night/droughted + malate 0.19; day/droughted + malate 0.38; night/well-watered 0.3; day/well-watered 0.66. (P. Barnwell and N. Smirnoff, unpubl.)

This shows that the kinetic properties of PEPC are rapidly influenced by the hydration state of the leaf. It is not possible to determine the importance of the higher affinity for PEP in dehydrated leaves until more extensive measurements have been made, particularly through the remainder of the dark period. It is, however, clear that there is not a close correlation between kinetic properties with

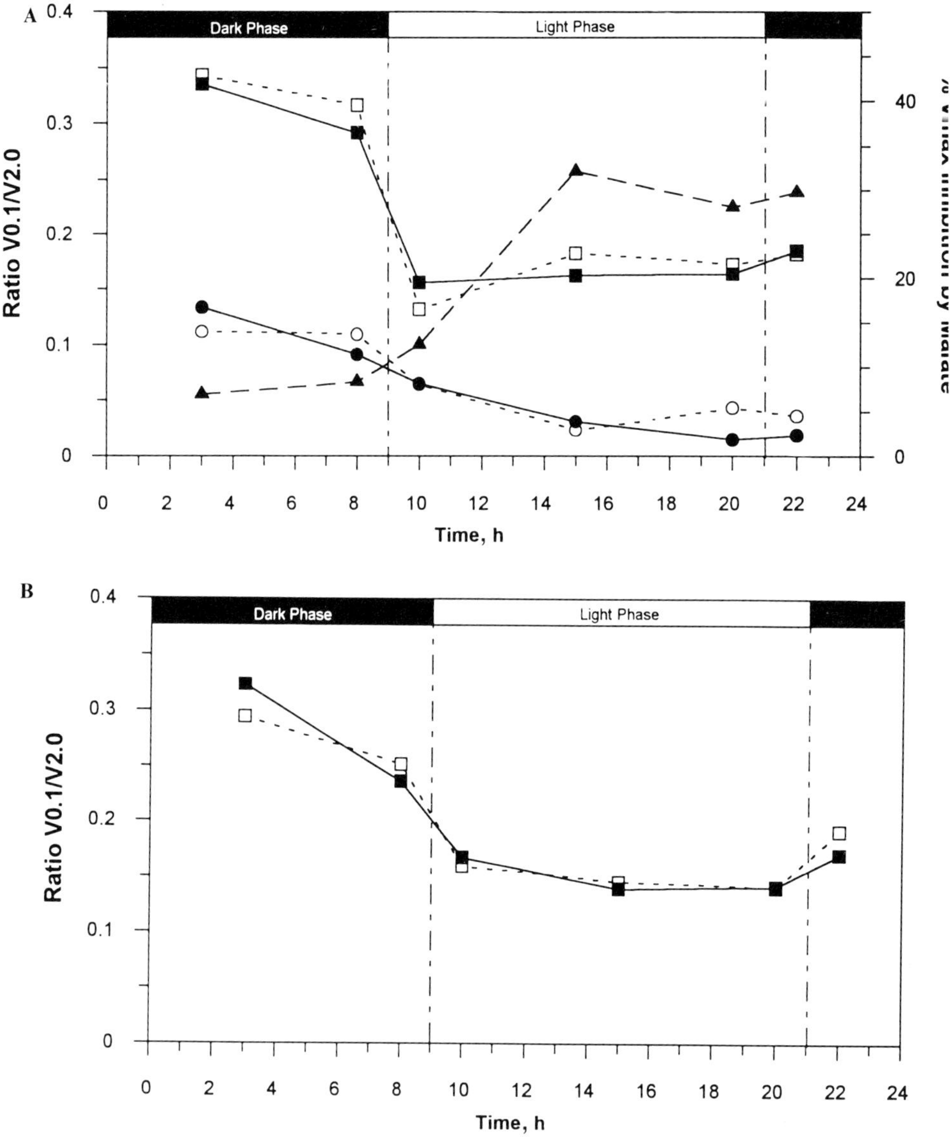

Fig. 11.6A,B. Comparison of the properties of PEPC from (**A**) long-term droughted and (**B**) well-watered *Sedum telephium* leaves throughout a day-night cycle. See Fig. 11.5 for details of plants and PEPC extraction. The $V_{0.1}/V_{2.0}$ ratio (ratio between rates at 0.1 and 2.0 mM PEP) is shown for assays with (●, ○) and without (■, □) 2 mM malate; ▲ % inhibition of V_{max} by 2 mM malate. *Solid lines* are measured rates and *dotted lines* are calculated from the curves fitted to the kinetic data using Michaelis-Menten or allosteric models as appropriate. (P. Barnwell and N. Smirnoff, unpubl.)

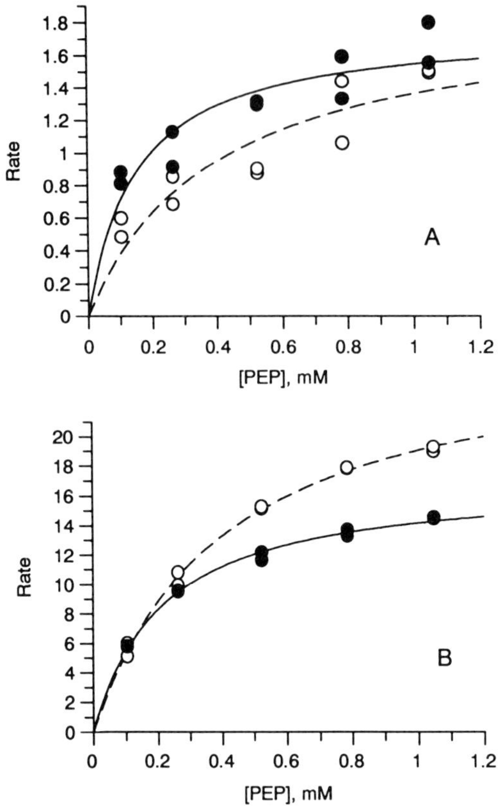

Fig. 11.7A,B. The effect of the hydration state of detached *Sedum telephium* leaves on activity versus PEP concentration of PEPC in desalted crude extracts. PEPC was extracted during the late dark period from leaves treated as described in Fig. 11.2. **A** Well-watered plants; **B** droughted plants; ○ hydrated leaves; ● dehydrated leaves. The K_m values (mM PEP $\pm$ SD) are: droughted plants/rehydrated leaves 0.41 ± 0.03; droughted plants/dehydrated leaves 0.22 ± 0.01; well-watered plants/rehydrated leaves 0.40 ± 0.16; well-watered plants/dehydrated leaves 0.15 ± 0.05. (N. Smirnoff, unpubl.)

respect to PEP and sensitivity to malate (Fig. 11.6). This suggests that PEPC can exhibit subtle variations in its properties, superimposed on day/night fluctuations. This is also supported by the observation that PEPC from plants grown at low photon flux densities and which accumulate less malate, has lower affinity for PEP and lower malate sensitivity (Borland and Griffiths 1992). The properties of PEPC are affected by exposure to malate at low PEP concentrations, protein concentration and the presence of compatible solutes such as glycerol (Wu and Wedding 1985, 1987; Manetas et al. 1986; Selinioti et al. 1987; Podesta and Andreo 1989). Exposure to varying malate concentrations in the leaves during extraction can affect subsequent malate sensitivity (Kluge et al. 1980), although inclusion of glycerol in the extraction medium should minimize this effect

(Podesta and Andreo 1989). The malate may promote dissociation into dimers, while glycerol promotes association into tetramers (Wu and Wedding 1985, 1987; Podesta and Andreo 1989). The sensitivity of PEPC to conditions is emphasized by the observation that the malate sensitivity of the night form of *Zea mays* PEPC is greater if it is not extracted but released directly into the assay mixture by introducing protoplasts which lyse and release PEPC in situ (Petropoulou et al. 1990). It is therefore not inconceivable that changes in leaf water status could influence PEPC by causing changes in concentration of protein, compatible solutes and effector molecules, which will then alter its properties in vivo and subsequently in vitro.

11.4 Conclusions and Speculations

The expression of CAM biochemistry in *S. telephium* is highly sensitive to leaf water status. The capacity for malate accumulation is related to both the maximum catalytic activity of PEPC and its regulatory properties, so that maximum accumulation can occur after a long period of water deficit which increases PEPC capacity. Short-term ($<$ 24 h) changes in water status modify the capacity for malate accumulation. How this is regulated is not understood. Based on the hypothesis of Lüttge (Lüttge et al. 1975, 1977), water status could alter transport of malate into the vacuole and thereby relieve or impose feedback inhibition on PEPC. Since the properties of PEPC also change in the short term, this hypothesis must be modified to incorporate either a direct or indirect effect on PEPC. A possibility which requires further investigation is that water status influences the activity of PEPC kinase, which is responsible for phosphorylating the enzyme at night and inducing a lower malate sensitivity (Carter et al. 1991).

To understand the triggering of malate accumulation fully will require more information on the regulation of storage carbohydrate mobilization and glycolysis. In particular, we need to know if control is exerted on PEP supply or on PEP utilization, or if it is more evenly distributed through the pathway. Starch is the major source of carbon for malate accumulation in *S. telephium* (Kenyon et al. 1985; Borland and Griffiths 1990). Water deficit is known to influence carbohydrate metabolism both by altering partitioning of photosynthate between starch and sucrose and possibly by stimulating starch breakdown (Oparka and Wright 1988; Bensari et al, 1990; Vassey et al 1991; Zrenner and Stitt 1991; Quick et al. 1992). Water deficit increases the activity of starch-degrading enzymes in the CAM plant *Aloë arborescens* (Verbücheln and Steup 1983). Increased PEP supply resulting from increased starch breakdown during water deficit could drive higher PEPC activity. Such an effect could be facilitated by the low sensitivity of phosphofructokinase to inhibition by PEP reported for one CAM plant (Sutton 1975). On the other hand, the rapid stimulation of dark carbon dioxide fixation in cell cultures by alkalinization of the medium and the corresponding increase in glycolysis are primarily controlled by PEPC and pyruvate kinase rather than by phosphofructokinase (Hatzfeld and Stitt 1991). They suggested that PEPC is rapidly activated by a combination of higher cytoplasmic pH, which decreases malate sensitivity, and increased malate transport into the vacuole. Neither of these was measured, so the hypothesis remains to be tested. In CAM plants there have been no direct measurements of cytoplasmic malate concentration. In vivo ^{13}C-nuclear magnetic resonance (NMR) spectroscopy is able to distinguish cytoplasmic and vacuolar pools in *Zea mays* root tips (Chang and Roberts 1989), but only a vacuolar pool is distinguishable in *Kalanchoë tubiflora* leaves, probably because of the small cytoplasmic volume (Stidham et al. 1983). However, with further technical developments, measurements may become possible in CAM leaves.

The other possible controlling factor, cytoplasmic pH, has been measured indirectly by labelled DMO uptake by leaves of *K. tubiflora* and other CAM plants. Although cytoplasmic pH may be generally higher at night, the oscillations are complex and perturbation by treatment with weak acids and bases favours control based on malate and proton transport at the tonoplast, rather than by pH effects on PEPC activity. For example, increased cytoplasmic pH in the dark inhibits malate accumulation, while an increase would be expected if the effect was mediated by PEPC (Marigo et al. 1982, 1983; Lüttge et al. 1982). Absolute pH was not measured and it would be of interest to measure this in relation to water status in *S. telephium*. ^{31}P-NMR can be used to determine cytoplasmic and vacuolar pH more precisely and preliminary measurements show that this is possible in *S. telephium* (R.G. Ratcliffe and N. Smirnoff, unpubl.). We have already shown with this technique that water deficit in *Zea mays* roots causes cytoplasmic alkalinisation (Spickett et al. 1992). It is possible that similar cytoplasmic pH changes brought about by changes in water status could alter the sensitivity of PEPC to malate and thereby modulate malate accumulation in *S. telephium*.

More extensive investigation of metabolism in *S. telephium* is needed before the basis of the switch between C_3 and CAM can be properly described, and this will require understanding of both the steps that contribute to regulation of malate metabolism and the means by which plant cells transduce changes in water status into biochemical responses. The present discussion has not touched on aspects related to malate decarboxylation and long-term induction of PEPC, which clearly also merit further investigation.

Acknowledgements. I am grateful to Paul Barnwell and Stefano Conti, who have allowed me to quote some of their unpublished results.

References

Alexandre J, Lassalles J-P (1991) Hydrostatic and osmotic pressure activated channel in plant vacuole. Biophys J 60: 1326–1336

Andreo CS, Gonzalez, DH, Iglesias AA (1987) Higher plant phosphoenolpyruvate carboxylase. Structure and regulation. FEBS Lett 213: 1–8

Bender M, Rouhani J, Vines HM, Black CC (1973) $^{13}C/^{12}C$ ratio changes in crassulacean acid metabolism plants. Plant Physiol 52: 427–430

Bensari M, Calmes J, Viala G (1990) Répartition du carbone fixé par photosynthèse entre l'amidon et le saccharose dans la feuille de soja. Influence d'un déficit hydrique. Plant Physiol Biochem 28: 113–121

Blanchard A (1991) Molecular and physiological studies on the drought response in *Sedum telephium* L. and *Delosperma nubigenum*. L. PhD Thesis, University of Exeter, Exeter

Bohnert HJ, Vernon DM, DeRocher EJ, Michalowski CB, Cushman JC (1992) Biochemistry and molecular biology of CAM. In: Wray JL (ed) Inducible plant proteins. Their biochemistry and molecular biology. Cambridge University Press, Cambridge, pp 113–137

Borland AM, Griffiths H (1990) The regulation of CAM and respiratory recycling by water supply and ligh regime in the C_3-CAM intermediate *Sedum telephium*. Funct Ecol 4: 33–39

Borland AM, Griffiths H (1992) Properties of phosphoenolpyruvate carboxylase and carbohydrate accumulation in the C_3-CAM intermediate *Sedum telephium* L. grown under different light and watering regimes. J Exp Bot 43: 353–361

Brulfert J, Vidal J, Le Marechal P, Gadal P, Queiroz O, Kluge M, Krüger I (1986) Phosphorylation-dephosphorylation process as a probable mechanism for the diurnal regulatory changes of phosphoenolpyruvate carboxylase in CAM plants. Biochem Biophys Res Commun 136: 151–159

Brulfert J, Kluge M, Güçlü S, Queiroz O (1988) Combined effects of drought, daylength and photoperiod on rapid shifts in the photosynthetic pathways of *Sedum spectabile*, a CAM species. Plant Physiol Biochem 26: 7–16

Brulfert J, Güçlü S, Kluge M (1991) Effects of abrupt or progressive drought on the photosynthetic mode of *Crassula sieberiana* cultivated under different daylengths. J Plant Physiol 138: 685–690

Carter PJ, Nimmo HG, Fewson CA, Wilkins MB (1991) Circadian rhythms in activity of a plant protein kinase. EMBO J 10: 2063–2068

Chang K, Roberts JKM (1989) Observation of cytoplasmic and vacuolar malate in maize root tips by ^{13}C-NMR spectroscopy. Plant Physiol 89: 197–203

Conti S, Smirnoff N (1992) Time course of CAM induction and response to different levels of water deficit in *Sedum telephium*. Ann Fac Sci Agrar Univ Napoli 26: 27–39

Ding JP, Pickard BG (1993) Mechanosensory calcium-selective cation channels in epidermal cells. Plant J 3: 83–110

Groenhof AC, Daniel PP, Bryant JA (1985) Adaptive changes in the properties of phosphoenolpyruvate carboxylase in water stressed plants. Biochem Soc Trans 13: 1154

Groenhof AC, Bryant JA, Etherington JR (1986) Photosynthetic changes in the inducible CAM plant *Sedum telephium* following the imposition of water stress. I. General characteristics. Ann Bot 57: 689–695

Groenhof AC, Bryant JA, Etherington JR (1988) Photosynthetic changes in the inducible CAM plant *Sedum telephium* following the imposition of water stress. II. Changes in the activity of phosphoenolpyruvate carboxylase. Ann Bot 62: 187–192

Groenhof AC, Smirnoff N, Bryant JA (1990) The appearance of a new molecular species of phosphoenolpyruvate carboxylase (PEPC) and the rapid induction of CAM in *Sedum telephium* L. Plant Cell Environ 13: 437–445

Hatzfeld W-D, Stitt M (1991) Regulation of glycolysis in heterotrophic cell suspension cultures of *Chenopodium rubrum* in response to proton fluxes at the plasmalemma. Physiol Plant 81: 103–110

Höfner R, Vazquez-Moreno L, Abou-Mandour AA, Bohnert HJ, Schmitt JM (1989) Two isoforms of phosphoenolpyruvate carboxylase in the facultative CAM plant *Mesembryanthemum crystallinum*. Plant Physiol Biochem 27: 803–810

Jones R, Wilkins MB, Coggins JR, Fewson CA, Malcolm ADB (1978) Phosphoenolpyruvate carboxylase from the crassulacean plant *Bryophyllum fedtschenkoi* Hamet et Perrier. Purification, molecular and kinetic properties. Biochem J 175: 391–406

Kenyon WH, Severson RF, Black CC (1985) Maintenance carbon cycle in crassulacean acid metabolism plant leaves. Source and compartmentation of carbon for nocturnal malate synthesis. Plant Physiol 77: 183–189

Kluge M, Böcher M, Jungnickel G (1980) Metabolic control of crassulacean acid metabolism: evidence for diurnally changing sensitivity against inhibition by malate of PEP-carboxylase in *Kalanchoë tubiflora* Hamet. Z Pflanzenphysiol 97: 197–204

Knopf O, Kluge M (1979) Properties of phosphoenolpyruvate carboxylase in *Sedum* species in relation to crassulacean acid metabolism. Plant Cell Environ 2: 73–78

Lawlor DW (1970) Absorption of polyethylene glycols by plants and their effects on plant growth. J Exp Bot 38: 834–841

Lee HSJ, Griffiths H (1987) Induction and repression of CAM in *Sedum telephium* L. in response to photoperiod and water stress. J Exp Bot 38: 187–192

Lüttge U, Kluge M, Ball E (1975) Effects of osmotic gradients on vacuolar malic acid storage. A basic principle in oscillatory behaviour of crassulacean acid metabolism. Plant Physiol 56: 613–616

Lüttge U, Ball E, Greenway H (1977) Effects of water and turgor potential on malate efflux from leaf slices of *Kalanchoë daigremontiana*. Plant Physiol 60: 521–523

Lüttge U, Smith JAC, Marigo G (1982) Membrane transport, osmoregulation, and the control of CAM. In: Ting IP, Gibbs M (eds) Crassulacean acid metabolism. American Society of Plant Physiologists, Rockville, pp 69–91

Manetas Y (1982) Changes in properties of phosphoenolpyruvate carboxylase from the CAM plant *Sedum praealtum* D.C. upon dark/light transition and their stabilization by glycerol. Photosynth Res 3: 321–333

Manetas Y, Petropoulou Y, Karabourniotis G (1986) Compatible solutes and their effects on phosphoenolpyruvate carboxylase of C_4-halophytes. Plant Cell Environ 9: 145–151

Marigo G, Ball E, Lüttge U, Smith JAC (1982) Use of the DMO technique for the study of relative changes of cytoplasmic pH in leaf cells in relation to CAM. Z Pflanzenphysiol 108: 223–233

Marigo G, Lüttge U, Smith JAC (1983) Cytoplasmic pH and the control of crassulacean acid metabolism. Z Pflanzenphysiol 109: 405–413

Martin CE, Higley M, Wang W-Z (1988) Recycling of CO_2 via crassulacean acid metabolism in the rock outcrop succulent *Sedum pulchellum* Michx. (Crassulaceae). Photosynth Res 18: 337–343

Müller D, Kluge M (1983) Immunological evidence for a crassulacean acid metabolism specific phosphoenolpyruvate carboxylase in *Sedum* and *Kalanchoë* species. Physiol Vég 21: 919–926

Müller D, Kluge M, Gröschel-Stewart U (1982) Comparative studies on immunological and molecular properties of phosphoenolpyruvate carboxylase in species of *Sedum* and *Kalanchoë* performing crassulacean acid metabolism. Plant Cell Environ 5: 223–230

Nimmo GA, Nimmo HG, Hamilton ID, Fewson CA, Wilkins MB (1986) Purification of the phosphorylated night form and dephosphorylated day form of phosphoenolpyruvate carboxylase from *Bryophyllum fedtschenkoi*. Biochem J 239: 213–220

Nott DL, Osmond CB (1982) Purification and properties of phosphoenolpyruvate carboxylase from plants with crassulacean acid metabolism. Aust J Plant Physiol 9: 409–422

Oparka KJ, Wright KM (1988) Osmotic regulation of starch synthesis in potato tubers. Planta 174: 123–126

Petropoulou Y, Manetas Y, Gavalas NA (1990) Intact mesophyll protoplasts from *Zea mays* as a source of phosphoenolpyruvate carboxylase unaffected by extraction: advantages and limitations. Physiol Plant 80: 605–611

Pilon-Smits EAH, 't Hart H, van Brederode J (1990) Phosphoenolpyruvate carboxylase in *Sedum rupestre* (Crassulaceae): drought-enhanced expression and purification. J Plant Physiol 136: 155–160

Pilon-Smits EAH, 't Hart H, Meesterburrie JAN, Naber P, Krueler R, van Brederode J (1991) Variation in crassulacean acid metabolism within the genus *Sedum*: carbon isotope composition and water status dependent phosphoenolpyruvate carboxylase activity. J Plant Physiol 137: 342–346

Podesta FE, Andreo CS (1989) Maize leaf phosphoenolpyruvate carboxylase. Oligomeric state and activity in the presence of glycerol. Plant Physiol 90: 427–433

Quick WP, Chaves MM, Wendler R, David M, Rodrigues ML, Passaharinho JA, Pereira JS, Adcock MD, Leegood RC, Stitt M (1992) The effect of water stress on photosynthetic carbon metabolism in four species grown under field conditions. Plant Cell Environ 15: 25–35

Reinhold L, Seiden A, Volokita M (1984) Is modulation of proton pumping a key event in osmoregulation? Plant Physiol 75: 846–849

Schroeder JL, Hedrich R (1989) Involvement of ion channels and active transport in osmoregulation and signalling of higher plant cells. Trends Biochem Sci 14: 187–192

Selinioti E, Nikolopoulos D, Manetas Y (1987) Organic cosolutes as stabilisers of phosphoenolpyruvate carboxylase in storage: an interpretation of their action. Aust J Plant Physiol 14: 203–210

Spickett CM, Smirnoff N, Ratcliffe RG (1992) Metabolic response of maize roots to hyperosmotic shock. An in vivo ^{31}P nuclear magnetic resonance study. Plant Physiol 99: 856–863

Stidham MA, Moreland DE, Siedow JN (1983) ^{13}C nuclear magnetic resonance studies of crassulacean acid metabolism in intact leaves of *Kalanchoë tubiflora*. Plant Physiol 73: 517–520

Sutton BG (1975) Kinetic properties of phosphorylase and 6-phosphofructokinase of *Kalanchoë daigremontiana* and *Atriplex spongiosa*. Aust J Plant Physiol 2: 403–411

Teeri J (1982) Photosynthetic variation in the Crassulaceae. In: Ting IP, Gibbs M (eds) Crassulacean acid metabolism. American Society of Plant Physiologists, Rockville, pp 244–259

Vassey TL, Quick WP, Sharkey TD, Stitt M (1991) Water stress, carbon dioxide, and light effects on sucrose-phosphate synthase activity in *Phaseolus vulgaris*. Physiol Plant 81: 37–44

Verbücheln O, Steup M (1983) Water-stress induced increase of starch-degrading enzyme activities in the CAM plant *Aloë arborescens*. Plant Physiol 72 (Suppl): 36

Wedding RT, Black MK (1986) Malate inhibition of phosphenolpyruvate carboxylase from *Crassula*. Plant Physiol 82: 985–990

Winter K (1982) Properties of phosphoenolpyruvate carboxylase in rapidly desalted leaf extracts of the crassulacean acid metabolism plant *Mesembryanthenum crystallinum* L. Planta 154: 298–308

Wolswinkel P (1990) Recent progress in research on the role of turgor-sensitive transport in seed development. Plant Physiol Biochem 28: 399–410

Wu M-X, Wedding RT (1985) Diurnal regulation of phosphoenolpyruvate carboxylase from *Crassula*. Plant Physiol 77: 667–675

Wu M-X, Wedding RT (1987) Regulation of phosphoenolpyruvate carboxylase from *Crassula argentea*. Further evidence on the dimer tetramer interconversion. Plant Physiol 84: 1080–1083

Wyse RE, Zamski E, Tomos AD (1986) Turgor regulation of sucrose transport in sugar beet taproot tissue. Plant Physiol 81: 478–481

Zrenner R, Stitt M (1991) Comparison of the effect of rapidly and gradually developing water stress on carbohydrate metabolism in spinach leaves. Plant Cell Environ 14: 939–946

12 Putative Causes and Consequences of Recycling CO_2 via Crassulacean Acid Metabolism

C.E. MARTIN

12.1 Introduction

The fate of respiratory CO_2 released simultaneously with the absorption of atmospheric CO_2 at night during crassulacean acid metabolism (CAM) is inevitable; both sources of CO_2 will contribute to the nocturnal accumulation of malate. As a result, determinations of stoichiometric relationships between the amount of malate accumulated overnight and the net amount of CO_2 absorbed from the external atmosphere should yield values greater than unity. Surprisingly, this is seldom the case. Studies of numerous CAM species under optimal conditions report ratios of the amount of malate accumulated overnight to the net amount of atmospheric CO_2 absorbed (hereafter referred to as "malate/CO_2") consistently near one (Medina and Delgado 1976; Nobel and Hartsock 1978, 1983; Eickmeier 1979; Nobel et al. 1984; Winter et al. 1986; Virzo De Santo et al. 1987). Apparently, either the rate of respiration is too low to contribute substantially to the nocturnal pool of malate, or experimental variability in measurements of CO_2 assimilation and malate accumulation during the night might obscure small differences between these measures of CAM.

In contrast to the above, there are instances whereby malate/CO_2 ratios greatly exceed one during CAM (sensu lato). These deviations from the expected ratio of approximately one can be grouped into three categories. The first category comprises plants exhibiting "CAM-cycling" (Ting 1985). These plants seldom perform CAM (sensu stricto); instead they exhibit C_3 gas exchange patterns concomitant with CAM acid fluctuations (Fig. 12.1; Ting and Rayder 1982; Martin and Zee 1983; Ting 1985; Martin and Jackson 1986; Monson 1989). The potential value of CAM-cycling is unclear, but suggestions include conservation of CO_2 (Martin and Zee 1983), conservation of water (Cockburn 1985; Martin et al. 1988b), "preparation" for CAM-idling (Rayder and Ting 1981; Ting and Burk 1983; Sipes and Ting 1985), minimization of photoinhibition (Osmond et al. 1980; Osmond 1982), and possibly enhancement of water uptake from the plant's surroundings via the osmotic effect of high concentrations of malate (Lüttge 1986, 1987; Harris and Martin 1991b). The latter two hypotheses have

Department of Botany, University of Kansas, Lawrence, KS 66045-2106, USA

Ecological Studies, Vol. 114
Winter/Smith (eds) Crassulacean Acid Metabolism

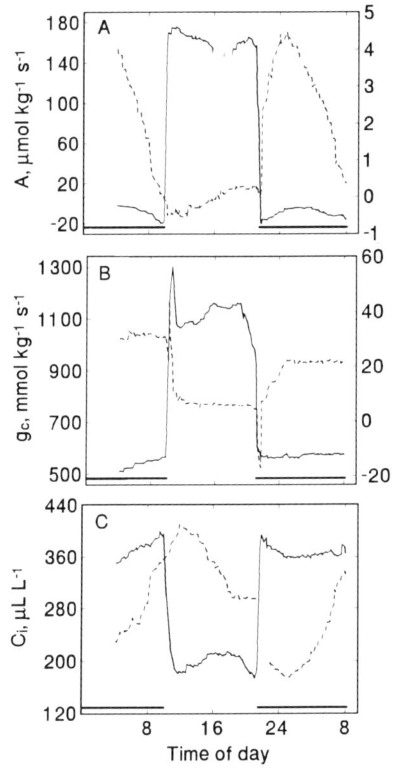

Fig. 12.1A–C. Net CO_2 exchange (*A*; panel **A**), shoot conductance to CO_2 (g_c; panel **B**), and shoot internal CO_2 concentration (C_i; panel **C**) of a representative CAM species (*Tillandsia setacea; dashed lines*) and a representative CAM-cycling species (*Talinum calycinum; solid lines*). *Thick black bars* along the time axes represent night periods. At the onset of the dark period, the CAM species exhibits absorption of water vapour; hence, conductances are negative (panel **B**). During this time, calculations of C_i were not possible (panel **C**). Gas exchange of both plants was measured under the same environmental conditions (see legend of Fig. 12.3). In panels **A** and **B**, the *right axes* (lower rates, same units) apply to the CAM species. Increases in leaf malate concentration over the second night were 110 mmol kg^{-1} (malate/CO_2 ratio = approx. 1.0) for the CAM plant and 380 mmol kg^{-1} (malate/CO_2 ratio = ∞) for the CAM-cycling species. All data are expressed on a dry-mass basis

not been tested in plants exhibiting CAM-cycling. Recycling respiratory CO_2 [the presumed substrate for acid accumulation at night when stomata are closed (Winter et al. 1986; Patel and Ting 1987)] by itself cannot benefit the plant any more than the fixation of atmospheric CO_2, as the same amount of energy must be expended to reduce CO_2 to carbohydrate whether the source of CO_2 is inside or outside the leaves. On the other hand, the daytime release of CO_2 from the decarboxylation of malate should effect an elevation of the internal CO_2 concentration of the leaf, resulting in partial stomatal closure and, consequently, reduced rates of transpiration (Martin et al. 1988b). Martin and coworkers have accumulated evidence in favour of this water-conservation hypothesis in studies of CAM-cycling in succulents growing on rock outcrops in the central U.S. (Martin et al. 1988a,b; Harris and Martin 1991a,b). The amounts of water conserved, however, were quite small, and, furthermore, no evidence for water conservation associated with CAM-cycling was obtained with several species of *Sedum* (Gravatt and Martin 1992).

Ting and coworkers have championed the suggestion that CAM-cycling may simply be a precursor to CAM-idling and presumably provides little benefit to the plant when well watered, i.e. during CAM-cycling (Rayder and Ting 1981; Ting and Burk 1983; Sipes and Ting 1985). Under severe drought stress, CAM plants, as well as most species exhibiting CAM-cycling, maintain stomatal closure day and night yet continue to undergo diel fluctuations in acidity typical of CAM (Szarek et al. 1973; Martin and Zee 1983; Ting 1985; Lee et al. 1989). This phenomenon, termed "CAM-idling" (Ting 1985), has not been experimentally investigated beyond descriptions of gas exchange and acid fluctuations. Szarek and colleagues (Szarek et al. 1973; Szarek and Ting 1974) first suggested that CAM-idling should maintain the plant in a metabolically active state during severe droughts allowing rapid recovery upon rehydration. Osmond and co-workers (Osmond et al. 1980) have also suggested that the elevation of tissue CO_2 concentrations during the day should minimize photoinhibition. Of course, these two hypotheses are not mutually exclusive. Although little experimental evidence exists for or against these hypotheses, few investigators question their validity.

During CAM-cycling and CAM-idling, stomatal conductances are low throughout the night yet malate accumulates in the photosynthetic tissue, presumably as a result of the fixation of respiratory CO_2 (Szarek and Ting 1974; Patel and Ting 1987). Thus, malate/CO_2 ratios in these two groups of plants are much greater than one; in fact, they are often infinity. One other group of plants exhibits malate/CO_2 ratios in excess of one. These plants, including numerous species in the Bromeliaceae, show CAM (sensu stricto) and, while simultaneously absorbing atmospheric CO_2, the apparent contribution of respiratory CO_2 to the accumulation of malate is sufficient to elevate the malate/CO_2 ratio (Sale and Neales 1980; Griffiths et al. 1986; Smith et al. 1986; Martin and Adams 1987; Griffiths 1988a; Borland and Griffiths 1990). Any potential benefits derived from recycling large amounts of respiratory CO_2 during CAM are difficult to envision upon consideration of the reduction in productivity necessitated by substituting respiratory CO_2 for atmospheric CO_2. Also, any savings in water resulting from elevated tissue internal CO_2 concentrations would presumably be offset by the investment of carbon (and the water expended during the acquisition of that carbon) in the "extra" respiratory and photosynthetic machinery associated with this CO_2 recycling.

Three hypotheses have been forwarded in an attempt to explain high levels of CO_2 recycling during CAM. First, respiration rates of photosynthetic tissue may be unavoidably high as a result of the warm climates characteristic of some CAM species that exhibit elevated malate/CO_2 ratios (Winter et al. 1986; Lüttge and Ball 1987; Fetene and Lüttge 1991). Second, contributions of respiratory CO_2 from nonphotosynthetic regions of the leaf may account for the malate accumulated in excess of atmospheric CO_2 uptake at night (Griffiths et al. 1986; Lüttge and Ball 1987). Third, plants exhibiting high levels of CO_2 recycling during CAM may be partially stressed (Griffiths 1988a,b; Lüttge 1990; Lüttge et al. 1991). With less stress, they might exhibit malate/CO_2 ratios of one; with more stress, they should undergo CAM-idling. Experimental studies testing these hypotheses are few. The first hypothesis seems untenable as not all species from warm climates exhibit elevated malate/CO_2 ratios (e.g., some species of *Tillandsia* in this study; see below and Loeschen et al. 1993), and some species from cool climates do exhibit malate/CO_2 ratios greater than one (Borland and Griffiths 1990). In an examination of gas-exchange characteristics of different tissues in leaves of several CAM species, Lüttge and Ball (1987) found no evidence for the second hypothesis. Finally, in studies of CAM in *Aechmea* species (Griffiths 1988a) and in *Bromelia humilis* (Fetene and Lüttge 1991), malate/CO_2 ratios varied from one to much greater than one, depending on environmental conditions, lending support to the third hypothesis.

It was the purpose of this study to investigate the ecophysiological consequences of CO_2 recycling during CAM-cycling as well as during CAM with an emphasis on species of *Talinum* (Portulacaceae) and *Tillandsia* (Bromeliaceae). Specifically, the goal was to examine further the water-conservation hypothesis regarding the potential benefit of CAM-cycling in *Talinum*, and to examine hypotheses about CO_2 recycling during CAM in *Tillandsia*.

12.2 Recycling of Respiratory CO_2 During CAM in *Tillandsia*

There are several reports of malate/CO_2 ratios in excess of one in various CAM species (Sale and Neales 1980; Griffiths et al. 1986; Borland and Griffiths 1990; Lüttge 1990; Fetene et al. 1991) but especially in epiphytic bromeliads (Griffiths et al. 1986; Smith et al. 1986; Martin and Adams 1987; Griffiths 1988a). In some cases, plants were examined under apparently nonoptimal conditions, suggesting that stress-induced stomatal closure restricted the uptake of atmospheric CO_2, thereby increasing the proportional contribution of respired CO_2 to overnight malate accumulation (Griffiths et al. 1986; Smith et al. 1986; Griffiths 1988a; Lüttge 1990). On the other hand, some species were examined under presumably optimal conditions. For example, Martin and Adams (1987) reported malate/CO_2 ratios of approximately two in well-watered individuals of *Tillandsia schiedeana* measured at 1400 $\mu mol\ m^{-2} s^{-1}$ PPFD, day/night temperatures of 25/20 °C, and day/night vapour-pressure deficits of 1.4/0.6 kPa. Not surprisingly, this ratio increased as plants were progressively drought-stressed.

Leaves of *T. schiedeana* contain large amounts of living, nonphotosynthetic water-storage parenchyma tissue (hereafter referred to as "hydrenchyma"). In a test of the hypothesis that respiration of nonphotosynthetic tissue contributes enough CO_2 to elevate the malate/CO_2 ratio, Loeschen et al. (1993) measured nocturnal CO_2 exchange and malate fluctuations in 12 species of *Tillandsia* with varying degrees of hydrenchyma. Although leaves of several species contained approximately 50% hydrenchyma, expressed as per cent of the cross-sectional area of a leaf, malate/CO_2 ratios were approximately one in all species except *T. schiedeana* (30% hydrenchyma). Thus, the presence of large amounts of living, nonphotosynthetic tissue adjacent to CAM-performing chlorenchyma tissue does not necessarily effect elevated malate/CO_2 ratios. Similar conclusions were reached by Lüttge and Ball (1987) with species of *Ananas*, *Aechmea*, and *Hechtia*.

The high malate/CO_2 ratios observed in *T. schiedeana*, as well as other CAM species, may be explained by one or both of the following: low stomatal conductances may restrict rates of CO_2 uptake at night, and/or tissue respiration rates may be unusually high. To test the latter, respiration rates of leaf slices of *T. schiedeana* were compared with those of five species of *Tillandsia* that exhibit malate/CO_2 ratios of one (Loeschen et al. 1993). Dark respiratory O_2 uptake was measured during the light period using an O_2 electrode to avoid complications from nocturnal CO_2 uptake. Rates of O_2 uptake in *T. schiedeana* were not significantly different from those of the other species (Fig. 12.2). Therefore, elevated tissue respiration rates cannot explain the high malate/CO_2 ratios measured in this species.

The accumulated evidence suggests that the capacity for fixation of atmospheric CO_2 during CAM in *T. schiedeana* is low, relative to other CAM species, while respiration rates are similar, hence increasing the relative contribution of respiratory CO_2 to the nocturnal accumulation of malate. Reduced capacity for fixation of atmospheric CO_2 may result from reductions in stomatal conductance. Indeed, maximal nocturnal CO_2 fixation rates and leaf conductances in *T. schiedeana* were lower than comparable values of two of the species used in the respiration experiment in spite of the fact that overnight accumulations of malate were similar (Fig. 12.3). In addition, although nocturnal increases in leaf malate concentrations in *T. ionantha* were nearly twice those in *T. schiedeana*, rates of CO_2 uptake and associated conductances in *T. ionantha* were approximately three times those in *T. schiedeana* (Fig. 12.3). Similarly, although gas exchange parameters of *T. balbisiana* and *T. setacea* were similar to those of *T. schiedeana*, nocturnal accumulations of malate were only half those of *T. schiedeana* (Fig. 12.3). One possible explanation for these findings is that only *T. schiedeana* suffered some degree of stress during the gas exchange and malate measurements in spite of the identical environmental conditions used during measurements with all species.

Based on the above evidence, the elevated malate/CO_2 ratios in *T. schiedeana* may result from stress-induced reductions in CO_2 uptake. With increasing stress,

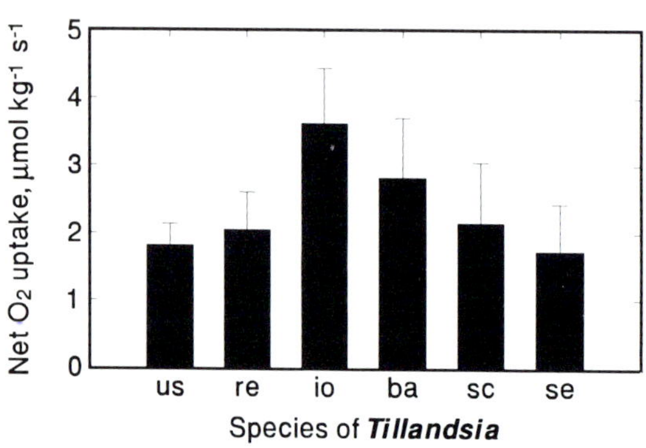

Fig. 12.2. Respiration rates of glasshouse-grown *Tillandsia usneoides* (*us*), *T. recurvata* (*re*), *T. ionantha* (*io*), *T. balbisiana* (*ba*), *T. schiedeana* (*sc*), and *T. setacea* (*se*). Oxygen consumption of leaf slices in water was measured at 20°C in the dark using a polarographic O_2 electrode and oxygen meter (see Martin and Harris 1993 for further details). Values are means of five plants, with one standard deviation shown. Rates are expressed on a dry-mass basis. There were no significant differences ($p > 0.05$; Kruskal-Wallis test) among the means

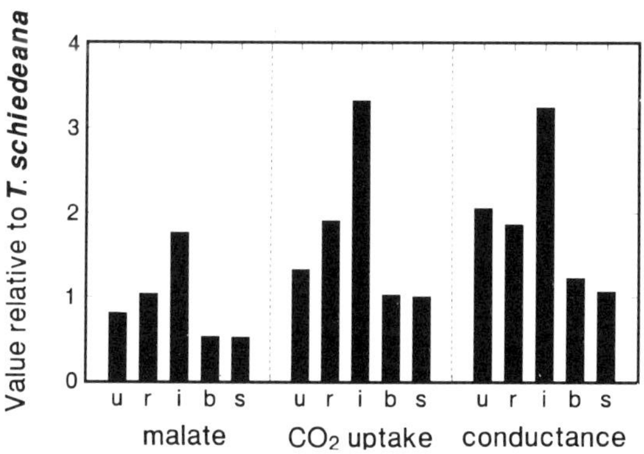

Fig. 12.3. Nocturnal accumulation of malate, maximum nocturnal rate of net CO_2 uptake, and maximum nocturnal shoot conductance to CO_2, expressed relative to the value for *Tillandsia schiedeana*, of five species of *Tillandsia* measured under identical environmental conditions (1000 to 1500 μmol m^{-2} s^{-1} PPFD for 12h, 30/20 °C day/night air temperatures, and 2.5/0.6 kPa vapour-pressure deficits). The species, abbreviations, and numbers of plants per mean value are: *T. usneoides* (*u*; 4), *T. recurvata* (*r*; 4), *T. ionantha* (*i*; 8), *T. balbisiana* (*b*; 4), *T. setacea* (*s*; 4), and *T. schiedeana* (4). In all three cases only *T. ionantha* means were significantly different ($p < 0.05$; Kruskal-Wallis and Mann-Whitney *U*-tests) from the others. Gas-exchange techniques and malate determinations are described in Loeschen et al. (1993)

e.g. increasing desiccation (Martin and Adams 1987), tissue malate/CO_2 ratios increase. When the fixation of atmospheric CO_2 eventually ceases as a result of severe stress, the plants undergo CAM-idling. Therefore, it seems logical to conclude that high levels of recycling of respiratory CO_2 simply reflect an intermediate stage of photosynthesis in a continuum from CAM under well-watered conditions to CAM-idling under severe stress. Similar conclusions were suggested by Griffiths (1988a, b, 1989), Lüttge (1990), and Fetene and Lüttge (1991).

12.3 Recycling of Respiratory CO_2 During CAM-Cycling in *Talinum*

Martin and coworkers estimated the potential savings in water associated with CAM-cycling in *Sedum pulchellum* (Martin et al. 1988a) and *Talinum calycinum* (Martin et al. 1988b) assuming that the amount of water conserved can be calculated by converting the amount of CO_2 recycled during nocturnal acidification into reductions in daytime transpiration using daytime values of water-use efficiency. The amounts of water potentially saved in these two species were low, less than 5% of normal daytime water loss, although higher values were obtained by using gas-exchange and acidity data from different individuals.

The water-conservation hypothesis is predicated on the assumption that the daytime decarboxylation of malate will elevate internal CO_2 concentrations in leaves of species exhibiting CAM-cycling (Martin et al. 1988b; Harris and Martin 1991a, b). This higher internal CO_2 concentration should effect partial stomatal closure during the day, at least in the morning when tissue acidity is highest and deacidification is underway (Cockburn et al. 1979). As is the case in CAM plants (sensu stricto), tissue CO_2 concentrations should be highest in the morning and lowest in the afternoon [often low enough to open the stomata in many CAM plants (Fig. 12.1; Kluge and Ting 1978; Winter 1985)]. Given that patterns of deacidification in plants with CAM-cycling are similar to those of CAM plants (compare Martin and Zee 1983, and Martin and Jackson 1986, with numerous CAM studies), the assumption that internal CO_2 concentrations in leaves of plants exhibiting CAM-cycling should be highest in the morning and then decrease throughout the day appears valid. Previous comparisons of rates of CO_2 uptake, leaf conductance, and internal CO_2 concentrations in *Talinum calycinum*, however, revealed no significant differences between morning and afternoon values of these gas-exchange parameters (Martin et al. 1988b; also see Fig. 12.1). This hypothesis was reevaluated using the same species under similar environmental conditions but correlating morning internal CO_2 concentrations and conductances to CO_2 with diurnal decreases in tissue malate content. The degree of nocturnal acidification did not correlate with either gas-exchange property (data not shown). Futhermore, when nocturnal acidification was inhibited by low temperature, leaf gas-exchange properties during the subsequent warm day differed little from plants undergoing normal levels of CAM-cycling (Fig. 12.4).

The above findings emphasize the insignificant effect of diel fluctuations in malate on gas exchange in photosynthetic tissues of plants exhibiting CAM-cycling. There are two possible explanations for this: either the diel fluctuations of

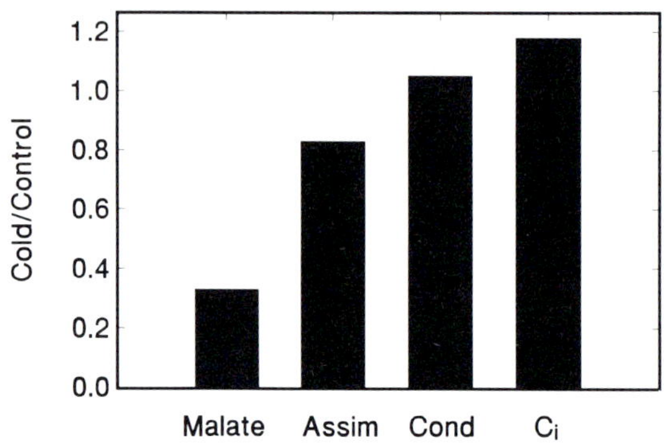

Fig. 12.4. Ratios of nocturnal malate accumulation and gas-exchange properties measured at noon in *Talinum calycinum* exposed to cold (5 °C) or control (20 °C) conditions the night before. Ratios represent comparisons of three cold-treated plants with 10 individuals monitored under control conditions for malate accumulations ("malate"; means were significantly different at $p < 0.05$; Mann-Whitney *U*-test), and comparisons of net CO_2 uptake ("assim"), conductance ("cond"), and internal CO_2 concentration ("C_i") at noon following a cold night with data collected at noon the previous day following a control night for three individuals. None of the gas-exchange properties was significantly different between treatments ($p < 0.05$; Mann-Whitney *U*-test). Gas-exchange techniques and malate determinations are described in Martin et al. (1988a,b) and Harris and Martin (1991a,b)

malate in plants with CAM-cycling are minor relative to CAM plants, or acid fluctuations may be similar in both types of plants, yet are proportionally inconsequential relative to rates of daytime CO_2 exchange and water loss in CAM-cycling species. If the latter were true, daytime rates of gas exchange in plants exhibiting CAM-cycling should greatly exceed the nocturnal rates of CAM plants and, presumably, the capacity of the latter for CO_2 fixation during the daytime. This is apparently the case. Comparisons of malate accumulations and gas-exchange rates between plants with CAM-cycling and selected CAM plants (sensu stricto) reveal that nocturnal malate accumulations in the two types of plants are not substantially different (Fig. 12.5), yet daytime rates of CO_2 uptake in CAM-cycling species are much greater than nocturnal gas-exchange rates in

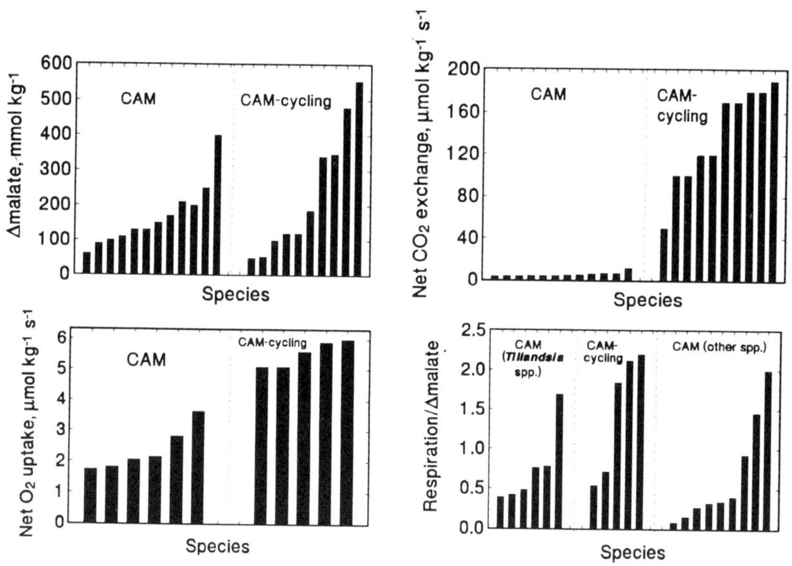

Fig. 12.5. Mean nocturnal increases in tissue malate concentrations (*upper left panel*) and mean maximum rates of net CO_2 uptake (*upper right panel*) at night for 12 species of *Tillandsia* ("CAM") and during the day for five species each of *Sedum* and *Talinum* ("CAM-cycling"). All plants (n = 4 or more) were well-watered (except one species of *Sedum*) and measured under similar environmental conditions (approx. 1500 µmol m^{-2} s^{-1} PPFD, 25 to 30 °C day temperature, 1.5 kPa day vapour-pressure deficit, 20 °C night temperature, and 0.5 kPa night vapour-pressure deficit). Data are expressed on a dry-mass basis and were obtained from Martin et al. (1988a), Harris and Martin (1991a), Gravatt and Martin (1992), and Loeschen et al. (1993)

Mean (n = 5) rates of dark respiration (*lower left panel*), determined with an O_2 electrode using sliced tissue in water, of six species of *Tillandsia* ("CAM"; same data as in Fig. 12.2) and five species of *Talinum* ("CAM-cycling"; data from Martin and Harris 1993) measured at 20 °C during the light period. Data are expressed on a dry-mass basis

Integrated nighttime respiration (based on O_2 uptake data) expressed as a fraction of the nocturnal accumulation of malate in the same species (*lower right panel*) for six species of *Tillandsia* ("CAM"; data from Fig. 12.2 and Loeschen et al. 1993), five species of *Talinum* ("CAM-cycling"; data from Martin and Harris 1993), and nine species of CAM plants ("other spp."). The latter data were selected from various published studies

CAM taxa (Figs. 12.1,12.5). Thus, although diel malate fluctuations during CAM-cycling can be substantial and similar to those in many CAM plants, the amount of respiratory CO_2 recycled during CAM-cycling appears inconsequential when compared to rates of daytime CO_2 uptake.

Given that nocturnal accumulations of malate in plants undergoing CAM-cycling can be equivalent to those of some CAM plants (Fig. 12.5), or even higher in CAM-cycling plants (Fig. 12.1), it may appear puzzling that, unlike CAM plants (Cockburn et al. 1979), the nocturnal fixation of CO_2 in the photosynthetic tissues does not lower internal CO_2 concentrations enough to effect stomatal opening. On the other hand, consideration of the high respiration rates in CAM-cycling species (Fig. 12.5) reveals that internal CO_2 concentrations most likely remain elevated in spite of fixation of CO_2 by PEPC. In other words, the rate of CO_2 release by respiration exceeds the rate of CO_2 uptake at night in photosynthetic tissues of plants with CAM-cycling. The same argument can be made during the day to explain why the CO_2 released from the decarboxylation of malate in photosynthetic tissues of these taxa does not result in stomatal closure. In this case, rates of daytime CO_2 fixation most likely greatly exceed rates of CO_2 release from malate decarboxylation. This would presumably have only a slight effect in decreasing stomatal conductance, explaining the low estimates obtained for water conservation potentially resulting from CAM-cycling (Martin et al. 1988a,b; Gravatt and Martin 1992).

The respiration rates of CAM-cycling species exceed those of selected CAM species; however, the difference is much less than predicted based on comparison of rates of atmospheric CO_2 uptake in the two groups of plants (Fig. 12.5). Furthermore, integration of these respiration rates implicates respiration as the sole source of CO_2 for acidification during CAM-cycling and, surprisingly, as a substantial source of CO_2 in many CAM species. The former is not surprising, but the latter is unexpected given the malate/CO_2 ratios of one reported for most of the CAM species included in this survey (Loeschen et al. 1993). This unexpected finding is not unique to data collected in this laboratory; respiration rates were high enough to account for a substantial proportion of nocturnal malate accumulation in several other CAM species examined in other laboratories (Fig. 12.5). Given the difficulty of measuring respiration rates in CAM plants, a result of interference from nocturnal CO_2 uptake, gas-diffusion problems in aqueous measuring systems, and potential wounding effects associated with measurements of tissue pieces, these findings may reflect large errors in estimating realistic respiration rates in CAM, but not CAM-cycling, species (see also Chap. 1). It is interesting to note in this regard that nocturnal malate accumulations in three CAM species measured in a CO_2-free environment were quite small relative to malate accumulation measured under normal air, implying that contributions of respiratory CO_2 to malate formation at night were minor (Winter et al. 1986). Perhaps this represents a more realistic approach in determining respiration rates in CAM tissues.

The apparently minor contribution of respiratory CO_2 to the physiology of plants undergoing CAM-cycling necessitates a reevaluation of the ecophysiological consequences of CAM-cycling (Martin et al. 1988b). It is difficult to envision any direct benefits to the plant by the relatively minor level of CO_2 recycling that occurs in these plants. Indeed, a more reasonable explanation is offered by Ting and coworkers, i.e. that CAM-cycling represents a precursor to CAM-idling but is not beneficial by itself (Rayder and Ting 1981; Ting and Burk 1983; Sipes and Ting 1985). Given that diel fluctuations of malate are as high as in many CAM species, and that large acid fluctuations are often maintained in CAM-cycling plants under severe drought stress, the physiology of CAM-cycling plants under severe drought is apparently identical to that of CAM plants under severe stress. Thus, the benefits of recycling respiratory CO_2 via CAM during severe stress, i.e., protection of the photosynthetic apparatus and maintenance of the plant in a metabolically "poised" state, should prove adaptive for CAM-cycling as well as CAM plants.

12.4 Concluding Remarks

The findings presented above argue against a direct benefit resulting from recycling large amounts of CO_2 in CAM plants and in plants exhibiting CAM-cycling. Instead, the results emphasize the potential importance of CAM-idling under conditions of severe stress. Thus, when CAM plants are stressed, CAM-idling may be an inevitable but potentially beneficial consequence of stomatal closure. Plants with CAM-cycling may best be viewed as C_3 plants when well-hydrated and CAM-idling plants when stressed. Given the apparent importance of CAM-idling, future research should be directed toward elucidation of the potentially beneficial consequences of this variation of CAM.

Acknowledgements. Special thanks go to numerous students in my laboratory over the past 13 years for helpful discussions and productive collaborations. One of those students, Valerie Loeschen, is gratefully acknowledged for outstanding contributions to my research efforts. I also thank Jan Elder and Judy Wiglesworth for their excellent word processing skills. Much of my research has been funded by the University of Kansas General Research Fund and the National Science Foundation.

References

Borland AM, Griffiths H (1990) The regulation of CAM and respiratory recycling by water supply and light regime in the C_3-CAM intermediate *Sedum telephium*. Funct Ecol 4: 33–39

Cockburn W (1985) Variation in photosynthetic acid metabolism in vascular plants: CAM and related phenomena. New Phytol 101: 3–24

Cockburn W, Ting IP, Sternberg LO (1979) Relationships between stomatal behavior and internal carbon dioxide concentration in crassulacean acid metabolism plants. Plant Physiol 63: 1029–1032

Eickmeier WG (1979) Eco-physiological differences between high and low elevation CAM species in Big Bend National Park, Texas. Am Midl Nat 101: 118–126

Fetene M, Lüttge U (1991) Environmental influences on carbon recycling in a terrestrial CAM bromeliad, *Bromelia humilis* Jacq. J Exp Bot 42: 25–31

Fetene M, Lüttge U, Ziegler H (1991) Flexibility in CO_2-fixation pathway of a highland *Kalanchoë, Kalanchoë petitiana* A. Rich. Bot Acta 104: 374–378

Gravatt DA, Martin CE (1992) Comparative ecophysiology of five species of *Sedum* (Crassulaceae) under well-watered and drought-stressed conditions. Oecologia 92: 532–541

Griffiths H (1988a) Carbon balance during CAM: an assessment of respiratory CO_2 recycling in the epiphytic bromeliads *Aechmea nudicaulis* and *Aechmea fendleri*. Plant Cell Environ 11: 603–611

Griffiths H (1988b) Crassulacean acid metabolism: a re-appraisal of physiological plasticity in form and function. Adv Bot Res 15: 43–92

Griffiths H (1989) Carbon dioxide concentrating mechanisms and the evolution of CAM in vascular epiphytes. In: Lüttge U (ed) Vascular plants as epiphytes. Evolution and ecophysiology. Springer, Berlin Heidelberg New York, pp 42–86

Griffiths H, Lüttge U, Stimmel K-H, Crook CE, Griffiths NM, Smith JAC (1986) Comparative ecophysiology of CAM and C_3 bromeliads. III. Environmental influences on CO_2 assimilation and transpiration. Plant Cell Environ 9: 385–393

Harris FS, Martin CE (1991a) Correlation between CAM-cycling and photosynthetic gas exchange in five species of *Talinum* (Portulacaceae). Plant Physiol 96: 1118–1124

Harris FS, Martin CE (1991b) Plasticity in the degree of CAM-cycling and its relationship to drought stress in five species of *Talinum* (Portulacaceae). Oecologia 86: 575–584

Kluge M, Ting IP (1978) Crassulacean acid metabolism. Analysis of an ecological adaptation. Springer, Berlin Heidelberg New York

Lee HSJ, Lüttge U, Medina E, Smith JAC, Cram WJ, Diaz M, Griffiths H, Popp M, Schäfer C, Stimmel K-H, Thonke B (1989) Ecophysiology of xerophytic and halophytic vegetation of a coastal alluvial plain in northern Venezuela. III. *Bromelia humilis* Jacq., a terrestrial CAM bromeliad. New Phytol 111: 253–271

Loeschen VS, Martin CE, Smith M, Eder SL (1993) Leaf anatomy and CO_2 recycling during crassulacean acid metabolism in twelve epiphytic species in the genus *Tillandsia* (Bromeliaceae). Int J Plant Sci 154: 100–106

Lüttge U (1986) Nocturnal water storage in plants having crassulacean acid metabolism. Planta 168: 287–289

Lüttge U (1987) Carbon dioxide and water demand: crassulacean acid metabolism (CAM), a versatile ecological adaptation exemplifying the need for integration in ecophysiological work. New Phytol 106: 593–629

Lüttge U (1990) Nocturnal citrate accumulation and its response to environmental stress in the CAM plant *Kalanchoë pinnata* (Lam.) Pers. Plant Cell Environ 13: 977–982

Lüttge U, Ball E (1987) Dark respiration of CAM plants. Plant Physiol Biochem 25: 3–10

Lüttge U, Ball E, Fetene M, Medina E (1991) Flexibility of crassulacean acid metabolism in *Kalanchoë pinnata* (Lam.) Pers. I. Response to irradiance and supply of nitrogen and water. J Plant Physiol 137: 259–267

Martin CE, Adams WW III (1987) Crassulacean acid metabolism, CO_2-recycling, and tissue desiccation in the Mexican epiphyte *Tillandsia schiedeana* Steud (Bromeliaceae). Photosynth Res 11: 237–244

Martin CE, Harris FS (1993) Nocturnal respiration rates and malate accumulation in five species of *Talinum* (Portulacaceae) during CAM-cycling. J Plant Physiol 141: 762–764

Martin CE, Jackson JL (1986) Photosynthetic pathways in a midwestern rock outcrop succulent, *Sedum nuttallianum* Raf. (Crassulaceae). Photosynth Res 8: 17–29

Martin CE, Zee AK (1983) C_3 photosynthesis and crassulacean acid metabolism in a Kansas rock outcrop succulent, *Talinum calycinum* Engelm. (Portulacaceae). Plant Physiol 73: 718–723

Martin CE, Higley M, Wang W-Z (1988a) Recycling of CO_2 via crassulacean acid metabolism in the rock outcrop succulent *Sedum pulchellum* Michx. (Crassulaceae). Photosynth Res 18: 337–343

Martin CE, Higley M, Wang W-Z (1988b) The ecophysiological significance of CO_2-recycling via crassulacean acid metabolism in *Talinum calycinum* Engelm. (Portulacaceae). Plant Physiol 86: 562–568

Medina E, Delgado M (1976) Photosynthesis and night CO_2 fixation in *Echeveria columbiana* v. Poellnitz. Photosynthetica 10: 155–163

Monson RK (1989) On the evolutionary pathways resulting in C_4 photosynthesis and crassulacean acid metabolism (CAM). Adv Ecol Res 19: 57–110

Nobel PS, Hartsock TL (1978) Resistance analysis of nocturnal carbon dioxide uptake by a crassulacean acid metabolism succulent, *Agave deserti*. Plant Physiol 61: 510–514

Nobel PS, Hartsock TL (1983) Relationships between photosynthetically active radiation, nocturnal acid accumulation, and CO_2 uptake from a crassulacean acid metabolism plant, *Opuntia ficus-indica*. Plant Physiol 71: 71–75

Nobel PS, Lüttge U, Heuer S, Ball E (1984) Influence of applied NaCl on crassulacean acid metabolism and ionic levels in a cactus, *Cereus validus*. Plant Physiol 75: 799–803

Osmond CB (1982) Carbon cycling and stability of the photosynthetic apparatus in CAM. In: Ting IP, Gibbs M (eds) Crassulacean acid metabolism. American Society of Plant Physiologists, Rockville, pp 112–127

Osmond CB, Winter K, Powles SB (1980) Adaptive significance of carbon dioxide cycling during photosynthesis in water-stressed plants. In: Turner NC, Kramer PJ (eds) Adaptation of plants to water and high temperature stress. John Wiley, New York, pp 139–154

Patel A, Ting IP (1987) Relationship between respiration and CAM-cycling in *Peperomia camptotricha*. Plant Physiol 84: 640–642

Rayder L, Ting IP (1981) Carbon metabolism in two species of *Pereskia* (Cactaceae). Plant Physiol 68: 139–142

Sale PJM, Neales TF (1980) Carbon dioxide assimilation by pineapple plants, *Ananas comosus* (L.) Merr. I. Effects of daily irradiance. Aust J Plant Physiol 7: 363–373

Sipes DL, Ting IP (1985) Crassulacean acid metabolism and crassulacean acid metabolism modifications in *Peperomia camptotricha*. Plant Physiol 77: 59–63

Smith JAC, Griffiths H, Lüttge U, Crook CE, Griffiths NM, Stimmel K-H (1986) Comparative ecophysiology of CAM and C_3 bromeliads. IV. Plant water relations. Plant Cell Environ 9: 395–410

Szarek SR, Ting IP (1974) Seasonal patterns of acid metabolism and gas exchange in *Opuntia basilaris*. Plant Physiol 54: 76–81

Szarek SR, Johnson HB, Ting IP (1973) Drought adaptation in *Opuntia basilaris*. Significance of recycling carbon through crassulacean acid metabolism. Plant Physiol 52: 539–541

Ting IP (1985) Crassulacean acid metabolism. Annu Rev Plant Physiol 36: 595–622

Ting IP, Burk JH (1983) Aspects of carbon metabolism in *Welwitschia*. Photosynth Res 32: 279–285

Ting IP, Rayder L (1982) Regulation of C_3 to CAM shifts. In: Ting IP, Gibbs M (eds) Crassulacean acid metabolism. American Society of Plant Physiologists, Rockville, pp 193–207

Virzo De Santo A, Fioretto A, Bartoli G, Alfani A (1987) Gas exchange of two CAM species of the genus *Cissus* (Vitaceae) differing in morphological features. Photosynth Res 13: 113–124

Winter K (1985) Crassulacean acid metabolism. In: Barber J, Baker NR (eds) Photosynthetic mechanisms and the environment. Elsevier, Amsterdam, pp 329–387

Winter K, Schröppel-Meier G, Caldwell MM (1986) Respiratory CO_2 as carbon source for nocturnal acid synthesis at high temperatures in three species exhibiting crassulacean acid metabolism. Plant Physiol 81: 390–394

13 Ontogenetic Development of Crassulacean Acid Metabolism as Modified by Water Stress in *Peperomia*

I.P. Ting, A. Patel, S. Kaur, J. Hann, and L. Walling

13.1 Introduction

Peperomia species (Piperaceae) may show C_3 photosynthesis, CAM photosynthesis, or CAM-cycling in which stomata and gas exchange occur primarily during the day with a diurnal fluctuation of acidity typical of CAM (Holthe et al. 1992). In CAM-cycling, enzymes of CAM are present (Sipes and Ting 1985; Ting 1985; Ting and Sipes 1985). In *Peperomia* species, there is a division of photosynthetic labor among the various tissues of the leaf (Nishio and Ting 1987, 1993). The upper multiple epidermis functions as a window leaf filtering out much of the incident radiant energy and has about 5% of the leaf chlorophyll (Nishio and Ting 1987). The middle palisade mesophyll has most of the C_3 photosynthetic activity including ribulose 1,5-bisphosphate carboxylase/oxygenase (RUBISCO) and the light-dependent CO_2 fixation. The lower spongy mesophyll has the appearance of CAM tissue, and along with the multiple epidermis contains most of the phospho*enol*pyruvate carboxylase (PEPC) and dark CO_2 fixation (Nishio and Ting 1987). The lower spongy mesophyll is enriched in PSI relative to PSII, evidently related to the extra ATP requirement of this tissue (Nishio and Ting 1993). Stomata are restricted to the abaxial epidermis. Malic acid accumulates during the night. During the subsequent day period, decarboxylation keeps CO_2 high in the vicinity of the photosynthetic palisade cells, thereby minimizing photorespiration (Nishio and Ting 1987).

In this chapter, we discuss the normal ontogenetic development of CAM in *Peperomia*. We also report how water stress alters CAM and results in an enhancement of PEPC activity during the expression of CAM in *Peperomia*.

13.2 Experimental Plant Material

A survey of 93 *Peperomia* taxa was conducted to assess the presence of C_3 photosynthesis or CAM photosynthesis (Holthe et al. 1992). Plants were obtained from the University of California, Riverside, plant collection; the University of California, Berkeley, Botanic Gardens; from KDB Custom Farming, Fresno, CA;

Department of Botany and Plant Sciences, University of California, Riverside, CA 92521, USA

Ecological Studies, Vol. 114
Winter/Smith (eds) Crassulacean Acid Metabolism

Glasshouse Works, Stewart, OH; from Ms. Anita Baudean, New Orleans, LA; and from the Department of Botany, Ohio Wesleyan University, Delaware, OH. All plants were propagated by cuttings and grown in a glasshouse under standard conditions of water and nutrition. PPFD did not exceed 135 μmol m^{-2} s^{-1}, except for those species known to be sun plants. The latter were exposed to about 500 μmol m^{-2} s^{-1} PPFD. CAM was assessed on the basis of gas exchange, diurnal organic acid fluctuation, and the presence and quantities of PEPC, phosphofructokinase (PFK), and pyrophosphate: fructose 6-phosphate phosphotransferase (PFP) activities (Holthe et al. 1992). Plants were categorized as C_3 if they met the following criteria: PEPC and PFP activities were low, no diel organic-acid fluctuation, and daytime gas exchange. Plants were categorized as CAM if PEPC and PFP acitivities were high, there was measurable diel organic-acid fluctuation, and if there was nocturnal gas exchange. Plants were categorized as CAM-cycling if PEPC and PFP activities were high and there was measurable diel organic-acid fluctuation, but only daytime gas exchange (Ting and Sipes 1985).

For the developmental studies reported here, clones of *Peperomia scandens* Ruiz and Pavón and *P. camptotricha* Miq. were propagated by vegetative cuttings and grown in 150-mm pots in a glasshouse at Riverside, CA. *P. scandens* is a shade epiphyte that grows naturally on rock outcrops in Mexico and parts of the Caribbean and was grown under shadecloth with an average PPFD of about 150 μmol m^{-2} s^{-1}. *P. camptotricha* is an epiphyte native to Mexico that grows in full sun and was grown in the glasshouse with an average PPFD of about 500 μmol m^{-2} s^{-1}. Plants were watered whenever necessary to maintain turgor except during water-stress experiments, which were conducted by withholding all water.

Figure 13.1 shows a drawing of the successive leaves of *Peperomia scandens* as they mature and are numbered 1 through 4 from the youngest that do not show attributes of CAM to the older, more mature leaves with CAM (see Figs. 13.2 to 13.6). Samples were taken for analysis from the leaves numbered 1 to 4. In *P. camptotricha*, 6 to 8 leaves of similar age are arranged in successive false whorls. Sampling was done by taking all leaves in successive false whorls numbered 1 (youngest) to 8 (older).

13.3 CAM in *Peperomia*

13.3.1 Distribution Among Species

Peperomia is a pantropical genus of more than 600 species most common in the Americas, the Caribbean, Hawaii, Polynesia, and the Malay Peninsula (Burger 1977). Plants occur mostly as epiphytes or grow on rocks and shallow soils. Nearly all have succulent leaves. In a survey of 93 taxa, we found 52% to have attributes associated with CAM. Of these, approximately two-thirds showed only daytime gas exchange, but still showed high PEPC activity and a diel fluctuation of acidity, i.e., were CAM-cycling rather than full CAM species. Representative

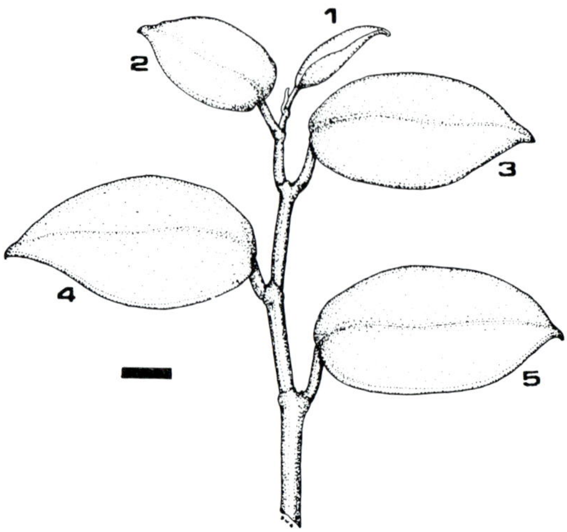

Fig. 13.1. Drawing of *Peperomia scandens* branch showing leaf arrangement. Leaves numbered *1* through *4* of increasing age were used in the experiments reported here. *Bar* = 10mm

Table 13.1. Mean values for PEPC activity, PFP activity, diel acid fluctuation, and occurrence of nocturnal gas exchange for 45 species of *Peperomia* surveyed experimentally (20 C_3 species, 22 CAM-cycling species, and 3 CAM species). Data from Holthe et al. (1992)

	C3	CAM-cycling	CAM	$P < F$[a]
PEPC[b]	51.4	119.9	211.0	0.09
PFP[b]	54.5	134.6	143.2	0.04
Acid[c]	1.6	29.2	23.8	0.001
Nocturnal Gas exchange	NO	NO	YES	–

[a]One-way ANOVA conducted by pooling data from CAM and CAM-cycling.
[b]Enzyme activities in nmol min^{-1} cm^{-2}.
[c]Diel organic-acid fluctuation in μmol g^{-1} fr. wt.

values for the enzyme activities and diel acid fluctuations for the C_3, CAM-cycling and full CAM species are shown in Table 13.1. Thus, the evidence to date indicates that CAM or the potential for CAM is widespread in this genus.

13.3.2 Ontogenetic Expression of CAM

In those species of *Peperomia* that show CAM or CAM-cycling (Ting and Sipes 1985), during normal development of leaves there is a progressive expression of CAM (Holthe et al. 1987). In the youngest leaves that can be sampled, all the attributes of C_3 photosynthesis are present. Stomata are open only during the daytime, all gas exchange occurs during the daylight period, there is no diel fluctuation of organic acids, and the enzymes usually associated with CAM are deficient or absent (Holthe et al. 1987, 1992). As leaves mature, there is first the expression of CAM enzyme activites [e.g. PEPC, PEP carboxykinase (PEPCK), and NADP-malic enzyme] accompanied by the appearance of a diel fluctuation of organic acids. At this time, virtually all gas exchange is during the day. By definition, the plants are showing CAM-cycling (Ting and Sipes 1985) and the diel fluctuation of acidity is largely the result of the refixation of respiratory CO_2 (Patel and Ting 1987). After the expression of CAM-cycling, there is a shift from daytime stomatal opening and gas exchange to nighttime stomatal opening and gas exchange such that the plants are showing full CAM (Holthe et al. 1987). Our model to depict the shift from C_3 photosynthesis to CAM is as follows:

$$C_3 \longrightarrow \text{CAM-cycling} \longrightarrow \text{CAM}$$

The rationale for this model is twofold. First, it is consistent with previously published data (Sipes and Ting 1985; Holthe et al. 1987). Second, there are some species of *Peperomia* that appear only to show C_3 photosynthesis, there are some that only show CAM-cycling, and some that develop to full CAM (Holthe et al. 1992). All of our evidence indicates that the expression of CAM in *Peperomia* is a normal developmental event that is largely independent of environmental perturbations such as water stress.

Our model may be rather general for many groups of succulent plants. For example, there is more PEPC activity in older leaves than in younger leaves of *Bryophyllum* (= *Kalanchoë*) *calycinum* (Nishida 1978) and *Mesembryanthemum crystallinum* (von Willert et al. 1976). And in *Kalanchoë daigremontiana* (Amagasa 1982) and *Mesembryanthemum crystallinum* (von Willert et al. 1976), the extent of diel organic-acid fluctuation is greater in older leaves than in younger leaves. Similarly, it is known that there is greater nocturnal gas exchange in older leaves than younger leaves of *Bryophyllum* (= *Kalanchoë*) *fedtschenkoi* (Jones 1975) and in *Senecio medley-woodii* (Eller et al. 1988). These observations and our observations with *Peperomia* are consistent with a developmental shift from C_3 to CAM in succulents. At least the *Peperomia* species that shift from C_3 to CAM-cycling or CAM appear different from *M. crystallinum* in that the latter species will stay CAM for extended periods of time (von Willert et al. 1976). The expression of CAM in *Peperomia* species appears to be independent of environmental signals, but water stress, light, and temperature modify the extent of the CAM activity (Patel 1992).

13.3.3 Modification of CAM Expression by Water Stress

Plants stressed by withholding water lost about 2 to 2.5% fresh weight per day. After rewatering, recovery to the control turgidity level occurred within 1 to 2 days. In *Peperomia*, stomata tend to respond to water stress by closing very rapidly (Hanscom and Ting 1978). Evidently, because of stomatal closure restricting exogenous CO_2 uptake, by 3 to 6 days of water stress diel organic-acid accumulation began to decrease. Rewatering resulted in a return to the control level within 9 days (data from leaf 4). In these same leaves, water stress resulted in a linear increase in PEPC activity up to about 9 days. Upon rewatering, activity returned to the well-watered control level within 6 days (data not shown).

When well-watered, successive leaves of *P. scandens* showed an increase in titratable acidity starting at less than 5 μmol cm^{-2} in the very youngest leaves tested (Fig. 13.2). However, in the youngest leaves, there was no evidence for diurnal acid fluctuation. As leaves matured, there was an enhanced fluctuation of acidity reaching as much as 18 to 20 μmol cm^{-2} (Fig. 13.2). When water stressed for 12 days, however, a slight day/night fluctuation of acidity was observed in the very youngest leaves, suggesting an earlier expression of CAM (Fig. 13.2). In the older leaves, the diel fluctuation of acidity was damped, evidently because of stomatal closure in response to the water stress. Experiments with *P. camptotricha* also showed increasing expression of CAM activity with leaf ageing as measured by diel acid fluctuation (Table 13.2). Young leaves from the youngest false whorls contained about 5 μmol cm^{-2} of acidity, but with virtually no diel fluctuation. There was a progressive increase in diel acid fluctuation until a

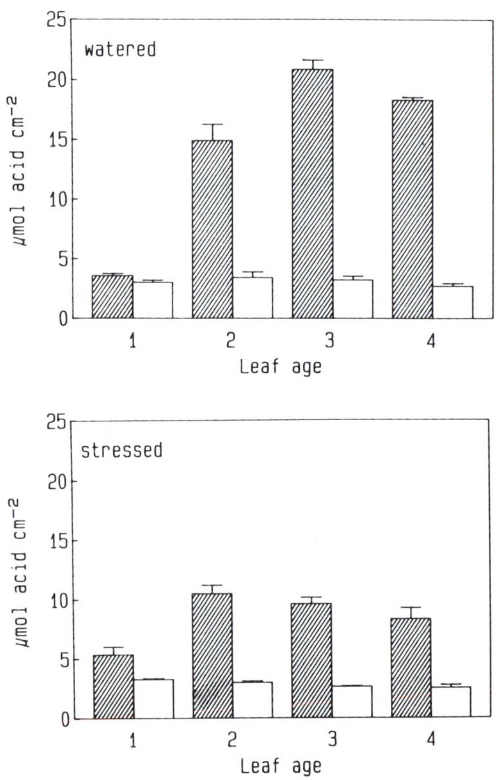

Fig. 13.2. *Upper panel* Diel fluctuation of titratable acidity in *P. scandens* with increasing leaf age (leaves *1–4*) determined with well-watered plants; *hatched bars* = end of dark period; *open bars* = end of light period. *Lower panel* Same as upper panel except plants under water stress. Values were determined by titration as previously described (Holthe et al. 1992) and are expressed as means ±SD of 3 samples

maximum of 18 μmol cm^{-2} was reached in the leaves of false whorls 7 and 8 (Table 13.2).

During the development of *P. scandens* leaves, PEPC activity increased reaching a maximum at about leaf age 3 (Fig. 13.3). There was a good correlation between the extent of diel acidity fluctuation and the maximum activity of PEPC in two experiments (Exp. 1: $r = 0.94$, $n = 3$; Exp. 2: $r = 0.88$, $n = 3$). When water stressed for 12 days, there was an increase in PEPC activity, even in the very youngest leaves tested (Fig. 13.3).

In *P. camptotricha*, PEPC activity increased with leaf maturation (Table 13.2) resulting in a linear correlation between acid accumulation and PEPC activity computed from the leaves of whorl 1 to the leaves of whorl 6 ($r = 0.997$). During water stress, there was about a 1.5-fold increase in extractable PEPC activity by leaf 4 (data not shown). Thus, *P. scandens* and *P. camptotricha* act similarly with respect to CAM expression in developing leaves and in their response to water stress.

Table 13.2. Relative changes in titratable acidity, PEPC activity, PEPC protein, and PEPC mRNA in successive leaves of *Peperomia camptotricha*. Data from Patel (1992) and Ting et al. (1993)

Whorl number[a]	Titratable acidity[b]	PEPC activity[c]	PEPC protein[d]	PEPC mRNA[d]
1–2	1.5	0.03	0.48	0.12
3–4	9.5	0.10	0.83	0.28
5–6	14.0	0.29	0.93	0.82
7–8	18.0	0.43	0.82	1.05

[a]Data for successive pairs of whorls were combined because of insufficient leaf material for protein and mRNA assay.
[b]μmol acid cm^{-2} diel fluctuation.
[c]μmol min^{-1} cm^{-2} PEPC activity.
[d]Relative units for PEPC protein and PEC mRNA determined by laser densitometry.

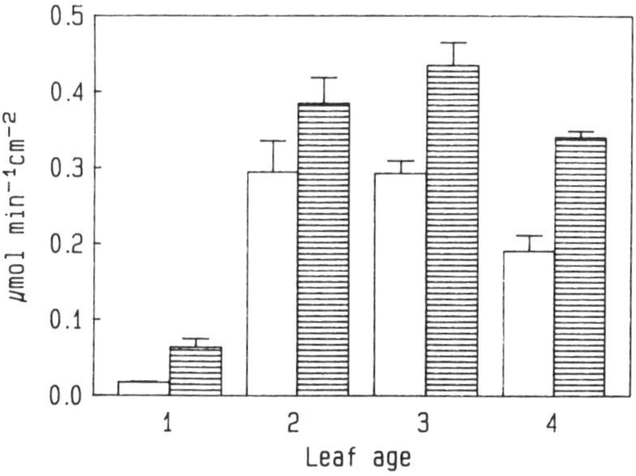

Fig. 13.3. PEPC activity in *P. scandens* with increasing age of leaves from *1* to *4*; *open bars* well-watered controls; *hatched bars* water-stressed plants. Assays were conducted as previously described (Sipes and Ting 1989) and data were statistically obtained from rate curves (Canellas and Wedding 1980). Values are means ± SD of three samples

PEPC protein in *P. scandens* also increased with increasing age of leaves (Fig. 13.4) consistent with the PEPC activity increase (Fig. 13.3). The specific activity increased between leaf 1 and leaves 2 to 4. In the latter three leaves, the specific activity was relatively constant at 0.087 ± 0.009 μmol min^{-1} cm^{-2} (PEPC protein $unit)^{-1}$. When water stressed for 12 days, there was only slight evidence of an increase in the accumulation of PEPC protein. In the very youngest leaves tested (leaf age 1), there was an increase in PEPC protein. Because of an increase in apparent PEPC activity in older leaves with little protein change, the calculated PEPC specific activity in leaves 2 to 4 was higher under the water stress conditions [0.14 ± 0.006 μmol min^{-1} cm^{-2} (PEPC protein $unit)^{-1}$].

In *P. camptotricha*, the PEPC-protein accumulation in developing leaves parallels the increase in PEPC activity (Table 13.2). There was a progressive increase in PEPC specific activity as well. Similarly to *P. scandens*, water stress in this species did not bring about significant enhancement in PEPC protein (data not shown), thereby resulting in an apparent increase in specific activity.

PEPC mRNA in *P. camptotricha* increased with ageing of the leaves in parallel with the increase in PEPC activity and PEPC protein (Table 13.2). In this experiment, PEPC mRNA increased in all leaves relative to the controls in response to 10 days water stress (Patel 1992). Because of limited plant material for the experiments reported here with *P. scandens*, we were unable to measure PEPC mRNA in the youngest leaf tested (i.e. leaf 1), but there was a clearly defined increase in PEPC transcripts in response to water stress in the older leaves (Fig. 13.5).

13.3.4 Recovery of Full CAM Expression After Rewatering

Figure 13.6 shows the increase in PEPC activity (V_{max}) to 10 days water stress in successive leaves of *P. scandens*, and the reversible nature of the increase when plants are rewatered. Similarly to the previous experiments, as leaves aged the

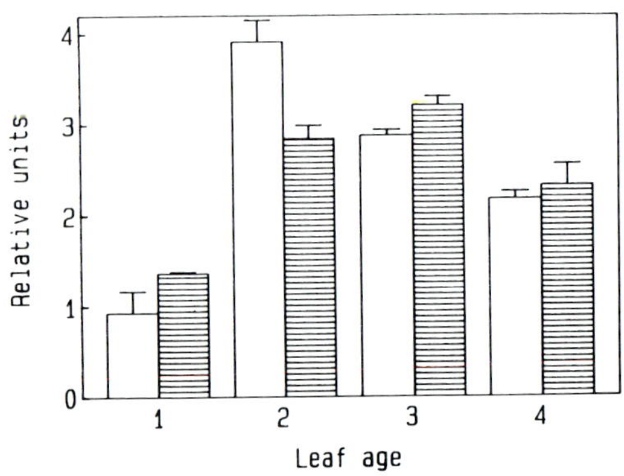

Fig. 13.4. PEPC protein in *P. scandens* with increasing age of leaves from *1* to *4*; *open bars* well-watered controls; *hatched bars* water-stressed plants. Data were obtained from Western blots (Pratt et al. 1986) hybridized with antibodies prepared against a CAM form of PEPC isolated from *P. camptotricha*. Values are means ± SD of three laser-densitometer traces

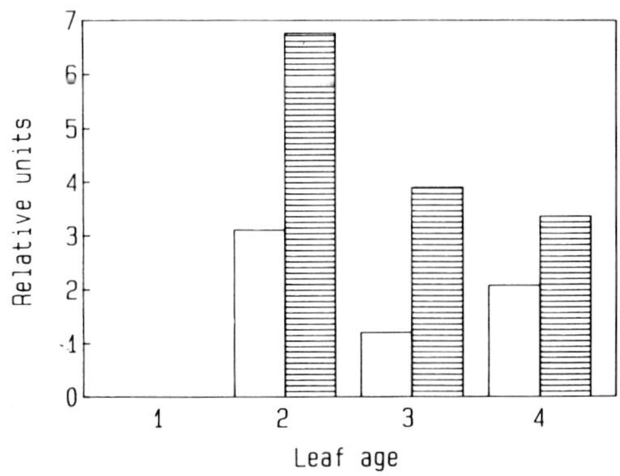

Fig. 13.5. PEPC mRNA in *P. scandens* from leaves *2* through *4*; *open bars* well-watered controls; *hatched bars* water-stressed plants. Data were obtained from Northern blots (Lehrach et al. 1977) hybridized against a cDNA probe made from the CAM PEPC of *P. scandens* (Patel 1992). Values are from laser-densitometer traces

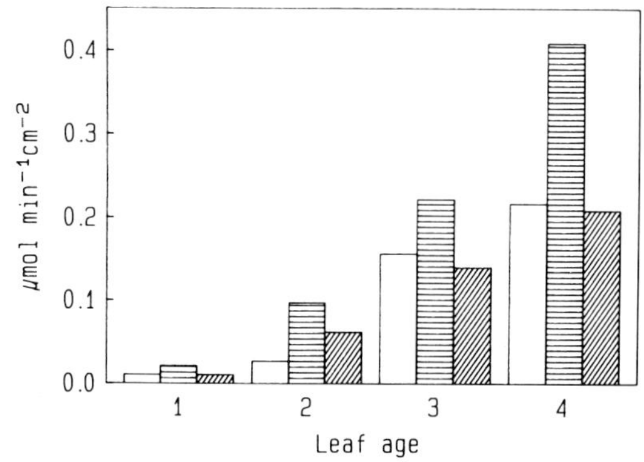

Fig. 13.6. PEPC activity in *P. scandens* leaves from well-watered control plants (*open bars*), water-stressed plants (*horizontally hatched bars*), and rewatered plants (*diagonally hatched bars*) in leaf ages *1* to *4*. Activity was determined as in Fig. 13.3

water stress resulted in an enhancement in PEPC activity above the normal ontogenetic increase as CAM is expressed. Rewatering resulted within 3 days in a reduction in extractable PEPC activity to the normal CAM levels (Fig. 13.6). There was a return to the normal diel fluctuation of acidity within 9 days.

13.4 Discussion of Water-Stress-Induced CAM Expression

13.4.1 General Effects

There are several apparent, easily measurable changes in *Peperomia* in response to water stress including: (1) stomatal closure restricting gas exchange; (2) reduction in the extent of diel organic-acid fluctuation evidently related to the stomatal closure; and (3) increased levels of PEPC mRNA and an increase in PEPC activity. It is also known that succulent plants respond to water stress by production of ethylene gas and abscisic acid (ABA) (Rayder and Ting 1983; Thomas et al. 1992). Although not proven, it is highly likely that the stomatal closure in response to water stress is related to the increase in tissue ABA, since the stomata of CAM plants are known to respond to ABA (Ting 1987).

13.4.2 PEPC mRNA

We can only speculate on the mechanism associated with the increase in PEPC mRNA in response to water stress. It is well known that water stress results in increased PEPC activity in many CAM plants and that in *Mesembryanthemum crystallinum* it is the result of de novo synthesis (see Chaps. 9 and 10). Possible factors regulating PEPC mRNA in the experiments reported here include CO_2, which would tend to be altered in leaves upon stomatal closure, an increase in tissue ethylene or ABA, and a decrease in organic acids including malic acid. CO_2 seems an unlikely factor, since in a previous experiment with *Portulacaria afra* we were unable to alter PEPC activity when grown at 40 ppm or 950 ppm CO_2 (Huerta and Ting 1988). The only measurable effect of CO_2 was that 40 ppm reduced the accumulation of organic acids. Similarly, Winter (1979) was unable to induce PEPC in *M. crystallinum* with high or low CO_2. However, in most of these experiments only PEPC activity was measured and not PEPC mRNA.

There is an increasing amount of evidence that ABA may play an important role in the expression of CAM when plants are under water stress. When *Portulacaria afra* is treated with ABA, CAM is expressed and PEPC activity increases similarly to the response to water stress (Ting 1981). Similarly, ABA treatment of *Mesembryanthemum crystallinum* results in the enhanced expression of CAM including the expression of PEPC (Chu et al. 1990) and the expression of PEPC mRNA (Cushman et al. 1989; but see also Thomas et al. 1992). In addition, the 5'-flanking region of the PEPC gene in *Mesembryanthemum* contains the same ABA-responsive elements as genes responsive to ABA (Cushman and Bohnert 1992). The possible significance of ABA in CAM induction and stress responses is further considered in Chapters 8, 9 and 10.

In previous experiments with *Portulacaria afra*, we were not able to show an enhancement of CAM or increased expression of PEPC activity by treatment with ethylene gas (A. Huerta and I.P. Ting, unpubl.). Thus, we do not believe that ethylene gas is a factor in CAM expression.

The increased accumulation of PEPC transcripts in response to water stress when plants are already performing CAM may be a compensatory response to the reduced tissue levels of malate; however, no mechanism is known. We can also speculate that water stress results in a reduction in PEPC protein, bringing about the production of more PEPC mRNA to compensate. Here it is visualized that the level of PEPC protein regulates mRNA levels.

13.4.3 PEPC Activity

We do not know why there is an increase in PEPC activity in response to water stress that does not appear to be a result of an increase in PEPC protein. However, diel changes in PEPC activity are well known (Kluge and Ting 1978) and the enzyme is known to be highly regulated (Winter 1980; see also Chaps. 3 and 11.). Thus, it is possible that water stress directly results in an increase in PEPC-specific activity.

In *Mesembryanthemum crystallinum*, salt stress induces the expression of one of the two PEPC isogenes associated with CAM (Cushman et al. 1989; see also Chap. 9). Similarly, water stress in *Peperomia* may result in the expression of one form of PEPC mRNA that codes for a protein that we do not detect with our antibody. However, this appears unlikely, since our PEPC antibody should be less specific than our cDNA probe and should detect any increase in PEPC protein.

13.4.4 Reversibility of the Water-Stress Response

The PEPC response to water stress is apparently fully reversible. In the experiments reported here, PEPC activity increased in response to water stress, reaching its maximum after about 9 days; upon rewatering, the activity returned to the control levels within 3 days (Fig. 13.6). Organic-acid fluctuation began to decrease within 3 to 6 days after stress, reaching a low level by 9 days. Recovery of organic-acid fluctuation occurred after 9 days. Thus, the CAM-related synthesis of organic acids seemed to decrease before the increase in PEPC activity, but PEPC activity was restored to the control level before titratable acidity returned to the control level. These results may be similar to those of *Mesembryanthemum crystallinum*, in which salt stress induces an increase in PEPC transcripts (Ostrem et al. 1987), which decline after the relief of the stress (Vernon et al. 1988; see also Chaps. 9 and 10).

This reversible expression of PEPC in response to salt and/or water stress in succulent plants may be a rather general phenomenon. As stated previously, it occurs with *Mesembryanthemum crystallinum* (Vernon et al. 1988) and we have observed a similar phenomenon when CAM-*Xerosicyos danguyi* and CAM-*Clusia rosea* are water stressed, forcing them into CAM-idling. Accompanying the transition from CAM to CAM-idling is an increase in PEPC activity, PEPC protein, and PEPC mRNA that is fully reversible upon rewatering (unpubl.).

13.5 Concluding Remarks

Peperomia species are excellent plants for studying the shift from C_3 photosynthesis to CAM. The genus is represented by species which show C_3, CAM-cycling, and CAM. The progression from C_3 to CAM-cycling or CAM appears to be a normal developmental phenomenon. However, this progression from C_3 to CAM during development can be modified with environmental perturbations such as water stress, light, and temperature (Patel 1992). Water stress results in a damping of organic-acid fluctuation, but an increase in PEPC mRNA and PEPC activity. Rewatering results in a return to the normal CAM level of activity.

Acknowledgements. This research was supported in part by US NSF grants DCB-8807860 (IPT) and DCB-8910462 (LLW). We thank Ms. Christine Peters for assistance in obtaining the data for Fig. 13.6.

References

Amagasa T (1982) The influence of leaf age on the diurnal changes of malate and starch in the CAM plant *Kalanchoë daigremontiana* Hamet et Perr. Z Pflanzenphysiol 108: 93–96

Burger WC (1977) The Piperales and the monocots--alternate hypothesis for the origin of monocotyledonous flowers. Bot Rev 43: 345–393

Canellas PF, Wedding RT (1980) Substrate and metal ion interactions in the NAD^+ malic enzyme for cauliflower. Arch Biochem Biophys 199: 259–264

Chu C, Dai Z, Ku MSB, Edwards GE (1990) Induction of crassulacean acid metabolism in the facultative halophyte *Mesembryanthemum crystallinum* by abscisic acid. Plant Physiol 93: 1253–1260

Cushman JC, Bohnert HJ (1992) Salt stress alters A/T-rich DNA-binding factor interactions with the phosphe*nol*pyruvate carboxylase promoter from *Mesembryanthemum crystallinum*. Plant Mol Biol 20: 411–424

Cushman JC, Meyer G, Michalowski CB, Schmitt JM, Bohnert HJ (1989) Salt stress leads to differential expression of two isogenes of phosphoe*nol*pyruvate carboxylase during crassulacean acid metabolism induction in the common ice plant. Plant Cell 1: 715–725

Eller BM, Ruess BR, Sharma S (1988) Carbon gain, water conservation and expression of CAM during leaf development of *Senecio medley-woodii*. J Plant Physiol 133: 304–309

Hanscom Z, Ting IP (1978) Responses of succulents to plant water stress. Plant Physiol 61: 327–330

Holthe PA, Sternberg LSL, Ting IP (1987) Developmental control of CAM in *Peperomia scandens*. Plant Physiol 84: 743–747

Holthe PA, Patel A, Ting IP (1992) The occurrence of CAM in *Peperomia*. Selbyana 13: 77–87

Huerta AJ, Ting IP (1988) Effects of various levels of CO_2 on the induction of crassulacean acid metabolism in *Portulacaria afra* (L.) Jacq. Plant Physiol 88: 183–188

Jiao JA, Chollet R (1991) Posttranslational regulation of phosphoenolpyruvate carboxylase in C_4 and crassulacean acid metabolism plants. Plant Physiol 95: 981–985

Jones MB (1975) The effect of leaf age on leaf resistance and CO_2 exchange of the CAM plant *Bryophyllum fedtschenkoi*. Planta 123: 91–96

Kluge M, Ting IP (1978) Crassulacean acid metabolism. Analysis of an ecological adaptation. Springer, Berlin Heidelberg New York

Lehrach H, Diamond D, Wozney JM, Boedtker H (1977) RNA molecular weight determinations by gel electrophoresis under denaturing conditions, a critical evaluation. Biochemistry 16: 4543–4751

Nishida K (1978) Effect of leaf age on light and dark $^{14}CO_2$ fixation in a CAM plant, *Bryophyllum calycinum*. Plant Cell Physiol 19: 935–941

Nishio JN, Ting IP (1987) Carbon flow and metabolic specialization in the tissue layers of the crassulacean acid metabolism plant, *Peperomia camptotricha*. Plant Physiol 84: 600–604

Nishio JN, Ting IP (1993) Photosynthetic characteristics of the palisade mesophyll and spongy mesophyll in the CAM/C_4 intermediate plant, *Peperomia camptotricha*. Bot Acta 106: 120–125

Ostrem JA, Olson SW, Schmitt JM, Bohnert HJ (1987) Salt stress increases the level of translatable mRNA for phosphoenolpyruvate carboxylase in *Mesembryanthemum crystallinum*. Plant Physiol 84: 1270–1275

Patel A (1992) Expression of photosynthetic genes during leaf development in *Peperomia* species. PhD Dissertation, University of California, Riverside, 123 pp

Patel A, Ting IP (1987) Relationship between CAM and respiration in *Peperomia camptotricha*. Plant Physiol 84: 640–642

Pautot V, Holzer F, Walling LL (1991) Differential expression of tomato proteinase inhibitor I and II genes during bacterial pathogen invasion and wounding. Mol Plant Microbe Interact 4: 284–292

Pratt LH, McCurdy W, Shimazaki Y, Cordonnier MM (1986) Immunodetection of phytochrome: Immunocytochemistry, immunoblotting, and immunoquantitation. In: Linskens HF, Jackson JF (eds) Immunology in plant sciences. Springer, Berlin Heidelberg New York, pp 50–74

Rayder L, Ting IP (1983) Shifts in the carbon metabolism of *Xerosicyos danguyi* H. Humb. (Cucurbitaceae) brought about by water stress. I. General characteristics. Plant Physiol 72: 606–610

Sipes D, Ting IP (1985) Crassulacean acid metabolism and crassulacean acid metabolism modifications in *Peperomia camptotricha*. Plant Physiol 77: 59–63

Sipes D, Ting IP (1989) Kinetic properties of phosphoenolpyruvate carboxylase in *Peperomia camptotricha*. Plant Physiol 91: 1050–1055

Thomas JC, McElwain EF, Bohnert HJ (1992) Convergent induction of osmotic stress-responses. Plant Physiol 100: 416–423

Ting IP (1981) Effects of abscisic acid on CAM in *Portulacaria afra*. Photosynth Res 2: 39–48

Ting IP (1985) Crassulacean acid metabolism. Annu Rev Plant Physiol 36: 595–622

Ting IP (1987) Stomata in plants with crassulacean acid metabolism. In: Zeiger E, Farquhar GD, Cowan IR (eds) Stomatal function. Stanford University Press, Stanford, pp 353–366

Ting IP, Sipes D (1985) Metabolic modifications of crassulacean acid metabolism – CAM-idling and CAM cycling. In: Burris RH, Ludden PW, Burris JE (eds) Nitrogen fixation and CO_2 metabolism. Elsevier, New York, pp 371–378

Ting IP, Hann J, Sipes D, Patel A, Walling LL (1993) Expression of P-enolpyruvate carboxylase and other aspects of CAM during the development of *Peperomia camptotricha* leaves. Bot Acta 106: 313–319

Vernon DM, Ostrem JA, Schmitt JM, Bohnert HJ (1988) PEPCase transcript levels in *Mesembryanthemum crystallinum* decline rapidly upon relief from salt stress. Plant Physiol 86: 1002–1004

Von Willert DJ, Kirst GO, Treichel S, von Willert K (1976) The effect of leaf age and salt stress on malate accumulation and phosphoenolpyruvate carboxylase activity in *Mesembryanthemum crystallinum*. Plant Sci Lett 7: 341–346

Winter K (1979) Effect of different CO_2 regimes on the induction of crassulacean acid metabolism in *Mesembryanthemum crystallinum* L. Aust J Plant Physiol 6: 589–594

Winter K (1980) Day/night changes in the sensitivity of phosphoenolpyruvate carboxylase to malate during crassulacean acid metabolism. Plant Physiol 65: 792–796

Wu MX, Wedding RT (1985) Regulation of phosphoenolpyruvate carboxylase from *Crassula* by interconversion of oligomeric forms. Arch Biochem Biophys 240: 655–662

14 Crassulacean Acid Metabolism in Leaves and Stems of *Cissus quadrangularis*

A. VIRZO DE SANTO and G. BARTOLI

14.1 Introduction

The genus *Cissus* (Vitaceae) is found throughout both the wet and arid tropics of the world. Surveys of $\delta^{13}C$ values for plants collected in their natural habitats, as well as studies in controlled conditions (Schütte et al. 1967; Milburn et al. 1968; Virzo De Santo et al. 1978, 1980, 1983, 1987a; Ting et al. 1983; Olivares et al. 1984) suggest that CAM is a widespread feature in the genus *Cissus* that might have facilitated the spread of the genus from the wet into the arid tropics.

The different species of the genus *Cissus* so far investigated exhibit different degrees of CAM expression (Table 14.1). *Cissus quadrangularis* L. is a species with succulent carbon-assimilating stems bearing small, short-lived leaves (Table 14.1). Great plasticity in the expression of CAM is found within this species, with marked differences between stems and leaves (Ting et al. 1983; Virzo De Santo et al. 1987b). Differences in CAM expression are found within the same organ, influenced mainly by water availability (Virzo De Santo et al. 1981, 1987a,b). The pattern of CO_2 fixation in the stem ranges from CO_2 uptake during both the night and the day in well-watered plants, to CO_2 fixation exclusively during the night with poor water availability, to CAM operating entirely using recycled respiratory CO_2 (CAM-idling) in droughted plants. The leaves, which normally show CAM-cycling, convert to CAM under conditions of moderate water stress.

In this chapter, the photosynthetic pattern of leaves and stems, their relationships and the influence of water shortage are discussed. The data derive from long-term studies made outdoors as well as under controlled conditions. Several components of CAM, such as gas exchange, nocturnal accumulation of malic acid, succulence and water relations have been quantified.

14.2 Main Characteristics of Leaf and Stem

The green succulent stem of *C. quadrangularis* exhibits a thick cuticle, a bilayered epidermis of small cells with about 50 stomata per mm^2, a cortex showing several

Dipartimento di Biologia Vegetale, Università di Napoli "Federico II", Via Foria 223, 80139 Napoli, Italy

Ecological Studies, Vol. 114
Winter/Smith (eds) Crassulacean Acid Metabolism

Table 14.1. A list of *Cissus* species known to perform CAM, including the main characteristics of leaf and stem, and the habitat

	Leaf	Stem	Habitat	Reference
C. gongylodes Burch.	Large, thin CAM-cycling	Slight succulence	Rain forests; South America	[d]
C. cactiformis Gilg	Small, deciduous CAM	Cactus-like CAM	Tropical xeric woods; Africa	[d]
C. hypoleuca Harv.		Succulent, CAM		[b]
C. quadrangularis L.	Small, deciduous CAM-cycling	Cactus-like CAM	Tropical, subtropical xeric woods; Africa, Arabia, Madagascar, East India, Moluccas	[cd]
C. quinquangularis Chiov.	Long-lived, succulent, CAM	Green CAM	Tropical xeric woods; Africa	[e]
C. rotundifolia Blume	Long-lived, succulent, CAM	Green, lignified	Tropical xeric woods; Africa	[e]
Cissus sp.	Small, thin deciduous CAM	Slight succulence CAM	Semi-arid environments; Kenya	[d]
C. trifoliata L.	Succulent, deciduous, CAM facultative	Lignified, climbing on columnar cacti	Semi-arid region; South America	[a]

[a]Olivares et al. (1984). [b]Schütte et al. (1967). [c]Ting et al. (1983). [d]Virzo De Santo et al. (1983). [e]Virzo De Santo, unpubl.

layers of photosynthetic parenchyma (cell dimensions about 53 × 40 μm), and a substantial pith consisting of water-storing parenchyma cells with few chloroplasts (Fig. 14.1). Stem succulence in the sense of Delf (1912) is about 1.3 kg H_2O m^{-2}, while the succulence at the cellular level (mesophyll succulence of Kluge and Ting 1978) is about 2.2 g H_2O mg^{-1} chlorophyll.

The leaves are small and fleshy and are produced only under well-watered conditions. Their life span is 2 to 4 months. The rather loose mesophyll (Fig. 14.1) is differentiated into palisade (cell dimensions about 29 × 70 μm) and spongy parenchyma (diameter about 45 μm). Stomata occur both on the adaxial and the abaxial surface with a frequency of 45 and 57 mm^{-2}, respectively. The values of succulence (0.24 kg m^{-2}) and mesophyll succulence (0.6 g H_2O mg^{-1} chlorophyll) of the leaf are lower than those of the stem.

14.3 Gas Exchange

14.3.1 Stem Gas Exchange

Figure 14.2A illustrates the characteristic 24-h pattern of stem net CO_2 exchange of a well-watered plant. The pattern is typical of CAM. Nighttime fixation is

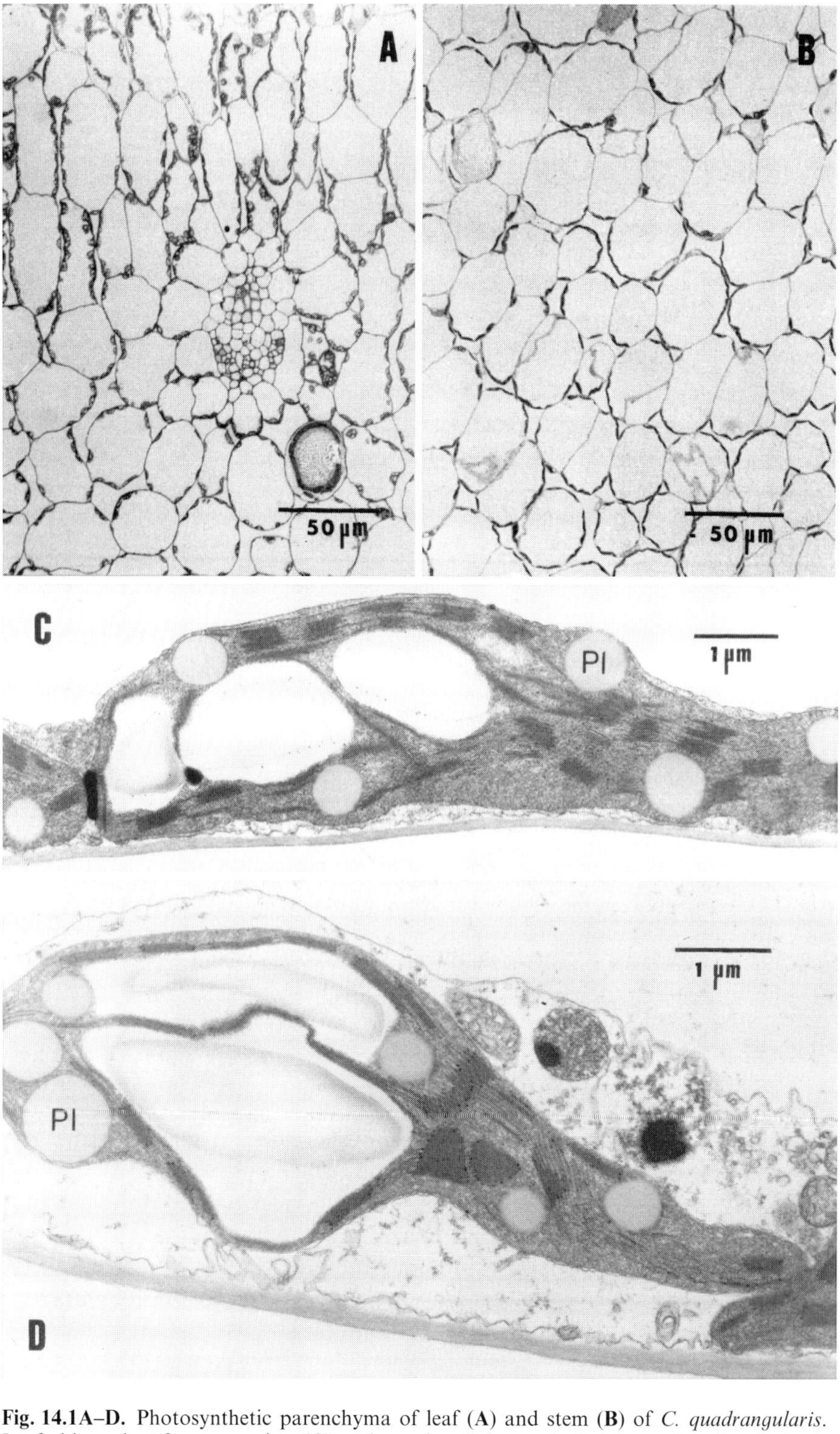

Fig. 14.1A–D. Photosynthetic parenchyma of leaf (**A**) and stem (**B**) of *C. quadrangularis*. Leaf chloroplast from morning (**C**) and evening (**D**) sample from a well-watered plant. Unlike stem chloroplasts (Virzo De Santo et al. 1984), plastosomes (*Pl*) and starch deposits are nearly equally abundant at the beginning and at the end of the light period

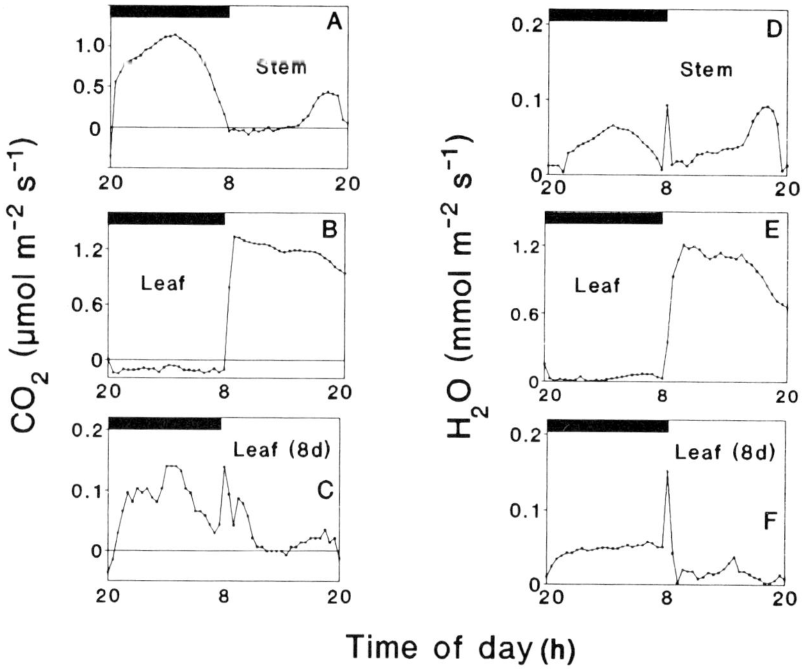

Fig. 14.2A–F. Daily time course of CO_2 exchange and transpiration of stem and leaf of *C. quadrangularis* under well-watered conditions and under water shortage (**C** and **F**; *8 d* number of days after induction of stress). Gas exchange and environmental parameters were measured in an open IRGA system using the GK 022 climate unit with flanged gas-exchange cuvette (Walz, Effeltrich, Germany). Experimental conditions were as in Table 14.2 (3 and 6). Plants were kept under the experimental conditions for at least 2 days before readings were taken. To induce water shortage, the stem was cut so as to leave two internodes below the one (and/or below the one bearing the leaf) enclosed in the cuvette

predominant over daytime fixation under any environmental condition (Fig. 14.3A–C; Virzo De Santo et al. 1987a). The highest rate of CO_2 uptake is reached in the first half of the dark period; afterwards CO_2 uptake declines. Transpiration shows the same pattern (Fig. 14.2D). The four phases of CO_2 exchange described by Osmond (1978) are evident in well-watered plants, although the contribution to total daily carbon gain of CO_2 fixation during the light period Phases II and IV varies with air temperature and humidity (Virzo De Santo et al. 1987a). Low ambient humidity seems to inhibit both daytime and nighttime stem CO_2 fixation of well-watered plants (Fig. 14.3A–C).

Due to the higher evaporative demand during the light period, water loss is higher during the light period than during the dark period. Ambient humidity influences the magnitude of stem transpiration much more than CO_2 fixation (Fig. 14.3A–F).

Water-use efficiency (WUE), expressed as mmol CO_2 mol^{-1} H_2O, is higher during nighttime CO_2 fixation compared to daytime CO_2 fixation (Table 14.2).

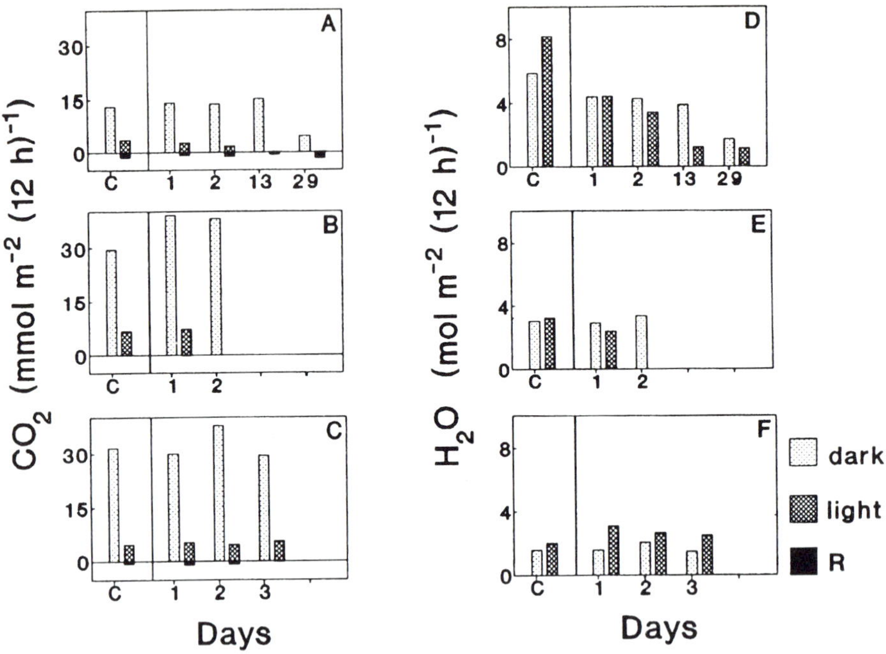

Fig. 14.3A–F. Integrated daytime and nighttime stem CO_2 uptake and transpiration under well-watered conditions (*C*) and for days after induction of drought at three different day/night regimes of relative air humidity. Integrated values of CO_2 release (*R*) during daytime are also shown. Experimental conditions are reported in Table 14.2 (**A** and **D** as indicated in 1; **B** and **E** as indicated in 2; **C** and **F** as indicated in 3). Water shortage was induced as described in Fig.14.2

Both daytime and nighttime WUE decrease with decreasing air relative humidity. Nighttime values of WUE are similar to those reported for other stem-succulent CAM plants (Szarek and Ting 1975; Nobel 1976, 1977; Osmond et al. 1979), whereas daytime values are somewhat lower.

14.3.2 Stem Gas Exchange Under Water Shortage

In response to water shortage and under low ambient humidity, daytime gas exchange is reduced in a few days (Fig. 14.3A,D). Nighttime CO_2 fixation may be unaffected over many days after the induction of stress, or somewhat increased (Fig. 14.3A,B,C). At a night/day RH of 34/23%, and temperature of 20/26 °C, Phase IV appears to be the most sensitive to water shortage (Fig. 14.4A). Moreover, WUE is lowest in Phase IV (nighttime, Phase II and Phase IV values of WUE are 2.4, 1.1 and 0.9 mmol CO_2 mol^{-1} H_2O, respectively). This is consistent with a prevailing C_3 carboxylation pattern in Phase IV (Winter 1985; Griffiths et al. 1986).

Table 14.2. Water-use efficiency (WUE) of well-watered and drought-stressed (in parentheses) stem and leaf of *Cissus quadrangularis*. WUE was calculated as mmol CO_2 uptake per mol H_2O lost

	Experimental conditions			mmol CO_2/mol H_2O		
	Night/day T (°C)	Night/day RH (%)	PPFD (μmol $m^{-2}s^{-1}$)	Night	Day	24 h
Stem						
1	20/26	34/23	400	2.2 (2.7)	0.4 (0.1)	1.2 (1.7)
2	20/26	50/35	400	9.8 (13.3)	2.1 (3.0)	5.8 (8.7)
3	20/26	63/43	400	19.8 (19.9)	2.4 (2.2)	10.0 (8.8)
Leaf						
4	18/27	43/28	800	1.4 (4.5)	1.6 (0.6)	1.5 (1.3)
5	20/26	50/35	800	2.1 (3.3)	2.1 (0.2)	2.1 (1.2)
6	21/27	59/43	800	– (2.1)	1.2 (0.2)	– (1.0)

The contribution of nighttime fixation to total 24-h CO_2 fixation increases with drought to 100% within a few weeks. However, the absolute amount of CO_2 uptake per 24 h decreases, and within 2 to 3 months of water shortage the gas-exchange pattern shifts towards CAM-idling, i.e. to a state in which CAM operates entirely using recycled respiratory CO_2 (Virzo De Santo et al. 1981, 1987a; see also Chaps. 12 and 13).

The magnitude of both daytime and nighttime CO_2 fixation (Fig. 14.5D,E) increases with tissue relative water content. However, only daytime CO_2 fixation and tissue RWC are significantly correlated ($r = 0.94$, $P < 0.001$). The relative contribution of nocturnal fixation increases with decreasing tissue RWC (Fig. 14.5F).

Rewatering of droughted plants induces a fast recovery of both daytime and nighttime CO_2 uptake (Fig. 14.4B). However, the relative increase in daytime CO_2 fixation is more than one order of magnitude greater than that in nocturnal fixation.

14.3.3 Leaf Gas Exchange

Leaves of well-watered plants of *C. quadrangularis* show a typical C_3 gas-exchange pattern, with net atmospheric CO_2 uptake occurring only during the day coupled with a low rate of net release of CO_2 at night (Fig. 14.2B). Considering that the leaves accumulate malic acid during the night (see below), it is reasonable to assume that the synthesis of malic acid is due to the recycling of respiratory CO_2. This type of photosynthetic pattern (CAM-cycling) has already been assessed in the leaves of *C. quadrangularis* by Ting et al. (1983; see also Chaps. 12 and 13).

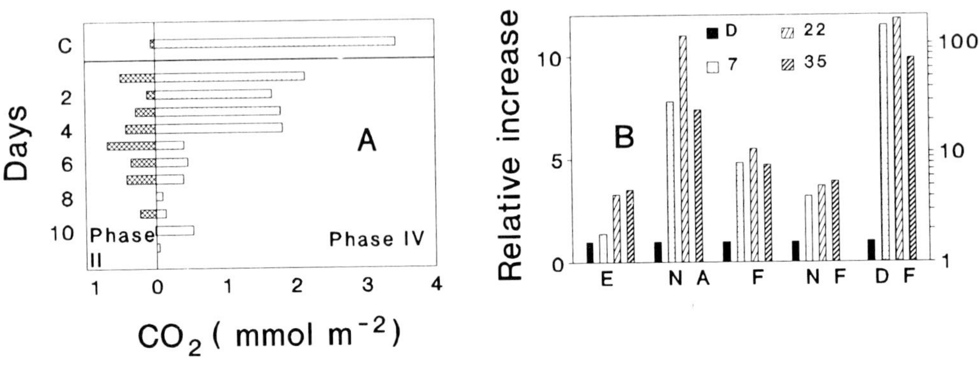

Fig. 14.4. A Stem-integrated CO_2 fixation during Phases II and IV under well-watered conditions (*C*) and at different time intervals (days) after induction of water shortage. Temperature 20/26°C, RH 34/23% night/day, 400 μmol photons m^{-2} s^{-1}. **B** Relative increases, compared with plants unwatered for 3 months, of evening stem concentration (*E*) and nocturnal accumulation (*NA*) of malic acid, and CO_2 fixation during daytime (*DF*), nighttime (*NF*) and the whole 24-h cycle (*F*), assessed 7, 22 and 35 days after rewatering. The values for droughted plants (*D*) are taken as 1. The logarithmic scale on the *right-hand axis* in (**B**) refers to *DF*

Air humidity influences leaf gas exchange, with the highest measured CO_2 uptake being observed in leaves at the highest RH (Fig. 14.6C). At low relative humidity, leaves of well-watered plants exhibit net CO_2 uptake both during nighttime and daytime (Fig. 14.6A,B). However, the amounts of CO_2 fixed at night are very small (0.6 to 5%)compared with daytime fixation. In well-watered plants, leaf transpiration at the higher ambient RH is one order of magnitude greater than stem transpiration (Figs. 14.3F, 14.6F). Moreover, about 98% of the water loss occurs during daytime, whereas in the stem nighttime transpiration is similar to daytime transpiration (Fig. 14.3F).

During dark fixation, leaf water-use efficiency (WUE) is lower than stem WUE (Table 14.2). In contrast, the WUE of leaf and stem are similar during light fixation.

Under similar environmental conditions, the magnitude of CO_2 fixation of well-watered plants is higher in the leaf than in the stem (Figs. 14.3A,B,C; 14.6A,B,C), consistent with the higher photosynthetic capacity of the leaf (D'Ambrosio and Virzo De Santo 1992).

According to the different patterns of gas exchange, daytime leaf transpiration is higher than stem transpiration, whereas nighttime leaf transpiration is lower than stem transpiration.

14.3.4 Leaf Gas Exchange Under Water Shortage

When water supply is withdrawn, leaf CO_2 exchange shifts from a C_3-like towards a CAM-like pattern (Fig. 14.2C). The time taken to shift towards the

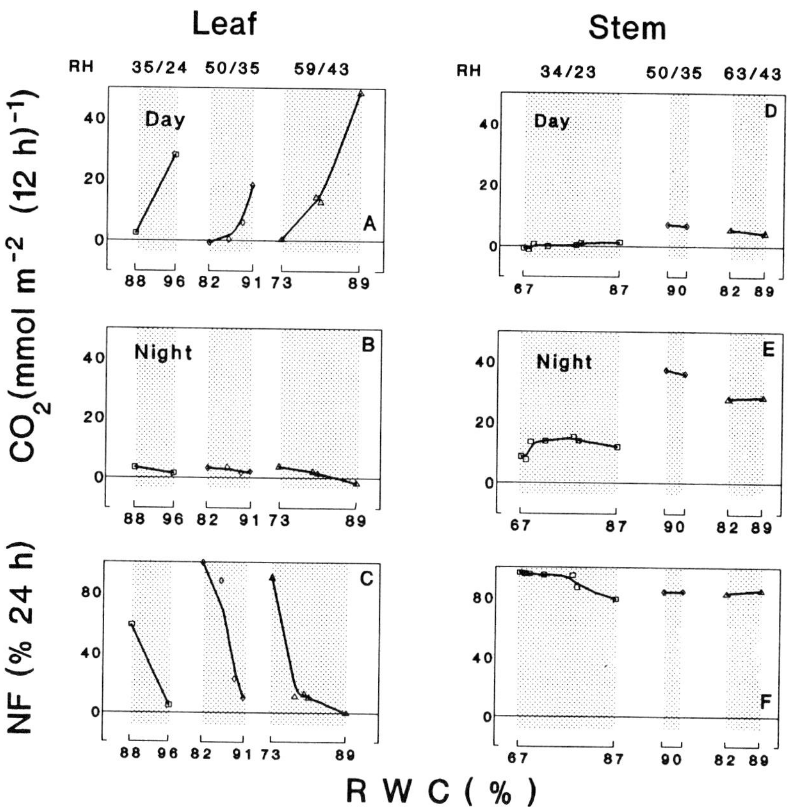

Fig. 14.5A–F. Integrated day and night CO_2 uptake of leaf and stem as a function of stem relative water content (*RWC*). Nighttime fixation (*NF*) as a percentage of fixation during the 24-h cycle is also shown. The measurements were made under three different night/day regimes of relative air humidity (values in %) as indicated *at the top*. RWC was measured as described in Virzo De Santo et al. (1984) in triplicate on segments taken from the second internode below the one (and/or below the one bearing the leaf) enclosed in the gas-exchange cuvette

CAM pattern is longer at higher relative humidity than at lower relative humidity (Fig. 14.6A,B,C).

Daytime CO_2 fixation is highly sensitive to water stress and is rapidly and drastically reduced by water shortage (Fig. 14.6A,B,C). This response parallels the response of transpiration to water shortage (Fig. 14.6D,E,F). Nocturnal CO_2 fixation increases progressively with drought, but its rate never exceeds 10 to 25% of that measured during the daytime with good water supply (Fig. 14.6 A,B,C).

Leaf gas exchange is highly influenced by plant RWC (measured on stem segments). While daytime CO_2 fixation decreases with decreasing RWC, an opposite trend is evidenced for nighttime CO_2 fixation (Fig. 14.5A,B) depending on the shift towards the CAM pattern of gas exchange (Fig. 14.2C; 14.6A,B,C).

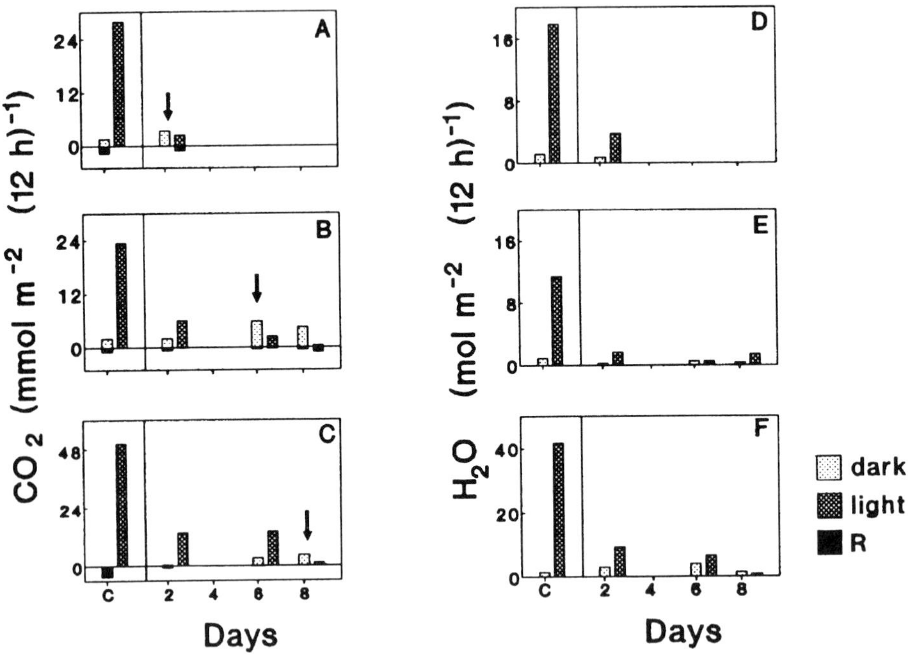

Fig. 14.6A–F. Integrated daytime and nighttime leaf CO_2 uptake and transpiration under well-watered conditions and under drought at three different regimes of relative air humidity. Experimental conditions are reported in Table 14.2 (**A** and **D** as indicated in 4; **B** and **E** as indicated in 5; **C** and **F** as indicated in 6). *Arrows* show the shift from the C_3 to the CAM mode of CO_2 exchange in response to water shortage. All other abbreviations as in Figs. 14.2 and 14.3

Accordingly, the nocturnal proportion of CO_2 fixation increases with decreasing RWC (Fig. 14.5C).

Leaf nocturnal WUE increases with the shift to the CAM pattern of gas exchange, although it never reaches values as high as those measured in the stem (Table 14.2). In contrast, daytime WUE markedly decreases with drought indicating that CO_2 fixation is reduced more than transpiration.

14.4 Nocturnal Accumulation of Malic Acid in Leaf and Stem

The magnitude of CAM activity in *C. quadrangularis* is highly influenced by environmental conditions (Virzo De Santo et al. 1980; Alfani et al. 1987). Stem nocturnal accumulation of malic acid in plants kept outdoors in the Naples Botanic Garden (Fig. 14.7A) was maximal in July when maximum/minimum temperatures averaged 32/19 °C and the photoperiod was about 14 h. Minimum acid accumulation occurred in November and March when maximum/minimum temperatures averaged 18/8 °C and the photoperiod was about 9 h and 12 h, respectively. Moreover, CAM activity was lower in droughted plants (Fig. 14.7A).

The leaves of *C. quadrangularis* exhibit nocturnal accumulation of malic acid under all environmental conditions, but to a lesser extent than the stem (Ting et al. 1983; Virzo De Santo et al. 1987b and unpubl.). In well-watered plants, nocturnal accumulation of malic acid, measured under a range of environmental conditions, averaged 37 and 29 μmol g^{-1} fr. wt. in stem and leaf, respectively (Virzo De Santo et al. 1987b). In plants grown outdoors under different levels of photosynthetic photon flux density (PPFD), nocturnal accumulation of malic acid was enhanced in the leaf by increasing PPFD, but no effect was found on the CAM activity of the stem (Virzo De Santo et al. 1987b).

14.4.1 Nocturnal Accumulation of Malic Acid and Water Shortage

Water shortage results in a decrease in stem RWC and a concomitant reduction in nocturnal malic-acid accumulation. Nocturnal accumulation of malic acid correlates positively with tissue RWC ($r = 0.85$; $P < 0.01$). In contrast, nocturnal accumulation of malic acid in the leaf is increased in response to a moderate drought (Figs. 14.7B; 14.8A; Virzo De Santo 1987b and unpubl.). By withholding water, the stem RWC of plants kept under controlled conditions decreased rapidly, whereas leaf RWC changed only slightly (Fig. 14.7B). We argue that the maintenance of high RWC in the leaf is connected with water transport from the succulent stem that serves as an effective water-storage buffer. Malic-acid fluctuations in the leaf were enhanced after 15 and 22 days without water, whereas no effect was observed in the stem. Similarly, measurements made in the Naples Botanic Garden on plants under natural day/night cycles showed that by

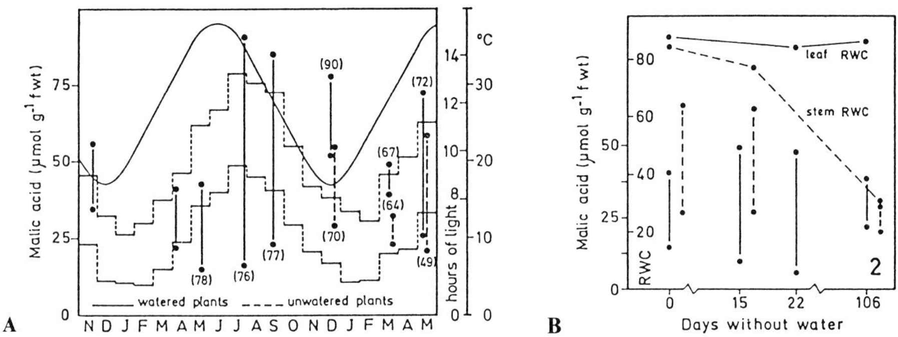

Fig. 14.7. A Seasonal variations of CAM activity in the stem of *C. quadrangularis* growing outdoors in the Naples Botanic Garden. Monthly means of maximum and minimum temperatures and daylength are reported. Values of RWC are given *in parentheses*. The *continuous curve* represents hours of daylight. (From Alfani et al. 1987 with permission). **B** Changes of CAM activity and tissue RWC in leaf and stem in response to water shortage (growth chamber: 27/16 °C, 55/80% RH day/night regime; photoperiod 9 h, PPFD: 150 μmol m^{-2} s^{-1}). (From Virzo De Santo et al. 1987b with permission). Upper and lower points refer to morning and evening values, respectively

watering the plants less frequently a decrease in CAM activity is induced in the stem together with a simultaneous increase of CAM activity in the leaf (Fig. 14.8A). The findings are consistent with the observed shift from the C_3 to the CAM pattern of leaf gas exchange under water shortage (Figs. 14.2C; 14.6A,B,C). However, a long period of drought reduces nocturnal accumulation of malic acid both in the stem and in the leaf; nevertheless, the reduction is lower in the leaf than in the stem (Fig. 14.8B).

It is worthwhile noting that the evening level of malic acid in the stem tends to decrease with water stress and to increase with rewatering (Figs. 14.4B; 14.8C,D). In contrast, in *Opuntia basilaris* the evening level of titrable acidity in irrigated

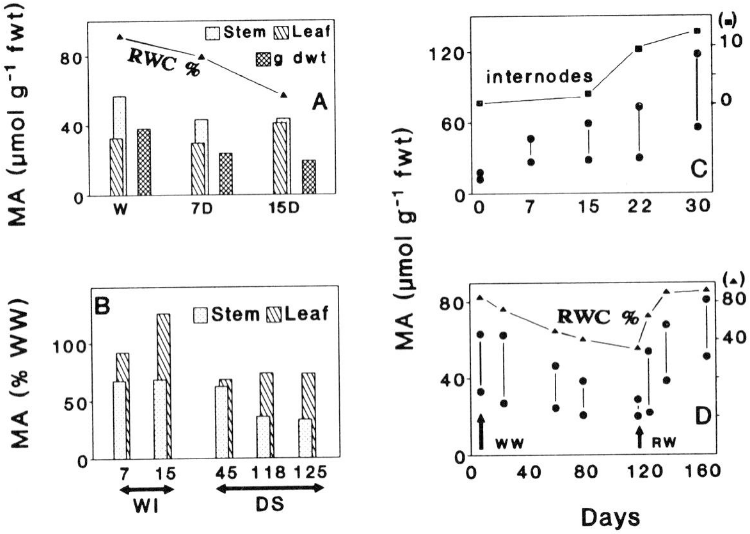

Fig. 14.8. **A** Nocturnal accumulation of malic acid (*MA*) in stem and leaf under different watering regimes (*W* plants watered every day; *7D* and *15D* plants watered once a week and twice a month, respectively) measured outdoors during a hot bright day in September 1989 (maximum day temperature 36 °C, minimum night temperature 17 °C; day/night RH 30/100%). Biomass (as g dry wt accumulated in the period July to September) and stem *RWC* of the three groups of plants are also given; **B** Changes in nocturnal accumulation of malic acid in stem and leaf in response to different watering regimes (*WI* frequency of watering in days) and to drought (*DS* days without water). Values are percentages of malate accumulation in well-watered (*WW*) plants. Measurements on droughted plants were performed outdoors in July (maximum day temperature 41°C, minimum night temperature 24°C; day/night RH 30/100%; photoperiod about 15 h; **C** Evening (*lower points*) and morning (*upper points*) stem concentrations of malic acid and growth (as number of new internodes) on recovery from drought stress after 135 days without water. Experimental conditions: 12 h photoperiod, PPFD 180 µmol m^{-2} s^{-1}; 25/14°C, 70/100% RH day/night regime; **D** Changes in evening and morning stem concentrations of malic acid and RWC in response to drought and to rewatering of droughted plants; *WW* water withheld; *RW* rewatered. Experimental conditions: 12 h photoperiod, PPFD 180 µmol m^{-2} s^{-1}, 27/16°C and 55/80% RH day/night regime. Other explanations as in Fig.14.7

plants was practically the same as in non-irrigated plants (Hanscom and Ting 1978). Similarly, the evening level of malic acid remained unchanged throughout the year in *Kalanchoë beharensis*, in which the quantity of CO_2 fixed during the night is relatively insensitive to changes in water availability (Kluge et al. 1992).

14.4.2 Plant Growth and Recovery from Water Stress

Restricted watering reduces plant growth (Fig. 14.8A), in spite of the increase in CAM activity in the leaves, which is unable to offset the decrease of CAM activity in the stem. Moreover, senescence causes the short-lived leaves to be shed. Rewatering of droughted plants induces rapid tissue rehydration and the recovery of stem CAM activity, which is accompanied by a rapid production of new internodes (Fig. 14.8C,D). Furthermore, growth is enhanced by production of new leaves which significantly increases carbon fixation.

Virzo De Santo et al. (1984, 1985) have demonstrated that cellular integrity of the stem chlorenchyma is maintained even at a RWC as low as 30%, which is similar to the lower limit for cell viability in cacti and agaves (Nobel 1988). Maintenance of the cellular integrity of stem chlorenchyma accounts for the ability of *C. quadrangularis* to recover quickly from water stress following rewatering.

Measurements of gas exchange and nocturnal accumulation of malic acid, monitored during a 35-d period following the rewatering of a droughted plant, indicate that both nocturnal accumulation of malic acid and net CO_2 uptake in the stem increase with watering. However, whereas malate accumulation increases by a factor of 8, net CO_2 uptake increases only by a factor of 3, suggesting that the recycling of respiratory CO_2 is high in such conditions. On the other hand, in well-watered plants the estimated amount of malic acid potentially resulting from nocturnal CO_2 uptake accounts for about 50% of the measured accumulation of malic acid under similar experimental conditions.

14.5 The Value of the Leaf

Based on the results presented we shall conclude by speculating about the value of the leaf for carbon gain in *C. quadrangularis*. Leaves of *C. quadrangularis* (1) contribute only about 19% (Fig.14.9) to the whole photosynthetic surface of the plant, (2) are formed only under conditions of abundant water supply, and (3) have a low water-use efficiency compared with the stem (Table 14.2), principally because the leaves are more C_3-like. This leads to the question of how the costs of producing the leaf are met by the interest gained from the investment. Under well-watered conditions, the leaf contribution to plant daytime CO_2 uptake is greater than that of the stem (Fig.14.9). Although this advantage is offset by the ability of the stem to fix larger amounts of atmospheric CO_2 during the night, leaf contribution with respect to total 24-h plant carbon gain is still greater than expected on the basis of the leaf contribution to plant photosynthetic surface

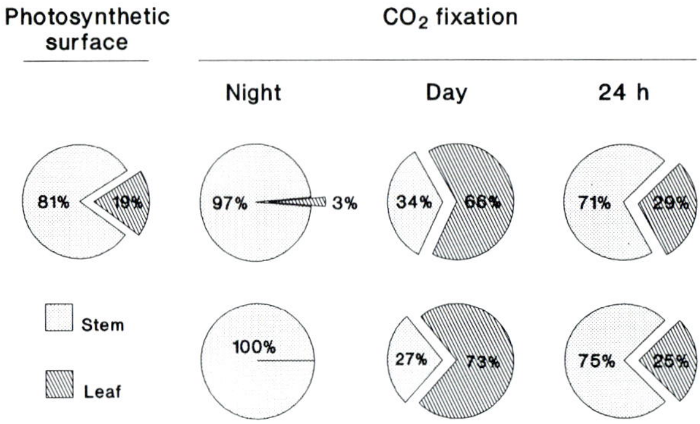

Fig. 14.9. Leaf and stem contribution to the photosynthetic surface and to nighttime, daytime and daily 24-h carbon-dioxide fixation of individual plants (based on measurements made on five plants). Carbon-dioxide fixation was measured at low (*upper row*) and at high (*lower row*) relative humidity (values in %)

(Fig.14.9). Moreover, we have shown that under water shortage stem CAM activity is reduced relatively more than leaf CAM activity, which is maintained longer under limiting conditions. *C. quadrangularis* invests most of its resources in the growth of the green succulent stem, thereby contributing to further carbon assimilation and, at the same time, to an increased water-storage capacity, the latter being a fundamental characteristic of adaptation to aridity. Nonetheless, leaves under conditions of high water availability enhance plant productivity through their higher photosynthetic rate.

Acknowledgements. Financial support by CNR and MURST is gratefully acknowledged. Most of the results have been obtained from research carried out in collaboration with A. Alfani and A. Fioretto.

References

Alfani A, Fioretto A, Ligrone R, Virzo De Santo A (1987) Relation between CAM and photoperiod in *Cissus quadrangularis* L., a preliminary report. G Bot Ital 121: 191–192

D'Ambrosio N, Virzo De Santo A (1992) Emissione di fluorescenza in foglie e fusti della specie CAM *Cissus quadrangularis* L. S. It. E. Atti 15: 705–706

Delf EM (1912) Transpiration in succulent plants. Ann Bot 26: 409–440

Griffiths H, Lüttge U, Stimmel KH, Crook CE, Griffiths NM, Smith JAC (1986) Comparative ecophysiology of CAM and C_3 bromeliads. III. Environmental influences on CO_2 assimilation and transpiration. Plant Cell Environ 9: 385–393

Hanscom Z III, Ting IP (1978) Irrigation magnifies CAM-photosynthesis in *Opuntia basilaris* (Cactaceae). Oecologia 33: 1–15

Kluge M, Ting IP (1978) Crassulacean acid metabolism. Analysis of an ecological adaptation. Springer, Berlin Heidelberg New York

Kluge M, Razanoelisoa B, Ravelomanana D, Brulfert J (1992) In situ studies of crassulacean acid metabolism in *Kalanchoë beharensis* Drake Del Castillo, a plant of the semi-arid southern region of Madagascar. New Phytol 120: 323–334

Milburn TR, Pearson DJ, Ndgewe NA (1968) Crassulacean acid metabolism under natural tropical conditions. New Phytol 67: 883–897
Nobel PS (1976) Water relations and photosynthesis of a desert CAM plant, *Agave deserti*. Plant Physiol 58: 576–582
Nobel PS (1977) Water relations and photosynthesis of a barrel cactus, *Ferocactus acanthodes*, in the Colorado desert. Oecologia 27: 117–133
Nobel PS (1988) Environmental biology of agaves and cacti. Cambridge University Press, Cambridge
Olivares E, Urich R, Montes G, Coronel I, Herrera A (1984) Occurrence of crassulacean acid metabolism in *Cissus trifoliata* L. (Vitaceae). Oecologia 61: 358–362
Osmond CB (1978) Crassulacean acid metabolism: a curiosity in context. Annu Rev Plant Physiol 29: 379–414
Osmond CB, Ludlow MM, Davis R, Cowan IR, Powles SB, Winter K (1979) Stomatal responses to humidity in *Opuntia inermis* in relation to control of CO_2 and H_2O exchange patterns. Oecologia 41: 65–76
Schütte KH, Steyn R, van der Westhuizen M (1967) Crassulacean acid metabolism in South African succulents: a preliminary investigation into its occurrence in various families. J S Afr Bot 33: 107–110
Szarek SR, Ting IP (1975) Photosynthetic efficiency of CAM plants in relation to C_3 and C_4 plants. In: Marcelle R (ed) Environmental and biological control of photosynthesis. Dr W Junk, Dordrecht, pp 289–297
Ting IP, Sternberg LO, DeNiro MJ (1983) Variable photosynthetic metabolism in leaves and stems of *Cissus quadrangularis* L. Plant Physiol 71: 677–679
Virzo De Santo A, Fioretto A, Alfani A (1978) Resistenza alla siccità di *Cissus quadrangularis* L. vitacea succulenta con CAM. Delpinoa 20: 37–44
Virzo De Santo A, Alfani A, Greco L, Fioretto A (1980) Environmental influences on CAM activity of *Cissus quadrangularis* L. J Exp Bot 31: 75–82
Virzo De Santo A, Bartoli G, Russo G, Fioretto A (1981) La fissazione della CO_2 in funzione della disponibilità idrica in *Cissus quadrangularis* L. G Bot Ital 115: 127–128
Virzo De Santo A, Alfani A, Russo G, Fioretto A (1983) Relationship between CAM and succulence in some species of Vitaceae and Piperaceae. Bot Gaz 114: 342–346
Virzo De Santo A, Ligrone R, Alfani A, Fioretto A, Russo G (1984) CAM activity and day/night changes in the ultrastructure of stem chlorenchyma of *Cissus quadrangularis* L. as influenced by drought. Plant Cell Environ 7: 105–112
Virzo De Santo A, Fioretto A, Ligrone R, Alfani A (1985) Ultrastructural changes and levels of CAM activity on recovery from water stress in *Cissus quadrangularis* L. G Bot Ital 119: 190–191
Virzo De Santo A, Fioretto A, Bartoli G, Alfani A (1987a) Gas exchange of two CAM species of the genus *Cissus* (Vitaceae) differing in morphological features. Photosynth Res 13: 113–124
Virzo De Santo A, Fioretto A, Alfani A (1987b) Effect of environmental conditions and age on CAM activity of the leaves in *Cissus quadrangularis* L. G Bot Ital 121: 213–215
Winter K (1985) Crassulacean acid metabolism. In: Barber J, Baker NR (eds) Photosynthetic mechanisms and the environment. Elsevier, Amsterdam, pp 329–387

15 Variations in the Phases of Crassulacean Acid Metabolism and Regulation of Carboxylation Patterns Determined by Carbon-Isotope-Discrimination Techniques

A.M. Borland and H. Griffiths

15.1 Introduction

The four phases of gas exchange, defined for CAM plants by Osmond in 1978, have provided the framework for the interpretation of physiological and biochemical studies on the regulation of this pathway in both constitutive and facultative CAM plants. Whilst the main CAM processes of dark uptake of CO_2 and organic-acid decarboxylation occur during Phases I and III, respectively, the daytime Phases II and IV are often considered merely as transitions between C_3 and C_4 carboxylation. However, uptake of external CO_2 during the day, particularly during Phase IV, contributes a substantial proportion of the assimilates used for the growth of consitutive CAM plants. In addition, daytime photosynthesis is essential for maintaining a positive carbon balance during the C_3-CAM transition in facultative plants.

The aim of this chapter is to consider some of the biochemical and physiological mechanisms underlying the regulation of carbon fluxes during Phases II and IV and to assess the relative contributions from daytime C_3 and C_4 carboxylation. Moreover, net CO_2 uptake during decarboxylation has been frequently observed for C_3-CAM intermediates and we propose that such observations may be reconciled with the high capacity of RUBISCO and the substantial rates of electron transport characteristic of many CAM plants in the light.

15.2 Phases II and IV: General Characteristics

15.2.1 Expression of Phases II and IV

The duration of Phases II and IV and the contribution that each make to net CO_2 uptake over 24 h is determined by environmental and physiological factors and is highly species-dependent. Table 15.1 lists examples for a number of constitutive and facultative CAM plants that show net dark fixation of CO_2 (i.e. Phase I gas exchange) and illustrates how the expression of Phases II and IV may be affected

Department of Agricultural and Environmental Science, Ridley Building, The University, Newcastle upon Tyne NE1 7RU, UK

Ecological Studies, Vol. 114
Winter/Smith (eds) Crassulacean Acid Metabolism

Table 15.1. The duration and contribution to net CO_2 uptake over 24 h of photosynthesis during Phases II and IV for a range of constitutive and facultative CAM plants growing under different environmental conditions

Species	Phase II (h)	% CO_2 fixed	Phase IV (h)	% CO_2 fixed	Conditions	Reference
Kalanchoë daigremontiana	1.5	10	4	30	12 h photoperiod, 27/15 °C	Kluge et al. (1981a)
	4	ND	ND	ND	Leaf temperature 15 °C	Winter and Tenhunen (1982)
	3	ND	ND	ND	Leaf temperature 20 °C	
	1.5	ND	ND	ND	Leaf temperature 25 °C	
	0.5	ND	ND	ND	Leaf temperature 30 °C	
Kalanchoë blossfeldiana	1.5	5	2.6	8	16-h photoperiod, 27/17 °C	Kluge et al. (1981b)
Kalanchoë tubiflora	2	ND	ND	ND	16-h photoperiod, 25/15 °C	Fischer and Kluge (1984)
	4	ND	ND	ND	CO_2-free air overnight	
Dudleya farinosa	6	41	6	39	Field, start of dry season	Bartholomew (1973)
	4	22	6	35	3 months later	
Agave deserti	3	3	8	72	25 days old	Nobel (1988)
	3	7	8	45	100 days old	
	3	10	4.5	10	400 days old	
Mesembryanthemum crystallinum	4	37	5	40	12-h photoperiod, 25/15 °C	Winter (1974)
	4	42	3.2	13		Whitehouse et al. (1991)
Clusia rosea	3 to 4	20 to 40	–	–	Field	Ball et al. (1991)
Clusia major	1.6	16	4	12	Simulated field	Franco et al. (1990)

Table 15.1 (*Contd.*)

Species	Phase II (h)	% CO_2 fixed	Phase IV (h)	% CO_2 fixed	Conditions	Reference
Clusia minor	Continuous uptake of CO_2 over 24 h				Simulated field	Franco et al. (1990)
Clusia minor	4	24 to 30	4	15 to 20	Field	Borland et al. (1993)
Clusia minor	5	58	2	13	12-h photoperiod, VPD 0.62 kPa	Schmitt et al. (1988)
	3	30	2.5	18	VPD 1.31 kPa	
Clusia minor	3.4	26	5	43	12-h photoperiod, 30/15 °C	Haag-Kerwer et al. (1992)
	3.4	37	4.5	36	30/20 °C	
	Continuous uptake of CO_2				30/25 °C	
Peperomia scandens	3.5	11	9.6	49	16-h photoperiod, 60 days old	Holthe et al. (1987)
	Continuous uptake of CO_2				75 days old	
Portulacaria afra	Continuous uptake of CO_2				Glasshouse + H_2O	Guralnick et al. (1992)
	2	20	4	4	− H_2O	
Sedum telephium	Continuous uptake of CO_2				12-h photoperiod, 20/12 °C	Borland and Griffiths (1990)

by temperature, water availability, photoperiod, humidity and leaf age. Patterns of gas exchange during the day may be influenced by the magnitude of CO_2 fixation and malic-acid accumulation during the previous night (Medina and Delgado 1976; Fischer and Kluge 1984). If CO_2 uptake during the night is high, subsequent daytime photosynthesis may be relatively low, and vice versa (Kluge and Ting 1978).

In terms of leaf carbon balance, the CO_2 fixed during Phase IV has been recognized as contributing a substantial proportion of the carbon skeletons required for growth (Allaway et al. 1974; Winter 1985; Borland et al. 1994). Moreover, under well-irrigated field conditions which favour the maximum contribution of net CO_2 fixation in Phase IV, the growth rates of pineapple crops may approach the lower limit of C_3 crop growth rates (Bartholomew and Kadzimin 1977). Although Phase II is usually considered as a brief light-stimulated burst of photosynthesis, gas exchange during this period can, in some instances, continue for up to 6 h and can make a substantial contribution to diel carbon balance, a fact that is often overlooked (see Table 15.1).

15.2.2 Physiological Regulation of Phases II and IV

By manipulating environmental parameters to modulate the phases of gas exchange, a number of studies have illustrated that the magnitude of Phases II and IV in situ are determined by a complex interplay of biochemical and physiological responses. Experiments where the amount of organic acids accumulating overnight have been controlled by administering CO_2-free air overnight, or by altering the length of the dark period, have suggested that the storage capacity of the vacuole will determine the magnitude of Phase II (Winter and Tenhunen 1982; Fischer and Kluge 1984). This may be considered in terms of the vacuolar capacity per se or via regulation of tonoplast transport processes and the energy requirements for H^+ plus mal^{2-} transport into the vacuole (see Chap. 4). The length of time that stomata remain open during Phase II also appears to be determined by the intercellular CO_2 partial pressure (p_i) (Kluge et al. 1981b), which in turn depends upon the onset of malic-acid efflux from the vacuole (Winter and Tenhunen 1982). Although little is known about the diel regulation of the decarboxylation enzymes or RUBISCO in CAM plants, the subsequent generation and consumption of CO_2 will obviously be important for determining p_i, the timing of stomatal closure at the end of Phase II, and the duration of Phase III. However, under conditions of drought, stomatal movements appear to be determined primarily by leaf water potential and, as in *Kalanchoë daigremontiana*, Phase IV may be eliminated (Smith and Lüttge 1985). Stomata may also respond directly to changes in humidity as shown by laboratory-based studies on *Clusia minor* (Schmitt et al. 1988).

15.2.3 Regulation of C_3/C_4 Carboxylation During Phases II and IV

Evidence for the transitional nature of Phases II and IV has been provided via a number of experimental approaches. Research into the biochemical regulation of PEPC has indicated that the enzyme is rapidly down-regulated during Phase II following malic-acid efflux from the vacuole, with PEPC sensitized to malate inhibition during the day (Winter 1982; Nimmo et al. 1987; see also Chap. 3). Malic acid is subsequently decarboxylated to generate CO_2, which is refixed by RUBISCO during Phase III when stomata are usually closed. The inhibition of PEPC is eventually relieved and C_4 carboxylation may increase towards the end of Phase IV once stomata have reopened. The in vitro studies of PEPC regulation during Phases II and IV have been largely confirmed by labelling studies using $^{14}CO_2$ and $^{13}CO_2$ (Osmond and Allaway 1974; Fischer and Kluge 1984; Ritz et al. 1986; Osmond et al. 1988). Together, these experiments support the suggestions that CO_2 is initially fixed via PEPC at the start of Phase II and that C_3 carboxylation via RUBISCO increases throughout Phase II. The shift from C_4 to C_3 carboxylation has also been confirmed by measurements of the O_2-sensitivity of CO_2 fixation during Phase II (Winter and Tenhunen 1982).

The reverse C_3/C_4 transition occurs during Phase IV and substantial $^{14}CO_2$ fixation into malic acid has been observed during Phase IV, particularly under low light intensities (Kluge et al. 1975). The non-invasive technique of on-line carbon-isotope discrimination has also demonstrated that "double carboxylation", involving both PEPC and RUBISCO, occurs during the transient phases of CAM in vivo (Griffiths et al. 1990; see Sect. 15.5).

15.3 Regulation of Daytime Photosynthesis in Facultative CAM Plants

During the C_3-CAM transition, a variety of gas-exchange characteristics may be observed outside the conventional pattern of four phases. These patterns include (1) daytime photosynthesis with respiratory CO_2 refixed at night (i.e. "CAM-cycling"), (2) continuous net uptake of CO_2 over 24 h, (3) late-afternoon (Phase IV) plus dark uptake of CO_2, (4) dark CO_2 uptake alone, and (5) continuous closure of stomata in cases of severe drought (i.e. "CAM-idling"). Where CO_2 is taken up continuously over the day/night cycle, it is sometimes difficult to distinguish different phases of daytime gas exchange, even though there may be a slight depression in photosynthetic rate during the middle of the day. Under these conditions CO_2 fixation reflects a combination of gross photosynthetic fluxes by night (respiration + net uptake) and by day (decarboxylation + net uptake). It is therefore of interest to determine when the midday depression of photosynthesis (often observed in C_3 plants growing in exposed arid habitats: Tenhunen et al. 1987) changes to the "midday depression" characteristic of Phase III in constitutive CAM. Three contrasting examples will be considered.

15.3.1 *Mesembryanthemum crystallinum*

During the C_3-CAM switch in *M. crystallinum*, slight depressions in the photosynthetic rate at midday gradually intensify to result in complete stomatal closure over the middle part of the day, with peaks in net photosynthesis at the start and latter part of the photoperiod (Winter and Gademann 1991; see Table 15.1). However, the decrease in stomatal conductance at midday is determined primarily by a decrease in leaf water potential and turgor pressure resulting from the salt stress, since p_i remains low during the early stages of the C_3-CAM transition (Winter and Gademann 1991). It is now thought that this reduction in leaf turgor may trigger the molecular mechanisms underlying CAM induction in *M. crystallinum*, since dark uptake of CO_2 develops subsequent to the midday depression of photosynthesis (Winter and Gademann 1991). Once dark fixation of CO_2 is induced after 7 to 10 days of salt stress, leaf turgor pressure may increase to a level higher than that before the salt treatment (Winter and Gademann 1991). However, even in these circumstances p_i still decreases during the middle of the day, unlike the findings with the constitutive CAM plant *K. daigremontiana*, where p_i increases shortly before the midday closure of stomata (Kluge et al. 1981b).

15.3.2 *Sedum telephium*

Under high PPFD, the imposition of drought stress serves mainly to enhance CAM activity in plants of *S. telephium*, since well-watered plants also accumulate organic acids overnight (Borland and Griffiths 1990). Such findings suggest that the machinery for CAM is always present in leaves of this species (Groenhof et al. 1990; Borland and Griffiths 1992; see also Chap. 11), thus distinguishing it from the true "induction" observed in *M. crystallinum* (see Chaps. 8,9 and 10). However, under low PPFD, CAM is only expressed in *S. telephium* after the imposition of drought (Borland and Griffiths 1990). Rates of daytime photosynthesis decline with increasing drought stress, and net dark uptake of CO_2 is only observed for severely droughted plants grown under higher PPFD (Fig.15.1; Borland and Griffiths 1990). Refixation of respiratory CO_2 therefore helps to maintain a positive carbon balance throughout the C_3-CAM transition and also in the full CAM state in this species, which may be described as predominantly exhibiting CAM-cycling (Harris and Martin 1991; Guralnick et al. 1992; Gravatt and Martin 1992; see also Chap. 12).

Even after the relative water content (RWC) of leaves drops to around 50%, CO_2 is still fixed continuously over the day with no apparent midday depression of photosynthesis (Fig.15.1). Whilst the relatively small amounts of organic acids accumulated overnight may help to maintain turgor at low relative water contents (A.M. Borland and H. Griffiths, unpubl.), the benefits for CO_2 supply can also be calculated by integrating CO_2 uptake from gas exchange and malate accumulation (Table 15.2). Under well-watered conditions, net daytime CO_2 uptake during the day is well in excess of that generated from any respiratory

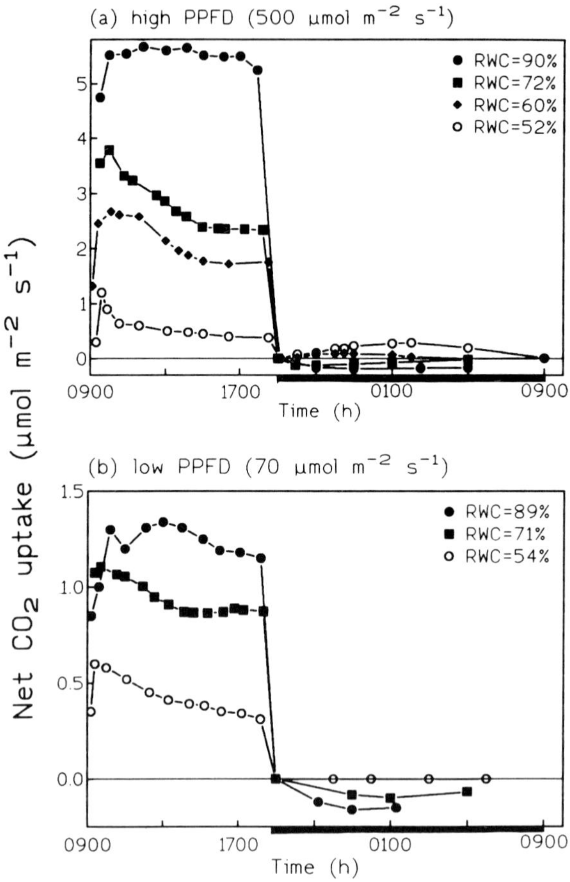

Fig. 15.1a,b. Rates of net CO_2 uptake by leaves of *Sedum telephium* grown under (**a**) high (500 μmol m^{-2} s^{-1}) or (**b**) low (70 μmol m^{-2} s^{-1}) PPFD, 20/13°C day/night temperature and 75/85% day/night relative humidity. Plants were subjected to varying degrees of water stress as reflected by relative water contents. The *solid bar* on the *x-axis* indicates the period of darkness

CO_2 refixation. However, when maximal CAM activity is induced (52 to 54% RWC), net daytime uptake equals that at night from net CO_2 uptake and/or respiratory refixation as indicated by nocturnal malate accumulation (Table 15.2). Thus, the supply of CO_2 in the light is doubled as a result of enhanced CAM activity and electron transport is maintained even under severe drought stress.

15.3.3 *Clusia minor*

In contrast to *M. crystallinum*, the C_3-CAM switch in *Clusia minor* is rapid and reversible, with two leaves on the same node capable of expressing either C_3 or

Table 15.2. Integrated net daytime CO_2 uptake by plants of *S. telephium* grown under high (500 μmol m^{-2} s^{-1}) or low (70 μmol m^{-2} s^{-1}) PPFD and subjected to increasing drought stress as reflected by leaf relative water contents (RWC). The amount of CO_2 generated from the decarboxylation of malate accumulated overnight is also shown

Growth conditions		Integrated net daytime CO_2 uptake (mmol m^{-2})	Δ malate
High PPFD	RWC 90%	238	6
	RWC 70%	117	16
	RWC 52%	22	23
Low PPFD	RWC 90%	53	0
	RWC 69%	38	6
	RWC 50%	17	16

CAM characteristics depending on leaf-air water-vapour pressure difference (Schmitt et al. 1988). Moreover, the magnitude of CAM inducible in *C. minor* is considerable. The nighttime accumulation of 1.4 molar titratable protons by *C. minor* in Trinidad is the highest ever reported for any facultative or constitutive CAM plant (Borland et al. 1992).

Studies on the expression of CAM in *C. minor* growing under natural conditions in Trinidad have illustrated that CAM is rapidly induced in terrestrial plants as the dry season progresses (Borland et al. 1992; 1993). Changes in gas-exchange patterns during the C_3-CAM transition in the field have demonstrated that during the wet season net uptake of CO_2 only occurs during the day, although a low level of acidification can result from the recapture of respiratory CO_2 at night (Borland et al. 1993). During the wet season there may also be a depression in photosynthetic rate after midday in exposed leaves. During the dry season, however, gas-exchange patterns for both exposed and shaded leaves are typical of CAM plants, with net uptake of CO_2 at night and stomatal closure during the middle of the day (Borland et al. 1993). The higher rates of photosynthesis observed during the dry season were probably related to the higher PPFDs encountered at this time (Borland et al. 1993).

The progression from C_3 to CAM gas-exchange characteristics in *C. minor* growing in Trinidad is comparable to those observed in laboratory-based studies on *M. crystallinum* (Winter and Gademann 1991; see Sect. 15.3.1). The magnification of the midday depression with increasing CAM in *C. minor* appears to be related primarily to changes in leaf water status responding to the increase in leaf-air vapour-pressure difference, as well as to the higher light intensities encountered around the middle of the day during the dry season (see also Winter et al. 1992).

The decarboxylation of substantial quantities of organic acids in leaves of *C. minor* might also be expected to lead to an increase in p_i. Leaves of *C. minor* can accumulate both citric and malic acids overnight. Although the decarboxylation

Table 15.3. Integrated net daytime CO_2 uptake in naturally exposed and shaded leaves of *C. minor* and the potential generation of CO_2 from organic-acid decarboxylation, assuming 1 mol of CO_2 is released from every mol of malate plus either 6 or 1 mol of CO_2 from each mol of citrate

Habitat	Net CO_2 uptake (mmol m^{-2})	Δ mal (mmol m^{-2})	Δ cit (mmol m^{-2})	Decarboxylation (mmol CO_2 m^{-2}) 6 mol CO_2/cit	Decarboxylation (mmol CO_2 m^{-2}) 1 mol CO_2/cit
Wet season					
Exposed	73.9	6.2	53.0	324	59
Shaded	62.1	0	0	0	0
Dry season					
Exposed	120.4	62.0	34.7	270	97
Shaded	52.6	50.6	10.4	113	61

of 1 mol of malic acid will release 1 mol of CO_2, the pathway of citric-acid degradation in CAM plants is not known. It has been suggested that, in theory, citric acid could be completely broken down during the day to release 6 mol CO_2 per mol citric acid (Haag-Kerwer et al. 1992). If this were the case, the amount of CO_2 generated internally would be at least twice as great as the net amount of CO_2 taken up during the day (Table 15.3). However, if we assume that only 1 mol of CO_2 is released from the decarboxylation of citric acid, the amount of CO_2 generated internally would be close to the integrated value for daytime photosynthesis.

15.4 Balance of C_3/C_4 Carboxylation in Facultative CAM Plants

From the examples described above, it would appear that patterns of daytime gas exchange exhibited during the C_3-CAM transition largely reflect changes in plant water status. However, during the shift towards CAM, day/night intermediary metabolism and carbon fluxes must also be regulated at the cellular level. This is particularly vital for C_3/CAM intermediates like *C. minor*, which can rapidly switch between the two modes of carbon acquisition.

Evidence for the transitions in C_3/C_4 carboxylation that occur during Phases II and IV (Sect. 15.2.3) has largely been gleaned from constitutive CAM plants or from facultative plants such as *M. crystallinum* or *K. blossfeldiana*, which show "molecularly definable" induction of CAM. Regulation of day/night carboxylation pathways in plants such as *S. telephium*, which are always endowed with the metabolic capacity for CAM and which continue to fix CO_2 continuously over the day even when CAM is enhanced, is largely unknown (see also Chap. 11). One way in which the activities of PEPC and RUBISCO may be reconciled with measurements of gas exchange is via measurements of short-term changes in carbon-isotope discrimination. This non-invasive technique can provide information on the extent of PEPC activity at the start and end of the day and indicate how shifts in the proportion of C_3 and C_4 carboxylation control daily carbon balance.

15.5 Instantaneous Discrimination of Carbon Isotopes

15.5.1 General Principles

Analysis of carbon-isotope ratios ($\delta^{13}C$) (see p. 5) has long been used to distinguish CAM plants and place the distribution of CAM in an ecological and phylogenetic context (Griffiths and Smith 1983; Griffiths 1992). For the constitutive CAM plant *K. daigremontiana*, only allowed to assimilate CO_2 at night, the $\delta^{13}C$ value of plant organic material (δ_p) was $-11‰$ (i.e. -0.011: Nalborczyk et al. 1975). In contrast, many C_3/CAM intermediates generally retain a predominantly C_3-like isotope composition (δ_p ranging from -25 to $-30‰$, i.e. -0.025 to -0.030), consistent with CAM operating mainly as a maintenance mechanism with little contribution to net carbon gain (Griffiths 1992). The relative proportions of C_3 and C_4 carboxylation processes will in general be integrated by the distribution of δ_p between these extremes.

More recently, it has been useful to distinguish the contribution made by the $\delta^{13}C$ of source CO_2 (δ_a) to plant organic material. Bulk air at 35 Pa currently has a δ_a of around $-8‰$ (i.e. -0.008), although the partial pressure and δ_a of source CO_2 can vary on a diel basis and with height in forest canopies (Broadmeadow et al. 1992). Thus the discrimination associated with biological processes can be derived using the relationship of Farquhar et al. (1989):

$$\Delta = \frac{\delta_a - \delta_p}{1 + \delta_p}. \tag{15.1}$$

Here, discrimination (Δ) is a positive term with the extent of depletion in ^{13}C directly related to the magnitude of discrimination processes. Typical values of Δ for the range of δ_p values quoted above would be $3.0‰$ (CAM; from Nalborczyk et al. 1976) and 17.4 to $22.7‰$ (C_3).

It is also possible to quantify the extent of C_3 and C_4 processes directly in real-time measurements of instantaneous discrimination and leaf gas exchange. Changes in the $^{13}C/^{12}C$ ratio of CO_2 in air as it passes over a leaf will reflect both diffusional resistance to CO_2 and carboxylation pathway (Evans et al. 1986). Instantaneous discrimination (Δ) during gas exchange of leaves is calculated from the difference in the carbon-isotope composition of the air leaving the chamber with (δ_o) and without (δ_e) a leaf enclosed (Evans et al. 1986):

$$\Delta = \frac{\xi(\delta_o - \delta_e)}{1 + \delta_o - \xi(\delta_o - \delta_e)}, \tag{15.2}$$

where $\xi = p_e/(p_e - p_o)$ and p_e and p_o are the CO_2 partial pressures of the air entering and leaving the chamber, respectively, when a leaf is enclosed.

15.5.2 On-Line Discrimination in *Tillandsia utriculata*

The isotopic signature of CO_2 exchange for each of the phases of CAM in the epiphytic bromeliad *T. utriculata* clearly demonstrated the transitions between

C_3 and C_4 carboxylation during Phases II and IV in this constitutive CAM plant (Griffiths et al. 1990). Thus the increase in instantaneous Δ from 10.5‰, measured shortly after lights came on, to 21.2‰ 30 min later, illustrated the rapid down-regulation of PEPC and the increase in RUBISCO activity during Phase II. During Phase III, decarboxylation of malic acid occurred with a small amount of CO_2 highly enriched in ^{13}C leaking from the leaf. During Phase IV, instantaneous discrimination decreased from 33.9 to 19.4‰ over the course of 5 to 6 h, demonstrating an increasing contribution from PEPC to CO_2 fixation towards the end of the photoperiod.

15.5.3 On-Line Discrimination in *Sedum telephium*

In C_3 plants, the magnitude of instantaneous Δ is largely determined by the ratio of intercellular to atmospheric partial pressure of CO_2 (p_i/p_a) that prevails when tissue carbon is assimilated. By measuring p_i/p_a it is therefore possible to predict Δ for plants fixing CO_2 via the C_3 pathway using the model of Farquhar et al. (1989):

$$\Delta = a + (b - a)\frac{p_i}{p_a}, \tag{15.3}$$

where a is the fractionation occurring due to diffusion in air (4.4‰) and b is the net fractionation caused by RUBISCO carboxylation (27‰; Farquhar and Richards 1984). By comparing Δ measured "on-line" with Δ predicted from p_i/p_a, it is thus possible to distinguish the effects of drought stress on stomatal limitation of photosynthesis from changes in carboxylation pathway that occur as CAM activity is enhanced in *S. telephium*.

Unlike the measurements for *T. utriculata*, where bottled CO_2 with δ_a of −43‰ was used as the source CO_2, for the experiments conducted on *S. telephium* (and *C. minor*, Sect. 15.5.4) source CO_2 was taken from atmospheric air which was stabilized by passage through a 1500-l mixing volume, with a regression equation relating natural variations in δ_a and p_a (Borland et al. 1993). The results, detailed in Fig. 15.2, illustrate that in well-watered plants of *S. telephium*, where measured and predicted Δ show close agreement, all CO_2 is fixed during the day with a Δ indicative of C_3 carboxylation via RUBISCO. Although there may be some refixation of respiratory CO_2 by PEPC at night (Fig. 15.1, Table 15.2), C_4 carboxylation had presumably ceased by the end of the dark period/beginning of the photoperiod, or was so low in comparison to RUBISCO activity that no significant decrease in Δ could be detected. As drought stress developed, measured and predicted Δ decreased, reflecting a decrease in stomatal conductance and an increase in water-use efficiency. Measured and predicted Δ also started to diverge at the beginning and end of the photoperiod (Fig. 15.2). The measured values of Δ, which were lower than those predicted from gas exchange, thus illustrate the increasing contribution from C_4 carboxylation as drought stress intensified.

These results indicate that C_4 carboxylation in *S. telephium* may operate for considerably longer during the day than was suggested by on-line measurements

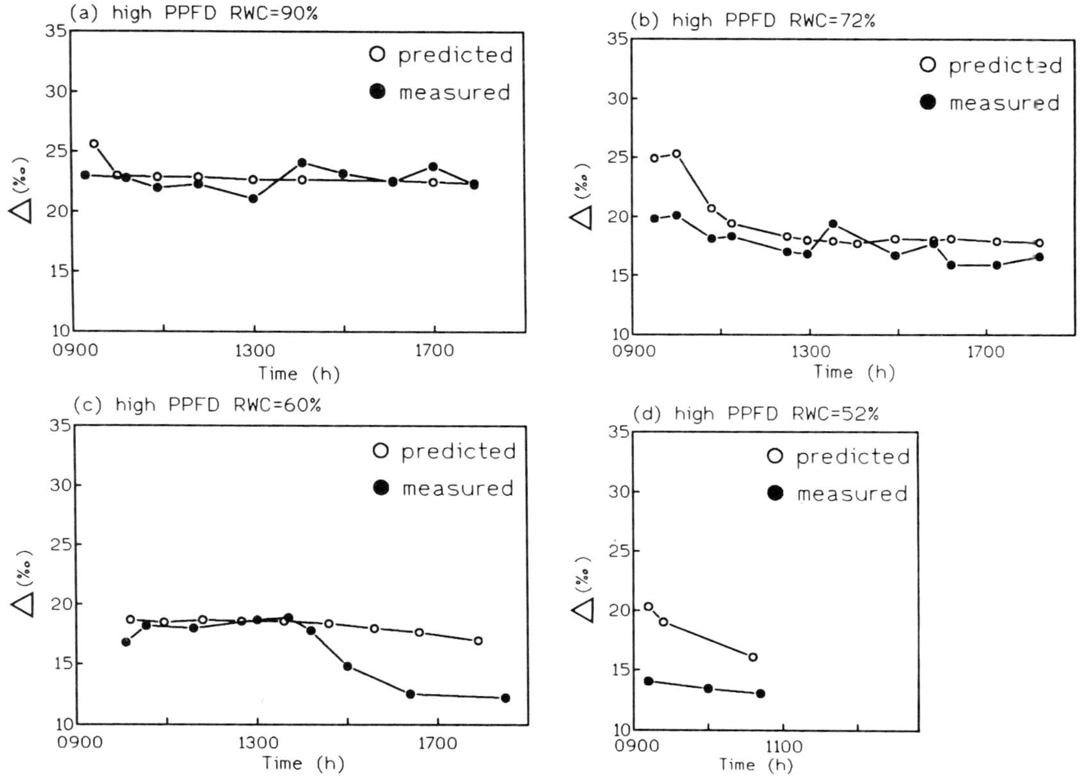

Fig. 15.2a–d. Diurnal changes in instantaneous ^{13}C discrimination (Δ) predicted from gas-exchange measurements of p_i/p_a and in Δ measured directly "on-line" for plants of *Sedum telephium* grown under a PPFD of 500 μmol m^{-2} s^{-1} and subjected to increasing drought stress as reflected by leaf relative water contents

for the constitutive CAM plant *T. utriculata* and by day/night changes in the kinetic properties of PEPC in plants performing pronounced CAM (Winter 1982; Nimmo et al. 1987). In vitro measurements of PEPC activity in *S. telephium* indicate that, in common with conventionally studied CAM plants, the enzyme extracted during the light period is more sensitive to malate inhibition than that extracted at night (Borland and Griffiths 1992; see also Chap.11). However "futile cycling" of CO_2 between PEPC and RUBISCO via a cycle of malic-acid synthesis and decarboxylation seems to occur early in Phase II and late in Phase IV for this species. Together with CO_2 taken up at night, it appears that daytime C_4 carboxylation can make a significant contribution to the fluxes of carbon in leaves of *S. telephium* during periods of stress when daytime gas exchange is reduced.

15.5.4 On-Line Discrimination in *Clusia minor*

During the wet season in Trinidad, values of instantaneous Δ for exposed and shaded leaves of *C. minor* indicated that most CO_2 was fixed via RUBISCO, with decreases in Δ over the latter part of the day largely reflecting a decrease in stomatal conductance (Fig.15.3). However, during the dry season, when the typical four phases of CAM were found in both exposed and shaded plants, values of instantaneous Δ measured over the first 3 h after dawn (6.4–9.1‰) indicated that C_4 carboxylation dominated CO_2 uptake for most of the morning when rates of net CO_2 uptake were at maximum (Fig.15.4). Thus under natural conditions it would appear that, in *Clusia* at least, the down-regulation of PEPC in Phase II occurs much more slowly than laboratory-based studies on other species have suggested (Nimmo et al. 1987; Griffiths et al. 1990). These results are supported by field observations on the rate of organic-acid decarboxylation in *C. rosea*, where acid breakdown only commenced some 2 to 3 h after dawn (Ball et al. 1991). The situation in *C. minor* is further complicated by the accumulation of substantial quantities of citric acid, presumably also in the vacuole. Citrate can chelate magnesium and, as such, could also be an effective inhibitor of cytosolic PEPC (Kluge and Osmond 1972; Kluge and Ting 1978).

Whilst most of Phase II carboxylation would appear to have been mediated by PEPC in exposed leaves during the dry season, there was a slight increase in C_3 carboxylation shortly before stomata closed, as evidenced by the rise in Δ (Fig.15.4a). However, in shaded leaves which had lower CAM activity, there was a more pronounced increase in C_3 carboxylation over the latter part of Phase II (Fig.15.4b). When stomata reopened later in the day, instantaneous Δ declined during Phase IV which, in shaded leaves, was partly due to a decrease in stomatal conductance (Borland et al. 1993). However, in exposed leaves stomata remained open throughout the transition from dusk to darkness, with the decrease in Δ reflecting the increase in PEPC activity (Fig.15.4).

Integration of carbon-isotope discrimination and net CO_2 exchange over the various phases of CAM indicated that C_4 carboxylation accounted for approximately 33% of carbon fixed during the day by exposed leaves and 18%

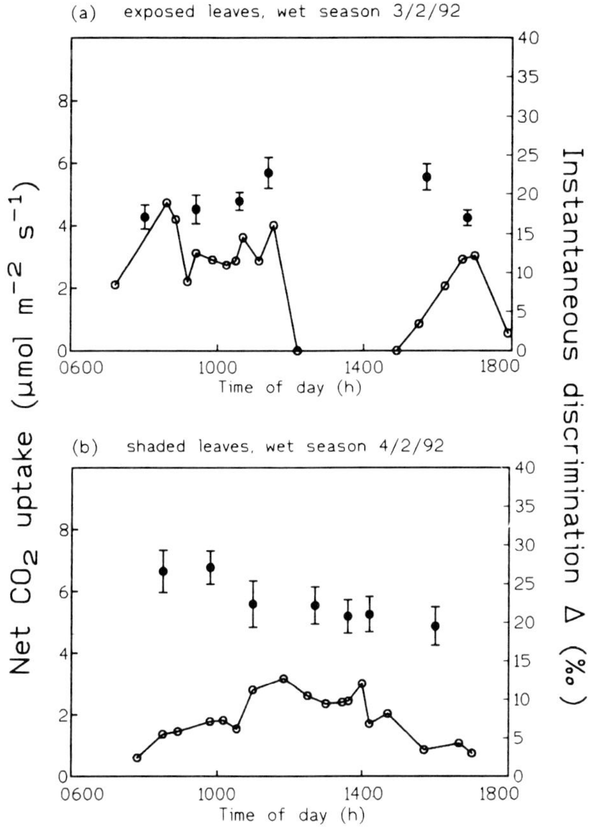

Fig. 15.3a,b. Gas exchange (○) and instantaneous ^{13}C discrimination Δ (●) measured during the day for (**a**) exposed and (**b**) shaded leaves of *Clusia minor* sampled during the wet season in Trinidad. Values for Δ are presented as a 3-to 4-point running mean obtained from five to eight replicates sampled at various times throughout the day

in shaded leaves (Borland et al. 1993). With dark uptake of CO_2 accounting for approximately 50% of the total carbon fixed over 24 hours by both shaded and exposed leaves, it is clear that during the dry season C_4 carboxylation contributes most of the carbon assimilated by *C. minor*. However, partitioning of C_3 photosynthates into a pool of transport sugars during Phase IV ensures that exported sucrose carries a predominantly C_3 isotopic signature, regardless of the degree of CAM (Borland et al. 1994).

15.5.5 Carbon-Isotope Discrimination in *Mesembryanthemum crystallinum*

The importance of CAM to the productivity and survival of *M. crystallinum* was implied from carbon-isotope-discrimination measurements for plants growing in the field and laboratory (Winter et al. 1978; Winter and Ziegler 1992). Flowering and

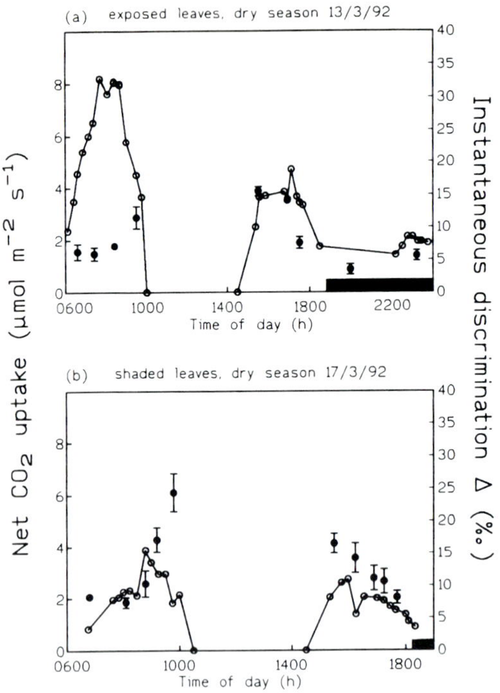

Fig. 15.4a,b. Gas exchange (○) and instantaneous ^{13}C discrimination Δ (●) for (**a**) exposed and (**b**) shaded leaves of *Clusia minor* sampled during the dry season in Trinidad. Values for Δ are presented as a 3- to 4-point running mean obtained from five to eight replicates sampled at various times throughout the day. The *solid bar* on the *x-axis* indicates the period of darkness

seed-set were reduced by a factor of 10 in plants deprived of CO_2 at night (Winter and Ziegler 1992). The $\delta^{13}C$ values of the seeds were less negative in the full CAM as opposed to the CO_2-free treatment (−16.4 and −20.7‰, respectively: Winter and Ziegler 1992). Thus, when calculated using Eq.(1) and a source CO_2 $\delta^{13}C$ of −8‰, discrimination (Δ) corresponds to 8.5‰ (CAM) and 13‰ (CO_2-free).

In the absence of on-line discrimination data for *M. crystallinum*, the values of Δ for organic material can be used to calculate the likely extent of daytime CO_2 fixation as a proportion of overall carbon gain. If plants were fixing CO_2 only at night, then the Δ value should be similar to that determined by Nalborczyk et al. (1975) for *K. daigremontiana* [i.e. $\delta^{13}C$ of −11‰, equivalent to a Δ of 3.0‰; Eq. (15.1)]. A similar value for Δ was measured directly for *C. minor* at night (Borland et al. 1993; Fig.15.4). By using the extreme values of Δ representing exclusively daytime (13.0‰) or nighttime (3.0‰) CO_2 uptake in *M. crystallinum*, 55‰ of carbon was gained during the day (13.0‰ × 0.55 = 7.14‰) and 45% during the night (3.0‰ × 0.45 = 1.36‰). This results in the observed Δ of (7.14 + 1.36) = 8.5‰ (Winter and Ziegler 1992) and corresponds well with published gas exchange data (Winter 1974; Whitehouse et al. 1991).

Additionally, the values of Δ can be used to calculate the average daily p_i/p_a and degree of stomatal limitation for the data of Winter and Ziegler (1992). For the C_3 form of *M. crystallinum*, with a $\delta^{13}C$ of $-29‰$ ($\Delta = 21.6‰$), p_i/p_a would on average have been 0.78; for the CAM form provided with CO_2 only in the light, p_i/p_a would on average have been 0.38. This compares closely with direct measurements of p_i/p_a where, after only 10 days of salt stress, p_i/p_a was reduced from 0.76 (C_3) to about 0.40 (Fig. 2 of Winter and Gademann 1991). Whether this reduction in stomatal conductance, which was higher during the middle of the light period, should be interpreted as the midday depression of photosynthesis or as the development of Phases II and IV remains to be determined. However, until measurements of on-line discrimination are made, it must be concluded that a diffusion-limited C_3 pathway can account for the Δ of organic material of *M. crystallinum* without any daytime contribution from PEPC activity.

15.6 Implications of Carbon Flow During Phases II and IV for C_3/CAM Intermediates

A brief survey of the literature illustrates that net uptake of CO_2 during Phases II and IV can make a substantial contribution to carbon cycling/turnover in a range of obligate and facultative CAM plants (Table 15.1). Moreover, measurements of on-line carbon-isotope discrimination for *S. telephium* and *C. minor* have illustrated that the cycling of carbon through PEPC forms a significant proportion of the carboxylation processes which occur during the day. Carbon cycling during the transitional phases of CAM appears particularly important under natural conditions, where the gradation of increasing and decreasing PPFD at dawn and dusk, respectively, leads to extended gas exchange in the light period.

For *M. crystallinum*, however, C_4 carboxylation appears to be largely, if not entirely, restricted to the dark period in plants in which CAM has been induced. Thus, the discrimination in seed material from plants of this species maintained in CO_2-free air at night could result solely from diffusion-limited carboxylation mediated by RUBISCO (Sect. 15.5.5).

The full induction of CAM in *M. crystallinum* translates into improved carbon gain and seed set on a seasonal basis (Winter and Ziegler 1992). However, for *S. telephium* and *C. minor*, the dark fixation of CO_2, together with the observed PEPC activity during Phases II and IV, does not result in substantial net carbon gain or productivity on a seasonal basis, since $\delta^{13}C$ values for organic material from both species fall within the range of C_3 plants (i.e. -23.2 to $-29.0‰$; Borland et al. 1992, Borland et al. 1992, 1994).

However, it can be considered that it is the flux of carbon through C_4 and C_3 products, rather than carbon gain, which is critical. Despite the extreme differences in the capacity for the induction of CAM in *S. telephium* and *C. minor*, regeneration of CO_2 from organic acids accumulated during Phase I contributes a significant proportion of daytime CO_2 fixation in both species. In *S. telephium*, the decarboxylation of malic acid (accumulated overnight from external and respiratory CO_2) can provide up to 50% of the carbon budget in the light, thus doubling gross CO_2 fixation although, in comparative terms, this is still less than 25% of the maximum net CO_2 uptake shown a few days earlier by the well-watered, C_3 form (Table 15.2). Since RUBISCO capacity is unlikely to have been approached in the droughted plants, p_i does not increase and stomata remain open. Consequently, CO_2 can be taken up continuously over 24 h during periods of drought stress. The high capacity of RUBISCO during Phase III may be related to the high photosynthetic capacities frequently encountered for CAM plants in the light when measured as rates of O_2 evolution (Adams and Osmond 1988; Adams et al. 1989; Borland and Griffiths 1989; Cote et al. 1989; Griffiths et al. 1989; Borland et al. 1992; Winter et al. 1992).

In *C. minor*, the duration of stomatal closure during "Phase III" increased in concert with enhanced CAM activity (Figs.15.3 and 15.4). Maximum CO_2 regeneration and high p_i therefore coincide with highest incident PPFD under natural conditions, and previous work has shown that the regeneration of CO_2 can alleviate photoinhibition in CAM plants (Osmond et al. 1980; Adams and Osmond 1988; Adams et al. 1989; Griffiths et al. 1989). It is also interesting to note that C_4 carboxylation predominated during Phases II and IV when CAM activity and net daytime CO_2 uptake rates were highest (Fig. 4). Given the high potential rates of electron transport (as determined from O_2 exchange: Borland et al. 1992) required to saturate RUBISCO capacity under saturating CO_2, perhaps futile cycling during Phase II serves as a useful mechanism to dissipate ATP and reductant derived when excess photons are absorbed, thus supplementing those mechanisms that down-regulate photochemical efficiency (see Chap.7). Therefore, under conditions of water stress and high PPFD encountered during the dry season, organic-acid pools may be conserved during Phase II in order that decarboxylation then coincides with maximum PPFD.

In conclusion, for obligate CAM plants and some facultative CAM plants such as *M. crystallinum*, C_4 carboxylation makes a significant long-term contribution to growth and productivity. However, in other facultative CAM plants where the switch to CAM is rapid and freely reversible, CAM may be viewed predominantly as a mechanism for maintaining photosynthetic integrity and a positive carbon balance during periods of stress.

References

Adams WW III, Osmond CB (1988) Internal CO_2 supply during photosynthesis of sun and shade grown CAM plants in relation to photoinhibition. Plant Physiol 86: 117–123

Adams WW III, Diaz M, Winter K (1989) Diurnal changes in photochemical efficiency, the reduction state of Q, radiationless energy dissipation and non-photochemical fluorescence quenching in cacti exposed to natural sunlight in northern Venezuela. Oecologia 80: 553–561

Allaway WG, Osmond CB, Troughton JH (1974) Environmental regulation of growth, photosynthetic pathway and carbon isotope discrimination ratio in plants capable of crassulacean acid metabolism. In: Bieleski RL, Ferguson AR, Cresswell MM (eds) Mechanisms of regulation of plant growth. Bull R Soc NZ 12: 195–205

Ball E, Hann J, Kluge M, Lee HSJ, Lüttge U, Orthen B, Popp M, Schmitt AK, Ting IP (1991) Ecophysiological comportment of the tropical CAM-tree *Clusia* in the field. II. Modes of photosynthesis in trees and seedlings. New Phytol 117: 483–491
Bartholomew DP (1973) Drought response in the gas exchange of *Dudleya farinosa* (Crassulaceae) grown under natural conditions. Photosynthetica 7: 114–120
Bartholomew DP, Kadzimin SB (1977) Pineapple. In: Alvim PT, Kozlowski TT (eds) Ecophysiology of tropical crops. Academic Press, New York, pp 113–156
Borland AM, Griffiths H (1989) The regulation of citric acid accumulation and carbon recycling during CAM in *Ananas comosus*. J Exp Bot 210: 53–60
Borland AM, Griffiths H (1990) The regulation of CAM and respiratory recycling by water supply and light regime in the C_3-CAM intermediate *Sedum telephium* L. Funct Ecol 4: 33–39
Borland AM, Griffiths H (1992) Properties of phosphoenol-pyruvate carboxylase and carbohydrate accumulation in the C_3-CAM intermediate *Sedum telephium* L. grown under different light and watering regimes. J Exp Bot 43: 353–361
Borland AM, Griffiths H, Maxwell C, Broadmeadow MSJ, Griffiths NM, Barnes JD (1992) On the ecophysiology of the Clusiaceae in Trinidad: expression of CAM in *Clusia minor* during the transition from wet to dry season and characterisation of three endemic species. New Phytol 122: 349–357
Borland AM, Griffiths H, Broadmeadow MSJ, Fordham MC, Maxwell C (1993) Short-term changes in carbon-isotope discrimination in the C_3-CAM intermediate *Clusia minor* L. growing in Trinidad. Oecologia 95: 444–453
Borland AM, Griffiths H, Broadmeadow MSJ, Fordham MJ, Maxwell C (1994) Carbon-isotope composition of biochemical fractions and the regulation of carbon balance in leaves of the C_3-crassulacean acid metabolism plant *Clusia minor* L. growing in Trinidad. Plant Physiol 106: 493–501.
Broadmeadow MSJ, Griffiths H, Maxwell C, Borland AM (1992) The carbon isotope ratio of plant organic material reflects temporal and spatial variations in CO_2 partial pressure and $\delta^{13}C$ within tropical forest formations in Trinidad. Oecologia 89: 435–441
Cote FX, André M, Folliot M, Massimino D, Daguenet A (1989) CO_2 and O_2 exchanges in the CAM plant *Ananas comosus* (L.) Merr. Plant Physiol 89: 61–68
Evans JR, Sharkey TD, Berry JA, Farquhar GD (1986) Carbon isotope discrimination measured concurrently with gas exchange to investigate CO_2 diffusion in leaves of higher plants. Aust J Plant Physiol 13: 281–292
Farquhar GD, Richards PA (1984) Isotopic composition of plant carbon correlates with water use efficiency of wheat genotypes. Aust J Plant Physiol 11: 339–352
Farquhar GD, Ehleringer JR, Hubick KT (1989) Carbon isotope discrimination and photosynthesis. Annu Rev Plant Physiol Plant Mol Biol 40: 503–537
Franco AC, Ball E, Lüttge U (1990) Patterns of gas exchange and organic acid oscillations in tropical trees of the genus *Clusia*. Oecologia 85: 108–114
Fischer A, Kluge M (1984) Studies on carbon flow in crassulacean acid metabolism during the initial light period. Planta 160: 121–128
Gravatt DA, Martin CE (1992) Comparative ecophysiology of five species of *Sedum* (Crassulaceae) under well-watered and drought-stressed conditions. Oecologia 92: 532–541
Griffiths H (1992) Carbon isotope discrimination and the integration of carbon assimilation pathways in terrestrial CAM plants. Plant Cell Environ 15: 1051–1062
Griffiths H, Smith JAC (1983) Photosynthetic pathways in the Bromeliaceae of Trinidad: relations between life-forms, habitat preference and the occurrence of CAM. Oecologia 60: 176–184
Griffiths H, Ong BL, Avadhani PN, Goh CJ (1989) Recycling of respiratory CO_2 during crassulacean acid metabolism: alleviation of photoinhibition in *Pyrrosia piloselloides*. Planta 179: 115–122
Griffiths H, Broadmeadow MSJ, Borland AM, Hetherington CS (1990) Short-term changes in carbon-isotope discrimination identify transitions between C_3 and C_4 carboxylation during crassulacean acid metabolism. Planta 181: 604–610
Groenhof AC, Smirnoff N, Bryant JA (1990) The appearance of a new molecular species of phosphoenolpyruvate carboxylase (PEPC) and the rapid induction of CAM in *Sedum telephium* L. Plant Cell Environ 13: 437–446

Guralnick LJ, Health RL, Goldstein G, Ting IP (1992) Fluorescence quenching in the varied photosynthetic modes of *Portulacaria afra* (L.) Jacq. Plant Physiol 99: 1309–1313

Haag-Kerwer A, Franco AC, Lüttge U (1992) The effect of temperature and light on the gas exchange and acid accumulation in the C_3-CAM plant *Clusia minor* L. J Exp Bot 43: 345–352

Harris FS, Martin CE (1991) Plasticity in the degree of CAM-cycling and its relationship to drought stress in five species of *Talinum* (Portulacaceae). Oecologia 86: 575–584

Holthe PA, Sternberg L da SL, Ting IP (1987) Developmental control of CAM in *Peperomia scandens*. Plant Physiol 84: 743–747

Kluge M, Osmond CB (1972) Studies on phosphoenolpyruvate carboxylase and other enzymes of crassulacean acid metabolinsm of *Bryophyllum tubiflorum* and *Sedum praealtum*. Z Pflanzenphysiol 66: 97–105

Kluge M, Ting IP (1978) Crassulacean acid metabolism. Analysis of an ecological adaptation. Springer, Berlin Heidelberg New York

Kluge M, Bley L, Schmid R (1975) Malate synthesis in crassulacean acid metabolism (CAM) via a double CO_2 dark fixation? In: Marcelle R (ed) Environmental and biological control of photosynthesis. Dr W Junk, The Hague, pp 281–288

Kluge M, Böhlke C, Queiroz O (1981a) Crassulacean acid metabolism (CAM) in *Kalanchoë*: changes in intercellular CO_2 concentration during a normal CAM cycle and during cycles in continuous light or darkness. Planta 152: 87–92

Kluge M, Brulfert J, Queiroz O (1981b) Diurnal changes in the regulatory properties of PEP-carboxylase in crassulacean acid metabolism (CAM). Plant Cell Environ 4: 251–256

Medina E, Delgado M (1976) Photosynthesis and night CO_2 fixation in *Echeveria columbiana* v. Poellnitz. Photosynthetica 10: 155–163

Nalborczyk E, LaCroix LJ, Hill R (1975) Environmental influences on light and dark CO_2 fixation by *Kalanchoë daigremontiana*. Can J Bot 53: 1132–1138

Nimmo GA, Wilkins MB, Fewson CA, Nimmo HG (1987) Persistent circadian rhythms in the phosphorylation state of phosphoenol-pyruvate carboxylase from *Bryophyllum feldtschenkoi* leaves and its sensitivity to inhibition by malate. Planta 170: 408–415

Nobel PS (1988) Environmental biology of agaves and cacti. Cambridge University Press, Cambridge

Osmond CB (1978) Crassulacean acid metabolism: a curiosity in context. Annu Rev Plant Physiol 29: 379–414

Osmond CB, Allaway WG (1974) Pathways of CO_2 fixation in the CAM plant *Kalanchoë daigremontiana*. I. Patterns of $^{14}CO_2$ fixation in the light. Aust J Plant Physiol 1: 503–511

Osmond CB, Winter K, Powles SB (1980) Adaptive significance of carbon dioxide recycling during photosynthesis in water-stressed plants. In: Turner NC, Kramer PJ (eds) Adaptation of plants to water and high temperature stress. John Wiley, New York, pp 139–154

Osmond CB, Holtum JAM, O'Leary MH, Roeske C, Wong OC, Summons RE, Avadhani PN (1988) Regulation of malic-acid metabolism in crassulacean acid metabolism plants in the dark and light: in-vivo evidence from ^{13}C-labelling patterns after $^{13}CO_2$ fixation. Planta 175: 184–192

Ritz D, Kluge M, Veith HJ (1986) Mass-spectrometric evidence for the double-carboxylation pathway of malate synthesis by crassulacean acid metabolism plants in light. Planta 167: 284–291

Schmitt AK, Lee HSJ, Lüttge U (1988) The response of the C_3-CAM tree, *Clusia rosea*, to light and water stress. I. Gas exchange characteristics. J Exp Bot 39: 1581–1590

Smith JAC, Lüttge U (1985) Day-night changes in leaf water relations associated with the rhythm of crassulacean acid metabolism in *Kalanchoë daigremontiana*. Planta 163: 272–282

Tenhunen JD, Catarino FM, Lange OL, Oechel WC (eds) (1987) Plant response to stress. Functional analysis in Mediterranean ecosystems. Springer, Berlin Heidelberg New York

Whitehouse DG, Rogers WJ, Tobin AK (1991) Photorespiratory enzyme activities in C_3 and CAM forms of the facultative CAM plant *Mesembryanthemum crystallinum* L. J Exp Bot 42: 485–492

Winter K (1974) NaCl-induzierter Crassulaceen-Säurestoffwechsel bei der Salzpflanze *Mesembryanthemum crystallinum*. Oecologia 15: 383–392

Winter K (1982) Properties of phosphoenolpyruvate carboxylase in rapidly prepared, desalted leaf extracts of the crassulacean acid metabolism plant *Mesembryanthemum crystallinum* L. Planta 154: 298–308

Winter K (1985) Crassulacean acid metabolism. In: Barber J, Baker NR (eds) Photosynthetic mechanisms and the environment. Elsevier, Amsterdam, pp 329–387

Winter K, Gademann R (1991) Daily changes in CO_2 and water vapor exchange, chlorophyll fluorescence and leaf water relations in the halophyte *Mesembryanthemum crystallinum* during the induction of crassulacean acid metabolism in response to high NaCl salinity. Plant Physiol 95: 768–776

Winter K, Tenhunen JD (1982) Light-stimulated burst of carbon dioxide uptake following nocturnal acidification in the crassulacean acid metabolism plant *Kalanchoë daigremontiana*. Plant Physiol 70: 1718–1722

Winter K, Ziegler H (1992) Induction of crassulacean acid metabolism in *Mesembryanthemum crystallinum* increases reproductive success under conditions of drought and salinity stress. Oecologia 92: 475–479

Winter K, Lüttge U, Winter E, Troughton JH (1978) Seasonal shift from C_3 photosynthesis to crassulacean acid metabolism in *Mesembryanthemum crystallinum* growing in its natural environment. Oecologia 34: 225–237

Winter K, Zotz G, Baur B, Dietz K-J (1992) Light and dark CO_2 fixation in *Clusia uvitana* and the effects of plant water status and CO_2 availability. Oecologia 91: 47–51

Part C
Ecophysiology and Evolution of Crassulacean Acid Metabolism

Preface

The third part of this book follows on from the discussion of the phenotypic plasticity of CAM in selected examples to survey the great diversity of natural environments in which CAM plants have established an ecological niche. Although the stem-succulent cacti were the first CAM plants to be studied in detail in the field (see Nobel 1988), the scope of ecophysiological investigations on CAM plants has broadened enormously in the last few years. These investigations are providing new insights into the selective pressures that can lead to the evolution of CAM in different environments.

Even for such well-studied species as the desert cacti and agaves, there have been some remarkable new findings from field research in recent years. In Chapter 16, Nobel challenges the common view of CAM plants as slow-growing "curiosities" and demonstrates that the annual productivities of well-irrigated species of *Opuntia* and *Agave* can be similar to, or even exceed, those of C_3 crop plants in comparable climates. Chapter 17 goes on to consider the significance for desert succulents in arid environments with seasonal rainfall of the root system, which has been notoriously difficult to investigate. In these environments, the roots must allow rapid absorption of water from the soil when wet, but be able to minimize water loss from the plant to the soil during periods of drought.

A little more than a decade ago, it would have been hard to imagine a more unlikely group of plants in which to find CAM than submerged aquatic vascular cryptogams. Keeley's pioneering studies, reviewed in Chapter 18, show that the CAM pathway allows nocturnal assimilation of carbon by these and a number of angiosperm macrophytes in freshwater habitats where the availability of dissolved inorganic carbon is low during the daytime.

CAM has also been recently discovered in the tropical genus *Clusia* (Clusiaceae), which contains free-standing arborescent species as well as shrubby epiphytes and hemiepiphytes. Chapter 19 illustrates the considerable versatility in daily CO_2-exchange patterns shown by different species of *Clusia*; these can range from those typical of C_3 plants to those representing full CAM. A detailed case study is described in Chapter 20 of a single species, *C. uvitana*, in which CO_2 exchange was monitored in situ over an entire year under natural conditions in a moist tropical forest.

Many of the basic principles of CAM have been elucidated from laboratory studies with species of *Kalanchoë* (Crassulaceae). This large genus exhibits considerable morphological diversity, but we are only just beginning to appreciate the significance of differences in growth habit for the ecophysiology of these plants in their natural environments in Madagascar (Chap. 21). Measurements of carbon- and hydrogen-isotope discrimination are now well-established techniques that can provide important information on the assimilation of organic matter over the life span of photosynthetic organs. These techniques can also be applied to studies of carbon assimilation of plants sampled from contrasting natural habitats (Chap. 22), as well as for taxonomic purposes to study the evolution of CAM in particular groups such as the Crassulaceae (Chap. 23). In the last chapter of this section (Chap. 24), Raven and Spicer summarize our understanding of the present-day functional significance of the CAM pathway and consider how, in the absence of any known fossil evidence, it may nevertheless be possible to reconstruct plausible evolutionary scenarios for the origin of CAM in terrestrial and aquatic habitats.

Reference

Nobel PS (1988) Environmental biology of agaves and cacti. Cambridge University Press, Cambridge

16 High Productivity of Certain Agronomic CAM Species

P.S. Nobel

16.1 Introduction

Plant productivity varies with the vegetation type and the environmental conditions. For native species in deserts, the aboveground dry-weight productivity averages only 1 Mg hectare^{-1} year^{-1} (Lieth and Whittaker 1975), whereas annual crops usually have productivities of 6 to 20 Mg ha^{-1} yr^{-1} (Evans 1975; Loomis and Gerakis 1975; Beadle et al. 1985; Nobel 1991)[1]. About 92% of all vascular plant species use the C_3 photosynthetic pathway (Fig. 16.1a) in which the first stable product of photosynthesis is a 3-carbon compound (3-phosphoglyceric acid; Ting 1985; Winter 1985; Lüttge 1987; Nobel 1991; Salisbury and Ross 1991). The highest productivities for C_3 crops range from 34 to 45 Mg ha^{-1} yr^{-1} (Table 16.1). For plants with the C_4 pathway (Fig. 16.1b), the first products formed are C_4 acids (e. g. oxaloacetic acid and malic acid). The highest productivities are 47 to 88 Mg ha^{-1} yr^{-1} for C_4 crops and 51 to 94 Mg ha^{-1} yr^{-1} for C_4 floodplain grasses (Table 16.1). Although the values for floodplain grasses are initial estimates, the high values for sugarcane (*Saccharum officinarum*) have been confirmed by multiple studies.

CAM plants, which generally have low productivites, represent over 6% of the 260 000 species of vascular plants (Nobel 1991; see also p. 9). Their stomatal opening and CO_2 uptake occur primarily at night, leading to accumulation of 4-carbon organic acids such as malic acid in the vacuoles of their chlorenchyma cells (Fig. 16.1c). The lower air and plant temperatures at night reduce their transpiration rates for a given degree of stomatal opening compared with C_3 and C_4 plants. Gas exchange for CAM plants is generally low during the daytime, when CO_2 released within the plant by decarboxylation of the organic acids is fixed into photosynthetic products by the C_3 pathway.

The 1600 species in the family Cactaceae all utilize CAM, except for a few primitive species with leaves as their main photosynthetic organs (Gibson and Nobel 1986; Nobel 1988). All of the approximately 200 species of *Mammillaria*, the largest genus in the Cactaceae, are relatively slow-growing. Most of the nearly

Laboratory of Biomedical and Environmental Sciences, 900 Veteran Avenue, University of California, Los Angeles, CA 90024-1786, USA

[1] 1 Mg = 10^6 g ≡ 1 metric tonne; and 1 hectare ≡ 10 000 m^2.

Ecological Studies, Vol. 114
Winter/Smith (eds) Crassulacean Acid Metabolism

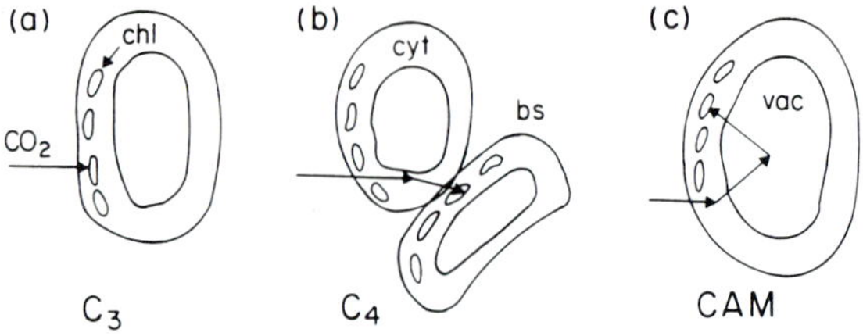

Fig. 16.1. Compartments involved with CO_2 fixation by the three photosynthetic pathways. For C_3 plants (*a*), the entire process occurs in the chloroplasts (*chl*) of mesophyll cells. For C_4 plants (*b*), the initial CO_2 fixation occurs in the cytosol (*cyt*) of mesophyll cells and the final CO_2 fixation occurs in chloroplasts in bundle-sheath (*bs*) cells. For CAM plants (*c*), the initial CO_2 fixation occurs in the cytosol; the resulting organic acids are stored overnight in the vacuole (*vac*) and are then subsequently decarboxylated during the daytime and refixed in the chloroplasts. Used by permission from Nobel (1991). (© *The New Phytologist*)

Table 16.1. Annual aboveground dry-weight productivity of the most productive C_3 and C_4 species

Type Species	Location	Maximal productivity (Mg ha^{-1} yr^{-1})	Reference
C_3 crops			
Beta vulgaris	California, USA	34	Loomis and Gerakis (1975)
Elaeis guineensis	Malaysia	40	Cooper (1975)
Manihot esculenta	Java; Madagascar; Sierra Leone	45	Loomis (1983)
C_3 trees			
Cryptomeria japonica	Japan	44	Kira (1975)
Eucalyptus globulus	Portugal	40	Jarvis and Leverenz (1983)
Eucalyptus grandis	South Africa	41	Schönau and Pennefather (1975)
C_4 crops			
Pennisetum purpureum	El Salvador; Puerto Rico	70 to 88	Loomis and Gerakis (1975); Loomis (1983)
Saccharum officinarum	Guyana; Hawaii, USA; Queensland, Australia	50 to 67	Evans (1975); Loomis and Gerakis (1975); Beadle et al. (1985)
Sorghum bicolor	California, USA	47	Loomis and Gerakis (1975)
C_4 floodplain			
Cyperus papyrus	Kenya	51	Jones (1986)
Echinochloa polystachya	Brazil	94	Piedade et al. (1991)

200 species of *Opuntia* also have low annual biomass productivity. However, certain species of *Opuntia* can have high productivities and are cultivated in about 25 countries worldwide for their fruit as well as for their cladodes (flattened stem segments) used as forage, fodder, and vegetables (Russell and Felker 1987; Nobel 1988). For instance, *Opuntia ficus-indica* and related species have long been

known to have aboveground dry-weight productivities of 18 to 20 Mg ha^{-1} yr^{-1} (Griffiths 1915; Monjauze and Le Houérou 1965; Flores Valdez and Aguirre Rivera 1979). The 140 species in the genus *Agave*, the largest genus in the family Agavaceae, apparently also use the CAM pathway. Until recently, the highest productivity reported for agaves was 25 to 26 Mg ha^{-1} yr^{-1} for *Agave mapisaga*, *A. salmiana*, and *A. tequilana*, species that are cultivated in Mexico for the production of alcoholic beverages (Nobel and Valenzuela 1987; Nobel 1988; García-Moya and Nobel 1990).

The primary purpose of this chapter is to assess the maximal productivity possible for certain agaves and cacti under ideal environmental conditions in the field. For this, the simultaneous effects of soil water status, air temperature, and light on net CO_2 uptake are considered. To maximize light interception for specific agaves and cacti, computer models predicting optimal spacing for plants in the field have been used for experimental plantings. Such predicted optimal spacing has been tested on plants in fertile soils where water was also not appreciably limiting. These conditions have led to biomass productivities of agaves and cacti that surpass those of nearly all other plants examined. To help understand the features underlying the high productivity of the CAM plants involved, gas exchange and energetics of CO_2 fixation by CAM plants will be compared with those of C_3 and C_4 plants.

16.2 Experimental Design for High Productivity

The simultaneous effects of environmental factors such as soil-water status, air temperature, and ambient light levels on net CO_2 uptake can be quantified using an Environmental Productivity Index (EPI; Nobel 1984; 1988):

$$\text{EPI} = \text{Water Index} \times \text{Temperature Index} \times \text{PPFD Index}, \qquad (16.1)$$

where the light absorbed by photosynthetic pigments is known as the photosynthetic photon flux density (PPFD, μmol photons m^{-2} s^{-1} from 400 to 700 nm, also often referred to as the photosynthetic photon flux, PPF). Each component index ranges from zero, when that environmental factor eliminates net CO_2 uptake over a 24-h period, to unity, when that factor is not limiting. The Water Index thus equals 1.00 for wet soil, the Temperature Index equals 1.00 for optimal temperatures, and the PPFD Index equals 1.00 for light saturation of daily net CO_2 uptake over 24-h periods.

To use EPI, the relationship between an environmental variable and net CO_2 uptake per unit shoot surface area over 24-h periods must be determined under carefully controlled conditions, usually in the laboratory. To quantify the influence of drought, defined as when the soil water potential decreases below the plant water potential, net CO_2 uptake is determined over 24-h periods during a course of soil drying. Nocturnal temperatures have more influence on daily net CO_2 uptake by CAM plants than do daytime temperatures, so their Temperature Index is usually based on average or minimal nighttime temperatures.

Temperature is usually not a major limiting factor for net CO_2 uptake by cultivated agaves and cacti over the course of a year, although episodic low temperatures can limit the distribution of various species. On the other hand, the availability of PPFD generally limits net CO_2 uptake for some parts of the shoots of agaves or cacti on a daily basis, especially when extensive shading occurs for the closely spaced plants that lead to high productivity per unit ground area.

Because CAM plants take up CO_2 primarily in the absence of light, net CO_2 uptake cannot be related to the instantaneous value of PPFD. Rather, net CO_2 uptake over 24-h periods is generally expressed as a function of the total daily PPFD, which is the integrated value of the instantaneous PPFD over the course of a day. The instantaneous PPFD can be calculated for any part of a shoot using a ray-tracing technique (Fig. 16.2). This technique determines the angle of the sunlight striking any surface and which surfaces are in shadows throughout the day, so the daily PPFD can be calculated for any surface. Hence, the PPFD Index [Eq. (16.1)], which is expressed per unit shoot surface area, can be predicted for any part of the shoot and for various spacings between plants along rows and between rows (Garcia de Cortázar and Nobel 1986). Incorporating the PPFD Index into EPI [Eq. (16.1)], net CO_2 uptake and hence productivity can be estimated under realistic field conditions.

The average PPFD Index is highest when the plants are far apart and hence do not shade each other. However, the highest productivity per unit ground area occurs when the plants are fairly close together. Close plant spacing is accompanied by a lower average PPFD Index due to some shading within the canopy but a much greater shoot area per unit ground area. The ray-tracing technique

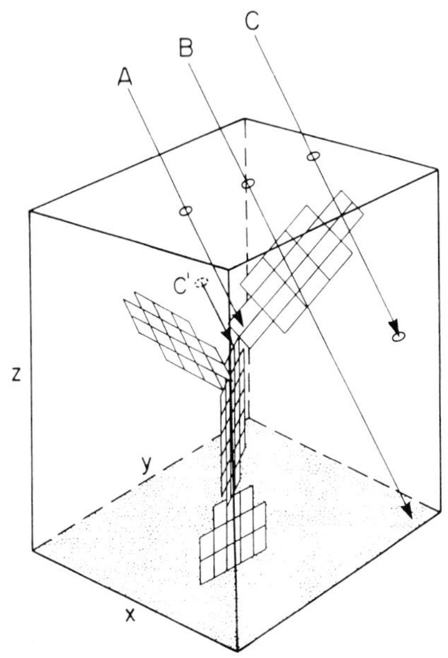

Fig. 16.2. Schematic illustration of the ray-tracing model used to calculate the interception of PPFD. Within a defined three-dimensional space, the surfaces of four cladodes of a platyopuntia are divided into many sub-surfaces. Sub-beam *A* is incident on one of these sub-surfaces, sub-beam *B* strikes the ground, and sub-beam *C* misses the plant and exits through the right-front plane. This latter sub-beam can be incident on an adjacent plant, which is geometrically equivalent to approaching the original plant at the opposite side but at the same height and at the same angle as for the exit of sub-beam *C* from the space. This is represented by sub-beam *C'*, which strikes a sub-surface of the central cladode. Used by permission from Garcia de Cortázar et al. (1985). (© Elsevier Science Publishers B.V.)

(Fig. 16.2) has been used to predict the spacing between plants that maximizes productivity for *Agave fourcroydes*, which is harvested for the fiber in its leaves. For this representative agave, maximal productivity occurs when its leaf area index (LAI), the total area for both sides of its leaves per unit ground area, is about 4.5 (Nobel 1991). Maximal productivity for a platyopuntia such as *Opuntia ficus-indica* has likewise been predicted, based on the ray-tracing technique, to occur at a similar value of its stem area index (SAI), the total area for both sides of its cladodes per unit ground area.

In the general case of nutrient limitations, another multiplicative factor can be incorporated into EPI [Eq. (16.1)] to quantify the effects of soil elements on net CO_2 uptake over 24-h periods. Such a Nutrient Index based on levels of various nutrients in the soil has been described for agaves and cacti (Nobel 1989). It incorporates nitrogen, phosphorus, potassium, sodium, and boron, the five elements that have the greatest demonstrated effects on net CO_2 uptake and growth for these species.

16.3 Productivity of Certain CAM Plants

To test for maximal productivity of agaves, approximately 6-year-old plants of *Agave mapisaga* Trel. var. *mapisaga* and of *Agave salmiana* Otto ex Salm. var. *salmiana* were selected. Both species can have substantial biomass productivity (García-Moya and Nobel 1990). Based on the ray-tracing model (Fig. 16.2), plants were chosen that had an LAI of about 4.5, indicating canopy closure with very little PPFD incident on the ground ($< 8\%$). Measurements were made in a year with substantial, well-distributed rainfall, so the Water Index [Eq. (16.1)] was close to 1.00 for nearly the entire year (Nobel et al. 1992). During the 1-year observation period, LAI increased to about 6 for both species, slightly decreasing productivity per unit ground area (Nobel et al. 1992).

Using the ray-tracing model to calculate the optimal SAI together with previous harvest data, plant spacing leading to maximal productivity of platy-opuntias was predicted. Cladodes of *Opuntia amyclea* Tenore and *Opuntia ficus-indica* (L.) Miller were therefore planted at a high density of 24 plants m^{-2} (240 000 plants ha^{-1}; Garcia de Cortázar and Nobel 1991; Nobel et al. 1992). Production of new daughter cladodes within a few months raised SAI to about 4, leading to canopy closure. Plants were subsequently pruned every 3 months to maintain SAI between 4 and 5, and the Water Index was maintained at 1.00 by daily drip irrigation.

During the 1-year observation periods, temperature was not a major limiting factor, as the average daily Temperature Index (Eq. (16.1)] was 0.84 to 0.90 for all species (Garcia de Cortázar et al. 1985; Nobel et al. 1992). In any case, temperature is an environmental variable not easily or inexpensively altered by agricultural practices, such as by changing the locations of the fields or using transparent covers (glass or plastic). Nutrients were also not a major limiting factor for the agaves and cacti considered, as all the sites were well fertilized.

To measure dry-weight productivity nondestructively, leaves of the agaves were harvested from neighbouring plants a few rows away from the monitored plants, and the periodically pruned cladodes from the platyopuntias were also harvested. Harvested material was dried at 80 °C in a forced-draft oven until no further weight change occurred (generally 3 d). Regressions were established between leaf dimensions and dry weight so that the dry weight of the monitored agaves could be estimated at 3-month intervals (Nobel 1984, 1988). The reported productivities (Table 16.2) for the agaves are based on such regressions, and those for the cacti are based on total harvests of the monitored plants at the end of 1-year or 2-year observation periods plus the dry weight of the cladodes periodically removed by pruning.

Annual aboveground dry-weight productivities under nearly optimal conditions of water, temperature, and light are extremely high for the agaves and cacti considered, 38 to 50 Mg ha^{-1} yr^{-1} (Table 16.2). These values exceed those of nearly all crops (Evans 1975; Loomis and Gerakis 1975; Beadle et al. 1985; Nobel 1991). For instance, the productivity of *O. ficus-indica* at sites in North America and in South America exceeds the productivity of the highest-producing C_3 crops and C_3 trees (Table 16.1). The values for the two agave species, which average 40 Mg ha^{-1} yr^{-1}, would have been even higher under irrigation during the winter and if the leaf area index had been a little lower during the second half of the observation year.

What is the maximal productivity possible for plants? Assuming no limitations other than the processing time for photons during photosynthesis, a maximal productivity of about 100 Mg ha^{-1} yr^{-1} could occur under year-round optimal conditions of all environmental factors and all soil factors (Nobel 1991). The maximal productivity for a generalized C_3 crop with early canopy closure and no limitation by drought or temperature can be 54 Mg ha^{-1} yr^{-1} in England (Monteith 1977). Indeed, 50 Mg ha^{-1} yr^{-1} is a realistic

Table 16.2. Annual aboveground dry-weight productivities of certain CAM plants under optimal spacing and wet conditions[a]

Species	Location	Productivity (Mg ha^{-1} yr^{-1})	Reference
Agave mapisaga	Tequexquinahuac, Mexico, Mexico	38 ± 2	Nobel et al. (1992)
Agave salmiana	Tequexquinahuac, Mexico, Mexico	42 ± 3	Nobel et al. (1992)
Opuntia amyclea	Saltillo, Coahuila, Mexico	45 ± 2	Nobel et al. (1992)
Opuntia ficus-indica	Santiago, Chile	47 to 50	Garcia de Cortázar and Nobel (1991, 1992)
O. ficus-indica	Saltillo, Coahuila, Mexico	47 ± 3	Nobel et al. (1992)

[a]Annual rainfall at Tequexquinahuac was 770 mm and the other plants were irrigated. Data are means $\pm$ SE ($n = 3$ or 4).

benchmark against which to assess maximal productivity among plants under any condition (Nobel 1991). In this regard, the six highest-producing C_3 species have an annual productivity averaging 41 Mg ha^{-1} yr^{-1} (Table 16.1), the CAM plant *O. ficus-indica* can average 48 Mg ha^{-1} yr^{-1} at various locations (Table 16.2), and the productivities of the highest-producing C_4 species can exceed these values.

Pineapple (*Ananas comosus*), another cultivated monocotyledonous CAM plant, has a predicted maximal productivity under optimal conditions of 35 Mg ha^{-1} yr^{-1} (Bartholomew and Kadzimin 1977). Its productivity as well as that of agaves and cacti will be lower under less than optimal environmental conditions. EPI [Eq. (16.1)] can be used to quantify the effect of such sub-optimal conditions of soil water, temperature, and light on net CO_2 uptake and hence on productivity. In an ecological context, environmental conditions are rarely close to ideal for productivity, and the daily net CO_2 uptake is usually less important than reproduction or protection from herbivory. For instance, the surfaces of many cacti are shaded by spines, which greatly reduces the PPFD available for photosynthesis and the resulting growth but can reduce herbivory and ameliorate thermal extremes, thereby promoting survival.

16.4 Gas Exchange and Biochemical Variations Among Photosynthetic Pathways

To help interpret the differences in maximal productivity among plants from the three photosynthetic pathways, the underlying gas-exchange differences must be examined. Because maximal productivity occurs under well-watered conditions, transpiration in general and limitations on gas exchange by stomata in particular are of secondary importance. The three photosynthetic pathways, however, do vary with regard to maximal rates of net CO_2 uptake per unit shoot area, which can affect productivity. Under well-watered conditions, optimal temperatures, and saturating PPFD, the maximal net CO_2 uptake rates average 48 μmol m^{-2} s^{-1} for the six C_3 species with the highest rates and 64 μmol m^{-2} s^{-1} for the six C_4 species with the highest rates (Nobel 1991). Such maximal CO_2 uptake rates are much lower for CAM plants, 29 to 34 μmol m^{-2} s^{-1} (Nobel 1988, 1991; Nobel et al. 1992). Thus, the greater maximal productivity of C_4 compared with C_3 plants is consistent with the greater maximal net CO_2 uptake rates of C_4 plants, but the argument based on maximal rates breaks down when CAM plants are considered.

Net CO_2 uptake over 24-h periods is of more relevance to assessing maximal productivity than is the maximal instantaneous rate of net CO_2 uptake. In this regard, CAM plants can have substantial net CO_2 uptake during the daytime in addition to their nocturnal net CO_2 uptake. On the other hand, C_3 and C_4 plants have no net CO_2 uptake during the night to supplement their daytime net CO_2 uptake. Maximal daily net CO_2 uptake under nearly optimal conditions averages 1100 mmol m^{-2} day^{-1} for *Agave mapisaga*, *A. salmiana*, and *O. ficus-indica* (Nobel and Hartsock 1983, 1984; Nobel et al. 1992). The highest values reported

Table 16.3. Energetics of the biochemistry of CO_2 fixation for C_3, C_4, and CAM plants[a]

	Energetic cost of pathway (kJ mol^{-1} CO_2)		
	C_3	C_4	CAM
Cost per mol CO_2 initially fixed	590	640 or 690	753 or 790
Cost per net mol CO_2 fixed, including photorespiration	867	640 or 690	753 or 790

[a]The energetic costs are calculated based on 3 ATP and 2 NADPH required per CO_2 fixed by the C_3 pathway, 1 or 2 more ATP for the C_4 pathway, and 3.25 or 4 more ATP by the CAM pathway; energies are 50 kJ mol^{-1} for ATP and 220 kJ mol^{-1} for NADPH; and the oxygenase activity of RUBISCO is assumed to be 25% of the carboxylase activity for C_3 plants, whereas photorespiration is assumed to be negligible for the C_4 and CAM pathways (Edwards and Walker 1983; Gregory 1989; Nobel 1991; Salisbury and Ross 1991; Taiz and Zeiger 1991; Chap. 25).

for C_3 and C_4 plants are similar, although high-quality field data on net CO_2 uptake over 24-h periods are limited for such species compared with data available for CAM plants, for which gas exchange is routinely measured over 24-h periods (Nobel 1988, 1991). Hence, data are currently inadequate to explain variations in maximal productivity among the photosynthetic pathways based on maximal daily net CO_2 uptake.

Net CO_2 uptake ultimately depends on the enzymatic processing of the CO_2 taken up. The three photosynthetic pathways differ in the sites of the initial CO_2 binding (Fig. 16.1), the daily timing and location of such binding, and the enzymes involved (Cockburn 1985; Gregory 1989). The C_3 photosynthetic carbon reduction cycle (Calvin-Benson cycle), which is part of all three pathways, takes place in the chloroplasts during the daytime and utilizes the enzyme ribulose 1, 5-bisphosphate carboxylase/oxygenase (RUBISCO) to bind CO_2. Besides binding CO_2, RUBISCO can also bind O_2 at the same site, thereby initiating photorespiration, a process that releases 0.5 CO_2 per O_2 bound. A considerable amount of energy, approximately 675 kJ per mol of O_2 bound, is required for the biochemical steps involved in photorespiration (Nobel 1991). For C_3 plants, the energetic cost for CO_2 fixation in terms of ATP and NADPH required per CO_2 initially bound is 590 kJ mol^{-1} (Table 16.3). The effective cost is raised substantially by the release of some of this CO_2 that occurs at the added energetic costs of photorespiration. At moderate temperatures, the oxygenase activity of RUBISCO is often about one-quarter of its carboxylase activity. This raises the cost per *net* mol of CO_2 fixed by 47% above that for the CO_2 initially fixed in C_3 plants (Nobel 1991).

For C_4 plants, the initial binding of CO_2 by phospho*enol*pyruvate carboxylase (PEPC) occurs in the cytosol of mesophyll cells, after which a 4-carbon organic-acid anion is shuttled to the bundle-sheath cells where CO_2 is released (Fig. 16.1b). This biochemical shuttle leads to a locally high concentration of CO_2 that overwhelms the oxygenase activity of RUBISCO, thereby essentially eliminating the costly release of CO_2 by photorespiration for C_4 plants. For CAM plants, the initial fixation of CO_2 by PEPC also takes place in the cytosol, as for C_4 plants, but the process occurs at night; the 4-carbon acid formed is actively transported into the large central vacuoles of the same cells for storage (Fig. 16.1c; see also Chap. 4). During the following daytime, the organic acids diffuse out of the vacuoles, are decarboxylated, and the CO_2 is released internally at a time when the stomates are closed. This decarboxylation can lead to concentrations of CO_2 high enough to overwhelm the oxygenase activity of RUBISCO, as also occurs for C_4 plants.

Energy is required to transport organic acids into the vacuoles of CAM plants (one ATP per malate; Ting 1985; Lüttge 1987), and energy is needed to manufacture the storage carbohydrate required for the initial CO_2 binding. Thus, CO_2 fixation is energetically more costly for CAM plants than for C_4 plants (Table 16.3). Both the C_4 and the CAM pathways can essentially avoid photorespiration, which causes C_3 plants to release about 20 to 30% of the CO_2 initially fixed at a considerable cost in terms of ATP and NADPH. The estimated mean cost per net CO_2 fixed, which can be 867 kJ mol^{-1} for C_3 plants when the oxygenase activity of RUBISCO is one-quarter of its carboxylase activity, can be 23% lower for C_4 plants and 11% lower for CAM plants (Table 16.3). Cellular and biochemical limitations therefore place CAM plants intermediate between C_3 and C_4 plants with regard to the costs of net CO_2 uptake. This energetic analysis at a cellular level of organization is consistent with variations in measured maximal productivities at a shoot level among the photosynthetic pathways, as the productivity of the most highly productive CAM plants is between the productivity of the most productive C_3 and C_4 plants (Tables 16.1 and 16.2).

16.5 Conclusions

Measuring and increasing the productivity of C_3 and C_4 plants was a major agricultural objective up to about 1975. After that, the major breeding emphasis for C_3 and C_4 plants shifted to enhancing the growth of the specific plant parts harvested, such as grain, to avoidance of lodging by shorter and sturdier stems, to control of pathogens, and to other such aspects. Comparable breeding attention has not occurred for CAM plants, which overall outnumber C_4 species more than fivefold. Increases in productivity for certain commercially important agaves and cacti can be expected due to future breeding efforts as well as to other agricultural improvements.

Studies with species of *Agave* and *Opuntia* (Table 16.2) demonstrate that certain CAM plants can have higher productivities under optimal conditions than do C_3 species (Table 16.1). The high productivity of these agaves and cacti can be explained at a cellular level by the great reduction of photorespiration by CAM plants, in which decarboxylation of organic acids can raise the CO_2 level such that the oxygenase activity of RUBISCO is substantially suppressed.

Stomatal limitations are apparently of little consequence when considering maximal productivity for plants of all three photosynthetic pathways. At the level of organization represented by shoot gas exchange, CAM plants can also take up CO_2 during the daytime, especially when water is not limiting, whereas C_3 and C_4 plants cannot take up CO_2 at night. Thus, the daily period of net CO_2 uptake can be longest for CAM plants, enhancing their productivity compared with that possible for C_3 and C_4 plants.

Agaves and cacti are perennials that can have substantial productivity year-round. Moreover, net CO_2 uptake and productivity by CAM plants can be achieved at a lower cost of water compared with C_3 and C_4 plants. By taking up CO_2 primarily at night when temperatures are lower and hence plant water loss is reduced, the water-use efficiency (CO_2 fixed per unit water lost) of CAM plants is generally about threefold higher than for C_4 plants and sixfold higher than for C_3 plants. The higher water-use efficiency of CAM plants can lead to a higher productivity per unit ground area. Consequently, various agaves and cacti can be successfully cultivated in arid and semi-arid regions where limited water availability greatly reduces the economic viability of C_3 and C_4 plants.

Acknowledgement. The research on agaves and cacti has been financially supported by the Environmental Sciences Division, Office of Health and Environmental Research, U.S. Department of Energy contract DE-FC03-87-ER60615 and the Research Committee of the UCLA Academic Senate.

References

Batholomew DP, Kadzimin SB (1977) Pineapple. In: Alvim P de T, Kozlowski TT (eds) Ecophysiology of tropical crops. Academic Press, New York, pp 113–156

Beadle CL, Long SP, Imbamba SK, Hall DO, Olembo RJ (1985) Photosynthesis in relation to plant production in terrestrial environments. Tycooly, Oxford

Cockburn W (1985) Tansley review no. 1. Variation in photosynthetic acid metabolism in vascular plants: CAM and related phenomena. New Phytol 101: 3–24

Cooper' JP (ed) (1975) Control of photosynthetic production in terrestrial systems. Photosynthesis and productivity in different environments. Cambridge University Press, Cambridge, pp 593–621

Edwards G, Walker D (1983) C_3, C_4: mechanisms, and cellular and environmental regulation, of photosynthesis. Blackwell, Oxford

Evans LT (ed) (1975) Crop physiology: some case histories. Cambridge University Press, Cambridge

Flores Valdez CA, Aguirre Rivera JR (1979) El nopal como forraje. Universidad Autónoma Chapingo, Chapingo, México

Garcia de Cortázar V, Nobel PS (1986) Modeling of PAR interception and productivity of a prickly pear cactus, *Opuntia ficus-indica*, at various spacings. Agron J 78: 80–85

Garcia de Cortázar V, Nobel PS (1991) Prediction and measurement of high annual productivity for *Opuntia ficus-indica*. Agric For Meteorol 56: 261–272

Garcia de Cortázar V, Nobel PS (1992) Biomass and fruit production for the prickly pear cactus, *Opuntia ficus-indica*. J Am Soc Hortic Sci 117: 558–562

Garcia de Cortázar V, Acevedo E, Nobel PS (1985) Modeling of PAR interception and productivity by *Opuntia ficus-indica*. Agric For Meteorol 34: 145–182

Garcia-Moya E, Nobel PS (1990) Leaf unfolding rates and responses to cuticle damaging for pulque agaves in Mexico. Desert Plants 10: 55–57

Gibson AC, Nobel PS (1986) The cactus primer. Harvard University Press, Cambridge

Gregory RPF (1989) Biochemistry of photosynthesis, 3rd edn. John Wiley, Chichester

Griffiths D (1915) Yields of native prickly pear in southern Texas. Bulletin 208, United States Department of Agriculture, Washington, DC

Jarvis JG, Leverenz JW (1983) Productivity of temperate, deciduous and evergreen forests. In: Lange OL, Nobel PS, Osmond CB, Ziegler H (eds) Physiological plant ecology IV, Encyclopedia of plant physiology, new series, vol 12D. Springer, Berlin Heidelberg New York, pp 233–280

Jones MB (1986) Wetlands. In: Baker NR, Long SP (eds) Photosynthesis in contrasting environments. Elsevier, Amsterdam, pp 103–138

Kira T (1975) Primary production of forests. In: Cooper JP (ed) Photosynthesis and productivity in different environments. Cambridge University Press, Cambridge, pp 5–40

Lieth H, Whittaker RH (eds) (1975) Primary productivity in the biosphere. Springer, Berlin Heidelberg New York

Loomis RS (1983) Productivity of agricultural systems. In: Lange OL, Nobel PS, Osmond CB, Ziegler H (eds) Physiological plant ecology IV. Encyclopedia of plant physiology, new series, vol 12D. Springer, Berlin Heidelberg New York, pp 151–172

Loomis RS, Gerakis PA (1975) Productivity of agricultural ecosystems. In: Cooper JP (ed) Photosynthesis and productivity in different environments. Cambridge University Press, Cambridge, pp 145–172

Lüttge U (1987) Tansley review no. 10. Carbon dioxide and water demand: crassulacean acid metabolism (CAM), a versatile ecological adaptation exemplifying the need for integration in ecophysiological work. New Phytol 106: 593–629

Monjauze A, Le Houérou HN (1965) Le rôle des *Opuntia* dans l'économie agricole Nord Africaine. Bull Ec Natl Super Tunisi 8–9: 85–164

Monteith JL (1977) Climate and the efficiency of crop production in Britain. Philos Trans R Soc Lond Ser B 281: 277–294

Nobel PS (1984) Productivity of *Agave deserti*: measurement by dry weight and monthly prediction using physiological responses to environmental parameters. Oecologia 64: 1–7

Nobel PS (1988) Environmental biology of agaves and cacti. Cambridge University Press, Cambridge

Nobel PS (1989) A nutrient index quantifying productivity of agaves and cacti. J Arid Environ 26: 635–645

Nobel PS (1991) Tansley review no. 32. Achievable productivities of CAM plants: basis for high values compared with C_3 and C_4 plants. New Phytol 119: 183–205

Nobel PS, Hartsock TL (1983) Relationships between photosynthetically active radiation, nocturnal acid accumulation, and CO_2 uptake for a crassulacean acid metabolism plant, *Opuntia ficus-indica*. Plant Physiol 71: 71–75

Nobel PS, Hartsock TL (1984) Physiological responses of *Opuntia ficus-indica* to growth temperature. Physiol Plant 60: 98–105

Nobel PS, Valenzuela AC (1987) Environmental responses and productivity of the CAM plant, *Agave tequilana*. Agric For Meteorol 39: 319–334

Nobel PS, García-Moya E, Quero E (1992) High annual productivity of certain agaves and cacti under cultivation. Plant Cell Environ 15: 329–335

Piedade MTF, Junk WJ, Long SP (1991) The productivity of the C_4 grass *Echinochloa polystachya* on the Amazon floodplain. Ecology 72: 1456–1463

Russell CE, Felker P (1987) The prickly-pears (*Opuntia* spp., Cactaceae): a source of human and animal food in semiarid regions. Econ Bot 41: 433–445

Salisbury FB, Ross CW (1991) Plant physiology, 4th edn. Wadsworth, Belmont

Schönau APG, Pennefather M (1975) A first account of profits at harvesting as a result of fertilizing *Eucalyptus grandis* at time of planting in southern Africa. S Afr For J 94: 29–35

Taiz L, Zeiger E (1991) Plant physiology. Benjamin/Cummings, Redwood City

Ting IP (1985) Crassulacean acid metabolism. Annu Rev Plant Physiol 36: 595–622

Winter K (1985) Crassulacean acid metabolism. In: Barber J, Baker NR (eds) Photosynthetic mechanisms and the environment. Elsevier, Amsterdam, pp 329–387

17 Features of Roots of CAM Plants

P.S. Nobel and G.B. North

17.1 Introduction

Plants utilizing crassulacean acid metabolism (CAM) typically grow in habitats with prolonged dry periods (Kluge and Ting 1978; Osmond et al. 1982; Winter 1985). Root access to water and nutrients is restricted by low soil moisture, such as in deserts and other seasonally arid environments, or by limited soil volume, such as in rock crevices and tree bark. For plants under these conditions, a root system that readily takes up water from a wet soil but limits water loss to a dry soil is particularly advantageous. Indeed, the root systems of CAM plants such as desert succulents have several features that facilitate absorption and conduction of water when it is available, reduce water loss when the soil is dry, and regain high water uptake rates when the soil is rewetted (Kausch 1965; Nobel and Sanderson 1984; Nobel and Cui 1992; North and Nobel 1991, 1992).

Because the ability of root systems to respond to variations in soil moisture is essential for nearly all CAM plants, this chapter will emphasize root hydraulic conductivity – a measure of the ability of roots to take up water – and its underlying anatomy under wet, dry, and rewetted soil conditions. The related subjects of root distribution, root: shoot ratios, root initiation and abscission, and root respiration will also be considered. Most of the data come from research on three species: *Agave deserti*, a monocotyledonous leaf succulent from the Sonoran Desert of the southwestern United States, and two dicotyledonous stem succulents, *Ferocactus acanthodes*, which is sympatric with *A. deserti*, and *Opuntia ficus-indica*, a highly productive species that is cultivated for fruits and fodder in warm regions worldwide. Consideration will also be given to other CAM species whose roots have been investigated, notably cultivated agaves, epiphytic cacti, and epiphytic orchids.

17.2 Anatomy and Morphology

The roots of most CAM plants lack the many morphological specializations of their shoots (Freeman 1969; Gibson and Nobel 1986). Yet such roots have

Laboratory of Biomedical and Environmental Sciences, 900 Veteran Avenue, University of California, Los Angeles, CA 90024-1786, USA

Ecological Studies, Vol. 114
Winter/Smith (eds) Crassulacean Acid Metabolism

anatomical and morphological features that help them to endure drought as well as to take advantage of intermittent rainfalls.

17.2.1 Monocotyledons–Agaves and Orchids

When young, the roots of *Agave deserti* resemble those of species of *Lilium*, *Zea mays*, and other monocotyledons. The vascular core (stele), including the relatively narrow protoxylem and early metaxylem vessels, is surrounded by a pericycle, an endodermis, a cortex, a single-layered exodermis, and an epidermis (Fig. 17.1A). Both the endodermis and the exodermis consist of suberized cells regularly interspersed with unsuberized cells, referred to as passage or transfer cells (Esau 1977). Such a dimorphic exodermis occurs in 41 angiosperm families, both monocotyledons and dicotyledons (Shishkoff 1987) and has been associated with other xeromorphic adaptations (von Guttenberg 1968). The unsuberized cells in the exodermis of aerial roots of CAM orchids often have tilosomes, fibrous bodies adjacent to their outer walls that may admit water when wet and restrict the exit of water when dry (Benzing et al. 1982). The pathway for water movement between orchid aerial roots and the atmosphere is further complicated by the velamen, which consists of one or more layers of nonliving, usually air-filled cells at the root surface (Benzing 1990). Radial channels in the outer walls of unsuberized exodermal cells in the CAM epiphyte *Hoya carnosa* may also behave like valves (Öleson 1978), although water permeability has not been measured for a dimorphic exodermis under wet *versus* dry conditions.

As roots of *A. deserti* become older, the late metaxylem vessels lose their crosswalls and thereby become fully conductive to water at about 150 to 200 mm from the root tip (Huang and Nobel 1992). Concurrently, cortical cells adjacent to the endodermis and exodermis become suberized and lignified (Fig. 17.1B). In 1-year-old roots, few layers of undifferentiated cortex remain between the heavily suberized layers (North and Nobel 1991). The many layers of suberized cells in the older roots of *A. deserti* are less permeable to water than are layers of parenchyma cells (Esau 1977; Schönherr and Ziegler 1980).

17.2.2 Dicotyledons–Cacti

The root anatomy of *Ferocactus acanthodes* and *Opuntia ficus-indica* is typical of most cactus species (Freeman 1969; Hamilton 1970; Gibson and Nobel 1986) and for most dicotyledons as well, at least when the roots are young (Esau 1977). For the first few weeks after initiation, their stele is composed of protoxylem and metaxylem interspersed with phloem and parenchyma cells, surrounded by a pericycle, endodermis, cortex, and epidermis. At about 2 weeks of age, secondary xylem begins to develop in the stele and consists of vessels, fibers, and rays of parenchyma cells. At the same time, which is earlier than for most dicotyledons, a periderm begins to form just outside the stele that is composed chiefly of rectangular cells that become increasingly suberized and lignified with age

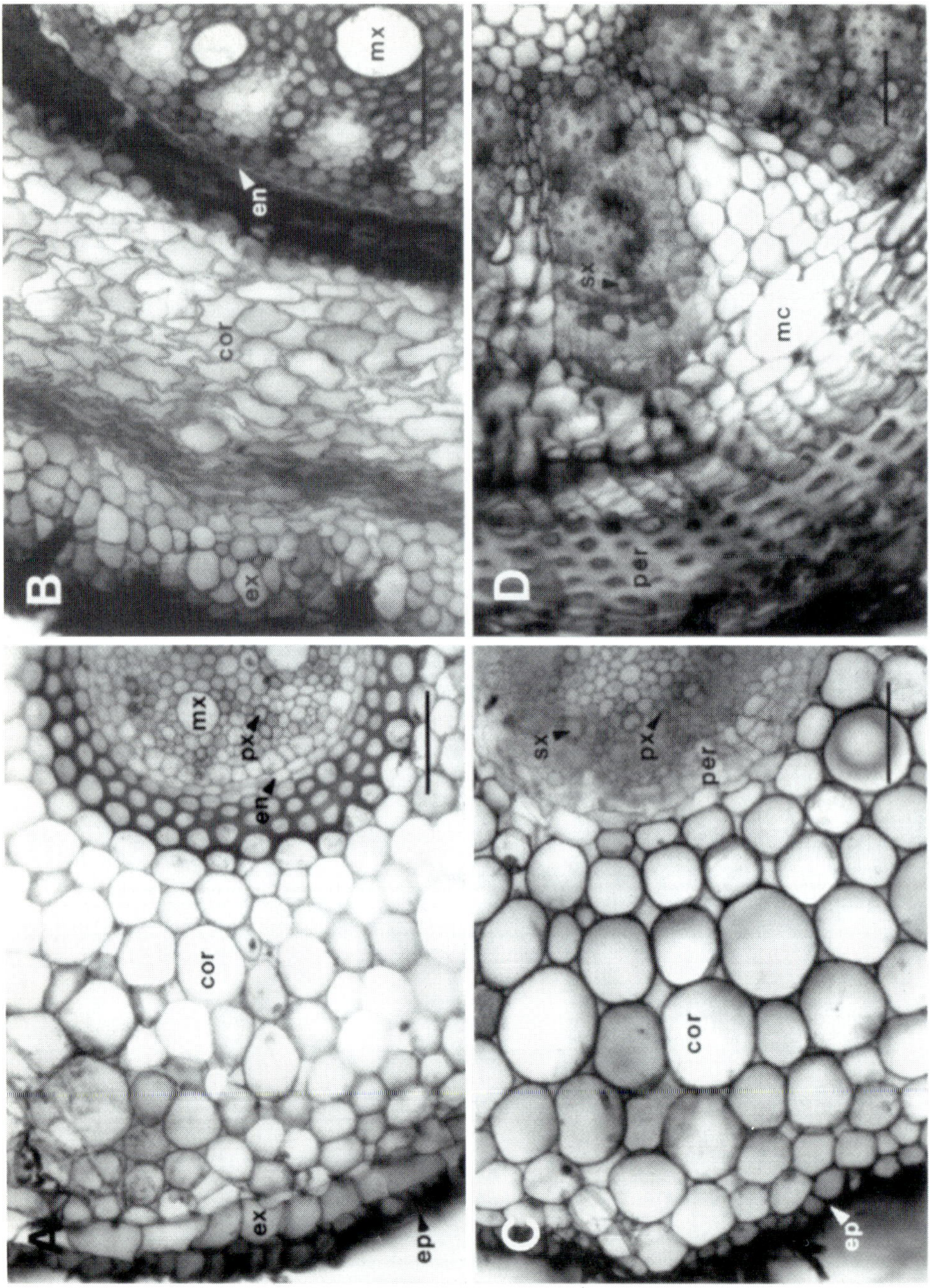
A
ep
ex
cor
en
px
mx
B
ex
cor
en
mx
C
ep
cor
per
px
sx
D
per
mc
sx

(Fig. 17.1C). Roots less than 8 to 10 weeks old are surrounded by a sheath composed of root hairs, mucilaginous secretions, and soil particles (North and Nobel 1992), similar to sheaths observed for roots of *Glycine max* (Sprent 1975), *Oryzopsis hymenoides* (Wullstein and Pratt 1981), and *Zea mays* (McCully and Canny 1988). For cacti, the sheath is particularly well developed for roots in drying soil, when it apparently influences water movement between the roots and the soil.

As the roots of *F. acanthodes* and *O. ficus-indica* grow older, the number of periderm layers increases, displacing the endodermis, cortex, and epidermis to the outside. At 3 to 4 months, roots of *F. acanthodes* and *O. ficus-indica* lose the cortex and epidermis, so the periderm then becomes the outermost tissue (Fig. 17.1D). One-year-old roots can have over 20 layers of periderm, with heavily suberized regions alternating with less suberized regions (North and Nobel 1992). Older roots of the tropical epiphytic cacti *Epiphyllum phyllanthus* and *Rhipsalis cassutha* have a similar multi-layered periderm, with regularly spaced radial channels that may lead to different water movements under wet versus dry soil conditions (G. B. North and P. S. Nobel, unpubl.).

In older roots of both desert and epiphytic cacti, the secondary xylem consists of wedges of vessels and fibers, separated by rays of parenchyma cells that often contain starch and, in exposed roots, even chloroplasts. Species of cacti with mucilage in the shoots frequently have large mucilage cells in the roots that are surrounded by secondary xylem parenchyma (Fig. 17.1D). Like the roots of most dicotyledons, cactus roots thicken with age due to secondary growth. Roots of epiphytic cacti are thinner than comparably aged roots of desert cacti and have smaller vessels with proportionately more fibers that can provide the mechanical strength needed to secure a hold in a tree canopy.

17.3 Distribution in Soil

CAM plants that grow in seasonally arid environments, such as deserts, typically have shallow roots, although columnar and arborescent species often have a few deeper, anchoring roots (Cannon 1911). The roots of most desert succulents occur mainly in the upper 200 mm of the soil (Rundel and Nobel 1991). The mean rooting depth for both *Agave deserti* and *Ferocactus acanthodes* is only 90 mm, and the maximum depth is less than 250 mm (Jordan and Nobel 1984; Hunt and Nobel 1987). The roots of most CAM epiphytes are also shallow, as indicated by the dimensions of soil mats dislodged from the forest canopy (Matelson et al. 1993). For CAM plants that grow in tree crevices or in pockets of soil on rocks, root systems are likely to be even shallower.

←

Fig. 17.1A–D. Light micrographs of transverse sections of roots stained with 0.1% (w/w) toluidine blue 0: *Agave deserti*, whose roots were 4 weeks **(A)** or 1 year **(B)** of age; and roots of *Opuntia ficus-indica* at 4 weeks **(C)** and 1 year of age **(D)**. Specific features identified are epidermis (*ep*), exodermis (*ex*), cortex (*cor*), endodermis (*en*) periderm (*per*), protoxylem (*px*), metaxylem (*mx*), secondary xylem (*sx*), and a mucilage cell (*mc*); *bars* = 100 μm

One consequence of shallow rooting is that the surrounding soil becomes wet or dry relatively quickly, thereby making rapid changes in the root hydraulic conductivity of CAM plants advantageous. Such changes are indicated by increases in transpiration or in water uptake a day or so after drought is terminated by rainfall, as have been observed for several desert succulents, including *Opuntia puberula* (Kausch 1965), *Opuntia basilaris* (Szarek et al. 1973), *A. deserti*, and *F. acanthodes* (Nobel 1976, 1977), as well as for *Mesembryanthemum pellitum* (Eller et al. 1991).

Even for shallow-rooted species, heterogeneities in soil moisture can arise due to rocks. In the northwestern Sonoran Desert, the soil water potential (Ψ_{soil}) within 10 mm of an exposed boulder or an underground rock is much higher (wetter) than at 100 mm away (Nobel et al. 1992a). Consistent with this difference in Ψ_{soil}, the main roots of three CAM succulents have substantially more lateral roots under rocks than at the same depth in soil without rocks (Table 17. 1). Averaged for the three species, the frequency of lateral branching is sixfold higher under rocks than in rock-free regions. Rocks can intercept the upward movement of water from deep moist layers during drought and also can provide cooler surfaces on their undersides where water may condense. In any case, the proliferation of roots of CAM plants near rocks (Table 17. 1) underscores the importance of soil water for root initiation and growth (Nobel et al. 1991, 1992a).

17.4 Root: Shoot Ratios

A direct corollary of shallow rooting is the low root : shoot ratio characteristic of most CAM plants. To be sure, a wide range of root : shoot ratios exists, particularly among epiphytes, from rootless *Tillandsia* species and virtually rootless tank bromeliads to essentially shootless orchids, such as *Campylocentrum pachyrrhizum*. Root : shoot ratios average about 1 among perennials, although the variation is great (Rundel and Nobel 1991). Shrubs in cold deserts of the Commonwealth of Independence States (formerly the USSR) and the United States have root: shoot ratios of 3.2 to 7.3. Shrubs in warm deserts of Australia, Israel, and the United States have root: shoot ratios of 0.3 to 1.4 (Rundel and

Table 17.1. Influence of rocks on the frequency of lateral roots for three succulents in the northwestern Sonoran Desert in southeastern California. Rocks were below the soil surface and averaged 160 mm in the horizontal direction. Data are means ± SE for 6 to 14 main roots. See Nobel et al. (1992a) for details

Species	Number of lateral roots per m of main root	
	No rock	Under rock
Agave deserti	5 ± 4	56 ± 11
Ferocactus acanthodes	23 ± 3	63 ± 9
Opuntia acanthocarpa	3 ± 1	54 ± 6

Nobel 1991). The average root: shoot ratio of tropical and temperate forest trees is only 0.3, the relatively low value reflecting the large amount of woody tissue in their shoots. The root: shoot ratio of agaves and cacti is much lower, averaging 0.11 (Table 17. 2).

The desert succulents *Agave deserti*, *Agave lechuguilla*, and *Ferocactus acanthodes* have an average root: shoot ratio of only 0. 09 (Table 17.2). This is less than the average root: shoot ratio of 0.13 for the more productive, cultivated species *Agave fourcroydes*, *Agave salmiana*, *Agave tequilana*, and *Opuntia ficus-indica* (Table 17.2). The root: shoot ratio for young plants of the epiphytic cactus *Epiphyllum phyllanthus* is about 0.06, similar to the ratio for *A. lechuguilla*. For older plants of the epiphyte, the ratio is probably somewhat lower, and indeed the root: shoot ratio of *A. lechuguilla* can be 0.04 under field conditions and 0.02 following nitrogen fertilization, the lowest value reported among all terrestrial plants (Nobel et al. 1989; Rundel and Nobel 1991).

Changes in root: shoot ratios with drought are varied, although shoot growth is often more inhibited than is root growth (Nobel 1988; Trewavas and Jones 1991). For some CAM species, root: shoot ratios decrease during prolonged drought due to the abscission of roots. The root: shoot ratio of *A. lechuguilla* growing in the Chihuahuan Desert increases by about 25% when plants are irrigated weekly (Nobel et al. 1989). This increase is associated with an increase in transpirational demand due to greater reliance on C_3 photosynthesis under well-watered conditions. In general, the low root: shoot ratios of CAM plants are consistent with their low rates of transpiration. CAM plants rely primarily on nocturnal CO_2 uptake when tissue temperatures are lower than daytime values, leading to less water loss and a three- to sixfold higher water-use efficiency (CO_2 fixed per water lost) than for C_3 or C_4 plants (Kluge and Ting 1978; Winter 1985; Nobel 1988, 1991a).

17.5 Respiration and Carbon Costs

An important consequence of low root: shoot ratios for CAM plants is a relatively low carbon investment in nonphotosynthetic tissue. Even for three cycles per year

Table 17.2. Average root:shoot ratios for mature agaves and cacti. Data are expressed on a dry-weight basis. See Nobel (1988) for details

Species	Root: shoot ratio
Agave deserti	0.11
Agave fourcroydes	0.15
Agave lechuguilla	0.06
Agave salmiana	0.11
Agave tequilana	0.12
Ferocactus acanthodes	0.09
Opuntia ficus-indica	0.14

of wet periods followed by droughts with the abscission of lateral roots, only about 11% of the shoot CO_2 uptake by *Agave deserti* is incorporated into the dry weight of its roots (Nobel 1988). Growth respiration rates for newly initiated roots are also low for *A. deserti, Ferocactus acanthodes*, and *Opuntia ficus-indica* in comparison with those of most other species (Nobel et al. 1992b). Indeed, the chief carbon cost for the growth of their new roots is represented by the structural carbon incorporated into the roots. Additional carbon, up to 1% of the carbon newly assimilated by the shoots, can be secreted as mucilage that helps to form soil sheaths for roots of *O. ficus-indica* (B. Huang, G.B. North, and P.S. Nobel, unpubl. observ.). Such a carbon cost can be offset by the improved water status and reduced water loss for the sheathed roots during drought.

Another carbon cost for root systems, maintenance respiration, is also low for some CAM plants. For 4-week-old roots of *A. deserti* and *F. acanthodes*, respiration rates per unit dry weight are about 7 μmol kg^{-1} s^{-1} (Palta and Nobel 1989a, b). However, maintenance respiration rates for two highly productive, cultivated species, *Agave salmiana* and *Agava mapisaga*, average about 40 μmol kg^{-1} s^{-1} (Nobel et al. 1993). These latter rates are similar to those for roots of most agronomic C_3 and C_4 plants (Nobel et al. 1993). High rates of root maintenance respiration do not always accompany high productivity for CAM succulents, as the rate for 4-week-old roots of the highly productive *O. ficus-indica* is only 5–8 μmol kg^{-1} s^{-1} (Palta and Nobel 1989b).

17.6 Water Uptake

Although CAM plants tend to have much higher water-use efficiencies than do C_3 or C_4 plants, water uptake is still necessary to support net CO_2 uptake and hence growth. A coefficient used to quantify water uptake is the root hydraulic conductivity, L_P (m s^{-1} MPa^{-1}), which relates the volumetric flux density of water into a root, J_v (m^3 m^{-2} s^{-1} = m s^{-1}), to a driving force, such as difference in water potential (Nobel 1991b):

$$J_v = L_P(\Psi_{surface} - \Psi_{xylem}), \qquad (17.1)$$

where $\Psi_{surface}$ is the water potential at the root surface and Ψ_{xylem} is that of the root xylem.

L_P can be determined for individual roots by measuring the J_v that is induced by applying a partial vacuum to the stele at the proximal end of an excised root placed in solution (Nobel et al. 1990). For the three CAM species considered, measurements were made on plants grown in large containers of soil in a glasshouse. Drought ($\Psi_{soil} < -0.5$ MPa) was imposed by withholding water (Ψ_{soil} became –5 MPa at 21 days), and rewetting raised Ψ_{soil} in the root zone to -0.01 MPa within 10 h.

17.6.1 Root Hydraulic Conductivity

Under wet conditions, L_P for young main roots and lateral roots of *Agave deserti*, *Ferocactus acanthodes*, and *Opuntia ficus-indica* averages 1.8×10^{-7} m s^{-1} MPa^{-1} (Fig. 17.2). This is within the range in L_P reported for other, mostly agronomic species (Fiscus 1977; Salim and Pitman 1984; Steudle et al. 1987). For young, non-aerial roots of the epiphytic CAM orchid *Vanilla planifolia*, L_P is 3.3×10^{-7} m s^{-1} MPa^{-1} (G. B. North and P. S. Nobel, unpubl.). Its roots are anatomically similar to those of *A. deserti*, but the endodermis for the orchid is less suberized than for roots of similar age for the agave. L_P for 1-year-old roots of *A. deserti* is much lower than that for 1-year-old roots of the two cactus species (Fig. 17.2), as a previous drought had increased suberization and consequently reduced the conductivity of the agave roots. Also, secondary thickening greatly increases the number of conducting xylem elements in a cross-section of the older cactus roots.

During 21 days of drying, L_P for young roots of *A. deserti* decreases by a factor of 10, partly because of the interruption of the radial pathway by lacunae formed by the collapse of cortical cells and partly because of embolism in the xylem (Fig. 17.2A; North and Nobel 1991). For young roots of the two cactus species, which are surrounded by soil sheaths, cortical cells remain relatively turgid and L_P decreases by only a factor of up to 2 during 21 days of soil drying (Fig. 17.2 B, C). L_P for 1-year-old roots of the agave does not change appreciably during soil

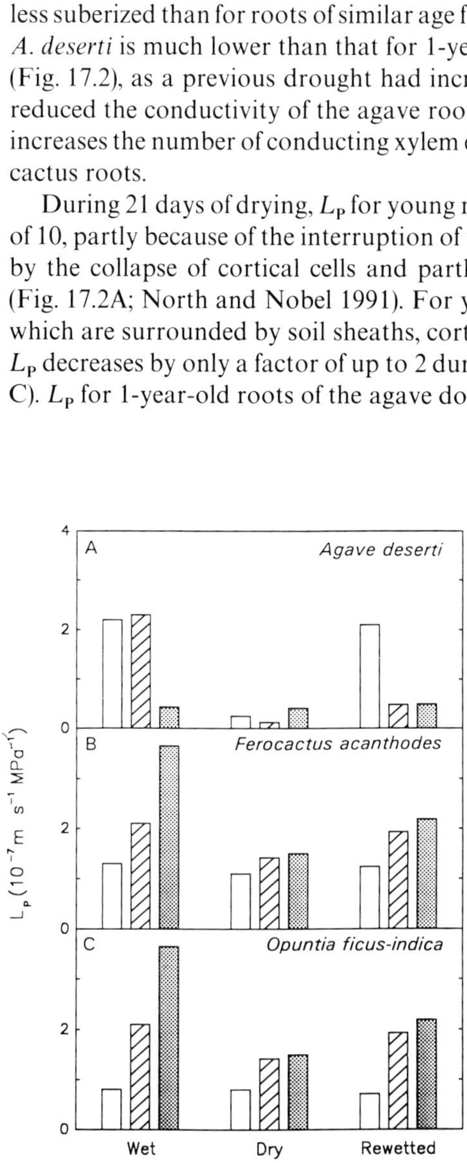

Fig. 17.2A–C. Effects of soil water status on root hydraulic conductivity (L_P) for *Agave deserti* (**A**), *Ferocactus acanthodes* (**B**), and *Opuntia ficus-indica* (**C**) for young main roots 4 to 6 weeks of age (*open left bar*), 4-week-old lateral roots occurring as branches on the main roots (*cross-hatched central bar*), and older main roots that averaged 1 year of age (*stippled right bar*) under wet, dry, and rewetted soil conditions. Under wet conditions and during rewetting, Ψ_{soil} was −0.1 MPa; at 21 days of drying, Ψ_{soil} was −7.7 MPa. Data are means for roots from four plants; standard errors averaged 9% of the means. See North and Nobel (1991, 1992) for further details

drying, whereas L_P of the cacti decreases more than 50%, although the number of suberized layers in the periderm does not increase significantly during drought (North and Nobel 1992). Apparently, the exposure of suberized layers to drying substantially decreases the permeability of the cactus roots, as occurs for other dicotyledons (Vogt et al. 1983).

After 7 days of rewetting, L_P for young main roots of all three species is restored to initial, pre-drought values (Fig. 17.2). The unsuberized cells in the exodermis of young roots of *A. deserti* may facilitate the rehydration of the cortical cells and the restoration of radial conductivity. Rewetting of the 1-year-old roots of *A. deserti*, which had appreciable suberization, has little effect on L_P (Fig. 17.2A). For 1-year-old roots of the two cactus species, after rewetting L_P increases to only 60% of its pre-drought level (Fig. 17.2B, C).

17.6.2 Axial Conductivity

L_P can be partitioned into a radial component from the root surface to the xylem and an axial component along the xylem. The volumetric flow of water per unit time in the root xylem, Q_v ($m^3\ s^{-1}$), is related to the axial hydraulic conductance per unit pressure gradient, K_h ($m^4\ s^{-1}\ MPa^{-1}$), as follows (Gibson et al. 1984):

$$K_h = Q_v/(\Delta P/l), \tag{17.2}$$

where the pressure drop ΔP (MPa) is applied along the length l (m) of a root segment. K_h was determined from Q_v measured by applying a partial vacuum to roots with their distal ends cut and barely immersed in solution (so that essentially no water entered radially). To assess the amount of embolism in the xylem, K_h was measured before and after root segments were pressurized in solution to remove air bubbles (Sperry 1986).

The ratio of axial conductivity measured before pressurization (K_h) to that measured afterwards (K_h^{max}) indicates the extent of air embolism in the xylem. Under wet conditions, K_h/K_h^{max} ranges from 0.82 to 1.01 for roots of the three species (Fig. 17.3). After 21 days of drying K_h/K_h^{max} is 0.76 for main roots of *A. deserti*, indicating relatively little embolism, but averages only 0.21 for the cacti, reflecting a more substantial loss of axial conductivity for the two cactus species. For lateral roots of the three species K_h/K_h^{max} averages 0.24 after 21 days of drying (Fig. 17.3).

For short root segments containing junctions between lateral roots and main roots, K_h/K_h^{max} decreases by about a factor of 10 for all three species during 21 days of drying, indicating more than 90% loss of conductivity due to embolism (North et al. 1992, 1993). The xylem at the main root – lateral root junctions of all three species is composed of relatively small tracheary elements that become more susceptible to embolism during drying because of degradation of their primary cell walls (North et al. 1993). Such preferential embolism at the junctions reduces the spread of embolism from lateral roots to main roots (North et al. 1992).

A similar role is performed by junctions between main roots and stems of *A. deserti* and *F. acanthodes* (Ewers et al. 1992).

For *A. deserti, F. acanthodes*, and *O. ficus-indica*, embolism is reversed and axial conductivity returns to its initial values after 7 days of rewetting (Fig. 17.3). This is probably the result of restored root pressure under the wet conditions (Tyree and Sperry 1989). The xylem of roots of the epiphytic cacti *Epiphyllum phyllanthus* and *Rhipsalis cassutha* embolizes much less during drought than for roots of the desert cacti, perhaps because the narrower vessels of the epiphytes are less susceptible to cavitation (Tyree and Sperry 1989). In addition, roots of the epiphytes recover fully from embolism within 4 days of rewetting.

17.6.3 Radial Conductivity

The other component of L_P, the radial hydraulic conductivity, L_R (m s^{-1} MPa^{-1}), equals the volume flux density of water at the root surface divided by the drop in water potential from the root surface to the root xylem. Using measured values of K_h and L_P together with the lengh (l, m) and the radius (r_{root}, m) of root segments, root radial conductivity averaged over the entire segment is calculated as follows (Landsberg and Fowkes 1978):

$$L_R = L_P \alpha l / \tanh (\alpha l), \qquad (17.3)$$

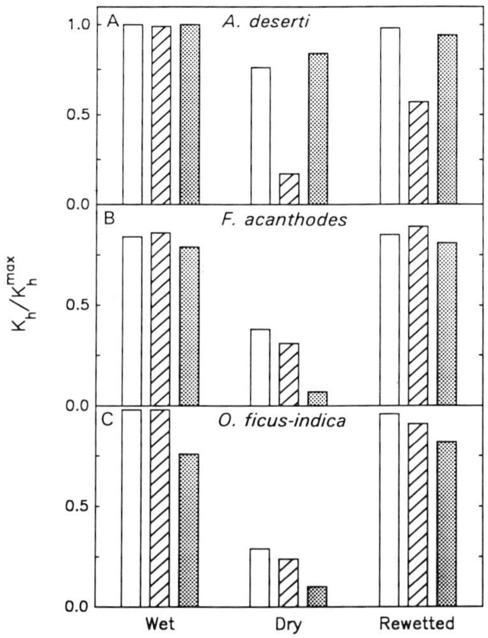

Fig. 17.3A–C. Effects of soil water status on root axial conductance per unit pressure gradient measured before (K_h) and after (K_h^{max}) pressurization to remove embolism for *Agave deserti* **(A)**, *Ferocactus acanthodes* **(B)**, and *Opuntia ficus-indica* **(C)** for young main roots (*open left bar*), lateral roots (*cross-hatched central bar*), and older main roots (*stippled right bar*). Data are means for roots from four plants; standard errors averaged 12% of the means. See Fig.17. 2 for conditions and references

where α equals $(2\pi\, r_{root} L_R/K_h)^{1/2}$. Equation (17.3) is solved by iteration, with L_R initially set equal to L_P and then gradually increased until convergence occurs.

For most roots, L_R is similar to L_P. When L_R is substantially higher than L_P, axial conductivity is the main limiter for L_P. For young roots of *A. deserti* under wet conditions, L_R is high due to the absence of heavily suberized tissues in the radial pathway (Fig. 17.4A). K_h is then limiting due to the relatively small diameters of the protoxylem and early metaxylem vessels and to the immaturity of the larger late metaxylem vessels (North and Nobel 1991; Huang and Nobel 1992). K_h is also limiting, although less so, for young roots of the two cactus species, which lack extensive secondary xylem (Fig. 17.4B, C; North and Nobel 1992).

During drying, the difference between L_P and L_R diminishes for young roots of *A. deserti* (Fig. 17.2A vs Fig. 17.4A), indicating that radial conductivity is becoming more limiting than axial conductivity. The development of cortical lacunae and especially the drying of the suberized layers contribute to the decrease in L_R. For roots of the two cactus species, changes in L_R during 21 days of drying are apparently less important than embolism in reducing L_P (Fig. 17.4B,C; North and Nobel 1992). Changes in L_R in response to rewetting parallel the changes in L_P for all three types of roots for all three species studied (Fig. 17.2 vs Fig. 17.4).

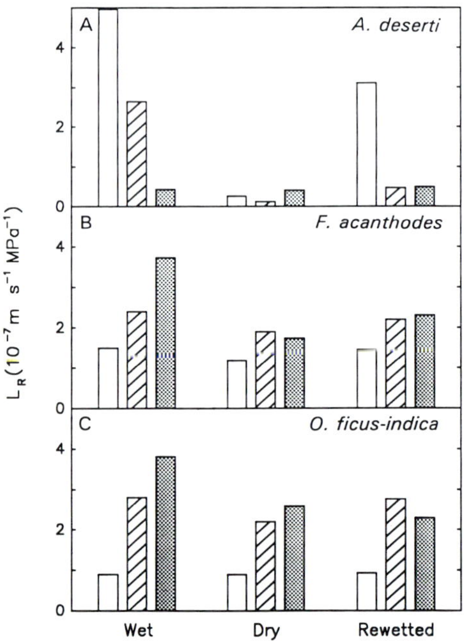

Fig. 17.4A–C. Effects of soil water status on root radial conductivity (L_R) for *Agave deserti* **(A)** *Ferocactus acanthodes* **(B)**, and *Opuntia ficus-indica* **(C)** for young main roots (*open left bar*), lateral roots (*cross-hatched central bar*), and older main roots (*stippled right bar*). L_R was calculated from root dimensions and mean values of L_P (Fig.17.3) and K_h for roots from four plants. See Fig 17.2 for conditions and references

17.6.4 Root Initiation and Abscission

In contrast to the decreases in L_R for all three species during drought, L_R for root segments containing main root–lateral root junctions actually increases during 21 days of drying (North et al. 1992). For *A. deserti*, whose lateral roots generally abscise during drought, an apoplastic pathway for radial water movement from the surface of the junction to the stele develops during 21 days of soil drying. On the other hand, *F. acanthodes* not only tends to retain its lateral roots but also initiates lateral root primordia during drying, which leads to rupturing of the periderm (North et al. 1993). Roots of the epiphytic cacti *E. phyllanthus* and *R. cassutha* also produce abundant lateral root primordia during 30 days of drying (G.B. North and P.S. Nobel, unpubl.). The heightened susceptibility to embolism caused by air entry near the junction during drying poses more of a problem for the monocotyledonous agave than for the dicotyledonous cacti, which can produce new vessels by secondary growth.

Main roots of *A. deserti* quickly initiate lateral roots in response to rewetting (North et al. 1993). For *F. acanthodes* and *O. ficus-indica*, lateral root primordia produced during drying will rapidly elongate upon rewetting, when additional primordia are also produced. The sites of lateral root initiation and development can substantially increase L_P within 24 h because of the proximity of a new root tip with high L_R to a main root stele with relatively high K_h. In contrast, sites of lateral root abscission become no more conductive to water than surrounding root regions 24 h after rewetting, indicating that adequate sealing has occurred (North et al. 1993).

17.7 Conclusions

The root systems of CAM succulents have evolved to limit water loss to a dry soil during drought yet readily take up water when soil moisture is replenished. They thus behave as rectifiers with respect to the movement of water between roots and soil (Nobel and Sanderson 1984; North and Nobel 1991, 1992). The decrease in L_P by a factor of 2 to 10 for roots of *Agave deserti*, *Ferocactus acanthodes*, and *Opuntia ficus-indica* is most important in the early stages of drought, when root hydraulic conductivity limits the overall conductivity of the root-soil pathway (Nobel and Cui 1992). During prolonged drought, the soil hydraulic conductivity becomes the limiting factor, preventing large water losses from succulent shoots through the roots to the soil (Nobel and Cui 1992). Upon rewetting, hydraulic conductivity is fully restored for young roots and increases for the root system due to the initiation of new roots. For most CAM plants, low root : shoot ratios not only limit water loss to the soil but also reduce root carbon costs. The relatively small biomass of roots of CAM plants implicitly testifies to their hydraulic efficiency, particularly in the crucial periods of rewetting after drought. Such water-uptake efficiency of their roots coupled with the high water-use efficiency of their shoots can lead to competitive advantages of CAM plants in an ecological context and relatively high productivity in an agronomic context.

Acknowledgements. The research on agaves and cacti has been financially supported by U.S. National Science Foundation grant DCB-90-02333, the Environmental Sciences Division, Office of Health and Environmental Research, U.S. Department of Energy contract DE-FC03-87-ER-60615, and an Andrew W. Mellon Foundation grant through the Smithsonian Institution.

References

Benzing DH (1990) Vascular epiphytes. Cambridge University Press, Cambridge

Benzing DH, Ott DW, Friedman WE (1982) Roots of *Sobralia macrantha* (Orchidaceae): structure and function of the velamen-exodermis complex. Am J Bot 69: 608–614

Cannon WA (1911) The root habits of desert plants. Publication 131, Carnegie Institution of Washington, Washington, DC

Eller BM, Ruess BR, Ferrari S (1991) Reestablishment of water uptake by succulents after drought: potometric field determinations in the Richtersveld (Cp., Rep. South Africa). Bot Helv 101: 259–265

Esau K (1977) Anatomy of seed plants, 2nd edn. John Wiley, New York

Ewers FW, North GB, Nobel PS (1992) Root-stem junctions of a desert monocotyledon and a dicotyledon: hydraulic consequences under wet conditions and during drought. New Phytol 121: 377–385

Fiscus EL (1977) Determination of hydraulic and osmotic properties of soybean root systems. Plant Physiol 59: 1013–1020

Freeman TP (1969) The developmental anatomy of *Opuntia basilaris* I. Embryo, root, transition zone. Am J Bot 56: 1067–1074

Gibson AC, Nobel PS (1986) The cactus primer. Harvard University Press, Cambridge

Gibson AC, Calkin HW, Nobel PS (1984) Xylem anatomy, water flow, and hydraulic conductance in the fern *Cyrtomium falcatum*. Am J Bot 71: 564–574

Hamilton MW (1970) The comparative morphology of three cylindropuntias. Am J Bot 57: 1255–1263

Huang B, Nobel PS (1992) Hydraulic conductivity and anatomy for lateral roots of *Agave deserti* during root growth and drought-induced abscission. J Exp Bot 43: 1441–1449

Hunt ER, Nobel PS (1987) Allometric root/shoot relationships and predicted water uptake for desert succulents. Ann Bot 59: 571–577

Jordan PW, Nobel PS (1984) Thermal and water relations of roots of desert succulents. Ann Bot 54: 705–717

Kausch W (1965) Beziehungen zwischen Wurzelwachstum, Transpiration und CO_2-Gaswechsel bei einigen Kakteen. Planta 66: 229–238

Kluge M, Ting IP (1978) Crassulacean acid metabolism. Analysis of an ecological adaptation. Springer, Berlin Heidelberg New York

Landsberg JJ, Fowkes ND (1978) Water movement through plant roots. Ann Bot 42: 493–508

Matelson TJ, Nadkarni NM, Longino JT (1993) Longevity of fallen epiphytes in a neotropical montane forest. Ecology 74: 265–269

McCully ME, Canny MJ (1988) Pathways and processes of water and nutrient in roots. Plant Soil 111: 159–170

Nobel PS (1976) Water relations and photosynthesis of a desert CAM plant, *Agave deserti*. Plant Physiol 58: 576–582

Nobel PS (1977) Water relations and photosynthesis of a barrel cactus, *Ferocactus acanthodes*, in the Colorado Desert. Oecologia 27: 117–133

Nobel PS (1988) Environmental biology of agaves and cacti. Cambridge University Press, Cambridge

Nobel PS (1991a) Tansley review 32. Achievable productivities of certain CAM plants: basis for high values compared with C_3 and C_4 plants. New Phytol 119: 183–205

Nobel PS (1991b) Physicochemical and environmental plant physiology. Academic Press, San Diego

Nobel PS, Cui M (1992) Hydraulic conductances of the soil, the root-soil air gap, and the root: changes for desert succulents in drying soil. J Exp Bot 43: 319–326

Nobel PS, Sanderson J (1984) Rectifier-like activities of roots of two desert succulents. J Exp Bot 35: 727–737

Nobel PS, Quero E, Linares H (1989) Root versus shoot biomass: responses to water, nitrogen, and phosphorus applications for *Agave lechuguilla*. Bot Gaz 150: 411–416

Nobel PS, Schulte PJ, North GB (1990) Water influx characteristics and hydraulic conductivity for roots of *Agave deserti* (Agavaceae). J Exp Bot 41: 409–415

Nobel PS, Loik ME, Meyer RW (1991) Microhabitat and diel tissue acidity changes for two sympatric cactus species differing in growth habit. J Ecol 79: 167–182

Nobel PS, Miller PM, Graham EA (1992a) Influence of rocks on soil temperature, soil water potential, and rooting patterns for desert succulents. Oecologia 92: 90–96

Nobel PS, Alm DM, Cavelier J (1992b) Growth respiration, maintenance respiration and structural-carbon costs for roots of three desert succulents. Funct Ecol 6: 79–85

Nobel PS, Huang B, Garcia-Moya E (1993) Root distribution, growth, respiration, and hydraulic conductivity for two highly productive agaves. J Exp Bot 44: 747–754

North GB, Nobel PS (1991) Changes in hydraulic conductivity and anatomy caused by drying and rewetting roots of *Agave deserti* (Agavaceae). Am J Bot 78: 906–915

North GB, Nobel PS (1992) Drought-induced changes in hydraulic conductivity and structure in roots of *Ferocactus acanthodes* and *Opuntia ficus-indica*. New Phytol 120: 9–19

North GB, Ewers FW, Nobel PS (1992) Main root-lateral root junctions of two desert succulents: changes in axial and radial components of hydraulic conductivity during drying. Am J Bot 79: 1039–1050

North GB, Huang B, Nobel PS (1993) Changes in structure and hydraulic conductivity for root junctions of desert succulents as soil water status varies. Bot Acta 106: 126–135

Öleson P (1978) Studies on the physiological sheaths in roots I. Ultrastructure of the exodermis in *Hoya carnosa* L. Protoplasma 94: 325–340

Osmond CB, Winter K, Ziegler H (1982) Functional significance of pathways of photosynthetic carbon assimilation. In: Lange OL, Nobel PS, Osmond CB, Ziegler H (eds) Physiological plant ecology II, Encyclopedia of plant physiology, new series, vol 12B. Springer, Berlin Heidelberg New York, pp 479–547

Palta JA, Nobel PS (1989a) Root respiration for *Agave deserti*: influence of temperature, water status, and root age on daily patterns. J Exp Bot 40: 181–186

Palta JA, Nobel PS (1989b) Influences of water status, temperature, and root age on daily patterns of root respiration for two cactus species. Ann Bot 63: 651–662

Rundel PW, Nobel PS (1991) Structure and function in desert root systems. In: Atkinson D (ed) Plant root growth, an ecological perspective. Blackwell, Oxford, pp 349–378

Salim M, Pitman MG (1984) Pressure-induced water and solute flow through plant roots. J Exp Bot 35: 869–881

Schönherr J, Ziegler H (1980) Water permeability of *Betula* periderm. Planta 147: 345–354

Shishkoff N (1987) Distribution of the dimorphic hypodermis of roots in angiosperm families. Ann Bot 60: 1–15

Sperry JS (1986) Relationship of xylem embolism to xylem pressure potential, stomatal closure and shoot morphology in the plant *Rhapis excelsa*. Plant Physiol 80: 110–116

Sprent JI (1975) Adherence of sand particles to soybean roots under water stress. New Phytol 74: 461–463

Steudle E, Oren R, Schulze ED (1987) Water transport of maize roots. Measurement of hydraulic conductivity, solute permeability, and of reflection coefficients of excised roots using the pressure probe. Plant Physiol 84: 1220–1232

Szarek SR, Johnson HB, Ting IP (1973) Drought adaptation in *Opuntia basilaris*. Significance of recycling carbon through crassulacean acid metabolism. Plant Physiol 23: 539–541

Trewavas AJ, Jones HG (1991) An assessment of the role of ABA in plant development. In: Davies WJ, Jones HG (eds) Abscisic acid: physiology and biochemistry. BIOS Scientific, Oxford, pp 169–188

Tyree MT, Sperry JS (1989) The vulnerability of xylem to cavitation and embolism. Annu Rev Plant Physiol Plant Mol Biol 40: 19–38

Vogt E, Schönherr J, Schmidt HW (1983) Water permeability of periderm membranes isolated enzymatically from potato tubers (*Solanum tuberosum* L.). Planta 158: 294–301

von Guttenberg H (1968) Der primäre Bau der Angiospermenwurzel. In Linsbauer K, Tischler G, Pascher A (eds) Handbuch der Pflanzenanatomie. Borntraeger, Berlin

Winter K (1985) Crassulacean acid metabolism. In: Barber J, Baker NR (eds) Photosynthetic mechanisms and the environment. Elsevier, Amsterdam, pp 329–387

Wullstein LH, Pratt SA (1981) Scanning electron microscopy of rhizosheaths of *Oryzopsis hymenoides*. Am J Bot 68: 408–419

18 Aquatic CAM Photosynthesis

J.E. KEELEY

18.1 Introduction

CAM photosynthesis in an aquatic plant was first suggested by the observation of marked diel (24 h) changes in malic acid in photosynthetic tissues, but not in nonphotosynthetic tissues, of *Isoetes howellii* (Lycophyta: Isoetaceae) (Keeley 1981). At the time everyone "knew" that CAM was an adaptation for increasing water-use efficiency in xeric adapted plants, and thus there was some reluctance to accepting the notion of an aquatic CAM plant. Phrases such as "CAM-like" or "aquatic acid metabolism (AAM)" were coined to avoid this apparent paradox (Cockburn 1983).

18.2 Evidence of CAM Photosynthesis

The evidence that crassulacean acid metabolism is present in the submerged aquatic *Isoetes howellii* is based on the following criteria.

1. CO_2 is assimilated in the dark and fixed into malic acid and remains in malic acid during the dark period (Fig 18.1).
2. Overnight acidification is followed by daytime deacidification resulting in a diel cycle of 100–300 mmol H^+ kg^{-1} fresh weight (Table 18.1).
3. Phospho*enol*pyruvate carboxylase (PEPC) activity is sufficient to account for the observed accumulation of malate (Table 18.2).
4. There is a 2:1 molar ratio between acidity and malate accumulation (Keeley 1981; Keeley and Busch 1984).
5. Dark fixation of CO_2 results in net CO_2 uptake in the dark (Keeley and Bowes 1982).
6. $^{14}CO_2$ fixed in the dark into malate is transferred in the light to phosphorylated compounds and accumulates in other soluble compounds such as sugars and insolubles such as starch (Fig. 18.2).
7. NADP-malic enzyme activity is sufficient to account for rates of decarboxylation (35 ± 8 mol kg^{-1} Chl h^{-1}, J.E. Keeley unpubl.).

Department of Biology, Occidental College, Los Angeles, CA 90041, USA

Ecological Studies, Vol. 114
Winter/Smith (eds) Crassulacean Acid Metabolism

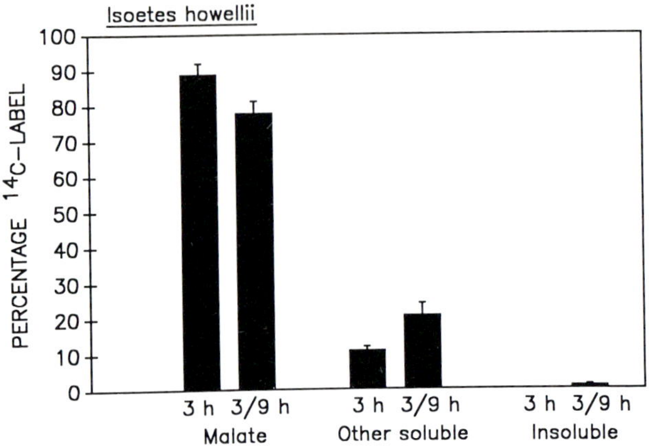

Fig. 18.1. Distribution of dark-fixed ^{14}C label after steady-state labelling for 3 h, and after labelling for 3 h followed by a chase of 9 h in ^{12}C medium in the dark for *Isoetes howellii*. (J.E. Keeley, unpubl.)

Table 18.1. Titratable acidity (TA) and malate (M) content in foliage of aquatic species. Values are means ± SD (number of samples in parentheses); FW = fresh weight

Species	(pH end-point	Titratable acidity (mmol H^+ kg^{-1} FW) Dusk	Dawn	Malate (mmol kg^{-1} FW) Dusk	Dawn	(n)	*t*-Test TA	M
Slight acid accumulation:								
Ascophyllum nodosum	(7.0)	23 ± 3	44 ± 8	2 ± 2	6 ± 2	(5)	**	*
Eleocharis acicularis	(8.3)	16 ± 2	36 ± 12	6 ± 2	12 ± 5	(4)	*	ns
Orcuttia viscida	(7.0)	7 ± 5	23 ± 6	11 ± 6	11 ± 6	(10)	**	ns
Vallisneria americana	(7.0)	11 ± 7	42 ± 22	–	–	(5)	**	–
Substantial acid accumulation:								
Isoetes howellii	(6.4)	12 ± 1	229 ± 19	21 ± 9	123 ± 5	(3)	**	**
Sagittaria subulata	(7.0)	8 ± 3	83 ± 26	10 + 2	45 ± 10	(6)	**	**
Crassula aquatica	(6.4)	0 ± 0	128 ± 13	15 ± 9	76 ± 5	(3)	**	**
Littorella uniflora	(6.4)	19 ± 22	112 ± 14	21 ± 14	66 ± 8	(4)	**	**

ns, $p > 0.05$; * $p < 0.05$; ** $p < 0.01$.
Values are from J.E. Keeley (publ. and unpubl.) except *Ascophyllum* (Johnston and Raven 1986).

8. Dark fixation contributes a substantial portion of the carbon gain (Keeley and Busch 1984).

In conclusion, it is apparent that *Isoetes howellii* is a submerged aquatic CAM plant.

Table 18.2. Phospho*enol*pyruvate carboxylase (PEPC) and ribulose 1,5-bisphosphate carboxylase (RUBISCO) activities in selected CAM and non-CAM aquatic species under submerged (Sub) and emergent (Emg) conditions (PEPC procedure after Jones et al. (1978) and RUBISCO after Lorimer et al. (1976)) compared with the terrestrial species *Spinacia oleracea* (C_3) and *Zea mays* (C_4). Data from J.E. Keeley (unpubl.)

		PEPC	RUBISCO	Ratio RUBISCO/PEPC
		(mol kg^{-1} Chl h^{-1})		
Isoetes howellii	(Sub)	36 ± 16	256 ± 68	7.1
	(Emg)	18 ± 10	553 ± 95	31.6
Crassula aquatica	(Sub)	178 ± 63	392 ± 161	2.2
	(Emg)	45 ± 5	854 ± 163	19.0
Eleocharis acicularis	(Sub)	84 ± 29	224 ± 941	2.7
	(Emg)	36 ± 7	675 ± 124	18.8
Orcuttia viscida	(Sub)	54 ± 25	285 ± 51	5.3
	(Emg)	168 ± 70	339 ± 58	2.0
Spinacia oleracea		54	865	16.0
Zea mays		842	462	0.5

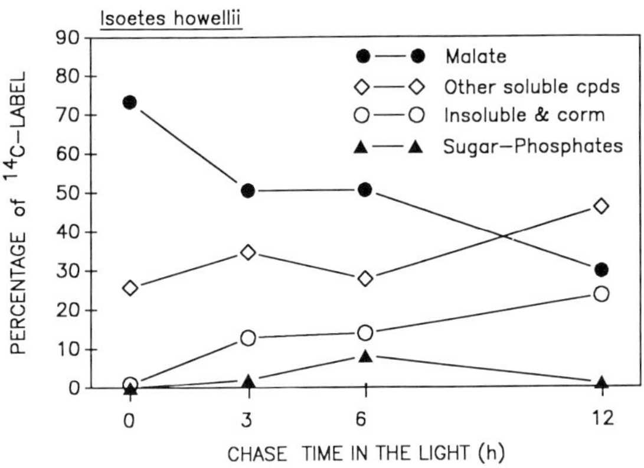

Fig. 18.2. Distribution of dark-fixed ^{14}C label after steady-state labelling in the dark for 3 h and a further 9 h incubation in the dark in ^{12}C medium, followed by a chase of 12 h in the light for *Isoetes howellii*. (J.E. Keeley, unpubl.)

18.3 Distribution of Aquatic CAM Plants

Detection of an overnight acidification/daytime deacidification cycle is the sine qua non of crassulacean acid metabolism and has been used routinely in surveys of aquatic plants for evidence of CAM. Detection of such an acid cycle is not conclusive evidence of CAM, as will be discussed below.

From the literature and unpublished data I can find reports of tests for overnight acid accumulation for 159 freshwater and marine species (Table 18.3).

Table 18.3. Distribution of detectable overnight acid accumulation in aquatic species. Substantial is > 50 mmol H^+ kg^{-1} fresh weight

	Overnight acid accumulation		
	None	Slight	Substantial
	(No. of species tested)		
[Cyanobacterial lichen]	1	0	0
Chlorophyta	13	0	0
Phaeophyta	11	8	0
Rhodophyta	9	0	0
Bryophyta	4	0	0
Lycophyta	0	0	33
Sphenophyta	1	0	0
Pterophyta	3	0	0
Anthophyta			
(Monocot)	28	8	1
(Dicot)	29	5	5
Total	99	21	39

Based on literature and unpublished data provided by myself, John Raven, Linda Handley, Barry Osmond, Jonathan Newman, Andrew Farmer and George Bowes.

Of these, 39 showed substantial (> 50 mmol H^+ kg^{-1} fresh weight) overnight accumulation of acidity, and another 21 have slight but statistically significant overnight acid accumulation (Table 18.3).

18.3.1 Aquatic CAM Species

The 39 species with substantial overnight acid accumulation are distributed in four genera, *Isoetes*, *Crassula*, *Littorella*, and *Sagittaria*.

Isoetes is worldwide a genus of more than 150 taxa, most of which are either lacustrine or amphibious (Tryon and Tryon 1982). All have a nearly identical growth form of a rosette of leaves arising from a corm (Fig. 18.3), and all but one or two are perennial. Of these, 33 have been tested for an acidification/deacidification cycle (Keeley 1982, 1987, unpubl.) and all exhibit an overnight accumulation of titratable acidity and malate of a similar magnitude to that observed for *Isoetes howellii* (Table 18.1). In two other species, the lacustrine *I. bolanderi* and the amphibious *I. orcuttii*, further studies, such as those described in Section 18.2, indicate that these taxa are quite similar to *I. howellii*. In particular, the patterns observed in Figs. 18.1 and 18.2 are nearly identical between these three species. The 33 *Isoetes* species thus far tested represent taxa from the range of known aquatic habitats (e.g. seasonal pools, oligotrophic lakes, freshwater tidal creeks, irrigation canals) and from many parts of the world including North America, Central America, South America, Europe, Africa, and Australia. Thus, it seems warranted to suggest that all aquatic *Isoetes* species will eventually prove to be CAM.

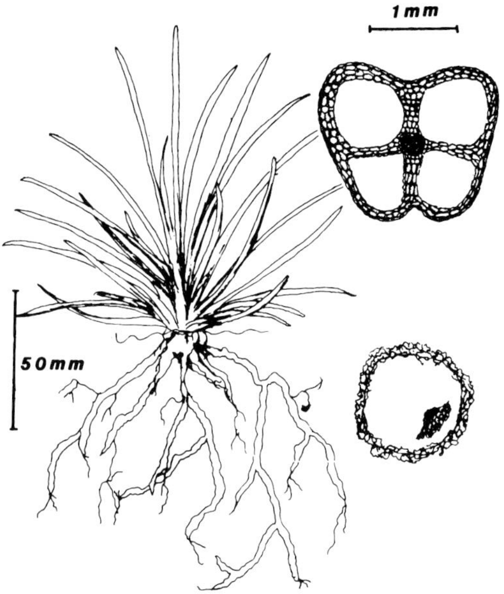

Fig. 18.3. Schematic illustration of *Isoetes* and cross-sections of a leaf and a root. (Keeley 1991)

Crassula (Crassulaceae) is a genus of 300 species, most of which are perennials. All aquatic species are annuals and very often co-occur with *Isoetes* in both oligotrophic lakes and seasonal pools. Four species (*C. aquatica*, *C. helmsii*, *C. natans*, *C. paludosa*) have been tested for CAM and all show an acidity cycle similar to that observed for *Crassula aquatica* (Table 18.1). Of the species tested, three were from seasonal pools in either North America, Africa, or Australia, and one was from oligotrophic lakes in the Andes of South America. *Crassula aquatica* has other features supporting the presence of CAM. Dark-fixed carbon is retained in malic acid throughout the night and is transferred during the day to phosphorylated compounds and accumulates in sugars and insoluble compounds. Additionally, PEPC activity is more than sufficient to account for dark fixation (Table 18.2).

The genus *Littorella* (Plantaginaceae) comprises two species, *L. americana* of North American high-latitude oligotrophic lakes, which has not been investigated for CAM, and the European *L. uniflora*, which is CAM; acidity and malate fluctuations are similar to other aquatic CAM species (Table 18.1) and the retention of dark-fixed carbon in malic acid during the night (such as shown in Fig. 18.1) and the transfer of dark fixed carbon in the light to phosphorylated sugars (such as shown in Fig. 18.2) are similar to *Isoetes* and *Crassula* taxa (J.E. Keeley; unpubl.). Also, the ratio of RUBISCO/PEPC (Farmer 1987) is quite similar to that reported for *Isoetes* (Table 18.2).

Andrew Farmer and George Bowes (unpubl.) have found that the freshwater tidal species *Sagittaria subulata* (Alismataceae) has significant overnight acid accumulation, and this has been verified (Table 18.1). Studies of the retention of dark-fixed carbon in malate during the night (such as in Fig. 18.1) and the transfer of dark fixed carbon in the light to phosphorylated sugars (such as in Fig. 18.2) are similar to *Isoetes* and *Crassula* taxa (J.E. Keeley, unpubl.). Unlike the previous three genera, there is documented evidence of lack of CAM in other species in this genus (Keeley and Morton 1982).

18.3.2 Ecological Distribution of Aquatic CAM Plants

In general, aquatic CAM plants are distributed in one of two habitats, oligotrophic lakes or mesotrophic shallow seasonal pools. The former habitats, being of low productivity, are widely distributed at high latitudes and at increasing latitudes with decreasing altitude. For example, *Isoetes* spp. are widely distributed in northern European lakes but are restricted to lakes above 3000 m at tropical and subtropical latitudes (J.E. Keeley, unpubl.). At low elevations in warm climates, permanent bodies of water are mesotrophic or eutrophic and CAM species are excluded from such habitats. Shallow seasonal pools at any altitude throughout the world are potential habitats for *Isoetes* and *Crassula* species.

This distribution is seemingly odd in the light of marked differences in the physical and chemical differences between such habitats (Fig. 18.4). Seasonal pools have a marked diurnal change in CO_2 availability due to the high photosynthetic demand resulting from the high temperatures, high light availability, and high plant biomass typical of such sites. Since these habitats are rain-fed, they are poorly buffered, and thus the pH is largely controlled by the inorganic carbon species; as CO_2 is consumed HCO_3^- predominates, and the pH may rise 2 to 4 units over a period of 4 to 6 h during the morning. The stress of such large pH changes may be tolerable due to the fact that the water is so poorly buffered relative to the cellular chemistry of the organisms that inhabit such environments.

Oligotrophic lakes, such as the Sierran lake illustrated in Fig. 18.4, show relatively little diurnal change in CO_2 availability. One characteristic they both share, however, is that during much of the day, CO_2 in the water is potentially limiting to photosynthesis. Although the level of CO_2 in almost all aquatic systems is higher than in air, the diffusional resistance of water is 4 orders of magnitude greater than in air, resulting in boundary-layer depletions around the plants (Raven et al. 1988). CAM-dominated lakes are more carbon limited than lakes lacking CAM species (Table 18.4).

There are a few exceptions to these distribution patterns. *Isoetes malinverniana* is restricted to drainage canals, often in the shade of trees, in northern Italy. *Sagittaria subulata*, and at least three *Isoetes* species, are restricted to freshwater tidal rivers where they alternate diurnally, according to the tides, between submergent and emergent conditions.

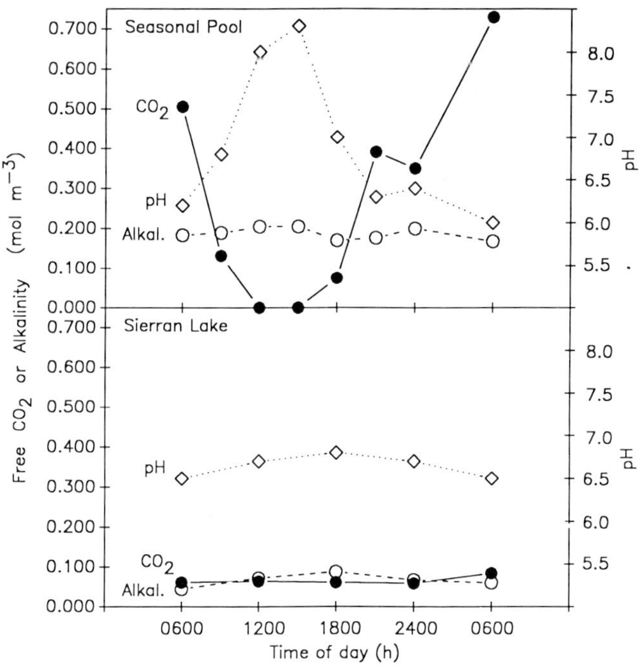

Fig. 18.4. Diel changes in water chemistry in a seasonal pool and a high-elevation oligotrophic Sierran lake (from Keeley et al. 1983; Keeley and Busch 1984; Alkal., Alkalinity (i.e. $HCO_3^- + CO_3^{2-}$)

Table 18.4. Comparison of lakes dominated by aquatic CAM species and lakes lacking CAM species. (Data gleaned from literature cited in text)

Lakes	pH	Total CO_2 (mol m^{-3})	Conductivity (μS m^{-1})
CAM-dominated	4 to 7	0.01 to 0.30	0.02 to 0.07
No CAM	7 to 10	2.5 to 5.0	2.0 to 7.0

18.3.3 Questionable Aquatic CAM Species

Slight but significant overnight acid accumulation has been observed in 21 species and examples of these taxa are shown in Table 18.1. It is questionable as to whether or not these acidity changes are related to CAM photosynthesis. Certainly in some species it is not. For example in *Orcuttia viscida* (Poaceae) it is apparent that carbon fixed in the dark does not remain in malate (Fig. 18.5), and the same is seen with the marine alga *Ascophyllum nodosum* (Fig 18.6), which was previously observed to have a CAM-like acid fluctuation (Johnston and Raven

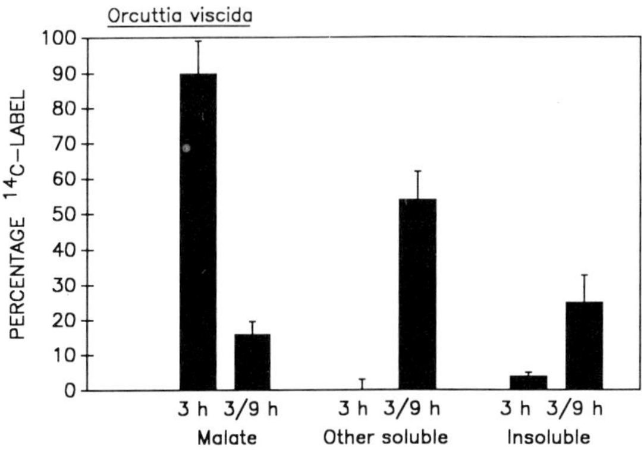

Fig. 18.5. Distribution of dark-fixed ^{14}C label after steady-state labeling for 3 h, and after labelling for 3 h followed by a chase of 9 h in ^{12}C medium in the dark for *Orcuttia viscida*. (J.E. Keeley, unpubl.)

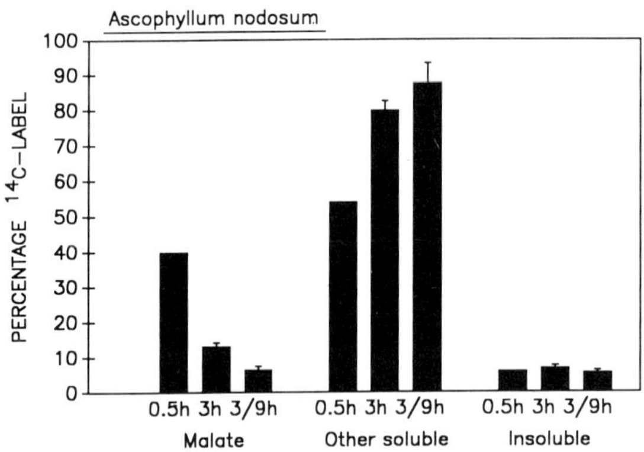

Fig. 18.6. Distribution of dark-fixed ^{14}C label after steady-state labelling for 0.5 h and 3 h, and after labelling for 3 h followed by a chase of 9 h in ^{12}C medium in the dark for *Ascophyllum nodosum*. (R. Montgomery and J.E. Keeley, unpubl.)

1986). The function of dark fixation and overnight acid accumulation in such species is unknown but several possibilities exist, e.g. it may play a role in nitrogen metabolism, as observed in some algae (Turpin et al. 1991).

18.4 Adaptive Significance of CAM in the Aquatic Environment

18.4.1 Seasonal Pool CAM Species

Under steady-state carbon conditions, *Isoetes howellii* shows net CO_2 uptake in the dark in excess of 20 mol kg^{-1} Chl h^{-1} at high CO_2 concentration, but uptake is markedly higher in the light (Fig. 18.7). However, under field conditions these

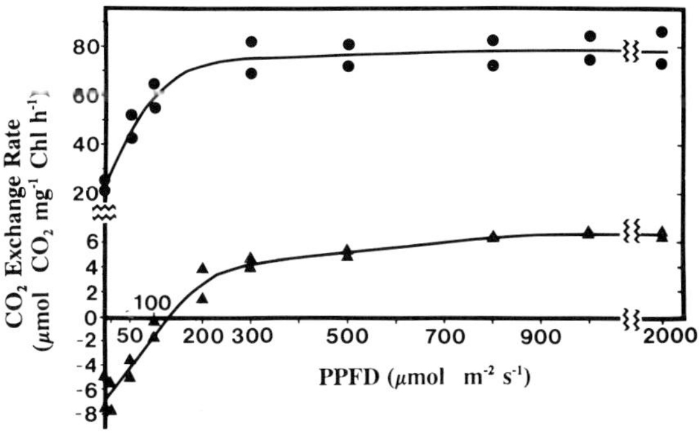

Fig. 18.7. Light response curves for *Isoetes howellii* leaves at 86.7 (circles) and 11.8 (triangles) µmol l^{-1} CO_2. (Keeley and Bowes 1982)

plants are not exposed to steady-state carbon conditions (Fig 18.4), and thus carbon uptake during the day is restricted to a brief window of time in the early morning (Fig. 18.8) In these environments, due to lower temperatures and respiration by the pool flora and invertebrate fauna, CO_2 builds up in the water overnight. For *Isoetes howellii*, dark CO_2 uptake under field conditions represents about 40% of the total 24-h carbon uptake (Keeley and Busch 1984). However, assimilation of ambient carbon during the night accounts for only about one-third of the total dark fixation (Keeley and Busch 1984). It would appear that much of the dark fixation and malic acid accumulation arises from re-fixation of respiratory CO_2 or CO_2 supplied by the roots. The latter is less likely in *Isoetes howellii* than in other *Isoetes* where it has been documented (Richardson et al. 1984). This is because the compact clay substrate would not lend itself to diffusion

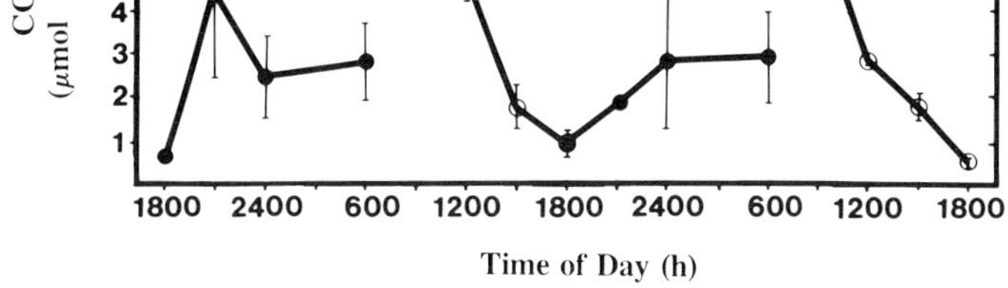

Fig. 18.8 Carbon-assimilation rates for submerged *Isoetes howellii* leaves in the field in southern California; *open circles* light uptake; *filled circles* dark uptake. (Keeley and Busch 1984)

of gases and the roots of this species are thin with dense cortex, distinctly unlike *Isoetes* from lacustrine environments where root uptake of carbon seems to be commonplace. Additionally, experiments with detached leaves sparged overnight with CO_2-free gas are still capable of overnight acid accumulation of about 150 mol H^+ kg^{-1} Chl (Keeley and Busch 1984).

Daytime depletion of acids in *Isoetes* foliage follows a pattern similar to the daytime depletion of CO_2 in the water, although it lags behind by about 3 h. It is also responsive to different ambient carbon conditions, being most rapid and complete under conditions of rapid ambient CO_2 depletion (Keeley and Busch 1984). These patterns are remarkably similar in *Isoetes orcuttii* and *Crassula aquatica*, which co-occur with *I. howellii* (J.E. Keeley, unpubl.).

The significance of CAM in seasonal pool plants is that it plays a major role in the carbon budget by allowing for the assimilation of carbon at night when it is abundant and by recapturing respiratory carbon that would otherwise be lost. Another way of looking at it is that these CAM plants compete for carbon at a time when other plants are not assimilating carbon. Ecologically, this may play an important role in promoting coexistence of these relatively diverse communities (Keeley and Sandquist 1991).

18.4.2 Lacustrine CAM Species

Like seasonal-pool CAM plants, Madsen (1987) has shown that *Isoetes lacustris* and *Littorella uniflora* are likewise capable of daytime CO_2 uptake; however, by extracting gas samples from the lacunae he has shown that under ambient CO_2 concentrations of 30 to 200 mmol m^{-3} the contribution of internal sources of carbon exceed external sources. It was suggested that decarboxylation of malic acid during the day was the source of internal CO_2, and the rate of decarboxylation was dependent upon ambient CO_2 levels (Fig. 18.9). It is also clear from these experiments, and others (Richardson et al. 1984; Raven et al. 1988), that substantial carbon is taken up from the roots. Similar patterns of dark and light fixation were also observed for the lacustrine *Isoetes bolanderii* (Sandquist and Keeley 1990) and the lacustrine *I. macrospora* (Boston and Adams 1986). In these species it was estimated that CAM contributed 45 to 55% of the annual carbon gain.

The adaptive significance of CAM for lacustrine species lies in the enhanced capacity for carbon assimilation by extending the period for carbon assimilation to 24 h. The general absence of CAM from eutrophic environments is related to the inefficiency of this pathway under high-carbon conditions. For example, Madsen (1987) demonstrated that, for both *Littorella uniflora* and *Isoetes lacustris*, the relative contribution of CAM to the total CO_2 uptake over 24 h declined from 95% at an external concentration of 30 mmol m^{-3} CO_2 to about 36% at 200 mmol m^{-3} CO_2. At 1.4 mol m^{-3} CAM ceased in *Littorella uniflora*, and this was tied to a decrease in PEPC activity (Hostrup and Wiegleb 1991). The more rapid growth rates of C_3 species under high inorganic nutrient eutrophic conditions undoubtedly play a role in the competitive exclusion of CAM plants from such

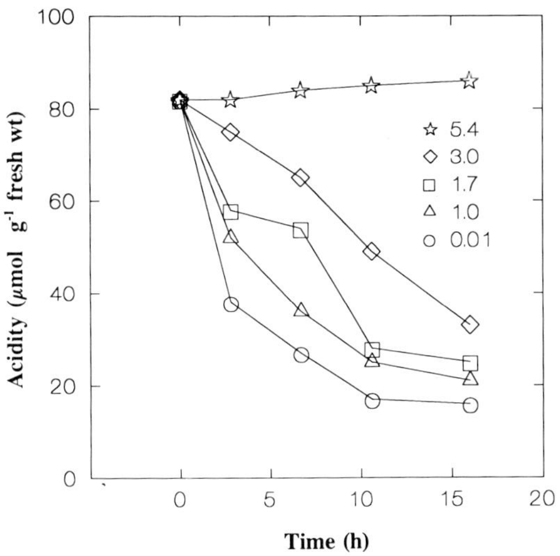

Fig. 18.9. Changes in titratable acidity in the light in leaves of *Littorella uniflora* incubated at 5.4, 3.0, 1.7, 1.0 and 0.01 mol m^{-3} free CO_2 in the water around the roots and with 0.01 mol m^{-3} free CO_2 in the water surrounding the leaves. (Madsen 1987)

sites, since CAM is suppressed under low PPFD (Robe and Griffiths 1990). Occasional reports of *Isoetes* from such eutrophic habitats (Seddon 1965) are often related to disturbance, which maintains the system in disequilibrium.

18.5 Aquatic CAM Plants in an Aerial Environment

All plants in seasonal pools become aerial at some time during their life cycle. *Isoetes howellii* loses CAM in aerial foliage, and this happens on a cell-by-cell basis while the bases of partially submerged plants retain CAM (Fig. 18.10). The same has been noted in CAM species from lacustrine habitats which become emergent at irregular intervals (Keeley et al. 1983; Aulio 1985).

The precise factors responsible for loss of CAM are unknown. The proximate answer seems to be that decreases in water potential at the leaf surface trigger the loss of CAM. This is supported by the observation that CAM is retained in *Isoetes howellii* under aerial conditions if maintained under high humidity, and is greatly reduced in submerged plants incubated in mannitol solutions with osmotic pressures > 0.1 MPa (equivalent to about 41 mol m^{-3} mannitol) (Table 18.5).

The ultimate answer for the loss of CAM is that in the aerial atmosphere diffusional resistances no longer limit CO_2 to photosynthetic organs, and a competitive advantage is likely gained by switching to strictly C_3 photosynthesis. The hypothesis is also supported by the absence of CAM in many terrestrial *Isoetes* species, which are likely to have derived from aquatic ancestors (Keeley 1983). The exception to this rule is the retention of CAM in the terrestrial *I. andicola* (= *Stylites andicola*) (Keeley et al. 1984) and other related terrestrials

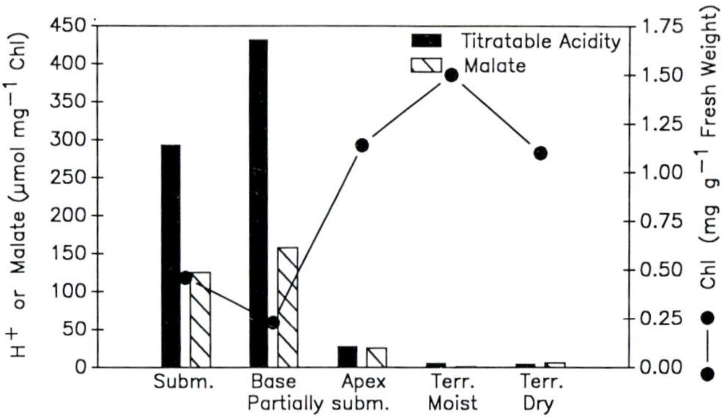

Fig. 18.10. Overnight accumulation of acidity and malate in *Isoetes howellii* foliage submerged (*Subm.*), partially submerged, and fully emergent (*Terr.*) under moist and dry soil conditions. (Keeley and Busch 1984)

Table 18.5. Titratable acidity and malate content in foliage of the aquatic CAM plant *Isoetes howellii* under submerged conditions for 1 month with mannitol or abscisic acid and under emergent conditions of differing humidities (from J.E. Keeley, unpubl.). Values are means ± SD (number of samples in parentheses)

	Titratable acidity (mmol H^+ kg^{-1} FW)		Malate (mmol kg^{-1} FW)		
	Dusk	Dawn	Dusk	Dawn	
Submerged:					
Control	16±11	143±25	14±5	71±17	(4)
ABA (0.01 mol m^{-3})	10±3	201±48	9±8	97±9	(4)
Mannitol					
80 mol m^{-3}	18±16	38±13	29±22	45±19	(3)
120 mol m^{-3}	3±1	6±5	1±1	12±12	(3)
240 mol m^{-3}	7±2	7±3	3±2	15±8	(3)
Emergent:					
RH < 65%	6+11	43±15	2±2	29±11	(4)
RH > 90%	16±11	277±25	5±3	128±13	(4)

above 4000 m in the Andes of South America (Keeley et al. 1994). These species are unique in that they lack stomata, and thus CAM seems to play a role in the recycling of CO_2 in these species hermetically sealed from the atmosphere.

18.6 Carbon-Isotope Discrimination

Carbon-isotope discrimination by aquatic CAM species is indistinguishable from co-occurring species of other photosynthetic pathways (Table 18.6). In general CAM species are more likely to be similar to C_3 species in the same

Table 18.6. Carbon-isotope ratios ($\delta^{13}C$) and discrimination (Δ) in aquatic plants (assumes free CO_2 is the only source of inorganic carbon) and emergent terrestrial (terr.) forms of two species (data from Keeley and Sandquist 1992). Values of Δ were calculated as described in Chapter 15 [Eq. (15.1), p. 239]

		$\delta^{13}C$ (‰)		
		Plant	Source	Δ (‰)
Fontinalis antipyretica	C_3	−26.9	−11.5	15.8
Isoetes bolanderi	CAM	−25.1	−11.5	14.0
Plagiobothrys undulatus	C_3	−27.4	−20.4	7.2
Isoetes howellii	CAM	−28.4	−20.4	8.2
	C_3 (terr.)	−29.4	−8.0	21.5
Orcuttia viscida[a]	C_4	−19.0	−11.0	8.2
	C_4 (terr.)	−12.9	−8.0	4.9

[a]The species shows a ^{14}C-labelling pattern in the light suggestive of C_4 photosynthesis but lacks Kranz anatomy (J.E. Keeley, unpubl.).

habitat than to CAM species in other habitats. The factors accounting for this pattern are described more fully in Keeley and Sandquist (1992), but are mainly related to the fact that isotope ratios are markedly affected by the signature of the source carbon and by the fact that discrimination by carboxylases may be eliminated by diffusional resistances in the aquatic environment.

18.7 Conclusions

CAM photosynthesis is one of the more remarkable examples of evolutionary convergence in the plant kingdom. Crassulacean acid metabolism is an evolutionary response to the stress of daytime CO_2 limitation. This comes about in arid habitats by daytime stomatal closure (selected to reduce water loss), whereas in aquatic habitats it comes about because ambient factors such as low carbon and high diffusional resistances limit the plant's access to CO_2. Terrestrial and aquatic CAM plants share the feature that both are largely restricted to relatively oligotrophic sites where there is a premium on surviving abiotic stresses. These oligotrophic sites limit plant growth, and thus there is little likelihood that the restricted growth rates of CAM plants will make them susceptible to competitive exclusion by neighbouring plants.

Acknowledegements. Much of the work reported here was supported by the National Science Foundation. I thank Kim Williams and Paulette Peckol for providing plant material.

References

Aulio K (1985) Differential expression of diel acid metabolism in two life forms of *Littorella uniflora* (L.) Aschers. New Phytol 100: 533–536

Boston HL, Adams MS (1986) The contribution of crassulacean acid metabolism to the annual productivity of two aquatic vascular plants. Oecologia 68: 615–622

Cockburn W (1983) Stomatal mechanism as the basis for the evolution of CAM and C_3 photosynthesis: a review. Plant Cell Environ 6: 275–279

Farmer AM (1987) Photosynthetic adaptation in *Lobelia dortmanna* and other isoetids. Photosynthetica 21: 51–55

Hostrup O, Wiegleb G (1991) The influence of different CO_2 concentrations in the light and the dark on diurnal malate rhythm and phosphoenolpyruvate carboxylase activities in leaves of *Littorella uniflora* (L.) Aschers. Aquat Bot 40: 91–100

Johnston AM, Raven JA (1986) Dark carbon fixation studies on the intertidal macroalga *Ascophyllum nodosum* (Phaeophyta). J Phycol 22: 78–83

Jones R, Wilkins MB, Coggins JR, Fewson DA, Malcolm ADB (1978) Phosphoenolpyruvate carboxylase from the crassulacean plant *Bryophyllum fedtschenkoi* Hamet et Perrier. Biochem J 175: 391–406

Keeley JE (1981) *Isoetes howellii*: a submerged aquatic CAM plant? Am J Bot 68: 420–424

Keeley JE (1982) Distribution of diurnal acid metabolism in the genus *Isoetes*. Am J Bot 69: 254–257

Keeley JE (1983) Lack of diurnal acid metabolism in two terrestrial *Isoetes* species. Photosynthetica 17: 93–94

Keeley JE (1987) The adaptive radiation of photosynthetic modes in the genus *Isoetes* (Isoetaceae). In: Crawford RMM (ed) Plant life in aquatic and amphibious habitats. Blackwell, Oxford, pp 113–128

Keeley JE (1991) Interactive role of stresses on structure and function in aquatic plants. In: Mooney HA, Winner WE, Pell EI (eds) Response of plants to multiple stresses. Academic Press, New York, pp 329–343

Keeley JE, Bowes G (1982) Gas exchange characteristics of the submerged aquatic crassulacean acid metabolism plant, *Isoetes howellii*. Plant Physiol 70: 1455–1458

Keeley JE, Busch G (1984) Carbon assimilation characteristics of the aquatic CAM plant *Isoetes howellii*. Plant Physiol 76: 525–530

Keeley JE, Morton BA (1982) Distribution of diurnal acid metabolism in submerged aquatic plants outside the genus *Isoetes*. Photosynthetica 16: 546–553

Keeley JE, Sandquist DR (1991) Diurnal photosynthesis cycle in CAM and non-CAM seasonal pool aquatic macrophytes. Ecology 72: 716–727

Keeley JE, Sandquist DR (1992) Commissioned Review. Carbon: freshwater plants. Plant Cell Environ 15: 1021–1035

Keeley JE, Walker CM, Mathews RP (1983) Crassulacean acid metabolism in *Isoetes bolanderi* in high elevation oligotrophic lakes. Oecologia 58: 63–69

Keeley JE, Osmond CB, Raven JA (1984) *Stylites*, a vascular land plant without stomata absorbs CO_2 via its roots. Nature 310: 694–695

Keeley JE, DeMasson D, Gonzalez R, Markham K (1994) Sediment based carbon nutrition in tropical alpine *Isoetes*. In: Rundel PW, Meinzer FC, Smith AP (eds) Tropical alpine environments: plant form and function. Cambridge University Press, Cambridge, pp 167–194

Lorimer GH, Badger MR, Andrews TJ (1976) The activation of ribulose-1, 5-bisphosphate carboxylase by carbon dioxide and magnesium ions. Biochemistry 15: 529–536

Madsen TV (1987) Interactions between internal and external CO_2 pools in the photosynthesis of the aquatic CAM plants *Littorella uniflora* (L.) Aschers and *Isoetes lacustris* L. New Phytol 106: 35–50

Raven JA, Handley LL, MacFarlane JJ, McInroy S, McKenzie L, Richards JH, Samuelsson G (1988) Transley Review No. 13. The role of CO_2 uptake by roots and CAM in acquisition of inorganic C by plants of the isoetid life-form: a review, with new data on *Eriocaulon decangulare* L. New Phytol 108: 125–148

Richardson K, Griffiths H, Reed ML, Raven JA, Griffiths NM (1984) Inorganic carbon assimilation in the isoetids, *Isoetes lacustris* L. and *Lobelia dortmanna* L. Oecologia 61: 115–121

Robe WE, Griffiths H (1990) Photosynthesis of *Littorella uniflora* grown under two PAR regimes: C_3 and CAM gas exchange and the regulation of internal CO_2 and O_2 concentrations. Oecologia 85: 128–136

Sandquist DR, Keeley JE (1990) Carbon uptake characteristics in two high elevation populations of the aquatic CAM plant *Isoetes bolanderi* (Isoetaceae). Am J Bot 77: 682–688

Seddon B (1965) Occurrence of *Isoetes echinospora* in eutrophic lakes in Wales. Ecology 46: 747–748

Tryon RM, Tryon AF (1982) Ferns and allied plants. Springer, Berlin Heidelberg New York

Turpin DH, Vanlerberghe GC, Amory AM, Guy RD (1991) The inorganic carbon requirements for nitrogen assimilation. Can J Bot 69: 1139–1145

19 *Clusia*: Plasticity and Diversity in a Genus of C_3/CAM Intermediate Tropical Trees

U. LÜTTGE

19.1 Diversity

Diversity is probably best considered as a genetic criterion. It refers to the question: which genes are there?

In phytosociology the question is that of floristic diversity, i.e. the richness of species in a given ecological unit. With respect to the genus of tropical trees and shrubs *Clusia*, we do not have very much information on how many different species may co-occur in a given habitat and thus may contribute to its diversity. In the karstic limestone hills of the Sierra San Luis in the State of Falcón in Venezuela we found about five sympatric species, two species that probably have to be separated from the *C. multiflora*-complex, one species belonging to the *C. rosea*-complex, *C. articulata*, and one as yet unidentified species (Franco et al. 1994, and unpubl. observ.). For the upper montane rain forest of Trinidad, Borland et al. (1992) in addition to *C. minor* report the three endemic species *C. aripoensis*, *C. intertexta* and *C. tocuchensis*. On Avila mountain in the coastal range of northern Venezuela near Caracas, we encounter the four species *C. minor*, *C. rosea*, *C. multiflora* and *C. articulata* (Steyermark and Huber 1978), but evidently they are distributed between different microsites.

A somewhat different question is that of taxonomic diversity, i.e. diversity within a given taxon. Diversity in the taxon of the genus *Clusia* is quite large. There are probably about 150 different species. Taxonomically, the genus is currently not well understood, and a revision is urgently needed.

In either case, i.e. in phytosociological ecology and in taxonomy, we may say that diversity is indeed a genetic question. Quantification of diversity is a matter of census with all its practical problems amplified by the generally very complex and diverse structure of tropical vegetation. Approaches of molecular biology may significantly promote progress in this area in the future. A pilot study has shown that, notwithstanding the considerable amounts of latex and phenolic compounds contained in the leaves of *Clusia*, genomic DNA can in fact be isolated in sufficient yield for DNA-amplification fingerprinting using the polymerase chain reaction. A selection of different primers has to be used to obtain

Institut für Botanik, Technische Hochschule Darmstadt, Schnittspahnstr. 3–5, D-64287 Darmstadt, Germany

Ecological Studies, Vol. 114
Winter/Smith (eds) Crassulacean Acid Metabolism

DNA-amplification fingerprints of appropriate simplicity and complexity so that genetic similarity and difference respectively can be assessed between different individuals of a given species and between different species (Böhm et al. 1993) This should be important in providing the basis for both the necessary identifications for ecophysiological purposes and further taxonomic revisions of the genus.

19.2 Plasticity

Plasticity is a functional or ecophysiological criterion referring to flexibility in response to environmental stress. In terms of molecular biology it refers to the question: which genes are turned on? It is a matter of intraspecific versatility, or in view of the genetic heterogeneity within a species, in the strictest sense it is a matter of flexibility of individuals or members of a clone. Since vegetative propagation of *Clusia* is easy in most cases, experimental studies of plasticity are facilitated.

Plasticity appears to be of particular adaptive value under conditions of medium stress or of variable stress over time and to interacting stress factors in non-extreme intensity ranges. Clearly under conditions of extreme stress imposed by one dominant factor, adaptation to that particular situation is needed and the selection is for specialists rather than for generalists. Conversely, with moderate and multifactorial stress varying with time, which is often typical for the tropical environment, flexibility appears to be an important adaptive advantage. As far as studies both in the field and in the laboratory show so far, *Clusia* ecophysiologically is one of the most flexible genera known. In this respect *Clusia* is possibly matched only by another tropical genus, namely *Peperomia* (Sipes and Ting 1985; Ting et al. 1985a; Holthe et al. 1987; see also Chap. 13).

19.2.1 Gas Exchange

Three modes of gas exchange are encountered in *Clusia*, in the laboratory as well as in the field, namely daytime CO_2 uptake as typical for C_3 photosynthesis, nocturnal CO_2 uptake as characteristic for crassulacean acid metabolism (CAM), and CO_2 uptake day and night. Shifts between these modes may occur rapidly in a given plant and are governed by various external factors, which may also interact (Fig. 19.1).

19.2.1.1 Availability of Water and Leaf-to-Air Water-Vapour Pressure Difference

Since CAM is generally considered to be an adaptation to limited supply of water, it is not surprising that reduced availability of water favours the expression of CAM in *Clusia*. Plants of *C. minor* droughted for 4 days in a growth chamber showed the typical four phases of CAM (Osmond 1978), with pronounced nocturnal CO_2 uptake (Phase I), day-time stomatal closure (Phase III), a

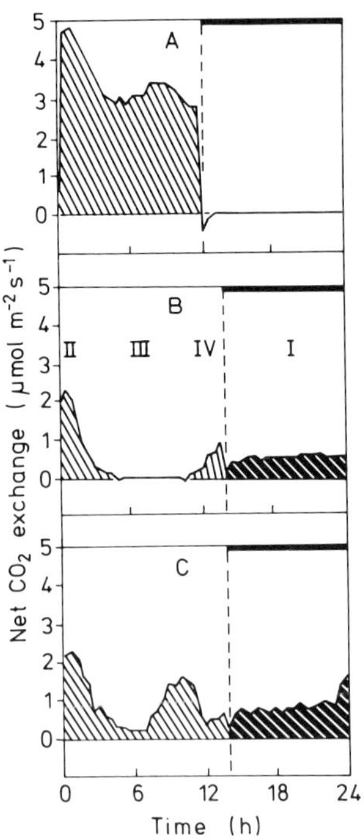

Fig. 19.1A–C. Three types of gas exchange in *Clusia minor* in the growth chamber. **A** C_3 photosynthesis; well-watered plants at a photosynthetic photon flux density of 1 700 μmol m^{-2} s^{-1} during the light period and a leaf-to-air water-vapour pressure difference (VPD) of 6.6 mPa Pa^{-1}; **B** CAM with Phases I to IV (Osmond 1978); droughted plants at a PPFD of 400 μmol m^{-2} s^{-1} and VPD of 13.5 mPa Pa^{-1}; **C** Net CO_2 uptake over 24 h; well-watered plants at a PPFD 400 μmol m^{-2} s^{-1} and VPD of 3.4 mPa Pa^{-1}. *Dark bars* indicate the dark period. (After Lee et al. 1989; modified from Lüttge 1991)

transitory peak of CO_2 uptake in the morning (Phase II), and some CO_2 uptake in the afternoon (Phase IV) (see also p. 4). When the plants were watered again in the morning of the fifth day, a leaf kept at a low leaf-to-air water-vapour pressure difference (VPD) of 6.2 mPa Pa^{-1} changed within a few hours to a significant CO_2 uptake in the light period and had a much suppressed CO_2 uptake in the following dark period (Fig. 19.2). However, the opposite leaf on the same node of the same plant at the same time kept at the higher VPD of 13.1 mPa Pa^{-1} continued to perform CAM (Schmitt et al. 1988).

19.2.1.2 Irradiance and Availability of Water

High irradiance often amplifies drought stress and therefore tends to enforce the expression of CAM. However, under sufficient supply of water, in *Clusia* the opposite effect may also be observed. In a leaf of *C. minor* originally kept at a photosynthetic photon flux density (PPFD) of 360 μmol m^{-2} s^{-1}, daytime CO_2 fixation almost doubled after transfer to 1200 μmol m^{-2} s^{-1} and nighttime CO_2 uptake was totally eliminated (Fig. 19.3; Schmitt et al. 1988). This effect only occurred in well-watered plants and not under drought stress. Evidently, with water not limiting, direct CO_2 fixation and assimilation with open stomata is

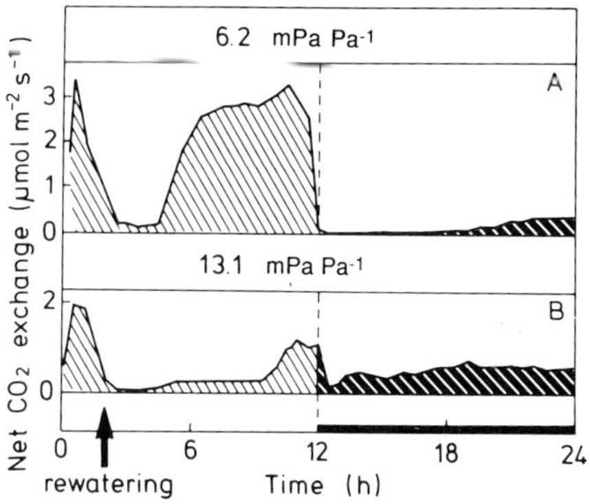

Fig. 19.2A–B. Net CO_2 exchange of two opposite leaves on the same node of *C. minor* in the growth chamber at different leaf-to-air water-vapour pressure differences (VPD). After 4 days of drought stress the plant was rewatered on the 5th day (*arrow*). **A** Leaf in a cuvette at a VPD of 6.2 mPa Pa^{-1}; **B** leaf at a VPD of 13.1 mPa Pa^{-1}. *Dark bars* indicate the dark period. (After Schmitt et al. 1988; from Lüttge 1991)

affordable and C_3 photosynthesis is more favourable in using high light intensities than CAM. In the opposite leaf on the same node, which continued to receive a PPFD of 360 µmol m^{-2} s^{-1}, such a shift was not observed and considerable dark CO_2 fixation was maintained.

19.2.1.3 Temperature

The day/night temperature regime plays a very important role in the expression of gas-exchange patterns of *C. minor*. A certain minimal day/night temperature difference is needed for net nocturnal CO_2 fixation. However, the behaviour is also modulated by light intensity supplied to the plants during growth. With plants grown at a PPFD of 260 to 300 µmol m^{-2} s^{-1}, highest nocturnal CO_2 uptake was obtained at day temperatures of 20 to 30 °C and a night temperature of 15 °C, while with plants kept at a PPFD of 30 to 50 µmol m^{-2} s^{-1} considerable nocturnal CO_2 uptake was only observed at a day/night temperature regime of 30/15 °C (see Fig. 4 in Haag-Kerwer et al. 1992). At low day/night temperature differences and/or higher night temperature, plants showed a C_3-like CO_2-exchange pattern, including nocturnal CO_2 release (Haag-Kerwer et al. 1992). The change in CO_2-exchange pattern occurs within a few days when the temperature regime is changed and is fully reversible (Fig. 19.4).

19.2.1.4 Survey of *Clusia* species

Most *Clusia* species so far studied in the laboratory and/or in the field show extraordinary flexibility in CO_2-exchange pattern (Table 19.1). It is not clear, however, if all *Clusia* species are C_3/CAM intermediates or if there are obligate C_3 or CAM plants among them. In our studies up to now, *C. multiflora* performed only C_3 photosynthesis and *C. alata* only CAM (Franco et al. 1994, and unpubl.).

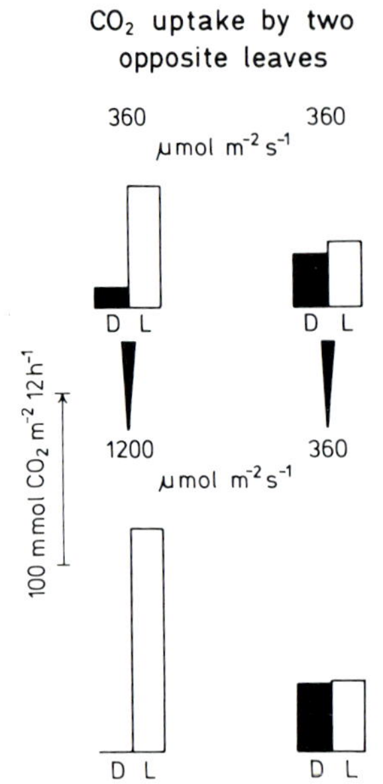

Fig. 19.3. Net CO_2 exchange of two opposite leaves on the same node of a well-watered plant of *C. minor* in the growth chamber at varied PPFD. One leaf was transferred from a PPFD of 360 µmol m^{-2} s^{-1} during the light period to 1200 µmol m^{-2} s^{-1}, while the other leaf was kept at 360 µmol^{-2} s^{-1}. *Columns* give integrated net CO_2 uptake for the light period (*L*) and dark period (*D*), respectively. (After data of Schmitt et al. 1988)

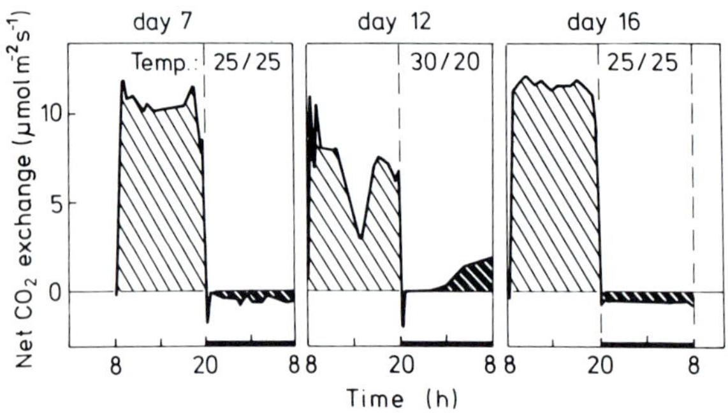

Fig. 19.4. Reversible changes between C_3-like CO_2 exchange of well-watered plants of *C. minor* in the growth chamber at day/night temperatures of 25/25 °C and CAM-like CO_2 exchange at 30/20 °C (PPFD 260 to 300 µmol m^{-2} s^{-1}). Plants were kept at 25/25 °C for 7 days followed by 30/20 °C up to day 12 and then returned to 25/25 °C on day 13. CO_2-exchange patterns measured on days 7, 12 and 16 are presented. *Dark bars* indicate the dark period. (After Haag-Kerwer et al. 1992, with permission of Oxford University Press)

19.2.2 Metabolism

Metabolic reactions in CAM-performing *Clusia* are also versatile. *Clusia* species can use both starch and soluble sugars (glucose, fructose) as precursors for nocturnal acid synthesis, and they can accumulate substantial amounts of citric acid in addition or alternatively to malic acid.

Table 19.1. *Clusia* species studied ecophysiologically

Species	Reference
Species showing C_3/CAM transitions	
C. aripoensis	Borland et al. (1992)
C. intertexta	Borland et al. (1992)
C. lanceolata	U. Lüttge and colleagues[a]
C. major	U. Lüttge and colleagues[a]
C. minor	U. Lüttge and colleagues[a]
C. rosea	U. Lüttge and colleagues[a]
	Sternberg et al. (1987)
	Ting et al. (1985b, 1987)
C. tocuchensis	Borland et al. (1992)
C. venosa	U. Lüttge and colleagues[a]
C. uvitana	Winter et al. (1992)
Oedematopus obovatus	U. Lüttge and colleagues[a]
Species with obligate C_3 photosynthesis	
C. multiflora	U. Lüttge and colleagues[a]
Species with obligate CAM	
C. alata	U. Lüttge and colleagues[a]

[a]The work of the present author's group was carried out with the groups of A.C. Franco and E. Medina.

19.2.2.1 Carbohydrates

In CAM-performing *C. rosea* in the field, day/night changes in carbohydrate levels on the basis of hexose units were about 1:1:0.5 glucose:fructose:starch (Fig. 19.5; Ball et al. 1991b). These carbohydrates are used for carbon skeletons needed for nocturnal synthesis of organic acids, and they must be resynthesized during the day, when the free hexoses are presumably stored in the vacuole. C.C. Black has analyzed several enzymes of glycolysis and gluconeogenesis in *Clusia* (see Chap. 2). He also has drawn our attention to the fact that there are no pathways in plants leading to the formation of free hexoses except splitting sucrose by invertase, so that in CAM plants with larger day-night oscillations of the levels of glucose and fructose invertase activity should be high, which in fact appears to be borne out in *Clusia* (see Chap. 2).

19.2.2.2 Organic Acids

Many different CAM plants may show small day/night oscillations in the levels of organic acids other than malic acid during CAM, especially of citric acid (Lüttge 1988). However, very substantial day/night changes in citric-acid levels have been observed so far only in CAM-performing *Clusia* species (Fig. 19.6; Popp et al.

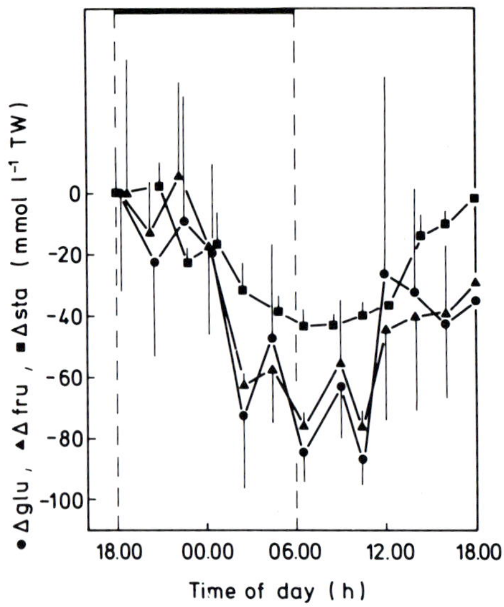

Fig. 19.5. Day/night changes in the levels of glucose (Δ*glu*), fructose (Δ*fru*) and starch in hexose equivalents (Δ*sta*) related to tissue water (*TW*) in leaves of *C. rosea* in the field. Changes in carbohydrate levels were related to the levels obtained at 18.30 h at the start of the measurements. The *dark bar* indicates the dark period. (Ball et al. 1991b, with permission)

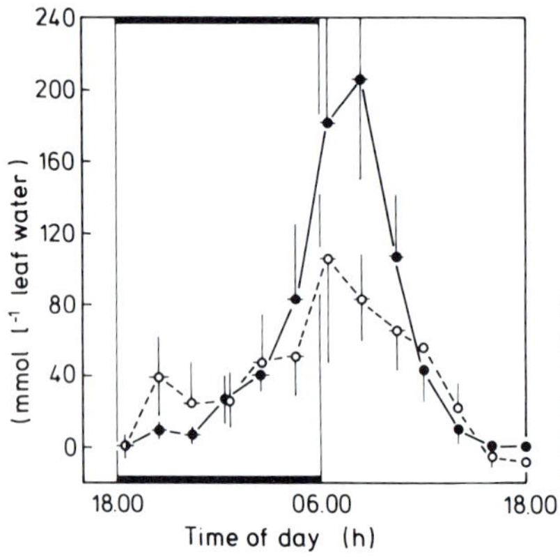

Fig. 19.6. Day/night changes in the levels of malate (●) and citrate (○) in leaves of *C. rosea* in the field related to leaf water. Changes were related to the levels observed at 18.30 h at the start of the measurements. *Dark bars* indicate the dark period. (Ball et al. 1991b, with permission)

1987; Franco et al. 1990, 1992, 1994; Ball et al. 1991b; Borland et al. 1992; Haag-Kerwer et al. 1992). $^{14}CO_2$-pulse/chase experiments with *C. minor* show that during CAM in the dark period label from external CO_2 first goes almost exclusively into malate and that subsequently, i.e. during a chase with normal air, label is transferred from malate into citrate (Fig. 19.7; Olivares et al. 1993). Since most likely malate is transported into the vacuole immediately after synthesis and citrate must be formed in the mitochondria, this underlines the problem of peripatetic paths of malate during the dark period of CAM with traffic between the compartments of vacuoles, cytosol and mitochondria as highlighted by Kalt et al. (1990). It also requires considerable export of citrate from mitochondria

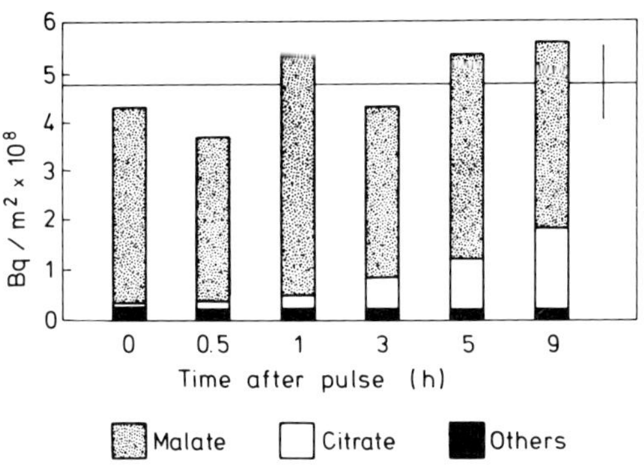

Fig. 19.7. Distribution of ^{14}C label in malate, citrate and other compounds after an exposure of detached leaves of *C. minor* to $^{14}CO_2$ during the last hour of the light period (0 h) and during a subsequent chase in normal air (0.5 to 9 h). *Horizontal line* (with SD error bar) gives average of total radioactivity in the samples taken for the times indicated. (After data of Olivares et al. 1993)

Table 19.2. Conditions leading to nocturnal accumulation of citrate alone (left column), malate alone (right column) and approximately similar amounts of citrate and malate (centre column) in *C. minor* in the growth chamber. Data for day/night changes of malate and citrate levels and for integrated nocturnal CO_2 gain (means ± SD (*n*)) are from Haag-Kerwer et al. (1992)

PPFD ($\mu mol\ m^{-2}\ s^{-1}$)	30 to 50	260 to 300	260 to 300
Day/night temperature (°C)	30/30	35/20	25/15
Δ Malate (mM)	−2	52	100
Δ Citrate (mM)	80	37	−10
Integrated nocturnal carbon gain ($mmol\ m^{-2}\ 12\ h^{-1}$)	−20.8 ±5.1 (6)	31.8 ±19.8 (5)	25.5 ±2.9 (5)

after uptake of pyruvate and malate. Studies with isolated spinach-leaf mitochondria have shown that such export does in fact occur. In C_3 plants in vivo, it mainly serves formation of 2-oxoglutarate via cytosolic aconitase and NADP-isocitrate dehydrogenase as a precursor for intermediates of the glutamine/glutamate cycle involved in N assimilation (Hanning and Heldt 1993). Hence, mitochondrial citrate transport may not be unusual, although it remains to be seen whether it is particularly highly expressed in *Clusia*.

The ratio of malic:citric acid accumulated during the dark period is highly affected by environmental factors like PPFD, day-night temperature regime and water availability (Franco et al. 1992; Haag-Kerwer et al. 1992). With varied day/night temperature regimes, the largest day/night changes of malate levels (Δmal) were obtained under conditions that also gave the largest rates of nocturnal fixation of external CO_2 (see Fig. 4 in Haag-Kerwer et al. 1992). The highest day/night changes of citrate levels were observed when nocturnal CO_2 uptake was low, there was even a release of CO_2, i.e. in plants grown at a PPFD of 260 to 300 $\mu mol\ m^{-2}\ s^{-1}$ at day/night temperatures of 20:20 °C and in plants kept at 30 to 50 $\mu mol\ m^{-2}\ s^{-1}$ PPFD with 30:30 °C (Haag-Kerwer et al. 1992). Thus, plants of *C. minor* can easily be conditioned to accumulate either malate or citrate alone, or both in about equal amounts (Table 19.2), or in other ratios.

19.3 The Ecophysiological Significance of Plasticity in *Clusia*

Plasticity should strengthen the competitive abilities of a species, especially under medium and variable stress (Sect. 19.2), and it should increase its ecological amplitude (Chazdon 1992).

19.3.1 Ecological Amplitude of *Clusia*: Habitats and Life Forms

The ecological amplitude of *Clusia* in fact is very large. *Clusia* may occur on coastal sand dunes and rocks, in savannas, in gallery forests, in lowland and montane rain forests and in cloud or fog forests (elfin forests) (Lüttge 1991; Haag-Kerwer and Lüttge 1993). More extravagantly, sites like epiphyte gardens or the tanks of epiphytic bromeliads (Schimper 1888; Ball et al. 1991a) and even termite mounds in asphalt lakes (Vareschi 1980) may be inhabited. Furthermore, *Clusia* species can occur as both terrestrial and epiphytic life forms. Seedlings of a given species may germinate either terrestrially or epiphytically. In the former case seedlings directly develop to independent trees. In the latter case two types of aerial roots are formed, establishing contact with the ground and strangling the host to establish a holdfast, respectively (Schimper 1888). Thus, these plants go through a hemiepiphytic stage; however, eventually they kill their host by girdling its phloem and by competition for light, and with its dead trunk rotting away they also become independent trees (Lüttge 1991). This high ecological versatility of *Clusia* most probably is not quite unrelated to its ecophysiological plasticity.

19.3.2 CO_2 Acquisition

The flexibility in CO_2 acquisition by light fixation via ribulose 1,5-bisphosphate carboxylase/oxygenase (RUBISCO) and by dark fixation via phospho*enol*-pyruvate carboxylase (PEPC), respectively, is the essential feature in the plasticity of *Clusia*. Gas-exchange measurements, both in the field and in the laboratory, however, only present part of the whole picture. They allow us to show which options are there and under which conditions they may possibly be expressed (Sect. 19.2.1). However, they do not show to what extent the plants really make use of these options over longer periods in their habitat (see Chap. 20).

The integrated evaluation that is needed is made possible by stable-isotope analysis. PEPC has a lower discrimination against ^{13}C during CO_2 fixation than RUBISCO, and hence stable carbon-isotope ratios ($\delta^{13}C$ values, defined as on p. 5) range from -9 to -18 ‰ for obligate CAM plants and from -22 to -32 ‰ for obligate C_3 plants (Troughton 1979; O'Leary 1981; Osmond et al. 1982). Many mature trees and young seedlings of *C. rosea* were sampled on St. John Island (US Virgin Islands, Lesser Antilles). Seedlings in general showed more negative, i.e. more C_3-like, $\delta^{13}C$ values than trees, suggesting that they made more use of direct CO_2 fixation via RUBISCO than trees (Fig. 19.8). On the other hand, the variability of $\delta^{13}C$ values in seedlings was larger than in trees, showing

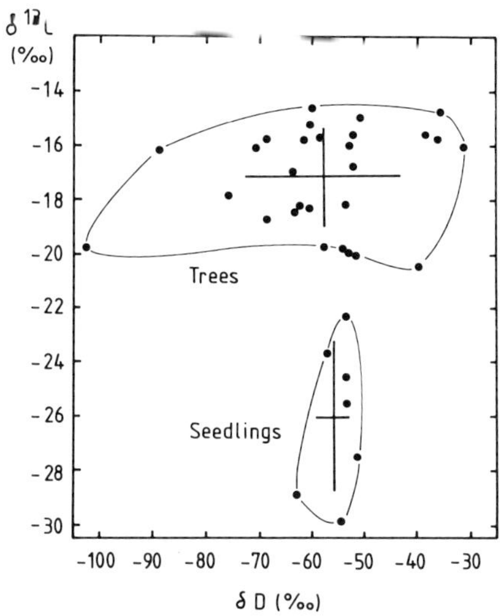

Fig. 19.8. Relation between $\delta^{13}C$ and δD values in seedlings and trees of *C. rosea* in the field with SD error bars (*crosses*) for the *encircled groups of points*. (Lüttge et al. 1993, with permission)

Table 19.3. Possible roles of day/night changes in the levels of malate (Δ mal) and citrate (Δ citr) during the CAM cycle. (After Franco et al. 1992)

	Δ mal	Δ citr
CO_2 acquisition	Yes	No
H_2O saving during CO_2 acquisition	Yes	No
Recycling, daytime	CO_2	CO_2
Recycling, nighttime	CO_2	C skeleton
Prevention of photoinhibition	Yes	Yes: more effectively
Osmotic changes	Yes	Limited
Buffer capacity	Low	High

that they made more extensive use of flexibility (Lüttge et al. 1993). Hydrogen-isotope ratios (δD values, defined as on p. 337) were also analyzed in the same material. There were no significant differences in δD values between trees and seedlings, but the variation was much larger in the trees than seedlings. Values of δD in dry plant matter are affected by a number of factors including the source of the water transpired (e.g. fresh precipitiation, water from the soil, groundwater, etc.) and the rate of transpiration (Rundel et al. 1989; see also Chap. 22). Thus, the larger variation of δD values in trees implies that trees have a more variable water budget in terms of supply and utilization than seedlings.

19.3.3 Accumulation of Organic Acids

The possible alternative accumulation of malic and citric acid, dependent on environmental conditions (Sect. 19.2.2), raises questions regarding the specific ecophysiological roles of the two acids (Table 19.3).

Much has been written about the ecophysiological implications of malic-acid accumulation. First of all, of course, it serves nocturnal storage of CO_2 after dark fixation of atmospheric CO_2 and drought adaptation related to this water-saving mechanism of CO_2 acquisition. After refixation of respiratory CO_2 it also serves internal recycling of CO_2, allowing maintenance of metabolism when stomatal conductance (and hence uptake of atmospheric CO_2) is increasingly reduced, even during the dark period as drought stress becomes more severe. This nocturnal malate accumulation and CO_2 storage with CO_2 recycling provides a source of CO_2 from decarboxylation of malate, which may be highly important during the subsequent day for the prevention of photoinhibition (Osmond 1982; Adams and Osmond 1988; Griffiths et al. 1989). Accumulation of malic acid also leads to an increased osmotic pressure in the vacuoles, not only if starch but also if free sugars are the precursors for malate synthesis, since two malate molecules may originate from one hexose unit. Nocturnal increases in osmotic pressure with osmotic storage of water may assist in the maintenance of the plant's water budget (Lüttge 1986).

Conversely, little is known about the possible ecophysiological role of citric-acid accumulation during the CAM cycle. Stoichiometry clearly shows that citric-acid accumulation is not accompanied by net CO_2 acquisition, and hence cannot serve a correlated water-saving function (Lüttge 1988). Osmotically it would only be active in increasing vacuolar osmotic pressure if starch were the precursor for citrate formation. In the case of vacuolar hexoses as precursors, as observed to a large extent in *Clusia* (Sect. 19.2.2), mobilization of the free hexoses from the vacuole and accumulation of citric acid would just cancel each other osmotically, because one hexose unit gives rise to one molecule of citric acid.

Since in contrast to malate formation there is stoichiometrically no net carbon gain in the formation of citrate following carbohydrate breakdown, citric acid accumulation also cannot serve net CO_2 recycling. However, citric-acid accumulation and remobilization does imply cycling of carbon. Energy conversion, i.e. formation of NADH and ATP, is occurring during nocturnal citrate synthesis (Lüttge 1988), so that carbon recycling via citrate, like CO_2 recycling via malate, would allow maintenance of metabolism behind stomata more or less closed day and night. Moreover, we may assume that daytime remobilization of citrate provides an even better protection against photoinhibition than that of malate. Malate is decarboxylated in the cytosol. Data of C.C. Black (Table 2.3 in Chap. 2) indicate that *Clusia rosea* is a PEP carboxykinase-type plant, where decarboxylation gives rise to 1 CO_2 for reassimilation in the light via RUBISCO and 1 PEP. The latter does not need to enter another organelle and can be used directly for gluconeogenesis in the cytosol. Even if we cannot generalize for all *Clusia* species that they are PEP carboxykinase-type plants, and if malate decarboxylation may also occur via malic enzyme, the resulting pyruvate would also not need to be transported into the mitochondria and could go via pyruvate, P_i dikinase towards gluconeogenesis. Conversely, citrate must enter the mitochondria to be broken down, and once in this compartment it may lead to as much as 6 CO_2 per molecule to be refixed via RUBISCO. This hypothesis can be checked by drought-stress experiments with *Clusia*, because when stomata become increasingly more closed day and night due to water-shortage stress, protection from photoinhibition by CO_2 derived from internally recycled metabolites should become increasingly important. Indeed, in various *Clusia* species the relative role of citrate increased, i.e. the ratio of malate:citrate during nocturnal acid accumulation decreased, during prolonged drought (Fig. 19.9; Franco et al. 1992).

Photosystem II chlorophyll-fluorescence quenching using the pulse-amplitude modulation technique (Schreiber and Bilger 1987) was studied in *C. minor* in the C_3 and CAM states. The CAM-state plants were obtained by treatment of plants grown at a PPFD of $\approx$ 400 μmol m^{-2} s^{-2} with a day/night temperature regime of 28/20 °C, conditions under which there is significant nocturnal citrate accumulation in addition to malate (see Fig. 4 in Haag-Kerwer et al. 1992). It is noteworthy with respect to the discussion of protection from photoinhibition that photochemical quenching, q_p, is much less affected by high temperatures above 40 °C in the CAM mode than in the C_3 mode (Fig. 19.10). Protection against photodamage and photodestruction may, therefore, be one of the keys for the understanding of citric-acid oscillations during CAM in *Clusia*, which certainly needs more study of exposed and shaded, droughted and well water-supplied plants in the field (see Franco et al. 1993).

Another important key may be the actual amount of organic acids that can oscillate. It is limited by the vacuolar buffering capacity. This affects the pH gradient between the cytosol and the vacuole interior, which in turn controls the tonoplast H^+-ATPase via the electrochemical proton gradient, $\Delta\mu_H{}^+$, against which it pumps protons into the vacuole. This primary active transport of protons is the driving force for electrophoretic transport of organic-acid anions and for organic-acid accumulation in the vacuole (Lüttge 1987; see also Chap. 4). We have shown that the tonoplast ATPase in CAM plants is a 2 H^+-pumping ATPase moving 2 H^+ per ATP hydrolyzed (Lüttge et al. 1981; Smith et al. 1982, 1984). This implies that $\Delta\mu_H{}^+$ at the tonoplast may not exceed half the free energy of ATP hydrolysis in the cytosol ($\approx$25 kJ mol^{-1}), and with a transtonoplast electrical potential gradient between zero and a few tens of millivolts positive,

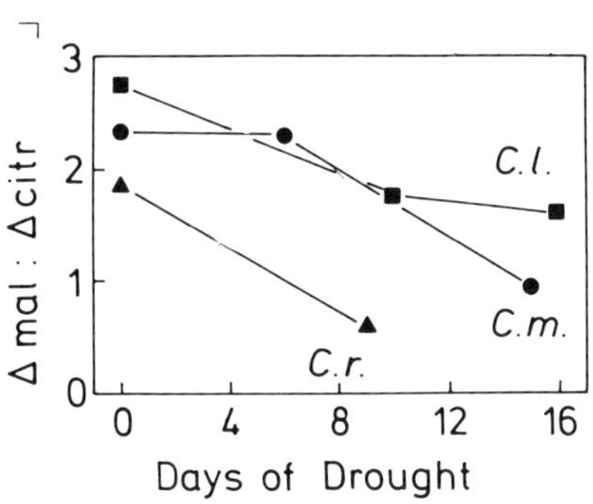

Fig. 19.9. Ratios of day/night changes of malate (Δ*mal*) and citrate (Δ*citr*) in leaves of *C. lanceolata* (*C. l.*), *C. minor* (*C. m.*) and *C. rosea* (*C. r.*) droughted for several days in the growth chamber. (After Franco et al. 1992)

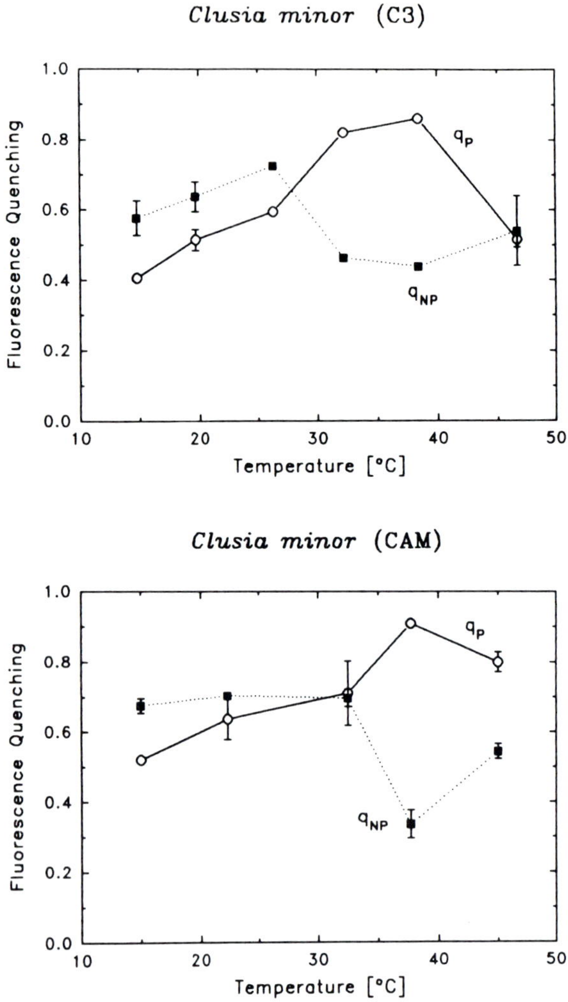

Fig. 19.10. Photochemical (q_P) and non-photochemical (q_{NP}) quenching of chlorophyll fluorescence in leaves of *Clusia minor* in the C_3 and CAM states. Plants were kept at a PPFD of 400 µmol m^{-2} s^{-1} at a constant day/night temperature of 25 °C and relative humidity of 70% for the C_3 state, and at day/night temperatures and relative humidities of 28/20 °C and 50/70%, respectively, for the CAM state. Measurements were performed during the light period (at the beginning of stomatal closure of Phase III for the CAM state) at a PPFD of 400 µmol m^{-2} s^{-1}. *Dots* give averages of triplicates with SD where larger than symbol size. (Measurements by T. Grams, Darmstadt)

vacuolar H^+-activity may not be much larger than that corresponding to a vacuolar pH of ≈ 3.0. Plants of *Clusia* have been shown to be the strongest acid accumulators known among CAM plants. In our studies *C. minor* and *C. rosea* reached up to ≈ 1200 mM titratable protons per night (Franco et al. 1992), and

Borland et al. (1992) even reported 1 410 mM for *C. minor* in the field in Trinidad. Nevertheless, vacuolar pH does not drop much below pH 3 and pumping of $2H^+$ per ATP remains thermodynamically possible (Ball et al. 1991b). It appears that the extraordinarily high total acid accumulation in *Clusia* is linked to citric-acid accumulation with its well-known strong buffering capacity (Beloserski and Proskwjakow 1956; Perrin and Dempsey 1974), so that buffering capacity also oscillates together with organic-acid levels.

Particularly high amplitudes of day/night oscillations of organic acids, of course, would imply high daytime refixation of CO_2 from the acids cycled and thus high protection from photoinhibition and photodamage. However, since citric acid does not contribute to C acquisition, it appears to be essential that there is flexibility in the use of citric and malic acid , as is observed in *Clusia* (Sect. 19.2.2). It must be important that this is regulated to serve the requirements given by variable stress, and that the ecophysiological advantages and disadvantages of either acid are optimized. There are clear hypotheses now to be evaluated by further ecophysiological tests in the field and by experiments in the phytotron.

19.4 Regulation of Plastic Responses

Plasticity may be regulated at the levels of transcription, translation and post-translational modulation of enzyme activity. A reasonable amount of information is now available regarding the "CAM enzymes" of glycolysis, organic-acid synthesis and gluconeogenesis (Bohnert et al. 1992; see Chap. 9) and the transport proteins of the tonoplast (Lüttge 1993; see Chap. 4) in the C_3/CAM-intermediate annual halophyte *Mesembryanthemum crystallinum*. However, little is known so far about *Clusia* in this respect. The only pertinent information appears to be an immunological study of PEPC in *C. uvitana*, showing that levels of this CAM enzyme increase and decrease within 10 days after the onset and alleviation of drought stress, respectively (Winter et al. 1992). Since C_3/CAM transitions in *Clusia* are often much more rapid than in *M. crystallinum*, it would be of particular interest to perform more work in this area in the future. The powerful methods of molecular biology with analyses of mRNA, in vitro translation and reverse transcription to cDNA should help to answer the question of plasticity posed in Section 19.2: which genes are turned on?

19.5 Plasticity and Diversity

In conclusion, with the question: are diversity and plasticity related to each other? we return to the outset. Clearly, if plasticity affects ecological amplitude and competitive abilities (Sect. 19.3) it should also affect floristic diversity. Moreover, with radiation, separation and segregation it also appears to be a basis of speciation, and hence evolution and taxonomic diversity. Ecophysiological studies on *Clusia* may contribute much to the understanding of these problems in the future (Lüttge 1993).

References

Adams WW III, Osmond CB (1988) Internal CO_2 supply during photosynthesis of sun and shade grown CAM plants in relation to photoinhibition. Plant Physiol 86: 117–123

Ball E, Hann J, Kluge M, Lee HSJ, Lüttge U, Orthen B, Popp M, Schmitt A, Ting IP (1991a) Ecophysiological comportment of the tropical CAM-tree *Clusia* in the field. I. Growth of *Clusia rosea* Jacq. on St. John, US Virgin Islands, Lesser Antilles. New Phytol 117: 473–481

Ball E, Hann J, Kluge M, Lee HSJ, Lüttge U, Orthen B, Popp M, Schmitt A, Ting IP (1991b) Ecophysiological comportment of the tropical CAM-tree *Clusia* in the field. II. Modes of photosynthesis in trees and seedlings. New Phytol 117: 483–491

Beloserski AN, Proskwjakow NI (1956) Praktikum der Biochemie der Pflanzen. Deutscher Verlag der Wissenschaften, Berlin

Böhm M, Löw R, Haag-Kerwer A, Lüttge U, Rausch T (1993) Evaluation of comparative DNA amplification fingerprinting for rapid species identification within the genus *Clusia*. Bot Acta 106: 448–453

Bohnert HJ, Vernon DM, DeRocher EJ, Michalowski CB, Cushman JC (1992) Biochemistry and molecular biology of CAM. In: Wray JL (ed) Inducible plant proteins. Cambridge University Press, Cambridge, pp 113–137

Borland AM, Griffiths H, Maxwell C, Broadmeadow MSJ, Griffiths NM, Barnes JD (1992) On the ecophysiology of the Clusiaceae in Trinidad: expression of CAM in *Clusia minor* L. during the transition from wet to dry season and characterization of the endemic species. New Phytol 122: 349–357

Chazdon RL (1992) Photosynthetic plasticity of two rain forest shrubs across natural gap transects. Oecologia 92: 586–595

Franco AC, Ball E, Lüttge U (1990) Patterns of gas exchange and organic acid oscillations in tropical trees of the genus *Clusia*. Oecologia 85: 108–114

Franco AC, Ball E, Lüttge U (1992) Differential effects of drought and light levels on accumulation of citric and malic acids during CAM in *Clusia*. Plant Cell Environ 15: 821–829

Franco AC, Olivares E, Ball E, Lüttge U, Haag-Kerwer A (1994) *In situ* studies of crassulacean acid metabolism in several sympatric species of tropical trees of the genus *Clusia*. New Phytol 126: 203–211

Griffiths H, Ong BL, Avadhani PN, Goh CJ (1989) Recycling of respiratory CO_2 during CAM metabolism: alleviation of photoinhibition in *Pyrrosia piloselloides*. Planta 179: 115–122

Haag-Kerwer A, Lüttge U (1993) Ecophysiological adaptation as a basis for high diversity in the genus *Clusia*. In: Barthlott W, Naumann CM, Schmidt-Loske K, Schuchmann KL (eds) Animal-plant interactions in tropical environments. Zoologisches Forschungsinstitut und Museum Alexander Koenig, Bonn, pp 115–123

Haag-Kerwer A, Franco AC, Lüttge U (1992) The effect of temperature and light on gas exchange and acid accumulation in the C_3-CAM plant *Clusia minor* L. J Exp Bot 43: 345–352

Hanning I, Heldt HW (1993) On the function of mitochondrial metabolism during photosynthesis in spinach (*Spinacia oleracea* L.) leaves. Plant Physiol 103: 1147–1154

Holthe PA, Sternberg LSL, Ting IP (1987) Developmental control of CAM in *Peperomia scandens*. Plant Physiol 84: 743–747

Kalt W, Osmond CB, Siedow J (1990) Malate metabolism in the dark after $^{13}CO_2$ fixation in the crassulacean plant *Kalanchoë tubiflora*. Plant Physiol 94: 826–832

Lee HSJ, Schmitt AK, Lüttge U (1989) The response of the C_3-CAM tree, *Clusia rosea*, to light and water stress. J Exp Bot 40: 171–179

Lüttge U (1986) Nocturnal water storage in plants having crassulacean acid metabolism. Planta 168: 287–289

Lüttge U (1987) Carbon dioxide and water demand: crassulacean acid metabolism (CAM), a versatile ecological adaptation exemplifying the need for integration in ecophysiological work. New Phytol 106: 593–629

Lüttge U (1988) Day-night changes of citric-acid levels in crassulacean acid metabolism: phenomenon and ecophysiological significance. Plant Cell Environ 11: 445–451

Lüttge U (1991) *Clusia*: Morphogenetische, physiologische und biochemische Strategien von Baumwürgern im tropischen Wald. Naturwissenschaften 78: 49–58

Lüttge U (1993) The role of crassulacean acid metabolism (CAM) in the adaptation of plants to salinity. New Phytol 125: 59–71
Lüttge U, Smith JAC, Marigo G, Osmond CB (1981) Energetics of malate accumulation in the vacuoles of *Kalanchoë tubiflora* cells. FEBS Lett 126: 81–84
Lüttge U, Ziegler H, Ting IP (1993) Ecophysiological comportment of the tropical C_3/CAM-intermediate tree *Clusia rosea* in the field as assessed by analyses of stable carbon- and hydrogen-isotope ratios. J Plant Physiol 142: 497–501
O'Leary MH (1981) Carbon isotope fractionation in plants. Phytochemistry 20: 553–567
Olivares E, Faist K, Kluge M, Lüttge U (1993) $^{14}CO_2$ pulse-chase labelling in *Clusia minor* L. J Exp Bot 44: 1527–1533
Osmond CB (1978) Crassulacean acid metabolism: a curiosity in context. Annu Rev Plant Physiol 29: 379–414
Osmond CB (1982) Carbon cycling and stability of the photosynthetic apparatus in CAM. In: Ting IP, Gibbs M (eds) Crassulacean acid metabolism. American Society of Plant Physiologists, Rockville, pp 112–127
Osmond CB, Winter K, Ziegler H (1982) Functional significance of different pathways of CO_2 fixation in photosynthesis. In: Lange OL, Nobel PS, Osmond CB, Ziegler H (eds) Physiological plant ecology II. Encyclopedia of plant physiology, new series, vol 12 B. Springer, Berlin Heidelberg New York, pp 479–547
Perrin DD, Dempsey B (1974) Buffers for pH and metal ion control. Chapman and Hall, London
Popp M, Kramer D, Lee H, Diaz M, Ziegler H, Lüttge U (1987) Crassulacean acid metabolism in tropical dicotyledonous trees of the genus *Clusia*. Trees 1: 238–247
Rundel PW, Ehleringer JR, Nagy KA (eds) (1989) Stable isotopes in ecological research. Springer, Berlin Heidelberg New York
Schimper AFW (1888) Botanische Mittheilungen aus den Tropen. II. Die epiphytische Vegetation Amerikas. G. Fischer, Jena
Schmitt AK, Lee HSJ, Lüttge U (1988) The response of the C_3-CAM tree, *Clusia rosea*, to light and water stress. I. Gas exchange characteristics. J Exp Bot 39: 1581–1590
Schreiber U, Bilger W (1987) Rapid assessment of stress effects on plant leaves by chlorophyll fluorescence measurements. In: Tenhunen JD, Catarino FM, Lange OL, Oechel WC (eds) Plant response to stress: functional analysis in mediterranean ecosystems. NATO ASI Series G: Ecological Sciences vol 15. Springer, Berlin Heidelberg New York, pp 27–53
Sipes DL, Ting IP (1985) Crassulacean acid metabolism and crassulacean acid metabolism modifications in *Peperomia camptotricha*. Plant Physiol 77: 59–63
Smith JAC, Marigo G, Lüttge U, Ball E (1982) Adenine-nucleotide levels during crassulacean acid metabolism and the energetics of malate accumulation in *Kalanchoë tubiflora*. Plant Sci Lett 26: 13–21
Smith JAC, Uribe EG, Ball E, Heuer S, Lüttge U (1984) Characterization of the vacuolar ATPase activity of the crassulacean-acid-metabolism plant *Kalanchoë daigremontiana*. Eur J Biochem 141: 415–420
Sternberg LSL, Ting IP, Price D, Hann J (1987) Photosynthesis in epiphytic and rooted *Clusia rosea* Jacq. Oecologia 72: 457–460
Steyermark JA, Huber O (1978) Flora del Avila. Soc Venezolana de Ciencias Naturales y Ministerio del Ambiente y de los Recursos Naturales Renovables, Caracas
Ting IP, Bates L, Sternberg LO, DeNiro MJ (1985a) Physiological and isotopic aspects of photosynthesis in *Peperomia*. Plant Physiol 78: 246–249
Ting IP, Lord EM, Sternberg LSL, DeNiro MJ (1985b) Crassulacean acid metabolism in the strangler *Clusia rosea* Jacq. Science 229: 969–971
Ting IP, Hann J, Holmbrook NM, Putz FE, Sternberg LSL, Price D, Goldstein G (1987) Photosynthesis in hemiepiphytic species of *Clusia* and *Ficus*. Oecologia 74: 339–346
Troughton JH (1979) $\delta^{13}C$ as an indicator of carboxylation reactions. In: Gibbs M, Latzko E (eds) Photosynthesis II. Photosynthetic carbon metabolism and related processes. Encyclopedia of plant physiology, new series, vol VI. Springer, Berlin Heidelberg New York, pp 140–149
Vareschi V (1980) Pflanzenökologie der Tropen. Ulmer, Stuttgart
Winter K, Zotz G, Baur B, Dietz KJ (1992) Light and dark CO_2-fixation in *Clusia uvitana* and the effects of plant water status and CO_2 availability. Oecologia 91: 47–51

20 Seasonal Changes in Daytime Versus Nighttime CO_2 Fixation of *Clusia uvitana* In Situ

G. ZOTZ and K. WINTER

20.1 Introduction

Since the discovery that *Clusia lundellii* is a CAM plant (Tinoco Ojanguren and Vásquez-Yánez 1983), species of *Clusia* have received considerable attention by plant ecophysiologists: first, because *Clusia* is one of the only genera amongst the Magnoliopsida (Dicotyledonae) that contain arborescent CAM species, and second, because some species of *Clusia* show substantial flexibility in the expression of CAM in response to the environment (Schmitt et al. 1988; Lee et al. 1989; Franco et al. 1990, 1991, 1992; Ball et al. 1991a,b; Borland et al. 1992; Haag-Kerwer et al. 1992; Winter et al. 1992) (see also Chaps. 15 and 19). In this chapter, we focus on *Clusia uvitana* Pitt., a hemiepiphytic species in the tropical moist forest of Barro Colorado Island (BCI), Panama (Croat 1978; Hammel 1986). Both epiphytic and hemiepiphytic plants, the latter possessing roots that reach the ground, can be found high up in the crowns of tall canopy trees, but also on rocks and tree trunks along the shoreline. *C. uvitana* is more abundant than C_3 hemiepiphytes such as species of *Ficus* (Todzia 1986). This finding probably reflects advantages of CAM in a climate characterized by a pronounced dry season from mid-December to April. During this time, monthly rainfall is frequently less than 20 mm, while the average monthly rainfall during the wet season usually exceeds 150 mm (Windsor 1990).

Studies with potted plants of *Clusia uvitana* under controlled conditions in the laboratory have demonstrated both slowly and rapidly occurring changes in CAM activity in response to changes in environmental conditions which are similar to responses found in *Clusia minor* (Chap. 19). Dark CO_2 fixation gradually increased over a period of several days as soil-water content declined (Winter et al. 1992). The increase in dark CO_2 fixation was accompanied by reversible increases in the level of PEPC protein (Winter et al. 1992). Upon rewatering, PEPC content and nocturnal carbon gain decreased, whilst daytime net carbon gain increased to its initial level before the onset of the stress treatment.

Nocturnal carbon gain rapidly increased (within 24 h) in response to reduced daytime carbon gain resulting from low ambient relative humidity and subatmospheric (17 Pa) CO_2 partial pressures during 12-h light periods (Winter

Smithsonian Tropical Research Institute, P.O. Box 2072, Balboa, Republic of Panama

Ecological Studies, Vol. 114
Winter/Smith (eds) Crassulacean Acid Metabolism

et al. 1992). CAM activity markedly decreased within 24 h when ambient CO_2 partial pressure was enhanced during the light period. The ratio nighttime: daytime net carbon gain also responded rapidly to changes in PPFD and temperature (Zotz and Winter 1993a). Low temperatures and low PPFDs resulted in a C_3- rather than a CAM-type pattern of CO_2 exchange.

In order to assess the ecological relevance of these laboratory findings, we explored the plasticity in the expression of CAM in *Clusia uvitana* in situ under natural tropical conditions. An open gas-exchange system with a climatized leaf cuvette was stationed in the canopy of a 47-m tall tree of *Ceiba pentandra* (L.) Gaertn., which was host to several *C. uvitana* (Zotz and Winter 1993a, 1994a,b,c). Diel (24-h) changes in CO_2 and water-vapour exchange, titratable acidity and leaf water status of exposed and shaded leaves of epiphytic and hemiepiphytic individuals were monitored over the course of an entire year. The role of leaf ontogeny in C_3-CAM relationships was also considered.

20.2 Seasonal Changes in the Expression of CAM

Figure 20.1 shows changes in diel (A_{24h}) and nocturnal (A_d) carbon balances of outer-canopy leaves of epiphytic and hemiepiphytic individuals of *C. uvitana*, growing side by side within the crown of a *Ceiba pentandra*, from January 1991 to January 1992. Also shown are daily levels of PPFD incident on the leaves and the weekly distribution of rainfall. Dark CO_2 fixation contributed significantly to diel carbon gain on most days, although CO_2 fixation during the day usually exceeded CO_2 fixation at night. The relationship between daily gas-exchange patterns and microclimatic conditions was analyzed by dividing the study year into four periods (Fig. 20.2). The division took into account the phenology of the host tree which was leafless from October 1991 to January 1992 (Zotz and Winter 1994c):

Period 1: dry season; *C. uvitana* shaded by the host tree (January to April 1991).
Period 2: wet season; *C. uvitana* shaded by the host tree (May to September 1991).
Period 3: wet season; *C. uvitana* not shaded by the host tree (mid-October to mid-December 1991).
Period 4: dry season; *C. uvitana* not shaded by the host tree (mid-December 1991 to mid-January 1992).

In hemiepiphytic individuals of *C. uvitana*, light was the main climatic factor determining the daily 24-h carbon gain and the expression of CAM. There was almost no seasonality in water potential and turgor pressure (derived from psychrometric measurements of water potential and osmotic pressure) of leaves (Zotz and Winter 1994b). The leaf-shedding of *Ceiba* (Periods 3 and 4) led to about twofold increases in daily PPFD and resulted in markedly increased

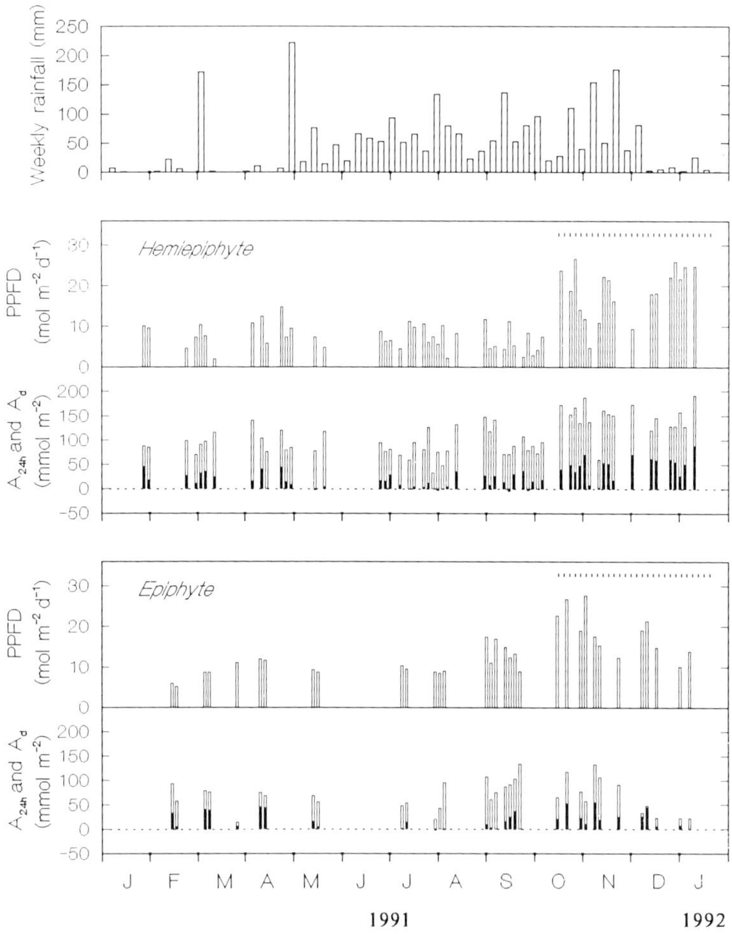

Fig. 20.1. Seasonal changes in weekly rainfall, daily integrated PPFD and CO_2 gas exchange in leaves of hemiepiphytic and epiphytic *C. uvitana*. A_{24h} daily (24-h) CO_2 balance (*open bars*); A_d nocturnal CO_2 balance (*closed bars*). The *dotted lines* in the panels showing PPFD indicate the time when the host tree was without foliage. (After Zotz and Winter 1994a,b)

carbon gain during both daytime and nighttime (Fig. 20.2). The leaf shedding also resulted in the decrease of minimum temperatures by about 2.5 °C. It is known that such temperature decreases at night may enhance net dark CO_2 fixation of CAM plants (Winter 1985). CAM activity was highest during Period 4 (early dry season, *C. uvitana* not shaded by *C. pentandra*), both in terms of nocturnal carbon gain (60 mmol CO_2 m^{-2} $night^{-1}$) and overnight organic acid accumulation (368 mmol H^+ m^{-2} $night^{-1}$) (Table 20.1). The ratio nighttime: daytime CO_2 uptake reached 1:1.4.

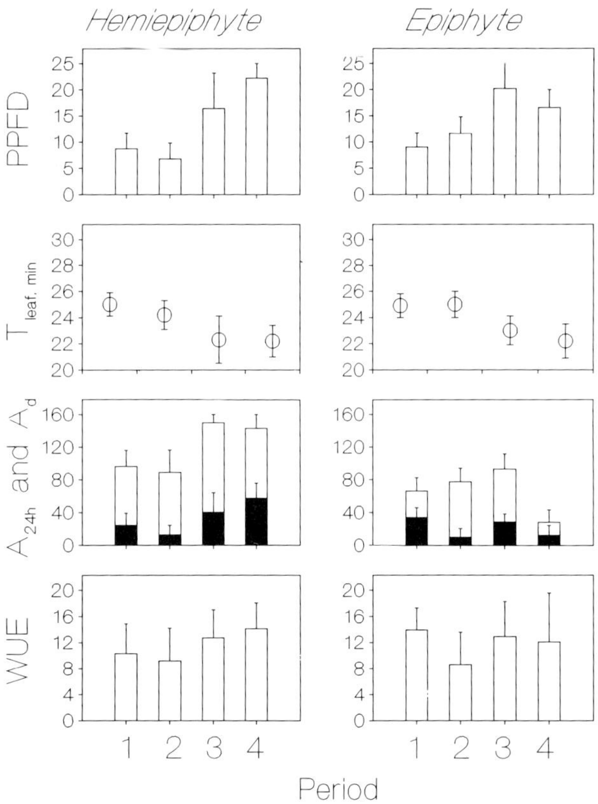

Fig. 20.2. Comparison of microclimatic conditions [*PPFD* (μmol m^{-2} s^{-1}) and $T_{leaf, min}$ minimum leaf temperature at night (°C)], 24-h (*open bars*) and nocturnal carbon gain (*closed bars*) (mmol m^{-2}), and 24-h water-use efficiency (*WUE* mmol CO_2 mol^{-1} H_2O) in leaves of epiphytic and hemiepiphytic *C. uvitana* during four periods from January 1991 through January 1992. The four periods are defined in the text (Sect. 20.2). Note that the duration of these periods differs substantially (period 1 = 4 months, period 2 = 5 months, period 3 = 2 months, period 4 = 1 month). Values are means ± SD. For the hemiepiphyte, n = 13, 26, 11, and 7 for periods 1, 2, 3, and 4, respectively. For the epiphyte, n = 7, 14, 7, and 5 for periods 1, 2, 3, and 4, respectively. (After G. Zotz and K. Winter 1994a, b; and previously unpubl. data)

In the epiphytic plants of *C. uvitana*, overall net carbon gain and nocturnal increase in organic acids were lower than in hemiepiphytic plants and CAM activity was clearly influenced by water availability. Turgor pressure occasionally decreased to close to zero after several weeks without rainfall during the dry season (Zotz and Winter 1994a). CAM was most pronounced during period 1, i.e. for much of the dry season except its early part (Fig. 20. 2). During period 1, mean values for nocturnal carbon gain and organic-acid accumulation were 30 mmol m^{-2} s^{-1} and 308 mmol H^+ m^{-2} night $^{-1}$, respectively, and the ratio nighttime: daytime CO_2 uptake was 1:1. Responses of diel patterns of net CO_2 exchange to

Table 20.1. Titratable acidity content (mmol $H^+ m^{-2}$) in hemiepiphytic and epiphytic *C. uvitana* from January 1991 to January 1992. Samples were collected at regular intervals during the indicated periods at dusk (H^+ 18.00 h) and the following dawn (H^+ 06.00 h). Nocturnal acidification (ΔH^+) was estimated as the difference. Data are means $\pm$ SD

	Hemiepiphyte				Epiphyte			
Period	n	H^+ 18.00	H^+ 06.00	ΔH^+	n	H^+ 18.00	H^+ 06.00	ΔH^+
(1) Jan–April 1991	6	122 ± 62	296 ± 71	174 ± 111	5	93 ± 77	401 ± 66	308 ± 114
(2) May–Sept. 1991	15	193 ± 105	303 ± 70	110 ± 99	8	240 ± 130	300 ± 88	59 ± 53
(3) Mid-Oct.–Mid-Dec. 1991	7	168 ± 86	421 ± 99	254 ± 86	3	238 ± 153	434 ± 48	195 ± 115
(4) Mid-Dec. 1991–Mid-Jan. 1992	5	121 ± 46	489 ± 56	368 ± 52	3	37 ± 12	273 ± 106	236 ± 116

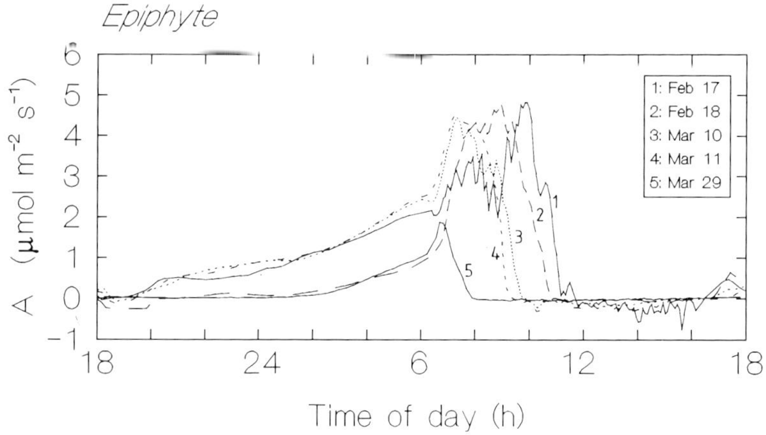

Fig. 20.3. Daily courses of net CO_2 exchange (A) of leaves of an epiphytic *C. uvitana* in response to progressive drought during the middle of the dry season, measured from 17 February to 29 March, 1991. After almost 3 weeks without rain (29 March), daytime net CO_2 uptake was limited to 1 h. (G. Zotz and Winter, previously unpubl. data)

increasing drought stress during the dry season are shown in Fig. 20.3. Between 17 February and 29 March, both the magnitude and duration of daytime CO_2 fixation gradually decreased. Daytime CO_2 fixation was eventually confined to a small peak shortly after dawn (March 29), whilst dark CO_2 fixation was much less affected by drought.

The CAM minimum occurred during the first part of the wet season (Period 2), when nocturnal net CO_2 uptake accounted for only, on average, 12.9% and 13.2% of 24-h carbon gain in hemiepiphytic and epiphytic *C. uvitana*, respectively. Throughout the whole study year, the day-to-day variation in dark versus light CO_2 fixation was small. Comparisons of nocturnal CO_2 uptake of individual leaves on 17 pairs of consecutive days showed that the mean difference in nocturnal carbon gain (A_d) between days was 15.4 $\pm$ 8.1 mmol CO_2 per night. The percentage change in the ratio A_d:A_{24h} averaged 12.8 $\pm$ 8.3% from one day to the next.

20.3 Short-Term Changes in the Expression of CAM

Extremely rapid changes between CAM and C_3 patterns of daily net CO_2 exchange were only observed once in the field, during an unusual rainfall event in the dry season 1991 (Zotz and Winter 1993a). Figure 20.4 shows 3 consecutive days of net CO_2 exchange of the leaf of an hemiepiphytic individual. The first 2 days were without precipitation, and net CO_2 uptake occurred at night, in the early morning and in the late afternoon. During 36 h of intense rainfall, temperatures declined and incident PPFDs stayed low, not exceeding 50 μmol

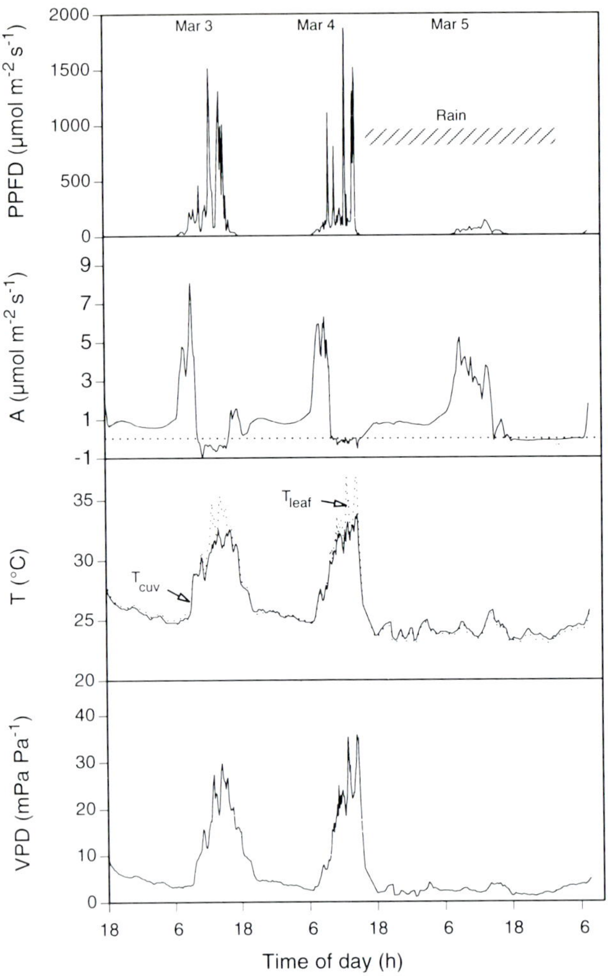

Fig. 20.4. Leaf gas exchange of a hemiepiphytic *C. uvitana* on 3 consecutive days beginning 18.00 h on 2 March 1991. Shown are daily changes in *PPFD*, net CO_2 exchange (*A*), temperature inside the leaf cuvette (= ambient air temperature), leaf temperature, and the leaf-to-air vapour pressure difference (*VPD*). (After Zotz and Winter 1993a; reprinted by permission of the American Society of Plant Physiologists)

m^{-2} s^{-1}. Nocturnal net CO_2 uptake was completely abolished and the diel pattern of gas exchange closely resembled that of a C_3 plant (Fig. 20.4, 5 March). We were able to reproduce a similar CAM-C_3 shift in the laboratory by subjecting leaves to similar decreases in light and temperature.

Like *Portulacaria afra* and certain other CAM species (Ting and Hanscom 1977; Ting and Rayder 1982), fully mature leaves of *C. uvitana* when exhibiting little CAM activity also contain high levels of organic acids day and night

(Table 20.1). There is a close negative correlation between the content of organic acids at dusk and the degree of nocturnal acidification (Zotz and Winter 1993a), i.e. leaves performing increased CAM activity do not necessarily show elevated levels of titratable acidity at dawn but lower levels at dusk. Conditions that reduce daily carbon gain by decreasing the availability of external CO_2 (e.g. high leaf-air water-vapour pressure differences, low ambient CO_2 partial pressures) probably initiate consumption of stored organic acids for photosynthetic CO_2 reduction and provide favourable conditions for dark CO_2 fixation during the following night. Conversely, low light and moderate temperature conditions as during the rainstorm on 5 March (Fig. 20.4) not only inhibited acid consumption during daytime; tissue acidity even increased slightly during the daytime, which made conditions unfavourable for dark CO_2 fixation. Thus, the organic acid status of the leaves during the light period seems to play a key role in regulating short-term changes in the magnitude of CAM in *C. uvitana*. These short-term changes may not involve PEPC synthesis and degradation.

20.4 The Effect of Leaf Ontogeny

The seasonal changes in CO_2 exchange patterns outlined above are experienced by individual leaves and are not the result of different sets of foliage. Outer-canopy leaves of hemiepiphytic individuals live for about 12 months (Zotz and Winter 1994b). Ontogenetic influences on leaf CO_2 exchange are mainly observed during the first 2 to 3 months of leaf development. Figure 20.5 depicts diel courses of net CO_2 exchange of a single leaf that was formed during the transition from wet to dry season in December 1990. It reached its final size in late January 1991 and was shed in early December 1991. Consistent with observations on other CAM species (Winter 1985), young leaves of *C. uvitana* show a C_3-type pattern of net CO_2 exchange. After 2 months (31 Jan., 1991), net CO_2 uptake still occurred exclusively during the daytime, although slight CAM activity was indicated by the absence of respiratory CO_2 loss for most of the night and by the small nocturnal increase in titratable acidity. In leaves older than 2 to 3 months, ontogenetic effects on CAM activity were negligible, and all changes in gas exchange were correlated with the environmental changes discussed in Section 20.1. After 5 months, at the end of the dry season (12 April, 1991), the leaf showed substantial rates of dark CO_2 fixation. Nocturnal carbon gain was markedly reduced during the first 6 months of the wet season, when *C. uvitana* was shaded by the leaves of *Ceiba* (19 July and 18 September, 1991). Thereafter (13 November), PPFD incident on *C. uvitana* increased owing to the leaf-shedding of *Ceiba* and dark CO_2 fixation was increased.

20.5 Correlation Between A_{max} and 24-h Carbon Gain

A previous study on diel courses of net CO_2 exchange in outer-canopy leaves of several C_3 species on Barro Colorado Island revealed a linear relationship

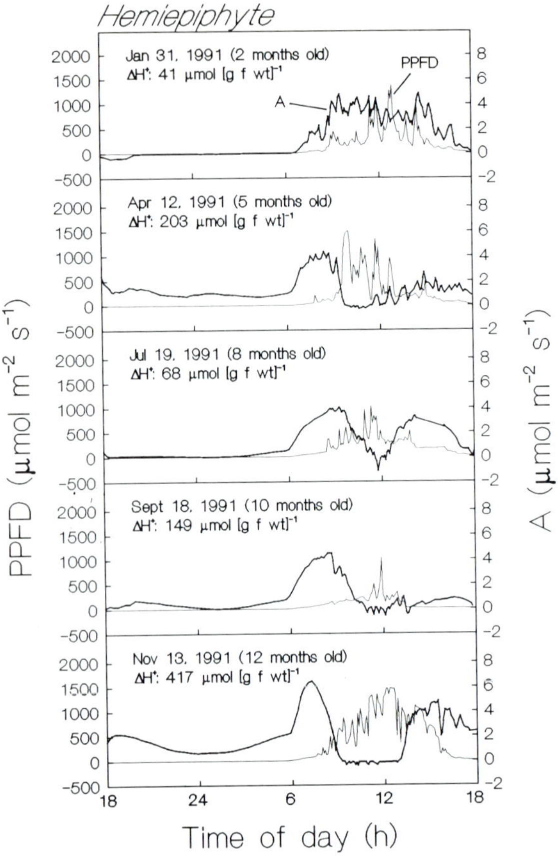

Fig. 20.5. Daily courses of net CO_2 exchange (A) and $PPFD$ over the lifetime of an outer-canopy leaf of a hemiepiphytic individual of *C. uvitana*. The same leaf was used for all measurements. It attained its full size in January 1991, and was shed about 12 months later in the first half of December 1991. Nocturnal acidification (ΔH^+ content of titratable acidity at dawn minus content of titratable acidity at dusk) was determined in the leaf opposite to the one used for gas-exchange measurements or on comparable leaves. (G. Zotz and K. Winter, previously unpubl. data)

between the daily carbon gain (A_{24h}) and the maximum rate of net CO_2 uptake (A_{max}): $A_{24h} = 19.2A_{max} - 7.7$ (Zotz and Winter 1993b). The same analysis for 127 24-h courses of gas exchange in epiphytic and hemiepiphytic *C. uvitana*, including outer-canopy and shaded inner-canopy leaves, yielded an essentially identical relationship: $A_{24h} = 19.8A_{max} - 12.9$ (Fig. 20.6). This is an extraordinary finding given that the slope of the A_{max}/A_{24h} relationship is influenced by the time per day a leaf operates at or near A_{max}. For example, cotton plants kept in the laboratory under 12-h light/12-h dark cycles and exposed to constant and high levels of PPFD during the light periods exhibited a much steeper slope of 24-h carbon

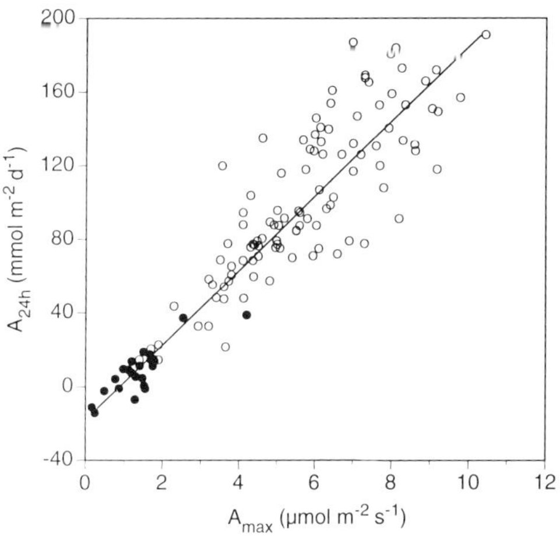

Fig. 20.6. Integrated 24-h carbon gain (A_{24h}) versus the maximum rate of net CO_2 assimilation (A_{max}) in epiphytic and hemiepiphytic *C. uvitana*. *Closed symbols* shade leaves; *open symbols* outer-canopy leaves. $A_{24h} = 19.8A_{max} - 12.9$. $r^2 = 0.82$, $p = 0.001$. (G. Zotz and K. Winter, previously unpubl. data)

gain versus A_{max} than any of the plants studied in situ on Barro Colorado Island (Königer and Winter 1993). Furthermore, a steeper slope than the one reported here was obtained in field investigations on a mediterranean shrub, *Arbutus unedo*, in Portugal (Beyschlag et al. 1987). In the latter study, leaves used for gas-exchange measurements were not maintained in their natural orientation but in a horizontal position, a treatment which may have increased 24-h carbon gain due to improved light interception. Compared to the C_3 plants on Barro Colorado Island, the period of CO_2 uptake at or near A_{max} in the C_3-CAM plant *C. uvitana* was generally shorter and confined to the early morning, but this disadvantage was evidently offset by the contribution of dark CO_2 fixation to 24-h carbon gain. Since A_{max} can be readily determined from instantaneous measurements of net CO_2 exchange (Zotz and Winter 1993b), this relationship should greatly facilitate attempts to scale from A_{max} measurements to daily and annual carbon balances of leaves and, eventually, whole canopies.

20.6 Summary and Conclusions

Under natural tropical conditions in the course of a year, fully developed outer-canopy leaves of the C_3-CAM plant *C. uvitana* derived, on 75% of 104 measuring days, between 5 and 50% of the daily carbon via dark CO_2 fixation. On 25% of the measuring days, the contribution of dark CO_2 fixation was either negligible or above 50%. The flexibility in engaging in the C_3 or CAM option of carbon assimilation is thus substantial. It remains to be seen, however, whether the physiological plasticity of *C. uvitana* and other species of *Clusia* with "C_3/CAM-intermediate" characteristics exceed that of terrestrial succulent-leaved CAM species from other genera. For example, in none of the species of

Kalanchoë, which have been classified as obligate CAM plants, has 24-h CO_2 exchange been thoroughly studied in situ throughout a whole year.

In hemiepiphytic individuals of *C. uvitana*, seasonal changes in PPFD markedly modulate total carbon gain and the expression of CAM, whilst in epiphytic individuals carbon gain and CAM activity are also strongly dependent on water availability. Rapid, environmentally induced changes between C_3- and CAM-type gas exchange patterns rarely occur under these natural tropical conditions. The lowest CAM activity is observed under conditions of reduced light levels during the wet season. The C_3 character of carbon assimilation is further increased at a whole-plant level, because about ten times more leaves per month develop during the wet than during the dry season (Zotz and Winter 1994b). During the first 2 months, these newly developed leaves exhibit net CO_2 uptake exclusively during daytime.

The linear relationship between A_{max} and diel carbon gain in the C_3-CAM plant *C. uvitana* extends previous findings with several rain-forest C_3 species (Zotz and Winter 1993b). These investigations with *C. uvitana* corroborate our earlier proposal that the A_{max}/A_{24h} relationship can be used to extrapolate from spot measurements of photosynthesis to daily and annual carbon balances and to scale up from leaves to canopies. Linear relationships between A_{max} and both diurnal and nocturnal carbon gain have also recently been found in the CAM plant *Kalanchoë pinnata* (Winter and Engelbrecht 1994).

Acknowledgements. We are grateful to T. Kursar and J. Wright for critical comments on the manuscript.

References

Ball E, Hann J, Kluge M, Lee HSJ, Lüttge U, Orthen B, Popp M, Schmitt A, Ting IP (1991a) Ecophysiological comportment of the tropical CAM-tree *Clusia* in the field. I. Growth of *Clusia rosea* Jacq. on St. John, US. Virgin Islands, Lesser Antilles. New Phytol 117: 473–481

Ball E, Hann J, Kluge M, Lee HSJ, Lüttge U, Orthen B, Popp M, Schmitt A, Ting IP (1991b) Ecophysiological comportment of the tropical CAM-tree *Clusia* in the field. II. Modes of photosynthesis in trees and seedlings. New Phytol 117: 483–492

Beyschlag W, Lange OL, Tenhunen JD (1987) Photosynthese und Wasserhaushalt der immergrünen mediterranen Hartlaubpflanze *Arbutus unedo* L. im Jahresverlauf am Freilandstandort in Portugal II. Charakteristische Gaswechselparameter von CO_2-Aufnahme und Transpiration. Flora 179: 399–420

Borland AM, Griffiths H, Maxwell C, Broadmeadow MSJ, Griffiths NM, Barnes JD (1992) On the ecophysiology of the Clusiaceae in Trinidad: expression of CAM in *Clusia minor* L. during the transition from wet to dry season and characterization of three endemic species. New Phytol 122: 349–357

Croat T (1978) Flora of Barro Colorado Island. Stanford University Press, Stanford

Franco AC, Ball E, Lüttge U (1990) Patterns of gas exchange and organic acid oscillations in tropical trees of the genus *Clusia*. Oecologia 85: 108–114

Franco AC, Ball E, Lüttge U (1991) The influence of nitrogen, light and water stress on CO_2 exchange and organic acid accumulation in the tropical C_3-CAM tree, *Clusia minor*. J Exp Bot 42: 597–603

Franco AC, Ball E, Lüttge U (1992) Differential effects of drought and light levels on accumulation of citric and malic acids during CAM in *Clusia minor*. Plant Cell Environ 15: 821–829
Haag-Kerwer A, Franco AC, Lüttge U (1992) The effect of temperature and light on gas exchange and acid accumulation in the C_3-CAM plant *Clusia minor* L. J Exp Bot 43: 345–352
Hammel BE (1986) New species of Clusiaceae from Central America with notes on *Clusia* and synonymy in the tribe Clusieae. Selbyana 9: 112–120
Königer M, Winter K (1993) Reduction of photosynthesis in sun leaves of *Gossypium hirsutum* L. under conditions of high light intensities and suboptimal leaf temperatures. Agronomie 13: 659–668
Lee HSJ, Schmitt AK, Lüttge U (1989) The response of the C_3-CAM tree, *Clusia rosea*, to light and water stress. II. Internal CO_2 concentration and water use efficiency. J Exp Bot 40: 171–179
Schmitt AK, Lee HSJ, Lüttge U (1988) The response of the C_3-CAM tree, *Clusia rosea*, to light and water stress. I. Gas exchange characteristics. J Exp Bot 39: 1581–1590
Ting IP, Hanscom Z (1977) Induction of acid metabolism in *Portulacaria afra*. Plant Physiol 59: 511–514
Ting IP, Rayder L (1982) Regulation of C_3 to CAM shifts. In: Ting IP, Gibbs M (eds) Crassulacean acid metabolism. American Society of Plant Physiologists, Rockville, pp 193–207
Tinoco Ojanguren C, Vásquez-Yánes C (1983) Especies CAM en la selva húmeda tropical de los Tuxtlas, Veracruz. Bol Soc Bot Méx 45: 150–153
Todzia C (1986) Growth habits, host tree species, and density of hemiepiphytes on Barro Colorado Island, Panama. Biotropica 18: 22–27
Windsor DM (1990) Climate and moisture variability in a tropical forest: long-term records from Barro Colorado Island, Panamá. Smithsonian Institution, Washington, DC
Winter K (1985) Crassulacean acid metabolism. In: Barber J, Baker NR (eds) Photosynthetic mechanisms and the environment. Elsevier, Amsterdam, pp 329–387
Winter K, Engelbrecht B (1994) Short-term CO_2 responses of light and dark CO_2 fixation in the crassulacean acid metabolism plant *Kalanchoë pinnata*. J Plant Physiol 144: 462–467
Winter K, Zotz G, Baur B, Dietz K-J (1992) Light and dark CO_2 fixation in *Clusia uvitana* and the effects of plant water status and CO_2 availability. Oecologia 91: 47–51
Zotz G, Winter K (1993a) Short-term regulation of crassulacean acid metabolism activity in a tropical hemiepiphyte, *Clusia uvitana*. Plant Physiol 102: 835–841
Zotz G, Winter K (1993b) Short-term photosynthesis measurements predict leaf carbon balance in tropical rain-forest canopy plants. Planta 191: 409–412
Zotz G, Winter K (1994a) Annual carbon balance and nitrogen use efficiency in tropical C_3 and CAM epiphytes, New Phytol 126: 481–492
Zotz G, Winter K (1994b) A one-year study on carbon, water and nutrient relationships in a tropical C_3-CAM hemi-epiphyte, *Clusia uvitana* Pittier. New Phytol 127: 45–60
Zotz G, Winter K (1994c) Photosynthesis of a tropical canopy tree, *Ceiba pentandra*, in a lowland rain forest in Panama. Tree Physiol 14: 1281–1301

21 Crassulacean Acid Metabolism in the Genus *Kalanchoë*: Ecological, Physiological and Biochemical Aspects

M. KLUGE[1] and J. BRULFERT[2]

> *"..It would be more accurate to regard the Kalanchoës as plants poorly equipped to struggle against other species, but which are very adaptable as regards the physical conditions of their environment..... individualistic and independent, they only subsist in hostile places for which no other plants would vie with them. It matters little to them if this hostility is due to a lack of water, a lack of humus, to elevation, or to the nature of the soil which does not hold together and erodes with the rain. .... When planted in a garden, they easily overtake the paths and the neighbouring flower-beds, but seldom make it over fences. This is because there is grass on the other side, against which they are helpless".*

P. Boiteau (extract from an unpublished manuscript on the genus *Kalanchoë*, translated by R. Greenberg)

21.1 Introduction

Crassulacean acid metabolism (CAM) was first described in a *Kalanchoë*[3] species (*K. pinnata* Pers. = *Bryophyllum calycinum* Salisb.; for a review see Kluge and Ting 1978). Moreover, many of the fundamental insights into the biochemistry and physiology of CAM have derived from studies conducted on species of this genus. Certain species of *Kalanchoë*, such as *K. daigremontiana* Hamet et Perr., *K. fedtschenkoi* Hamet et Perr. and *K. tubiflora* Hamet have been studied so intensively that they have formed the basis for many generalizations on the mechanisms and ecological significance of CAM.

[1]Institut für Botanik, Technische Hochschule Darmstadt, Schnittspahnstr. 10, D-64287 Darmstadt, Germany

[2]Laboratoire de Biochimie Fonctionelle des Membranes Végétales, CNRS, F-91198 Gif-sur-Yvette, France

[3]*Editors' note*: The genus *Kalanchoë* Adanson is usually considered to include *Bryophyllum* Salisbury, which is now reduced to the rank of a section or subgenus. The original diaeresis over the final letter in *Kalanchoë* – the name being derived from a Greek root – is permissible under Article 60.6 of the International Code of Botanical Nomenclature (Greuter et al. 1994) and can be retained

Ecological Studies, Vol. 114
Winter/Smith (eds) Crassulacean Acid Metabolism

However, the validity of generalizations based on a limited number of species should be critically assessed. Practical considerations may be important in using *Kalanchoë* species as model systems, such as the ease of cultivation and of obtaining genetically identical experimental plant material by vegetative propagation. But the genus *Kalanchoë* comprises not only terrestrial herbaceous leaf succulents like the species above, but also other very different growth forms (for example relatively thin-leaved plants, tall xeromorphic perennial bushes, epiphytes and climbers) that inhabit edaphically diverse biotopes. Thus, it is obvious that restricting studies to a few selected species will not lead to a complete picture. This holds particularly if evolutionary aspects and ecological implications of CAM are the main focus of interest.

Being aware of this problem, we have begun to investigate the distribution and characteristics of CAM in the genus *Kalanchoë* more systematically and on a comparative scale. In the present chapter we will review some recent results of this approach. Our investigations were greatly stimulated by three travels to Madagascar where most of the *Kalanchoë* species are naturally abundant and where we could directly experience the impressive diversity and ecological flexibility of these plants.

21.2 Results of $\delta^{13}C$ Surveys

21.2.1 CAM in Relation to Intragenic Taxonomy and Growth Forms

First we were interested in determining whether CAM is a phenomenon common to all species of the genus *Kalanchoë* and whether its characteristics coincide with taxonomic divisions within the various taxa of the genus. This type of question can be answered reliably by the analysis of the stable carbon-isotope ratios (Ehleringer and Rundel 1989). To the best of our knowledge none of the *Kalanchoë* species possesses Kranz anatomy. So the occurrence of C_4 photosynthesis can be ruled out in this genus. Thus, the entire range of $\delta^{13}C$ values observed can be discussed in terms of C_3 photosynthesis and CAM.

Neither amongst the 51 Malagasy nor the 25 African *Kalanchoë* species investigated have we detected an obligate C_3 plant (Kluge et al. 1991, 1993). Since our screening to date has comprised 76 of the approximately 130 known *Kalanchoë* species, it is reasonable to propose that all species in this genus are obligate or facultative CAM plants. Moreover, the considerable interspecific variability in carbon-isotope ratios we have found (Kluge et al. 1991, 1993) are consistent with the results from physiological studies (for reviews see for example Kluge and Ting 1978; Osmond 1978; Winter 1985; Lüttge 1987) showing in many *Kalanchoë* species a high phenotypic variability in CAM expression, with water relations, light intensity, photoperiod and the diel temperature regime being the major modulating factors.

In some species, CO_2 fixation proceeds entirely at night, independently of external factors, whereas other species show a high proportion of CO_2 uptake during the day, or even facultative CAM. In the Malagasy *Kalanchoë* species all types of CAM expression can be found, whereas in the species of continental Africa CAM with predominantly nocturnal CO_2 uptake is the rule (Kluge et al. 1993).

As is true for *Aeonium* (Pilon-Smits et al. 1992; see also Chap. 23), the interspecific variability of CAM expression in the genus *Kalanchoë* correlates with taxonomic divisions within the genus. According to P. Boiteau (unpubl.), the genus *Kalanchoë* can be divided into 3 sections (I. *Kitchingia*, II. *Bryophyllum*, III. *Eukalanchoë*) and 15 groups (Kluge et al. 1993). We found that the phenotypically most flexible modes of CAM, i.e. those with a high proportion of CO_2 uptake occurring during the day, can be attributed mainly to the phylogenetically more ancestral taxa, namely to all groups of the section *Kitchingia* and to certain groups of the section *Bryophyllum*. CAM with predominant dark CO_2 fixation, combined with low phenotypic flexibility, seems to be the typical mode of photosynthesis in the more derived taxa, i.e. in the majority of the groups in the sections *Bryophyllum* and *Eukalanchoë*.

There is also a clear correlation between the expression of CAM and plant morphology. Those *Kalanchoë* species showing CAM with a high degree of CO_2 fixation during the day are relatively thin-leaved and have generally more mesomorphic leaf anatomy, whereas full CAM is associated with pronounced succulence. Species of the latter type are often tall, bush-shaped perennial plants (Kluge et al. 1993). As shown by Queiroz and Brulfert (1982), short-day treatment can convert *K. blossfeldiana* Poelln. cv. Tom Thumb from a thin-leaved C_3 plant into a clearly leaf-succulent CAM plant.

21.2.2 Ecological Aspects

Madagascar provides excellent opportunities for studying the ecophysiological relevance of CAM for the genus *Kalanchoë*. The complex mosaic of climatic zones of this island, ranging from perhumid to arid areas, with associated gradients in temperature (Guillaumet 1984), leads to a similarly complex pattern of different ecological conditions.

Kalanchoë species occur in all climatic zones of Madagascar. Our survey of the long-term photosynthetic behaviour of these plants by $\delta^{13}C$ analysis (Kluge et al. 1991) revealed that the carbon-isotope ratios are clearly related to the climate of the individual sites. In the species from the high-rainfall regions, $\delta^{13}C$ values indicated that a large C_3 contribution to carbon gain was typical, whereas those from the arid regions indicated CAM with predominantly nocturnal CO_2 uptake.

The $\delta^{13}C$ values also mirror microclimatic differences. For instance, on the dry and hot "Inselberg" Mount Angavokely (Central Tableland of Madagascar) *K. campanulata* Baill. is abundant, a species with C_3-like $\delta^{13}C$ values (Kluge et al.

1991). This at first sight curious phenomenon can be explained by the fact that on this Inselberg the species is restricted to permanently shady and consequently humid microhabitats (gorges between the gneiss blocks, undergrowth of the relict forest), where the evaporative demand remains low and the stomata can be kept open during the day.

21.2.3 CAM Evolution in the Genus *Kalanchoë*

As outlined above, the pattern of CAM expression shown by a given species within the genus *Kalanchoë* mirrors its taxonomic position and can also be related to its growth form and habitat. This is further support for the view that the diversity of CAM found in the genus is not solely based on phenotypic flexibility, but rather has a strong genotypic component. The fact that in *Kalanchoë*, as in *Aeonium* (Pilon-Smits et al. 1992) and *Sedum* (Teeri 1982; see also Chap. 23), the occurrence of CAM together with CO_2 uptake mainly by the C_3 pathway correlates with the phylogenetically ancestral taxa, whereas in the more derived taxa CAM with predominantly nocturnal CO_2 uptake is common, supports the assumption that CAM has evolved from C_3 photosynthesis, presumably via intermediate modes of CAM performance (Medina 1974; Teeri 1982; Pilon-Smits et al. 1992). It is reasonable to assume that the evolution of full CAM from a C_3-dominated mode of photosynthesis was an important factor enabling species of *Kalanchoë* to occupy arid sites. This assumption is also consistent with the contention founded on taxonomic arguments (P. Boiteau, unpubl.) that the centre of adaptive radiation for the genus *Kalanchoë* is located in the moister regions of Madagascar, from where the plants have spread over the dry regions of the island and the eastern parts of continental Africa. The fact that most of the known *Kalanchoë* species are endemic to Madagascar (Rauh 1973; Jacobsen 1981) is further support for this view.

21.3 Experimental Approaches

21.3.1 Comparison of CAM Behaviour

The proposals on the ecophysiology of CAM derived from the $\delta^{13}C$ survey of plant material collected from natural habitats can be fully substantiated by measurements of gas exchange and other CAM parameters in the laboratory or in the field. This will be illustrated in the following by comparison of three species, namely *K. beharensis* Drake del Castillo, *K. miniata* Hilsenb. et Bojer, and *K. porphyrocalyx* Baillon.

Kalanchoë beharensis, a large bush-shaped perennial plant up to 3 m high, belongs to the section *Eukalanchoë* (P. Boiteau and R. Mannoni, quoted in Jacobsen 1981) and is endemic to the semi-arid south of Madagascar. The $\delta^{13}C$ values of $-12.80 \pm 2.4‰$ estimated in samples obtained from various sites suggest that this plant possesses CAM with CO_2 uptake proceeding entirely during the night (Kluge et al. 1992) and shows low phenotypic flexibility. Indeed, we observed that, regardless of whether the plants were naturally droughted for several months or had received 2 months of artificial daily watering, the CAM behaviour was identical: CO_2 uptake occurred entirely during the night and was quantitatively the same (Kluge et al. 1992). The amplitude of the diel acid fluctuation was also unaffected (Kluge et al. 1992). This is in contrast to the in situ behaviour of *Opuntia basilaris* (Szarek and Ting 1975), which after the onset of the dry season shows "CAM-idling", where no more CO_2 exchange with the atmosphere occurs and CAM operates exclusively at the expense of recycling respiratory CO_2 (Ting 1985). The ability of *K. beharensis* to maintain net CO_2 uptake may be connected with a relatively favourable water status of the photosynthetically active leaves (Kluge et al. 1992). As proposed by Kluge et al. (1992), the homeostasis of leaf water relations in *K. beharensis* may be achieved by massive water transfer out of the older leaves, which then wilt and are shed. Such water transfer has been observed in *Kalanchoë daigremontiana* (Schäfer and Lüttge 1987). It is reasonable to assume (but remains to be demonstrated) that the other bush-shaped *Kalanchoë* species abundant in the arid south of Madagascar (Jacobsen 1981) also behave like *K. beharensis*.

Compared with *K. beharensis*, the characteristics of CAM observed in *K. porphyrocalyx* and *K. miniata* are clearly different. These latter species belong to the first two taxonomic groups of the section *Bryophyllum* (P. Boiteau, unpubl.; see Kluge et al. 1993) and are thus members of the more ancestral taxa of the genus. They are both abundant on the Central Tableland of Madagascar, which is much less arid compared with the south where *K. beharensis* grows. The $\delta^{13}C$ values found in *K. porphyrocalyx* were $-23.7 \pm 4.4‰$ and in *K. miniata* $-19.3 \pm 6.4‰$ (see Kluge et al. 1991). Such values suggest that these species in their natural habitats may show considerable fixation of CO_2 uptake by the C_3 pathway. These predictions are supported by the results of comparative measurements of gas exchange and other CAM parameters in the laboratory (J. Brulfert and M. Kluge, unpubl.). As shown in Fig. 21.1, well-watered plants of both species take up external CO_2 mainly during the day, whereas nocturnal CO_2 uptake proceeds at low rates. In this state a slight diel fluctation of malic acid can be observed (data not shown). However, drought causes a strong suppression of CO_2 uptake during the day and causes an enhancement of nocturnal CO_2 fixation (Fig. 21.1) combined with a large increase in the amplitude of diel acid fluctuation. Upon rewatering, the former CO_2-exchange pattern is reestablished within a few hours, except that nocturnal CO_2 uptake remains somewhat higher compared with the level before application of water stress. The diel acid fluctuation is also reversed upon rewatering, but not phospho*enol*pyruvate carboxylase (PEPC) capacity, which remains higher than before the onset of drought (data not shown).

Even if at first sight *K. porphyrocalyx* and *K. miniata* seem to behave similarly with respect to their gas exchange, certain differences are apparent. Compared with *K. porphyrocalyx*, the gas-exchange transition in *K. miniata* upon drought

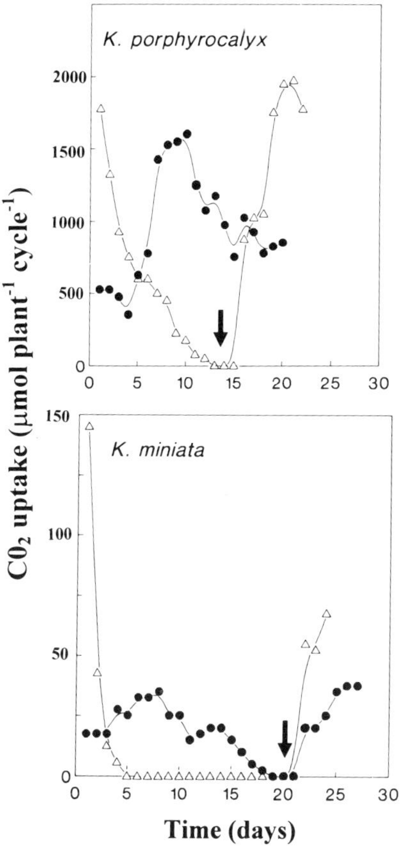

Fig. 21.1. Effects of drought on the CO_2 exchange of *K. porphyrocalyx* and *K. miniata* (data obtained by J. Brulfert at the phytotron of CNRS, Gif-sur-Yvette). External conditions: PPFD 280 μmol m^{-2} s^{-1}, 27 °C (day)/17 °C(night), relative air humidity 70% throughout the diel cycle, and daily watering. Drought was applied by withholding water and lowering the relative air humidity to 30% throughout the diel cycle. The diagrams show the changes in cumulative net CO_2 exchange during the night (*closed symbols*) and during the day (*open symbols*) of the diel cycles throughout the duration of the drought/reirrigation experiment. Onset of drought was at zero time. Time point of rewatering is indicated by the *arrows*. For experimental reasons the data were expressed on a per-plant basis

from the C_3-dominated to a CAM mode without CO_2 fixation during the daytime proceeds more rapidly and more completely (Fig. 21.1). This may be associated with differences in plant water relations. For instance, when droughted in a climate chamber under identical conditions, 15 days after cessation of watering *K. porphyrocalyx* still remains turgid, whereas *K. miniata* wilts. This lower dehydration resistance in *K. miniata* might be the result of epidermal properties, a view supported by the fact that detached leaves of this species lose water much more rapidly than those of *K. porphyrocalyx*: 2 days after detachment the relative leaf water content of *K. miniata* had dropped to about 50%,but remained at 80% in *K. porphyrocalyx*. Even in the strongly wilted state, *K. miniata* still performs CAM with net CO_2 uptake during the night (cf. Fig. 21.1). This implies that full turgor of the mesophyll cells is not a prerequisite for the performance of CAM. Indeed, leaves of *K. daigremontiana* will perform CAM when the mesophyll cells are at essentially zero turgor (Rygol et al. 1987).

The lower dehydration resistance of *K. miniata* was an unexpected finding. This species is a plant preferentially occupying dry and exposed microhabitats and growing frequently in shallow soils covering rocks. With such a habitat preference one might expect well-developed dehydration resistance in *K. miniata*. However, as outlined above, the contrary holds true in this species. This apparent paradox can be explained if the life cycle of the plant is taken into account. *K. miniata* retains its leaves and therefore grows mainly during the rainy season. CAM, especially in its highly flexible mode seen in *K. miniata*, may be helpful in maintaining net carbon gain during shorter periods of limited water availability that may occur during the rainy season. Towards the end of the moist summer, *K. miniata* enters the reproductive phase by developing flowers and seeds while most of the leaves are shed. The plant then survives the months of the dry season in the form of resting buds located near the ground, from which new shoots subsequently develop. Thus, *K. miniata* avoids the dangerous 6 to 7 months of the dry season by behaving analogously to geophytes.

K. porphyrocalyx shows a very different life form, growing as a shade-tolerant epiphyte in the forests of the mild humid mountain areas of Madagascar. In contrast to the quasi-annual *K. miniata*, *K. porphyrocalyx* is a true perennial that retains its full foliage throughout the year, even during the dry season when it may occasionally be exposed to water deficits. Thus, the superior dehydration resistance we observed in *K. porphyrocalyx* could be important for maintenance of favourable plant water status at times when rainfall is less frequent. In *K. porphyrocalyx*, CAM can be interpreted as an auxiliary mechanism allowing net carbon gain during periods of reduced water availability, whereas C_3 photosynthesis dominates the acquisition of external carbon during the time when water is freely available.

21.3.2 Regulation of CAM: Studies with Epiphytic Species

We have found that the epiphytes *K. porphyrocalyx* and *K. uniflora* are particularly suitable species for studying certain mechanisms behind the regulation of the diel CAM cycle. We will consider here a typical example.

Kalanchoë uniflora (Stapf) Hamet, an epiphytic species with highly flexible CAM (Schäfer and Lüttge 1986, 1988), has been used to investigate interrelationships between the diel fluctuations of PEPC properties and the actual patterns of carbon flow. This species develops hanging shoots that reach a length of several decimeters and bear many pairs of small leaves of similar size. Therefore it is easy to collect numerous very similar leaf samples from a single shoot. At high air humidity each leaf node develops minute functional roots. Thus, by cutting the shoot at the internodes, it is possible to obtain individual CAM-performing systems consisting of a node with rootlets and a single pair of leaves. This system is ideal, for instance, for applying radioactive tracers via the roots while monitoring gas exchange and other parameters.

Using this *K. uniflora* system, Petra Maier in the laboratory of M. Kluge (Maier 1992) has been able to measure simultaneously the diel courses of CO_2 exchange, malic-acid level, incorporation of $^{14}CO_2$ under illumination and in darkness (together with the resulting label distribution among metabolites), and the properties of PEPC (activity, sensitivity towards the allosteric inhibitor malate, phosphorylation status). This is the first example of such a comprehensive measurement of these variables in a single system throughout complete diel cycles. Figure 21.2 shows that the leaves on the detached nodes perform CAM, with nocturnal net CO_2 uptake accompanied by malate accumulation. As expected from previous studies (Kluge et al 1980; Winter 1980), PEPC changed its regulatory properties: the enzyme was most sensitive to inhibition by malate during the day and less so at night (Fig. 21.2C). There was also a diel change in phosphorylation of the PEPC protein (Nimmo et al. 1984; Brulfert et al. 1986), as indicated by the degree of ^{32}P labelling after tracer incorporation in vivo (Fig. 21.2D). Highest labelling occurred at the beginning of the night when PEPC began to lose gradually its sensitivity towards malate and was thereby changed into the physiologically active form. However, during the remainder of the night the ^{32}P label in PEPC began to decrease while the enzyme became increasingly insensitive to malate. Since it is now widely accepted that low sensitivity of PEPC at night towards inhibition by malate is the consequence of protein phosphorylation (Nimmo et al. 1987; Kluge et al. 1988; see also Chap. 3), this result was unexpected and requires further investigation.

As Fig. 21.2 shows, there was a close correlation between the gradual diel changes of the regulatory properties of PEPC on the one hand, and the mode of CO_2 fixation on the other. During the day, when the sensitivity of PEPC towards malate was high, $^{14}CO_2$ incorporation under illumination was considerably higher than in samples labelled at the same time in darkness, and labelling in the light led to tracer incorporation preferentially into end products of the C_3 pathway (mainly into carbohydrates and phosphorylated compounds; Fig. 21.2B). This means that at this time of the diel cycle CO_2 fixation in the light was mediated by the C_3 pathway, while the C_4 pathway was suppressed. However, during the night, when PEPC was in its physiologically active form, $^{14}CO_2$ incorporation in illuminated samples was quantitatively the same as in darkened samples, and light $^{14}CO_2$ fixation preferentially labelled malate and some other products of the C_4 pathway. This means that at this time CO_2 fixation in illuminated samples was mediated mainly by PEPC, while the C_3 pathway was suppressed.

P. Maier's comprehensive experiment with *K. uniflora* shows that the diel changes in PEPC properties have consequences for directing the metabolic carbon flow of CAM. Since the C_3 pathway apparently has access to CO_2 only if

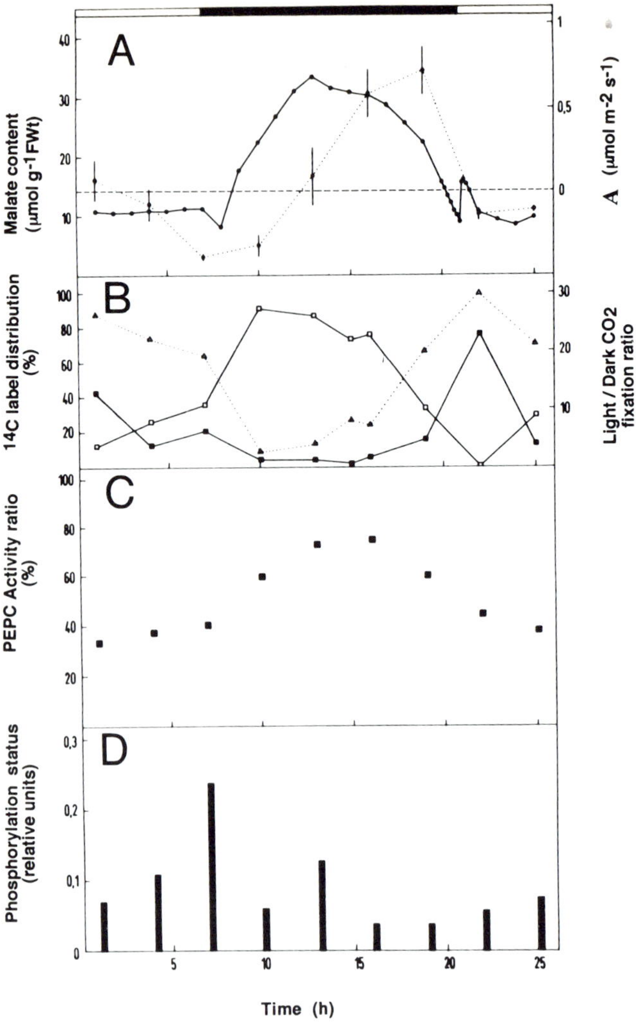

Fig. 21.2A–D. *K. uniflora*: Diel course of CAM in pairs of leaflets isolated from the donor plants by cutting the shoots between the nodes. All measurements were made at the same time from a population of samples derived from one donor plant. External conditions during the experiment: PPFD 250 μmol m^{-2} s^{-1}, 24 °C (day)/17 °C (night), relative humidity 70% (day and night). **A** Net CO_2 exchange (●—●) measured continuously by infrared gas analysis, and malate content per g fresh weight (*FWt*) of the leaves (●···●); **B** Results of $^{14}CO_2$ labelling. Four leaflets per sample were split with a razor blade parallel to the surface to overcome possible limitations of label uptake by stomatal resistance. At each time point, one sample was incubated with $^{14}CO_2$ in the light and a parallel sample in darkness. After 15 min of incubation the samples were killed, extracted and the total incorporation of ^{14}C together with its distribution among different fractions estimated according to Böcher and Kluge (1977). The data show the light/dark $^{14}CO_2$ fixation ratio (■) and label distribution after light $^{14}CO_2$ fixation (C_3-products △; C_4-products □); **C** Time course of activity ratio of PEPC. Extraction and assay of PEPC followed the procedure of Krüger and Kluge (1988). The enzyme activity was measured in parallel samples of the same crude but desalted extract by estimation of enzyme activity at pH 7.0 with 1 mM PEP in presence and absence of 0.1 mM L-malate. Activity ratio (%) = enzyme activity in the presence of 0.1 mM L-malate/enzyme activity in the absence of malate × 100 %; **D** Phosphorylation status of PEPC. The enzyme was labelled by uptake of $^{32}P_i$ into isolated nodes via the emergent rootlets. After 24 h of label incorporation, the leaflets were held for a further 24 h in unlabelled water. During the whole procedure, gas exchange was measured under the conditions described above. PEPC was extracted and analyzed for ^{32}P-incorporation as described by Kluge et al. (1988). The phosphorylation status (relative units) was evaluated by densitometric measurements of the PEPC bands on autoradiograms of the electrophoresis gels

PEPC is in its physiologically inactive form, this supports the view derived from many enzymological studies (see e.g. the reviews by Kluge and Ting 1978; Osmond 1978; Winter 1985; see also Chap. 3) that futile recycling of malate-derived CO_2 in CAM is prevented by regulation of PEPC.

21.4 Conclusions

Because of their biological flexibility, the members of the genus *Kalanchoë* have fascinated many plant scientists in addition to those involved in photosynthesis research. Among them the late Pierre Boiteau has to be mentioned, formerly Director of the Botanical Gardens Zimbazasa (Antananarivo, Madagascar) and one of the leading authorities on the Malagasy succulent flora. In an unfinished draft of a monograph of the genus *Kalanchoë* by P. Boiteau, we found an assessment describing the ecological diversity of these plants in a very illustrative manner, from which we quote a further extract here more extensively (translation from the French by R. Greenberg):

> "..... *It would be wrong to think that, by virtue of being succulent, the Kalanchoës are basically xerophytic plants; the number of Malagasy species undeniably adapted to life on arid land is by no means greater than those that are not. Nearly all species, no doubt, easily withstand some lack of water, even those which in their natural habitat never had to undergo this lack. .... Growing them is also the easiest thing in the world: all one has to do is to keep other plants away from them. They require hardly any other care and adapt to practically anything as long as this condition is respected. There are, however, a very small number of species that grow poorly and are thus tricky to cultivate. These are exceptions. Most Kalanchoë species are hard to bring down. When their flowers are cut and put into a vase, they develop roots and grow. They even continue to grow in herbarium sheets when they are thought to be dry. They have to be killed (by putting them into 5% formalin for example ...) if one wants to make dried specimens.....*".

Since in all *Kalanchoë* species the photosynthetic option is probably CAM, the wide inter- and intraspecific variability of CAM patterns found in the genus may be one of the reasons for their high biological adaptability. This is consistent with the interpretation of CAM as essentially a survival mechanism based on the superior water-use efficiency of photosynthetic carbon acquisition, rather than a mechanism to enhance photosynthetic productivity sufficient to overcome potential competitors (Kluge and Ting 1978).

Besides CAM, other adaptations also enable *Kalanchoë* species to colonize ecologically specialized habitats. The often very obvious and spectacular modes of vegetative propagation typical of many *Kalanchoë* species may be taken as an example of such an adaptation (Ravelomanana 1993). The different life-history strategies, discussed above for *K. porphyrocalyx* and *K. miniata*, are a further such example.

Lüttge (1987) has emphasized the great importance of comparative approaches for finding answers to the still-outstanding questions concerning CAM, particularly its ecological and evolutionary significance. It will be useful in the future to extend such comparative studies to many other genera and families of CAM plants. The availability of powerful screening methods, e.g. the estimation of stable-isotope ratios, or the techniques of molecular biology, can facilitate and thus stimulate such comparative research into CAM plants.

Acknowledgement. We thank Dr. Lucile Allorge-Boiteau for allowing us to quote from the unpublished manuscript of P. Boiteau.

References

Böcher M, Kluge M (1977) Der C4-Weg der C-Fixierung bei *Spinacia oleracea*. I. ^{14}C-Markierungsmuster suspendierter Blattstreifen unter dem Einfluß des Suspensionsmediums. Z Pflanzenphysiol 83: 347–361

Brulfert J, Vidal J, LeMarechal P, Gadal P, Queiroz O, Kluge M, Krüger I (1986) Phosphorylation-dephosphorylation process as a probable mechanism for the diurnal regulatory changes of phosphoenolpyruvate carboxylase in CAM plants. Biochem Biophys Res Commun 136: 151–159

Brulfert J, Güçlü S, Taybi T, Pierre J-N (1993) Enzymatic responses to water-stress in detached leaves of the CAM plant *Kalanchoë blossfeldiana* Poelln. Plant Physiol Biochem 31: 491–497

Ehleringer JR, Rundel PW (1989) Stable isotopes: history, units, and instrumentation. In: Rundel PW, Ehleringer JR, Nagy KA (eds) Stable isotopes in ecological research. Springer, Berlin Heidelberg New York, pp 1–15

Greuter W, Barrie FR, Burdet HM, Chaloner WG, Demoulin V, Hawksworth DL, Jørgensen PM, Nicolson DH, Silva PC, Trehane P, McNeill J (1994) International code of botanical nomenclature (Tokyo code). Koeltz Scientific Books, Königstein

Guillaumet J-L (1984) The vegetation: an extraordinary diversity. In: Jolly A, Oberle P, Albinac R (eds) Key environments: Madagascar. Pergamon Press, Oxford, pp 27–74

Jacobsen H (1981) Das Sukkulentenlexikon, 2nd edition. G. Fischer, Stuttgart

Kluge M, Ting IP (1978) Crassulacean acid metabolism. Analysis of an ecological adaptation. Springer, Berlin Heidelberg New York

Kluge M, Böcher M, Jungnickel G (1980) Metabolic control of crassulacean acid metabolism: evidence for diurnally changing sensitivity against inhibition by malate of PEP carboxylase in *Kalanchoë tubiflora* Hamet. Z Pflanzenphysiol 97: 197–204

Kluge M, Maier P, Brulfert J, Faist K, Wollny E (1988) Regulation of phosphoenolpyruvate carboxylase in crassulacean acid metabolism: in vitro phosphorylation of the enzyme. J Plant Physiol 133: 252–256

Kluge M, Brulfert J, Ravelomanana D, Lipp J, Ziegler H (1991) Crassulacean acid metabolism in *Kalanchoë* species collected in various climatic zones of Madagascar: a survey by $\delta^{13}C$ analysis. Oecologia 88: 407–414

Kluge M, Razanoelisoa B, Ravelomanana D, Brulfert J (1992) *In situ* studies of crassulacean acid metabolism in *Kalanchoë beharensis* Drake del Castillo, a plant of the semi-arid southern region of Madagascar. New Phytol 120: 323–334

Kluge M, Brulfert J, Lipp J, Ravelomanana D, Ziegler H (1993) A comparative study by $\delta^{13}C$-analysis of crassulacean acid metabolism (CAM) in *Kalanchoë* (Crassulaceae) species of Africa and Madagascar. Bot Acta 106: 320–324

Krüger I, Kluge M (1988) Diurnal changes in the regulatory properties of phosphoenolpyruvate carboxylase in CAM plants: are alterations in the quaternary structure involved? Bot Acta 101: 32–35

Lüttge U (1987) Carbon dioxide and water demand: crassulacean acid metabolism (CAM), a versatile ecological adaptation exemplifying the need for integration in ecophysiological work. New Phytol 106: 593–629

Maier P (1992) Die Regulation des Crassulaceen-Säurestoffwechsels: Untersuchungen zur Bedeutung der reversiblen Phosphorylierung der Phosphoenolpyruvat Carboxylase. Doctoral Dissertation, Technische Hochschule Darmstadt

Medina E (1974) Dark CO_2 fixation, habitat preference and evolution within the Bromeliaceae. Evolution 28: 677–686

Nimmo GA, Nimmo HG, Fewson CA, Wilkins MB (1984) Diurnal changes in the properties of phosphoenolpyruvate carboxylase in *Bryophyllum* leaves: a possible covalent modification. FEBS Lett 178: 199–203

Nimmo GA, Wilkins MB, Fewson CA, Nimmo HG (1987) Persistent circadian rhythms in the phosphorylation state of phosphoenolpyruvate carboxylase from *Bryophyllum fedtschenkoi* leaves and in its sensitivity to inhibition by malate. Planta 170: 408–415

Osmond CB (1978) Crassulacean acid metabolism: a curiosity in context. Annu Rev Plant Physiol 29: 379–414

Pilon-Smits EAH, 't Hart H, Maas JW, Meesterburrie JAN, Kreuler R, van Brederode J (1992) The evolution of crassulacean acid metabolism in *Aeonium* inferred from carbon isotope composition and enzyme activities. Oecologia 91: 548–553

Queiroz O, Brulfert J (1982) Photoperiod-controlled induction and enhancement of seasonal adaptation to drought. In: Ting IP, Gibbs M (eds) Crassulacean acid metabolism. American Society of Plant Physiologists, Rockville, pp 208–230

Rauh W (1973) Über die Zonierung und Differenzierung der Vegetation Madagaskars. Akademie der Wissenschaften und der Literatur, Mainz

Ravelomanana D (1993) Ecophysiologie des *Kalanchoë* de Madagascar: Le métabolisme acide des Crassulacées (CAM) en tant que processus physiologique adaptif. Thesis Docteur d'Etat, University of Antananarivo, Madagascar

Rygol J, Winter K, Zimmermann U (1987) The relationship between turgor pressure and titratable acidity in mesophyll cells of intact leaves of a crassulacean-acid-metabolism plant, *Kalanchoë daigremontiana* Hamet et Perr. Planta 172: 487–493

Schäfer C, Lüttge U (1986) Effects of water stress on gas exchange and water relations of a succulent epiphyte, *Kalanchoë uniflora*. Oecologia 71: 127–132

Schäfer C, Lüttge U (1987) Water translocation in *Kalanchoë daigremontiana* during periods of drought. Plant Cell Environ 10: 761–766

Schäfer C, Lüttge U (1988) Effects of high irradiances on photosynthesis, growth and crassulacean acid metabolism in the epiphyte *Kalanchoë uniflora*. Oecologia 75: 567–574

Szarek SR, Ting IP (1975) Physiological responses to rainfall in *Opuntia basilaris*. Am J Bot 62: 602–609

Teeri JA (1982) Photosynthetic variation in the Crassulaceae. In: Ting IP, Gibbs M (eds) Crassulacean acid metabolism. American Society of Plant Physiologists, Rockville, pp 244–259

Ting IP (1985) Crassulacean acid metabolism. Annu Rev Plant Physiol 36: 595–622

Winter K (1980) Day/night changes in the sensitivity of phosphoenolpyruvate carboxylase to malate during crassulacean acid metabolism. Plant Physiol 65: 792–796

Winter K (1985) Crassulacean acid metabolism. In: Barber J, Baker NR (eds) Photosynthetic mechanisms and the environment. Elsevier, Amsterdam, pp 329–387

22 Carbon- and Hydrogen-Isotope Discrimination in Crassulacean Acid Metabolism[1]

H. ZIEGLER

22.1 Introduction

Plants have different means of adapting their metabolic, especially their photosynthetic, activity to short- or long-lasting water stress. The CO_2-concentrating mechanism of C_4 photosynthesis, for example, allows carbon assimilation at low water cost. CAM in xerophytes – at least in plants with water-storage tissues – is an important adaptation in environments with intermittent water supply. The most effective protection against the detrimental effects of drought is the ability of plants ("resurrection plants", "poikilohydric plants") or parts of plants (spores, pollens, seeds, or fruits) to dry out without permanent injury and to restore their full metabolic capacity after rehydration. Interestingly, poikilohydric higher plants never show C_4 or CAM photosynthesis (Table 22.1).

The special application of the on-line monitoring of $^{13}C/^{12}C$ discrimination in CAM plants is discussed in Chap. 15. In the present chapter, possible relationships between the $^{13}C/^{12}C$ discrimination and the $^{1}H/^{2}H$ discrimination in CAM plants are considered.

22.2 Basic Principles

The basic source of deuterium in the organic material of an autotrophic plant (using water as a hydrogen source) is the medium water, and in terrestrial plants this means the soil water. The deuterium content of this soil water again depends mainly on its concentration in precipitation. Condensate from water vapour is enriched in deuterium, with the result that the remaining vapour will be increasingly depleted in deuterium as moisture is removed from the air. Since the absolute water content of the atmosphere increases with temperature, an increasing depletion of deuterium in precipitation will result as a consequence of decreasing temperature. This is termed the "climatic effect" (Schiegl 1970, 1972). The variations in deuterium content of natural waters caused by

---

Institut für Botanik und Mikrobiologie, Technische Universität München, Arcisstr. 21, D-80290 München, Germany

[1]Dedicated to Prof. Dr. W. Rüdiger, Munich, on the occasion of his 60th birthday

Ecological Studies Vol. 114
Winter/Smith (eds) Crassulacean Acid Metabolism

Table 22.1. Carbon-isotope ratios ($\delta^{13}C$ values) of resurrection plants

Species (Family)	Site	$\delta^{13}C$ (‰)
Myrothamnus flabellifolia Welw. (Myrothamnaceae)	Namibia, SE of Solitaire	−26.5
Chamaegigas intrepidus Dinter ex Heil (Scrophulariaceae)	Namibia, White Giraffe Farm	−26.1

condensation and evaporation lie between +100‰ and −400‰ with respect to the deuterium content of standard mean ocean water (SMOW; Craig 1961).[2]

As early as 1934 – 2 years after Urey et al. (1932) discovered deuterium – Washburn and Smith (1934 a,b) found gravimetrically with silica pycnometers that the leaf-tissue water of terrestrial plants was heavier than the soil water. It turned out later that the leaf-tissue water is enriched in deuterium, as well as in oxygen-18. This enrichment of deuterium in leaf-tissue water is not the result of a discrimination against HDO compared with H_2O during uptake by the roots or during transport of water in the xylem, but arises during transpiration from the leaves (Ziegler 1989). This is also true for CAM plants.

Bokhoven and Theeuwen (1956) were the first to describe a much lower deuterium content of organic plant material when compared with tissue water. As was shown by Estep and Hoering (1981) with microalgae, the deuterium content of the organic material of a plant depends on the deuterium content of the tissue (or medium) water only when plants are growing photoautotrophically. If the cells grow heterotrophically, the δD value of the organic material depends largely on the deuterium content of the food material. This is also true for heterotrophic tissues or organs in an autotrophic higher plant, e.g. for roots and seeds.

An additional fractionation occurs in metabolism when the products of pyruvate dehydrogenase (acetyl CoA) are involved. Lipids, for example, are always depleted in deuterium compared with source carbohydrates (Smith and Epstein 1970; Smith and Ziegler 1990). This has to be taken into account when lipid-rich plant material (e.g. some seeds, fruits, membrane-rich cells) is analyzed.

When we consider that the deuterium content of the organic material of a plant depends primarily on that of the tissue water of photosynthesizing organs, and further that the main discrimination against the heavier isotope probably occurs during light-mediated pyridine nucleotide reduction, it is not surprising that plants with different types of photosynthetic CO_2-fixation pathways differ in the deuterium contents of their organic material, since these different types show different water-use efficiencies (see review by Osmond et al. 1982).

[2]The relative deuterium content is expressed as:

$$\delta D(‰) = \left[\frac{(D/H)_{sample}}{(D/H)_{standard}} - 1\right] \times 1000.$$

Table 22.2. $\delta^{13}C$ values and δD values of the organic material (δD^{org}) and tissue water (δD^{H_2O}) in succulent plants at their natural sites[a]

Species	Site	δD^{org} (‰)	δD^{H_2O} (‰)	$\delta^{13}C$ (‰)
Carpobrotus acinaciformis (L.) L. Bolus (Aizoaceae)	Calabria, Lauria Italy	−101.0	+10.1	−27.7
– " –	Malia, Crete	−69.2	+10.4	−24.2
– " –	– " –	−63.9	+30.5	−22.5
– " –	– " –	−57.4	+19.9	−22.4
– " –	– " –	−54.9	+29.1	−21.6
Aeonium urbicum (C. Sm.) Webb et Berth.	Sicily, Realmonte	−36.9	−2.2	−14.6
(Crassulaceae) – " –	– " –	−29.6	−5.0	−14.1
Sedum coeruleum Vahl (Crassulaceae)	Sicily, Realmonte	−50.7	+2.6	−24.8
Sedum reflexum L. (Crassulaceae)	Sicily, Realmonte	−69.6	+2.2	−24.2
Umbilicus pendulinus DC. (Crassulaceae)	Sicily, Anapo Gorge	−103.8	−5.0	−27.2
Opuntia ficus-indica (L.) Miller (Cactaceae)	" " "	+2.9	+9.4	−12.6
Agave americana L. (Agavaceae)	Malia, Crete	+14.7	+23.0	−12.6

[a] cf. Ziegler (1979, 1989).

CAM plants generally show a much higher deuterium content in their organic material than C_3 or C_4 plants from the same biotope or under the same cultivation conditions (Ziegler et al. 1976; Sternberg et al. 1984; Ziegler 1989; Medina et al. 1991). This difference can be used to distinguish between CAM plants and C_4 plants at arid sites when $\delta^{13}C$ values do not provide a clear distinction.

CAM plants in their natural habitats also have a high deuterium content in their tissue water (Table 22.2). This was first reported by Schiegl (1970): *Opuntia* (CAM) tissue water had on average a 20‰ less negative δD value than *Populus* (C_3). At that time, there was no explantion for this difference.

The accumulation of heavy water in the tissues of CAM plants in dry environments is easily explained. Succulent CAM plants at dry sites are adapted to habitats where they fill up their large water-storing vacuoles occasionally after infrequent heavy rains and where the long dry periods occurring in between handicap potential non-CAM competitors. During the drought periods, these CAM plants transpire slowly but continuously (mainly at night), discriminating against the heavier isotope and therefore accumulating HDO in the remaining tissue water. Since metabolism in CAM plants involving cycles of malate synthesis and degradation continues even after long-lasting water stress (Szarek et al. 1973; Osmond 1975), and since this metabolism is also dependent on tissue water, the deuterium content in the organic material of CAM plants becomes progressively higher the more severe and long-lasting the water-stress period (Ziegler et al. 1976). It is not clear, however, whether the enrichment of deuterium in the tissue water can explain fully the relatively high deuterium content in the organic material of CAM plants.

22.3 Correlations Between $\delta^{13}C$ and δD Values?

Long-lasting drought should enhance dark CO_2 fixation via phospho*enol*-pyruvate carboxylase (PEPC) relative to light fixation directly via ribulose 1,5-bisphosphate carboxylase/oxygenase (RUBISCO), and consequently increase the $^{13}C/^{12}C$ ratio in the biomass (giving less-negative $\delta^{13}C$ values). As mentioned above, dry periods also lead to an increased deuterium content of the biomass (i.e. to less negative δD values).

Since the δD value depends on many variables (taxon-specific as well as environmental ones), there is little value in attempting to compare all the numerous plant species with very different systematic positions and from extremely diverse ecological habitats that we analyzed to date for $\delta^{13}C$ and δD values. Instead, as an example of particular life-forms, Fig. 22.1 shows the relations between the $\delta^{13}C$ and the δD values in the dry matter of different plants collected in Madagascar. Only in the case of epiphytic orchids at Manankazo (Fig. 22.1A) is there a clear-cut correlation between the two values. Fig. 22.1B indicates that even within the orchids this correlation is not always found. The difference in $\delta^{13}C$ values between the terrestrial and epiphytic orchids (the terrestrial orchids in general being C_3-type, the epiphytic ones mostly CAM, and among CAM species the epiphytic ones being heavier than the terrestrial) was also found in material from other regions of the world.

The reason for this lack in general correlation between $\delta^{13}C$ and δD values may be attributable to differences in the physical and chemical environment at the microsites of the plants. As with the question of systematic differences in δD values in the biomass of C_3 and C_4 species (Ziegler 1989), the problem can apparently only be solved by cultivation of a CAM species under constant conditions (especially with irrigation water of uniform δD value), with the amount of water supply as the only variable.

22.4 Differences in $\delta^{13}C$ and δD Values Between Different Organs and Tissues

The $\delta^{13}C$ and δD values can change in leaves (the most frequently analyzed organ) with their age and position (Table 22.3). There are also differences within a single organ: in leaves, for example between epidermis and mesophyll (Table 22.4), and in different layers of the stem (bark, cambium, wood, etc.; Table 22.4). Another example is differences between spines and stems in the Didiereaceae (Table 22.3; see also Winter 1979) and Cactaceae. A detailed $\delta^{13}C$ analysis of different tissues of the stem and roots of the Chilean cactus *Copiapoa cinerea* var. *columba-alba* is shown in Fig. 22.2. There is no

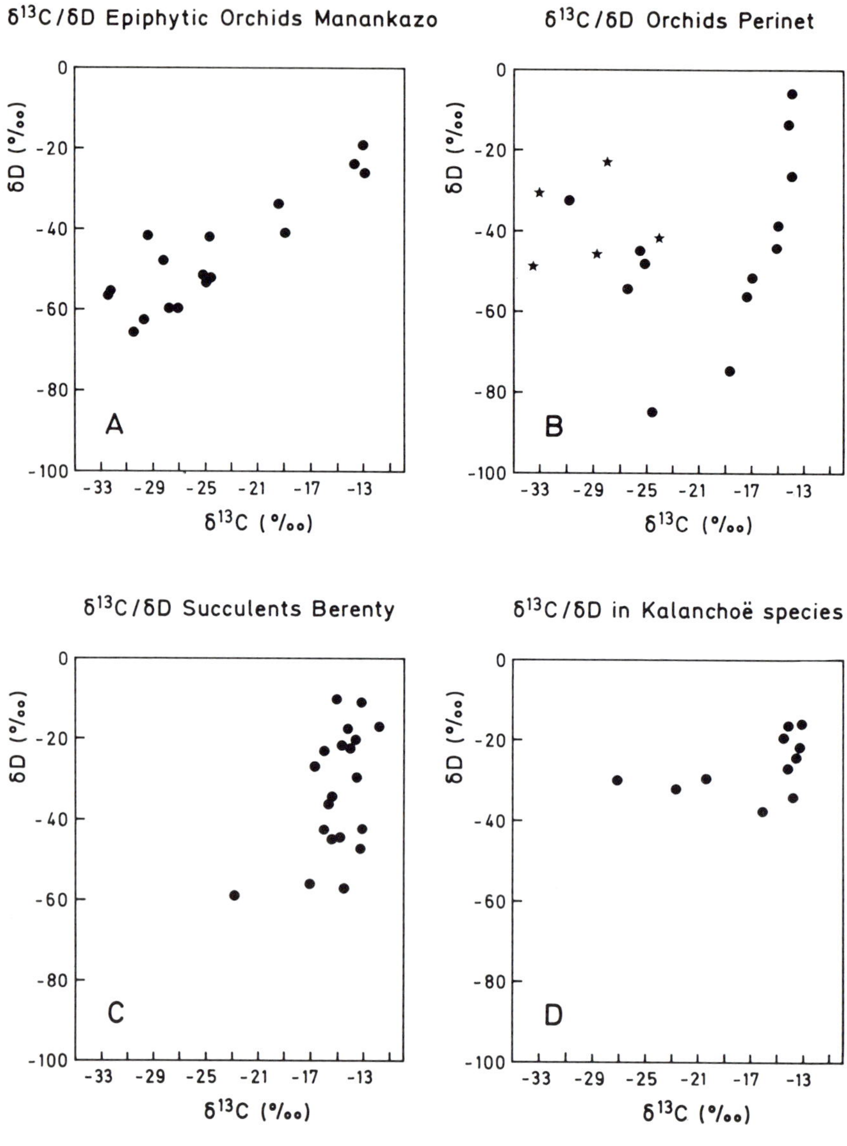

Fig. 22.1A–D. Correlations between the $\delta^{13}C$ and δD values in the biomass of C_3 and CAM plants in Madagascar. **A, B** Orchid species from Manankazo and Perinet, respectively; **C** succulent plants from Berenty; **D** *Kalanchoë* species. *Manankazo* Evergreen montane forest. *Perinet* Evergreen montane forest (maki reservate). *Berenty* Xerophilous bush. (Material collected by Dr. M. Kluge, Darmstadt)

clear difference in the $\delta^{13}C$ value of the N- and S-half of this phototropically curved stem, but there is considerable variation in the different tissues and between root and stem (Kraus 1993).

Table 22.3. $\delta^{13}C$ values of Didiereaceae, Madagascar (collected by M. Kluge, Darmstadt)

Species	$\delta^{13}C$ (‰)
Alluaudia montagnacii Rauh	
Leaves	−21.3
Spines	−15.6
Steam	−16.9
Alluaudia procera (Drake) Drake	
Leaves (young)	−22.7
″ (old)	−21.1
Spines	−16.3
Stem (young)	−22.2
″ (old)	−17.1
Alluaudia comosa Drake	
Leaves	−16.6
Spines	−14.5
Stem	−15.8
Alluaudia humbertii Choux	
Leaves	−18.4
Spines	−15.4
Stem	−15.8
Alluaudia dumosa Drake	
Stem	−16.6
Alluaudiopsis fiherenensis Humbert et Choux	
Leaves	−23.5
Spines	−22.3
Stem	−21.4
Alluaudiopsis marnieriana Rauh	
Leaves	−22.7
Spines	−20.0
Stem	−20.8

As a general rule, the $\delta^{13}C$ values in stems or petioles are less negative than those in leaves, but the associated δD values do not show a clear trend (Fig. 22.3).

The reason for the differences in the $\delta^{13}C$ values of different organs or tissues may be related to the following factors :

– Difference in chemical composition: high lipid or wax content (Table 22.4) presumes the action of pyruvate dehydrogenase, which discriminates strongly against ^{13}C-containing pyruvate compared with ^{12}C-pyruvate and gives rise to ^{13}C-poor material. On the other hand, low lipid content – e.g. in dead and membrane-deprived spines, or in parts of stems – will increase the $\delta^{13}C$ content, giving less negative $\delta^{13}C$ values.

In the different organs and tissues of CAM plants, the relative contributions of PEPC and RUBISCO to carbon gain and transport of carbon between tissues can also have some influence.

Table 22.4. $\delta^{13}C$ and δD values and ash content (% dry weight) in different tissues or organs of species of different photosynthetic types

Species	Site	Type	$\delta^{13}C$ (‰)	δD (‰)	Ash (%)
Kalanchoë daigremontiana Hamet et Perr. (Crassulaceae)	Cultivated	CAM			
Epidermis (average of 3 samples)			−20.6	−60.7	10.0
Mesophyll			−16.3	−15.2	26.0
Sarcocaulon mossamedense (Welw.) Hiern (Geraniaceae)	Namibia	CAM			
Leaves			−21.5		
Stem (wax)			−26.0		
Stem (wax, inner part)			−26.5		
Stem (cortex)			−25.6		
Stem (cambium)			−24.6		
Stem (wood)			−25.6		
Commelina benghalensis L. (Commelinaceae)	Cultivated	C_3			
Leaf (epidermis, abaxial)			−28.2	−75.7	20.9
Leaf (mesophyll)			−29.3	−99.3	16.4

As far as differences in δD values are concerned, there are the following possible explanations:

– The relative deuterium content decreases in organic material derived from the action of pyruvate dehydrogenase (e.g. lipids, waxes).
– Different organs and tissues may have different δD values in their tissue water, the starting point for the biosyntheses. The experimental demonstration of this fact is very difficult and up to now has not been performed for CAM plants. (For C_4 plants see Smith et al. 1991.)

The picture becomes even more complicated when the nitrogen supply to a CAM plant varies. As can be seen from Fig. 22.3, low nitrogen supply increases the ^{13}C discrimination ($\delta^{13}C$ values becoming more negative) in the leaves of *Crassula ovata* as well as in stems and petioles (but not in the roots), both with good water supply and during water stress (Widmann 1991).

22.5 $\delta^{13}C$ and δD Values of Parasites on CAM Plants

22.5.1 Holoparasites

Holoparasites are organisms that are not (or to a negligible degree only) able to produce organic material by their own photosynthesis. They obtain all

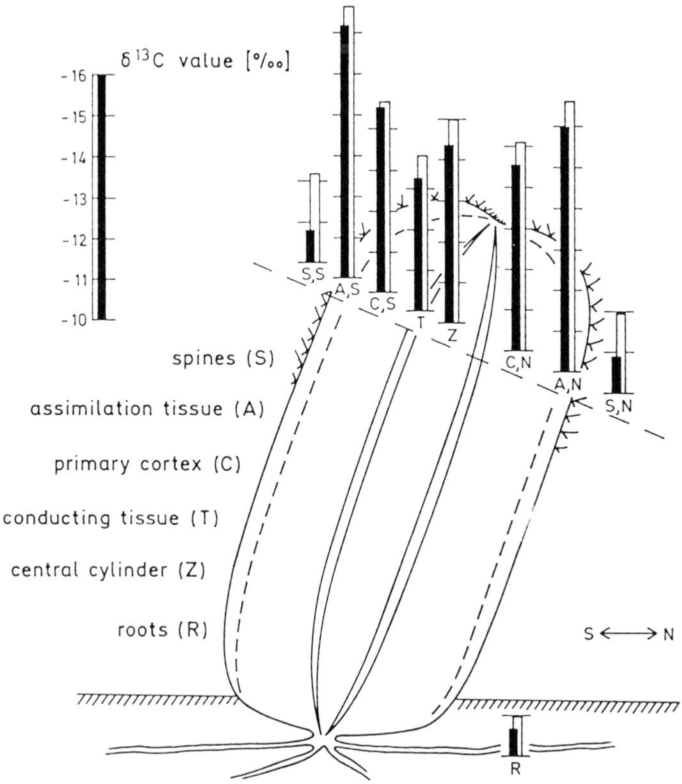

Fig. 22.2. $\delta^{13}C$ values of different tissues of a phototropically curved plant of *Copiapoa cinerea* var. *columba-alba* in the National Parque Pan de Azúcar, Chile. Cross-section in a N/S direction. *Black bar* Average value of all analyses; difference between *white bar* and *black bar* is one standard deviation. (Kraus 1993)

necessary compounds from their hosts. They usually have a direct connection to the phloem of their hosts.

It is apparent from the data in Table 22.5 that the $\delta^{13}C$ values of the parasites on CAM plants mirror those of their hosts. *Hydnora africana*, for example, has a similarly high ^{13}C content to its CAM-host *Euphorbia damarana* in its natural habitat in Namibia. The polyphagous *Cuscuta hyalina*, feeding on C_3, C_4 and CAM plants, always has $\delta^{13}C$ values resembling those of the different hosts. There was no statistical difference in the $\delta^{13}C$ values between host and parasite in 30 pairs of different photosynthetic types (C_3, C_4 and CAM) (Ziegler 1994).

Surprisingly, there is a significant ($p < 0.001$) difference between the deuterium content of the organic material of the host and its holoparasite: the parasite is always richer in deuterium.

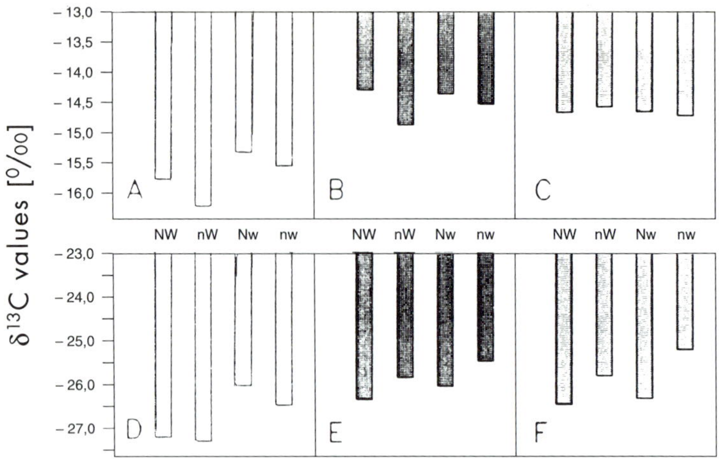

Fig. 22.3A–F. $\delta^{13}C$ values in leaves (**A, D**), stem/petiole (**B, E**), and roots (**C, F**) of the CAM plant *Crassula ovata* (**A–C**) and of the succulent C_3 plant *Peperomia obtusifolia* (**D–F**) at the end of the vegetation periods (averages of 3 replicates) *NW* Abundant nitrogen and water supply to the soil; *nW* low nitrogen, abundant water; *Nw* abundant nitrogen, low water; *nw* low nitrogen, low water. (Details in Widmann 1991)

Table 22.5. $\delta^{13}C$ and δD values in pairs of holoparasites and their CAM hosts

Species	$\delta^{13}C$(‰)	δD(‰)
Hydnora africana Thunb.	−11.4	−17.0
Euphorbia damarana Leach	−12.0	−49.2
Namibia		
Cuscuta hyalina Heyne ex Roth[a]	−17.9	−9.2
Sedum dendroideum Moçiño et Sessé ssp. *praealtum* (DC.) R.T. Clausen	−18.6	−24.2
Cuscuta hyalina Heyne ex Roth	−13.5	−9.3
Kalanchoë tubiflora Hamet	−14.2	−20.9
Cuscuta hyalina Heyne ex Roth	−16.5	−11.1
Kalanchoë daigremontiana Hamet et Perr.	−17.5	−25.6
Cuscuta hyalina Heyne ex Roth	−13.0	+9.3
Crassula portulacea Lam.	−14.3	−2.3

[a]The *Cuscuta* species and their host plants were cultivated in the glasshouse of the Institut für Botanik und Mikrobiologie, TU, München.

22.5.2 Hemiparasites

The best analyzed of the hemiparasites are the mistletoes (families Viscaceae and Loranthaceae). As true hemiparasites they are able to perform photosynthesis, and as far as is known this is C_3 photosynthesis. However, in some cases mistletoes also obtain a considerable amount of organic material from

Table 22.6. $\delta^{13}C$ and δD values of pairs of mistletoes and their CAM hosts

Species	$\delta^{13}C$	δD
Bakerella sp.	−26.5	−71.5
Didierea trollii Capuron et Rauh	−20.5	−47.1
Madagascar[a]		
Tapinanthus oleifolius (Wendl.) Danser	−24.5	−76.6
Euphorbia damarana Leach	−12.0	−49.2
Namibia		
Tapinanthus oleifolius	−17.6	−58.9
(Wendl.) Danser		
Aloë dichotoma Masson	−13.8	−42.8
Neuhof, Namibia		
Viscum capense L.f.	−25.9	−58.9
Euphorbia mauretanica L.	−18.4	−63.4
Karoo Botanic Garden		
Viscum crassulae Ecklon et Zeyh.	−26.2	6.2
Portulacaria afra Jacq.	−22.5	6.1
Karoo Botanic Garden		
Tristerix aphyllus (Miers ex DC.) Tieghem	−13.7	–
Eulychnia acida Philippi	−12.1	–
Tololo, Chile[b]		
Tristerix aphyllus (Miers ex DC.) Tieghem	−13.3	–
Echinopsis chilensis (Colla) Friedrich et G. Rowley	−12.2	–
Tololo, Chile[b]		
Tristerix aphyllus (Miers ex DC.) Tieghem	−13.7	–
Echinopsis skottsbergii (Backeb.) Friedrich et G. Rowley	−12.6	
Fray Jorge, Chile[b]		

[a]Collected by M. Kluge.
[b]Collected by R.G. Kraus.

their hosts (Ehleringer et al. 1985; Marshall and Ehleringer 1990; Schulze et al. 1991).

When we consider the δD values of misteltoes and their hosts, it should be emphasized that we analyzed the organic material rather than the tissue water of these associations. The average δD ($\pm SD$) of Australian host plants ($n = 42$; all C_3 plants) was $-72.8 \pm 19.6‰$, and of their mistletoes $-51.1 \pm 16.8‰$; this difference was highly significant ($p < 0.001$). In host plants from Africa and Madagascar ($n = 24$), the average δD value of the C_3 plants was $-67.8 \pm 10.7‰$, and for their parasites $-53.3 \pm 12.3‰$. In the CAM hosts from Africa shown in Table 22.6, the average δD value was $-41.7 \pm 21.4‰$, and for their mistletoes $-54.4 \pm 28.1‰$. This means that the mistletoe:host δD ratio is reversed in the pairs with CAM hosts compared with C_3 hosts. There was no significant correlation between the $\delta^{13}C$ and δD values of the Australian ($r = 0.160$) or African hosts ($r = 0.382$), nor between the $\delta^{13}C$ and δD values of the Australian ($r = 0.259$) or African parasites ($r = 0.073$).

22.5.3 The Isotopic Fate of Deuterium in Parasites

Differences in the δD values of hosts and parasites are much more difficult to explain than their $\delta^{13}C$ values. The enrichment of deuterium in the biomass of the holoparasites compared with the hosts is at present not understood (Ziegler 1986). The lower deuterium content in the biomass of some mistletoes on CAM hosts may be caused by the fact that mistletoes obtain their tissue water (used in their own photosynthesis) from the transpiration stream of the host, which is not much enriched in deuterium, while the CAM host also uses water from water-storage tissues or vacuoles, which can be considerably enriched in deuterium (Ziegler 1989 and above).

22.6 $\delta^{13}C$ and δD Values of Nectar

Since nectar-bearing plants are in general C_3 plants and since mixing of honey with sugarcane sugar or hydrolyzed corn starch (C_4 plants) has been detected, we proposed to use the determination of the $\delta^{13}C$ value of honey to check for falsifications (Ziegler et al. 1977). The δD values are useful for the identification of geographical origin, e.g. of *Citrus* honeys, because the more arid the site of origin the richer the deuterium content. It turns out that CAM plants can deliver not only a nectar rich in both ^{13}C and deuterium (Lüttge et al. 1985), but that there can also be a natural "heavy honey" derived from the CAM plant *Aloë* (honey from *A. marlothii*, S. Africa: $\delta^{13}C$ $-17.1‰$; δD $+39.6‰$).

22.7 Conclusions

CAM plants increase their relative ^{13}C and deuterium contents when they are exposed to drought; the expected correlation between the $\delta^{13}C$ and δD values, however, can be found in only a few cases. This may be due to the strong microhabitat influence on the δD value. The isotope ratios are further dependent on the organ and tissue used for analysis and also on mineral (nitrogen) nutrition. In holoparasites on CAM plants, the $\delta^{13}C$ value of the parasite biomass mirrors that of the host, while the parasite is always richer in deuterium. In hemiparasites, the mistletoe: host δD ratio is reversed (mistletoes more negative) in the pairs with CAM hosts compared with C_3 hosts. Nectar from CAM plants can be the source of a natural "heavy honey."

Acknowledgements. All isotope determinations in Munich were made in cooperation with W. Stichler, P. Trimborn and J. Lipp, Institut für Hydrologie der

Gesellschaft für Strahlen- und Umweltforschung. The work was supported by the Deutsche Forschungsgemeinschaft.

References

Bokhoven C, Theeuwen HHJ (1956) Deuterium content of some natural organic substances. Ned Akad Wet B 59: 78–83

Craig H (1961) Standard for reporting concentrations of deuterium and oxygen-18 in natural waters. Science 133: 1833–1834

Ehleringer JR, Schulze ED, Ziegler H, Lange OL, Farquhar GD, Cowan IR (1985) Xylem-tapping mistletoes: water or nutrient parasites? Science 227: 1479–1481

Estep MF, Hoering TC (1981) Stable hydrogen isotope fractionation during autotrophic and mixotrophic growth of microalgae. Plant Physiol 67: 474–477

Kraus RG (1993) Ökophysiologie chilenischer Kakteen. Doctoral Thesis, Technische Universität, München

Lüttge U, Stichler W, Ziegler H (1985) Isotope ratios ($\delta^{13}C$ and δD) of nectar in comparison to tissues in C_3 and CAM plants. Isr J Bot 34: 103–112

Marshall JD, Ehleringer JR (1990) Are xylem tapping mistletoes partially heterotrophic? Oecologia 84: 244–248

Medina E, Lüttge U, Leal F, Ziegler H (1991) Carbon and hydrogen isotope ratios in bromeliads growing under different light environments in natural conditions. Bot Acta 104: 47–52

Osmond CB (1975) Environmental control of photosynthetic options in crassulacean acid metabolism plants. In: Marcelle R (ed) Environmental and biological control of photosynthesis. Junk, The Hague, pp 299–309

Osmond CB, Winter K, Ziegler H (1982) Functional significance of different pathways of CO_2 fixation in photosynthesis. In: Lange OL, Noble PS, Osmond CB, Ziegler H (eds) Physiological plant ecology II. Encyclopedia of plant physiology, new series, vol 12 B. Springer, Berlin Heidelberg New York, pp 479–547

Schiegl WE (1970) Natural deuterium in biogenic materials. Influence of environment and geophysical applications. PhD Thesis, University of South Africa, Pretoria

Schiegl WE (1972) Deuterium content of peat as a paleoclimatic recorder. Science 175: 512–513

Schulze ED, Lange OL, Ziegler H, Gebauer G (1991) Carbon and nitrogen isotope ratios of mistletoes growing on nitrogen and non-nitrogen fixing hosts and on CAM plants in the Namib desert confirm partial heterotrophy. Oecologia 88: 457–462

Smith BN, Epstein S (1970) Biogeochemistry of the stable isotopes of hydrogen and carbon in salt marsh biota. Plant Physiol 46: 738–742

Smith BN, Ziegler H (1990) Isotopic fractionation of hydrogen in plants. Bot Acta 103: 335–342

Smith BN, Ziegler H, Lipp J (1991) Isotopic evidence for mesophyll reduction in *Zea mays*, a NADP-malic enzyme plant. Naturwissenschaften 78: 358–359

Sternberg LO, DeNiro MJ, Johnson HB (1984) Isotope ratios of cellulose from plants having different photosynthetic pathways. Plant Physiol 74: 557–561

Szarek SR, Johnson HB, Ting IP (1973) Drought adaption in *Opuntia basilaris*. Significance of recycling carbon through crassulacean acid metabolism. Plant Physiol 52: 539–541

Urey HC, Brickwedde FG, Murphy GM (1932) A hydrogen isotope of mass 2 and its concentration. Phys Rev 39: 164

Washburn EW, Smith ER (1934a) An examination of water from various sources for variation in isotopic composition. J Res Nat Bur Stand 12: 305–311

Washburn EW, Smith ER (1934b) The isotopic fractionation of water by physiological processes. Science 79: 188–189

Widmann K (1991) Vergleichende Untersuchungen zum Stickstoff- und Wasserhaushalt sukkulenter Pflanzen. Doctoral Thesis, Technische Universität München

Winter K (1979) $\delta^{13}C$ values of some succulent plants from Madagascar. Oecologia 40: 103–112

Ziegler H (1979) Diskriminierung von Kohlenstoff- und Wasserstoffisotopen: Zusammenhänge mit dem Photosynthesemechanismus und den Standortbedingungen. Ber Dtsch Bot Ges 92: 169–184

Ziegler H (1986) Control of photosynthesis by variation of diffusion resistance in mistletoes and their hosts. In: Marcelle R, Clijstra H, von Poucke M (eds) Biological control of photosynthesis. Martinus Nijhoff, Dordrecht, pp 171–185

Ziegler H (1989) Hydrogen isotope fractionation in plant tissues. In: Rundel PW, Ehleringer JR, Nagy KA (eds) Stable isotopes in ecological research. Springer, Berlin Heidelberg New York, pp 105–123

Ziegler H (1994) Deuterium content in organic material of hosts and their parasites. In: Schulze E-D, Caldwell M (eds) Ecophysiology of photosynthesis. Springer, Berlin Heidelberg New York, pp 393–408

Ziegler H, Osmond CB, Stichler W, Trimborn P (1976) Hydrogen isotope discrimination in higher plants: correlation with photosynthetic pathway and environment. Planta 128: 85–92

Ziegler H, Stichler W, Maurizio A, Vorwohl G (1977) Die Verwendung stabiler Isotope zur Charakterisierung von Honigen, ihrer Herkunft und ihrer Verfälschung. Apidologie 8: 337–347

23 Evolutionary Aspects of Crassulacean Acid Metabolism in the Crassulaceae

E.A.H. PILON-SMITS, H. 'T HART, and J. VAN BREDERODE

23.1 Introduction

The CAM pathway is taxonomically widespread and is believed to be used by about 16 000 species of vascular plants (see p. 9), a figure that includes more than 6% of all angiosperms. In addition, CAM has been reported in certain advanced groups of lower vascular plants (*Isoetes*; see Chap. 18) and in the specialized gymnosperm *Welwitschia mirabilis* (Schulze et al. 1976; Winter 1985).

Since CAM is mostly present in the more specialized groups of different families, the CAM pathway has probably evolved from C_3 photosynthesis relatively recently (Teeri 1982a; Griffiths 1989). The origin of CAM is most probably polyphyletic: it has evolved independently many times in different plant families (Osmond 1978; Lüttge 1987), and even within individual families (Smith et al. 1986).

Evolutionary studies on the Bromeliaceae, Cactaceae and Crassulaceae suggest an evolutionary direction from C_3 via intermediates, like CAM-cycling, to facultative or obligate CAM (Teeri 1982a; Griffiths 1988; Monson 1989). The evolution of CAM often runs parallel with specialization and the occupation of new, dry habitats (Medina 1974; Griffiths and Smith 1983; Smith 1989).

23.2 Taxonomy of the Crassulaceae

The Crassulaceae comprise about 1500 species and constitute a very diverse group, both cytologically and morphologically. The members of this family occupy predominantly dry habitats of the temperate and subtropical zones. In this chapter we will focus on two genera of the Crassulaceae: the genus *Sedum* of the subfamily Sedoideae, and the genus *Aeonium* of the subfamily Sempervivoideae.

Sedum is the largest genus (ca. 500 species) of the Crassulaceae and contains some taxa that are considered to bear resemblance to the hypothetical ancestor

Department of Plant Ecology and Evolutionary Biology, University of Utrecht, Padualaan 8, 3584 CH Utrecht, The Netherlands

Ecological Studies Vol. 114
Winter/Smith (eds) Crassulacean Acid Metabolism

of this family (Teeri 1982a,b; 't Hart 1991, 1992). Morphologically, *Sedum* is also the most diverse genus of the Crassulaceae and, apart from the small semi-aquatic genus *Tillaea* (subfam. Crassuloideae), has the widest distribution (Berger 1930; 't Hart 1982). Within *Sedum*, three largely allopatric groups are discerned: *S.* sect. *Sedum* (Europe), *S.* sect. *Americana*, and *S.* sect. *Asiatica*. From these sections the other genera of Crassulaceae – except for Crassuloideae – are thought to have evolved, as depicted in Fig. 23.1 (Uhl 1961, 1963; Teeri 1982a,b; 't Hart 1982). Consequently, *Sedum* is most probably paraphyletic (although derived from a common ancestor, it does not include all its descendants). The evolutionary relationships within *Sedum* have long

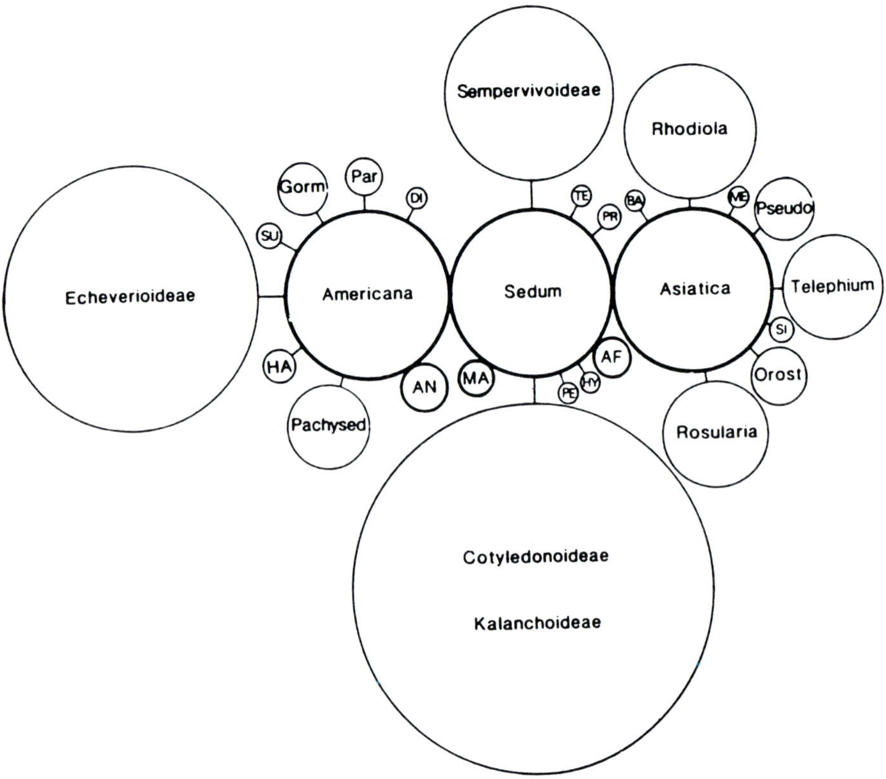

Fig. 23.1. Diagram of the classification of the Crassulaceae (from 't Hart 1982). The areas of the circles are proportional to the number of species of the taxa they represent. The six *touching circles with heavy contour lines* represent the sections of *Sedum* subgenus *Sedum* (AF sect. Africana; AN sect. Andina; MA sect. Macaronesia). The *circles* representing the other taxa of the subfamily Sedoideae are all connected to subgenus *Sedum* by *short lines*. *BA Balfouria*; *HA Hasseanthus*; *HY Hypagophytum*; *DI Diamorpha*; *Gorm Gormania*; *ME Meterostachys*; *Orost Orostachys*; *Pachysed Pachysedum*; *Par Parvisedum*; *PE Perrierosedum*; *PR Prometheum*; *Pseudo Pseudosedum*; *SI Sinocrassula*; *SU Sulcus*; *TE Telmissa*

been subject to investigation, but the relationships between most species are still largely unknown ('t Hart 1991).

Aeonium is one of the genera presumed to have evolved from the European Sedoideae. Together with *Sempervivum, Aichryson, Monanthes* and *Greenovia*, it forms the subfamily Sempervivoideae. *Aeonium* displays a large variety of growth-forms, ranging from small undifferentiated shrubs to rosettes and woody bushes (Lems 1960). Of the 37 species and subspecies of *Aeonium* (Liu 1989), 31 are endemic to the Canary Islands and two species to Madeira. Of the remaining four species – large woody shrubs – one occurs on the Cape Verde Islands, one in Morocco and, surprisingly, two in East Africa and Yemen. Despite the large morphological variation, the genetic divergence within *Aeonium* appears to be very small (T.H.M. Mes, unpubl. data). Therefore it is probably a young, monophyletic group.

23.3 Evolutionary Studies of CAM in the Crassulaceae

Comparative CAM studies have been performed on several taxa of the Crassulaceae. A significant degree of variation in CAM has been reported in *Sedum* (Knopf and Kluge 1979; Müller et al. 1980, 1982; Teeri 1982a). Based on both carbon-isotope ratios and PEPC immunological properties, the Mexican species (constitutive CAM) can be distinguished from those of North America or Europe (C_3 or facultative CAM). Teeri (1982a,b) has made a carbon-isotope survey of CAM in ten American genera. He proposed an evolutionary succession from (primitive) *Sedum* taxa, displaying low or intermediate CAM, to various advanced genera with less-negative carbon-isotope ratios. From a Central American *Sedum* ancestor ($\delta^{13}C = -22‰$, defined as on p. 5), several groups with strong CAM may have evolved (Mexican *Sedum, Echeveria, Lenophyllum, Pachyphytum, Graptopetalum*). The genera *Dudleya, Villadia* and *Cremnophila* show intermediate values. The genera *Diamorpha and Parvisedum* (weak CAM) may have evolved from a North American *Sedum* ancestor.

Another extensive carbon-isotope survey of CAM was made of the genus *Kalanchoë* (Kluge et al. 1991). There appears to be very large photosynthetic variation within this genus, with carbon-isotope ratios ranging from -10 to $-30‰$. The development of CAM in *Kalanchoë* correlates with both growth-form (stronger CAM in larger species) and habitat (stronger CAM in dry habitats).

Furthermore, an inventory of CAM was made of a limited number of species from the Sempervivoideae. Studies of *Aeonium* by Tenhunen et al. (1982) and Lösch (1984, 1990) indicate a substantial degree of variation in CAM within this genus. The other genera of the subfamily have been less extensively studied, but appear to have less photosynthetic plasticity. The genera *Greenovia, Monanthes* and *Aichryson* mainly have $\delta^{13}C$ values indicative of C_3 (Tenhunen et al. 1982; Lösch 1984, 1990). Seven *Sempervivum* species all showed values that indicate strong CAM, species with less negative values originating from drier habitats (Osmond et al. 1975).

A special case is formed by the aquatic CAM species *Tillaea aquatica* (= *Crassula aquatica*). Since CAM is absent in emerged leaves of this plant, it is believed that this species has evolved CAM as an adaptation to carbon-limiting aquatic conditions (Keeley and Morton 1982; see Chap. 18).

It is generally assumed that some form of CAM is present in all species of the Crassulaceae – with the possible exceptions of *Tillaea erecta* (= *Crassula erecta*, Keeley and Morton 1982) and *Sedum ternatum* (Teeri 1982b) – ranging from CAM-cycling to obligate CAM.

23.4 Evolution of CAM in *Sedum* and *Aeonium*

The two genera *Sedum and Aeonium* which differ greatly in numbers of species, geographical distribution and age, provide interesting groups for evolutionary CAM studies. Both genera are morphologically very variable and occupy a range of habitats, differing widely in humidity and temperature. We have made an inventory of the variation in CAM in these genera under controlled conditions. By comparison with the ecology (habitat) and phylogeny of the taxa, we investigated whether the variation in photosynthetic physiology is historically or environmentally determined, or both.

For the determination of CAM activity in *Sedum* and *Aeonium* the following parameters were used: (1) The activity of the CAM enzymes phosphoenolpyruvate carboxylase (PEPC) and NADP-malic enzyme (NADP-ME). (2) Leaf thickness, which is connected with succulence and known to be correlated with CAM activity (Teeri et al. 1981). (3) The carbon-isotope ratio. The $\delta^{13}C$ value, (Chaps. 15 and 22) is a very suitable CAM parameter for evolutionary studies since it is a long-term representation of the photosynthetic pathway in plants. It has been shown to be largely genetically determined (Teeri et al. 1981). These three parameters are rapidly assessable and therefore suitable for the screening of a large number of species. The results from these parameters are usually highly correlated, which indicates their reliability.

The plants were grown in a glasshouse under identical conditions. No supplementary light was supplied. In winter a minimum temperature of 10°C was maintained. Samples for all measurements were collected in summer. Carbon-isotope-ratio determinations were performed as described previously (Pilon-Smits et al. 1991a). Enzyme extraction as well as the determination of PEPC activity was performed as described previously (Pilon-Smits et al. 1990). NADP-ME activity was measured as described by Pilon-Smits et al. (1992). Protein determinations were made according to Bradford (1976).

The statistical analyses were performed using the Number Cruncher Statistical System (Dr. J.L. Hintze, 865 East 400 North, Kaysville, Utah 84037, USA; 801-546-0445). The clustering method is the K-means algorithm (Hartigan 1975).

23.4.1 *Sedum*

We have made a survey of the CAM activity in a group of 26 European *Sedum* species, representing the morphological variation present in the European *Sedum* flora, supplemented by some Mexican species. As parameters for CAM we used two criteria: (1) the carbon-isotope ratio and (2) the PEPC activity under normal, dry (1 week) or wet conditions, to distinguish facultative from obligate CAM. In addition, the leaf thickness was determined. The results are summarized in Table 23.1.

The three CAM parameters gave highly correlated results (Pilon-Smits et al. 1990). The Mexican *Sedum* species showed a high PEPC activity irrespective of water supply, with $\delta^{13}C$ values indicative of obligate CAM (−15 to −20‰). The PEPC activity in the European *Sedum* species forms a continuum with variable influence of water supply (C_3/facultative CAM). Incidentally, the photosynthetic plasticity of the species is not only reflected by the dry/wet PEPC ratio, but also by the (correlated) degree of seasonal fluctuation in PEPC activity (Pilon-Smits et al. 1991b). The carbon-isotope ratios of the European group (−25 to −32‰) indicate that a large portion of their carbon is fixed during the daytime by the C_3 pathway, supplemented to various extents by nocturnal fixation of respiratory CO_2 (CAM-cycling). The species from Madeira holds an intermediate position. This thick-leaved plant shows high, water-independent PEPC activity and a $\delta^{13}C$ value of −25‰.

The various types of CAM are functional in different circumstances. Obligate CAM, as exhibited by the Mexican species, is functional in stable, dry

Table 23.1. Summary of the characteristics of the *Sedum* species examined

Origin	Character[a] state	$\delta^{13}C$ (‰)	S.A.[b] PEPC	S.A. PEPC[b] dry/wet[c]	Leaf thickness (mm)
Mexico					
S. burrito	Advanced	−16.9	3120	1.8	8.0
S. dendroideum	Advanced	−22.2	1625	2.8	4.5
S. nussbaumerianum	Advanced	−15.0	2003	2.2	6.0
Madeira					
S. fusiforme	Primitive	−25.6	1233	0.9	5.0
Europe					
6 spp. average	Primitive	−30.1	352	2.8	1.0
18 spp. average	Advanced	−27.8	683	3.4	1.2
Extremes:					
S. aetnense	Primitive	−31.5	233	1.0	0.8
S. grisebachii	Primitive	−29.6	567	4.9	0.8
S. montanum	Advanced	−28.9	396	11.5	1.4
S. tymphaeum	Advanced	−25.5	1170	1.0	1.4

[a]Dichotomy based on morphological characters ('t Hart 1991).
[b]S.A. = specific activity (mU mg^{-1} protein).
[c]Ratio 1 week dry/1 week wet.

habitats. Facultative CAM, as displayed by most European *Sedum* species, enables plants to survive the dry season by switching to CAM, whereas in times of sufficient water supply they use the more productive C_3 pathway.

In *Sedum* the variation in CAM agrees to some extent with the major evolutionary trends within the genus (Table 23.1). More advanced *Sedum* species ('t Hart 1991) generally show a higher CAM activity than less specialized species. However, the species from the dry cliffs of southern Madeira, which has retained several primitive morphological characters, has developed substantial CAM activity, together with morphological adaptation to drought (succulence and secondary woodiness).

We propose that latent CAM is present (in the form of CAM-cycling) in all *Sedum* species, which can fairly easily evolve into facultative or obligate CAM. It may well be possible that CAM has evolved more than once in this genus.

23.4.2 *Aeonium*

Twenty-nine of the 37 species of *Aeonium* were studied. For most species, several plants originating from different locations were investigated. The variation in CAM activity was studied by measuring $\delta^{13}C$ values, as well as the specific activity of PEPC and NADP-ME.

A substantial variation in CAM appears to be present in *Aeonium*, as indicated by $\delta^{13}C$ values ranging from -13.6 to $-27.1‰$, and PEPC and NADP-ME activities ranging from 269 to 2251 and 103 to 585 mU mg^{-1} protein, respectively (Pilon-Smits et al. 1992). The values of the three CAM criteria are highly correlated.

To determine whether the variation in CAM represents an adaptation to different habitats, we have examined whether the three parameters were correlated with the climatic conditions of the natural habitats of the species and their growth- and life-forms, respectively. Four main ecological zones can be distinguished on the Macaronesian Islands (= Canary Islands, Cape Verde Islands, Madeira). These are (1) laurel forest (moist, cool); (2) north-facing cliffs (humid, cool); (3) south-exposed slopes (hot, dry); (4) (sub)alpine regions (cold, dry). These zones are fairly well separated, and the species are usually confined to a single zone (Lems 1960). Analysis of variance shows that the values of the three CAM parameters are each significantly related to the ecological zones. The species from dry habitats show strong CAM, whereas the others show C_3 or intermediate values.

In order to investigate the correlation of CAM activity with growth-form, we have arranged the different growth- and life-forms into the six classes schematically depicted in Fig. 23.2. These classes largely correspond with the sections distinguished by Lems (1960). The carbon-isotope ratio and NADP-ME activity are both significantly related to growth-form. The relation with PEPC activity, however, is only significant when enzyme activities based on fresh weight are used instead of protein content. In Fig. 23.2 the average CAM

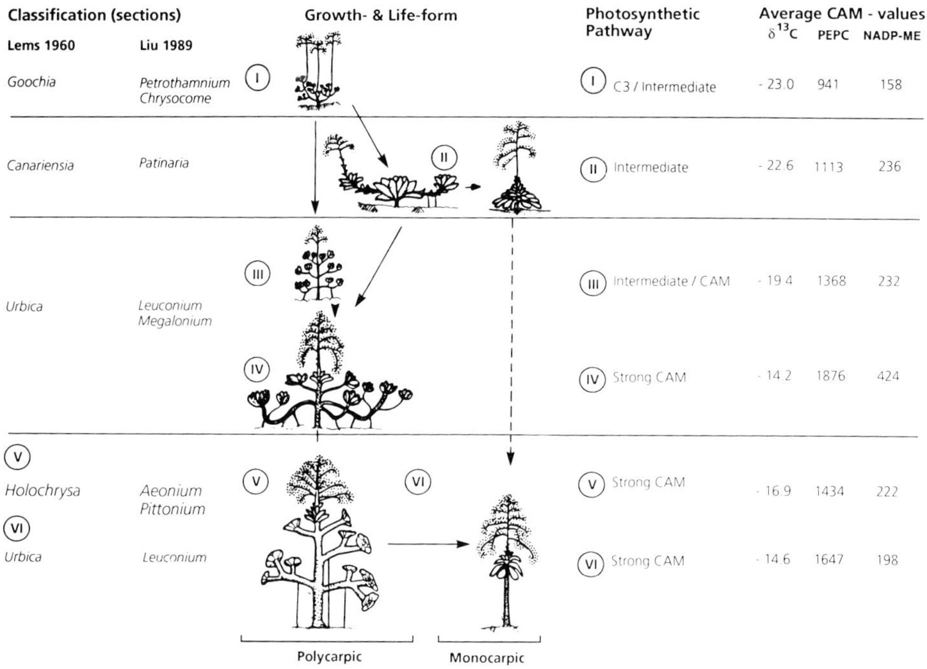

Fig. 23.2. Possible evolutionary succession of growth-forms in *Aeonium* (from Pilon-Smits et al. 1992). Corresponding classification is shown *on the left*, average CAM values and deduced photosynthetic pathway *on the right*. Illustrations were copied from Lems (1960)

values for the different growth-forms are given. Thus CAM in *Aeonium* appears to be correlated with habitat as well as growth-form. CAM and the large woody growth-forms have probably co-evolved as physiological and morphological adaptations, respectively, to dry habitats.

A cluster analysis was performed using the $\delta^{13}C$, PEPC and NADP-ME values. The five clusters obtained agree remarkably well with intrageneric classifications (Lems 1960; Liu 1989). This can be explained by the fact that these classifications are partly based on growth-form.

In view of the general assumption that CAM has evolved from C_3 photosynthesis, we propose an evolutionary succession within *Aeonium*, based on growth-form and CAM activity (Fig. 23.2). CAM probably evolved – along with the morphological differentiation – from small undifferentiated species, showing little or no CAM, via rosettes and/or candelabrum-shaped forms with intermediate or strong CAM, to large woody species showing strong CAM. This evolutionary direction agrees with the general assumption that *Aeonium* has evolved from *Sedum*, and is supported by Liu (1989), Thiede (1991), and by chloroplast DNA restriction fragment length polymorphism (RFLP) analysis

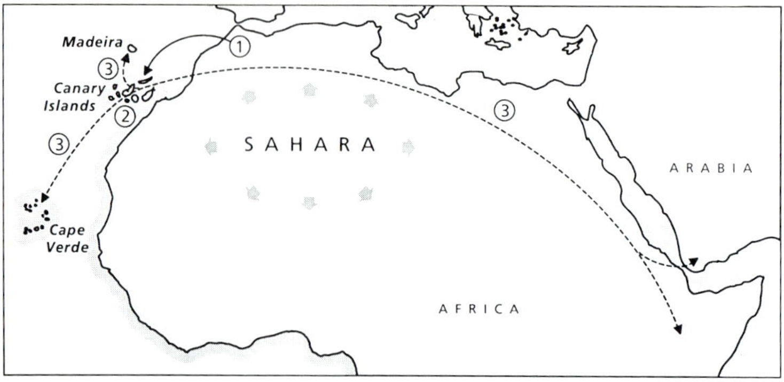

Fig. 23.3. Scenario for the evolution of *Aeonium*. *1 Aeonium* progenitor evolves from *Sedum* in Morocco and subsequently invades the Canary Islands; *2* The availability of many new niches leads to rapid speciation through adaptive radiation; *3* Large woody *Aeonium* form spreads over other Atlantic archipelagoes and over the African mainland, where its distribution is later disrupted by the expanding Sahara

(T.H.M. Mes, unpubl.). However, it does not agree with the evolutionary schemes proposed by Lems (1960) and Lösch (1990), who consider the large woody species to be the most primitive.

As a possible scenario for the evolution of *Aeonium*, we propose the following succession of events (Fig. 23.3). The progenitor of *Aeonium* evolved from a *Sedum* like ancestor, perhaps resembling the present-day *S. gattefossei* from Morocco (Uhl 1961). This progenitor of *Aeonium* (a small herb with C_3 photosynthesis) occupied the different newly available habitats, in the process giving rise to a wide range of different forms through adaptive radiation. One of the most specialized forms of these newly developed *Aeonium* species (large, woody, with strong CAM) then returned to the mainland. It spread through the whole of North Africa and the Cape Verde Islands. Finally the expansion of the Sahara desert disrupted its area, leading to the present-day disjunct distribution.

In view of the small genetic variation within the genus *Aeonium* (T.H.M. Mes, unpubl.) and the recent origin of some of the Canary Islands (Pleistocene: Schmicke 1976), we may conclude that CAM has evolved quite rapidly in this genus. The molecular basis for this rapid evolution of CAM in *Aeonium* – such as regulatory changes in promoter regions? – needs further study.

23.5 Conclusions

As a basis for evolutionary studies, inventories of CAM in different genera of the Crassulaceae have proven very useful. Those groups that occupy a wide range of habitats have offered especially interesting results and we hope will continue to do so in the future.

In the groups that have been studied so far, strong CAM is mainly found in the specialized taxa, which mostly occupy dry habitats and are often large and woody. Most probably CAM has evolved independently within *Kalanchoë*, *Aeonium*, *Sempervivum*, and in *Sedum* possibly even more than once.

Weak CAM in the form of CAM-cycling is probably present in all species of the Crassulaceae. Apparently CAM-cycling can evolve into strong CAM fairly rapidly under suitable selection pressures, as has been shown for *Aeonium*. Typically in the Crassulaceae, the occupation of new, arid habitats leads to morphological and physiological adaptation. This co-evolution gives rise to new, specialized species, showing strong CAM and the "Crassulaceae / CAM traits": succulence, a rosulate or semi-rosulate growth-form, and secondary woodiness.

Acknowledgements. The investigations were supported by the Netherlands Foundation for Biological Research (B.I.O.N.), with financial aid from the Netherlands Organization for the Advancement of Research (N.W.O.). We thank R. Kreuler and J.A.N. Meesterburrie for their help in determining the carbon-isotope ratios.

References

Berger A (1930) Crassulaceae. In: Engler A, Prantl K (eds) Die natürlichen Pflanzenfamilien 2. Aufl., 18A. Wilhelm Engelmann, Leipzig, pp 352–483

Bradford MM (1976) A rapid and sensitive method for the quantitation of microgram quantities of protein utilizing the principle of protein-dye binding. Anal Biochem 72: 248–254

Griffiths H (1988) Crassulacean acid metabolism: a re-appraisal of physiological plasticity in form and function. Adv Bot Res 15: 43–92

Griffiths H (1989) Carbon dioxide concentrating mechanisms and the evolution of CAM in vascular epiphytes. In: Lüttge U (ed) Vascular plants as epiphytes: evolution and ecophysiology. Springer, Berlin Heidelberg New York, pp 42–86

Griffiths H, Smith JAC (1983) Photosynthetic pathways in the Bromeliaceae of Trinidad: relations between life-forms, habitat preference and the occurrence of CAM. Oecologia 60: 176–184

Hartigan J (1975) Clustering algorithms. Wiley, New York

Keeley JE, Morton BA (1982) Distribution of diurnal acid metabolism in submerged aquatic plants outside the genus *Isoetes*. Photosynthetica 16: 546–553

Kluge M, Brulfert J, Ravelomanana D, Lipp J, Ziegler H (1991) Crassulacean acid metabolism in *Kalanchoë* species collected in various climatic zones of Madagascar: a survey by $\delta^{13}C$ analysis. Oecologia 88: 407–414

Knopf O, Kluge M (1979) Properties of phosphoenolpyruvate carboxylase in *Sedum* species in relation to crassulacean acid metabolism. Plant Cell Environ 2: 73–78

Lems K (1960) Botanical notes on the Canary Islands II. The evolution of plant forms in the islands: *Aeonium*. Ecology 41: 1–17

Liu H-Y (1989) Systematics of *Aeonium* (Crassulaceae). Special publications No 3, National Museum of Natural Science. 1 Kuan Chien Rd., Taichung, Taiwan, ROC

Lösch R (1984) Species-specific responses to temperature in acid metabolism and gas exchange performance of Macaronesian Sempervivoideae. In: Margaris NS,

Arianoustou-Farragitaki M, Oechel WC (eds) Tasks for vegetation science, vol 13. W Junk, The Hague, pp 117–126

Lösch R (1990) Funktionelle Voraussetzungen der adaptiven Nischenbesetzung in der Evolution der Makaronesischen Semperviven. Dissertationes Botanicae 146. Cramer, Berlin

Lüttge U (1987) Carbon dioxide and water demand: crassulacean acid metabolism, a versatile ecological adaptation exemplifying the need for integration in ecophysiological work. New Phytol 106: 593–629

Medina E (1974) Dark CO_2 fixation, habitat preference and evolution within the Bromeliaceae. Evolution 28: 677–686

Monson RK (1989) On the evolutionary pathways resulting in C_4 photosynthesis and CAM. Adv Ecol Res 19: 57–110

Müller D, Kluge M, Gröschel-Stewart U (1980) Immunological studies on phosphoenolpyruvate carboxylase in *Sedum* species with crassulacean acid metabolism or C_3 photosynthesis. Plant Cell Environ 3: 249–253

Müller D, Kluge M, Gröschel-Stewart U (1982) Comparative studies on immunological and molecular properties of phosphoenolpyruvate carboxylase in species of *Sedum* and *Kalanchoë* performing crassulacean acid metabolism (CAM). Plant Cell Environ 5: 223–230

Osmond CB (1978) Crassulacean acid metabolism: a curiosity in context. Annu Rev Plant Physiol 29: 379–414

Osmond CB, Ziegler H, Stichler W, Trimborn P (1975) Carbon isotope discrimination in alpine succulent plants supposed to be capable of CAM. Oecologia 18: 209–217

Pilon-Smits EAH, 't Hart H, van Brederode J (1990) Phosphoenolpyruvate carboxylase in *Sedum rupestre* (Crassulaceae): drought-enhanced expression and purification. J Plant Physiol 136: 155–160

Pilon-Smits EAH, 't Hart H, Meesterburrie JAN, Naber P, Kreuler R, van Brederode J (1991a) Variation in crassulacean acid metabolism within the genus *Sedum*: carbon isotope composition and water status dependent phosphoenolpyruvate carboxylase activity. J Plant Physiol 137: 342–346

Pilon-Smits EAH, 't Hart H, van Brederode J (1991b) Seasonal variation of phosphoenolpyruvate carboxylase specific activity in fifteen species exhibiting facultative or obligate crassulacean acid metabolism. J Plant Physiol 138: 581–587

Pilon-Smits EAH, 't Hart H, Maas JW, Meesterburrie JAN, Kreuler R, van Brederode J (1992) The evolution of crassulacean acid metabolism in *Aeonium* inferred from carbon isotope composition and enzyme activities. Oecologia 91: 548–553

Schmicke H-U (1976) The geology of the Canary Islands. In: Kunkel G (ed) Biogeography and ecology of the Canary Islands. Junk, The Hague, pp 67–184

Schulze E-D, Ziegler H, Stichler W (1976) Environmental control of crassulacean acid metabolism in *Welwitschia mirabilis* Hook. fil. in its range of natural distribution in the Namib desert. Oecologia 24: 323–334

Smith JAC (1989) Epiphytic bromeliads. In: Lüttge U (ed) Vascular plants as epiphytes: evolution and ecophysiology. Springer, Berlin Heidelberg New York, pp 109–138

Smith JAC, Griffiths H, Lüttge U (1986) Comparative ecopysiology of CAM and C_3 bromeliads. I. The ecology of the Bromeliaceae in Trinidad. Plant Cell Environ 9: 359–376

Teeri JA (1982a) Carbon isotopes and the evolution of C4 photosynthesis and crassulacean acid metabolism. In: Nitacki MH (ed) Biochemical aspects of evolutionary biology. University of Chicago Press, Chicago, pp 93–130

Teeri JA (1982b) Photosynthetic variation in the Crassulaceae. In: Ting IP, Gibbs M (eds) Crassulacean acid metabolism. American Society of Plant Physiologists, Rockville, pp 244–259

Teeri JA, Tonsor SJ, Turner M (1981) Leaf thickness and carbon isotope composition in the Crassulaceae. Oecologia 50: 367–369

Tenhunen JD, Tenhunen LC, Ziegler H, Stichler W, Lange OL (1982) Variation in carbon isotope ratios of Sempervivoideae species from different habitats of Teneriffe in the spring. Oecologia 55: 217–224 't Hart H (1982) The white-flowered European *Sedum*. 1. Principles of a phylogenetic classification of the Sedoideae (Crassulaceae) and the position of the white-flowered *Sedum* species. Proc K Ned Akad Wet Ser C 85: 663–675

't Hart H (1986) Hybridisation and speciation in *Sedum* from Europe. In: Urbanska K (ed) Abstr IOBP Symp differentiation and speciation in higher plants. Zürich, p 64

't Hart H (1991) Evolution and classification of the European *Sedum* species (Crassulaceae). Flora Mediterranea 1: 31–61

't Hart H (1992) The archetype of the Crassulaceae. Acta Bot Neerl 41: 107–108

Thiede J (1991) Untersuchungen zur Anatomie der Vegetationsorgane Makaronesischer Sempervivoideae (Crassulaceae). Diplomarbeit, Institut für Allgemeine Botanik, Universität Hamburg

Uhl CH (1961) The chromosomes of the Sempervivoideae (Crassulaceae). Am J Bot 48: 114–123

Uhl CH (1963) Chromosomes and phylogeny of the Crassulaceae. Cact Succulent J Am 25: 80–84

Winter K (1985) Crassulacean acid metabolism. In: Barber J, Baker NR (eds) Photosynthetic mechanisms and the environment. Elsevier, Amsterdam, pp 329–387

24 The Evolution of Crassulacean Acid Metabolism

J.A. Raven[1] and R.A. Spicer[2]

24.1 Introduction

This paper takes a wide view of the evolution of CAM; in particular it addresses the question why CAM does what it does and why other CAM-like possibilities have not been used.

In considering evolution of a metabolic trait such as CAM, we need to distinguish clearly between the selective significance of the trait in the context of the environment in which it evolved, and the mechanisms by which the genetic information specifying CAM could have been derived from pre-existing genetic determinants. Such considerations of the evolution of CAM need to be underpinned by geological evidence as to the environment (e.g. past bio-geochemical cycles and atmospheric composition), the timing of the occurrence of CAM taxa, the phylogeny of extant CAM taxa as shown by molecular genetic and other evidence, and of the molecular genetics of the parts of the CAM mechanism and of their putative precursors.

In conjecturing as to evolutionary pathways, we note that it is more important to devise a path in which each step was plausible and resulted in a functional product, rather than to minimize the number of steps to the end-product.

24.2 Selective Significance of CAM Traits in Terrestrial Plants in the Context of Climates and Palaeoclimates

For *terrestrial* CAM plants, major selective significance is attached to the very low ratio of water lost in transpiration per unit net C accumulated relative to what is found in C_3 and C_4 homoiohydric plants (Osmond 1978; Osmond et al. 1982; Lüttge 1987; Griffiths 1988; Winter and Ziegler 1992). This means that more biomass can be accumulated per unit external water input in CAM plants than in C_3 or C_4 plants, granted equal acquisition capacity by the plant and no differential competition from evaporation, drainage, or uptake by other

[1]Department of Biological Sciences, University of Dundee, Dundee, DD1 4HN, UK
[2]Department of Earth Sciences, The Open University, Milton Keynes MK7 6AA, UK

Ecological Studies Vol. 114
Winter/Smith (eds) Crassulacean Acid Metabolism

plants. These competitive processes may be more significant for CAM plants to the extent that they may exhibit lower specific growth rates in nature (but see Nobel 1991 and Chap. 16).

The timing of evolution of the terrestrial vascular CAM plants must have been later than the evolution of their C_3 vascular ancestors, which arose not later than the mid-Silurian (see Edwards and Seldon 1992; Raven 1993). Geochemical mass-balance considerations suggest that the CO_2 partial pressure in the atmosphere was substantially higher than it is today, while the partial pressure of O_2 was probably slightly lower (Fig. 24.1; Berner and Canfield 1989; Berner 1990; Berner 1993). The relevance of this to the ecophysiology of these early vascular plants is that the water (transpiration) cost of net C assimilation and biomass accumulation in C_3 plants depends, other things being equal, on the CO_2 and O_2 partial pressures in the atmosphere, granted appropriate stomatal behaviour. Higher CO_2 partial pressure and/or lower O_2 partial pressure increase the net CO_2 fixation rate by the PCRC + PCOC (photosynthetic carbon reduction cycle plus photosynthetic carbon oxidation cycle) at a given stomatal conductance and RUBISCO (ribulose 1,5-bisphosphate carboxylase-oxygenase) quantity and level of activation in C_3 plants. This increases the

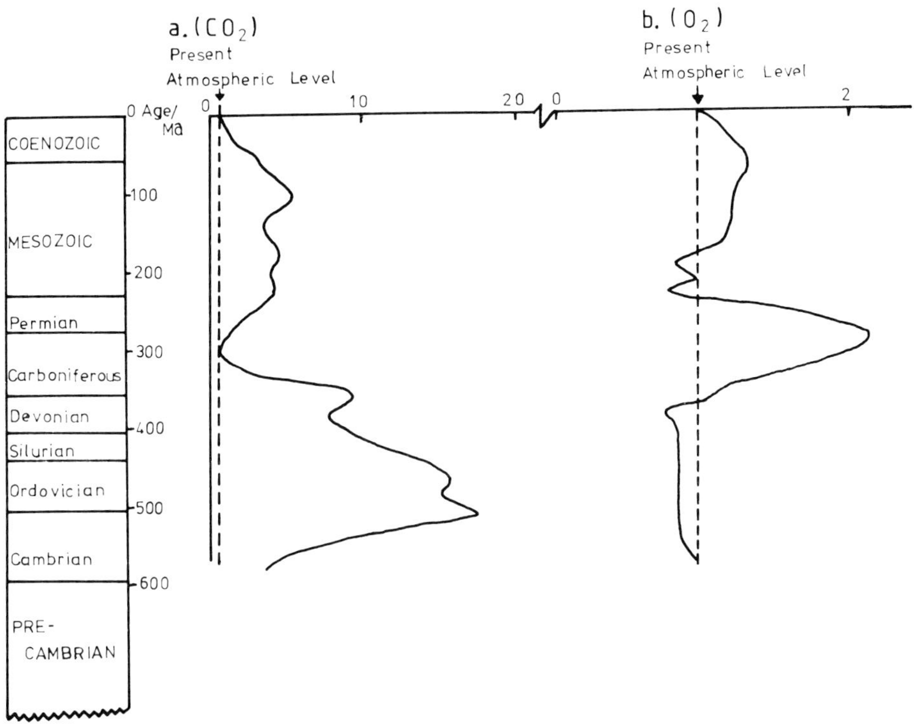

Fig. 24.1. Estimates of Phanerozoic atmospheric CO_2 (**a** after Berner 1990; see also Berner 1993) and O_2 (**b** after Berner and Canfield 1989) relative to the present level. *Scales* represent relative values based on present atmospheric levels of unity. *Ma* Million years

quantity of CO_2 fixed per unit of photosynthesizing material, while leaving the water loss over the same time period unchanged. If, as is sometimes the case, stomatal density and/or stomatal aperture is decreased at higher CO_2 partial pressures for growth and measurement (Raven 1993), the net CO_2 fixation rate in the light could be maintained at the rate attained under the lower CO_2 partial pressures, while the rate of water loss would be decreased. In either of these cases the water cost of carbon (and biomass) gain would be decreased; such arguments and conclusions are commonly used in discussions of the influence of the present anthropogenic increase in CO_2 partial pressure and are, of course, backed up by experimental data (Long 1991). Such arguments are qualitatively independent of the kinetic properties of RUBISCO involved, an important consideration in view of possible changes in RUBISCO kinetics during the evolution of C_3 land plants (see below).

CAM employs PEPC (phospho*enol*pyruvate carboxylase) for dark carboxylation; this carboxylase uses HCO_3^- as its effective "immediate" substrate (but see Jane et al. 1992) and, at prevailing values of cytosol pH (and hence of the equilibrium CO_2:HCO_3^- ratio: Holtum et al. 1984) and of stomatal conductance, is essentially saturated by extant CO_2 partial pressures. The refixation phase in the photoperiod, occurring as it does behind closed stomata, is at a higher CO_2 partial pressure and CO_2:O_2 ratio than obtain in air-equilibrium solutions today, thus increasing carboxylase and decreasing oxygenase activity of RUBISCO. Overall, this makes extant terrestrial CAM photosynthesis essentially independent of increases in CO_2 partial pressure above the present atmospheric level.

These arguments suggest that the relative advantage of CAM over C_3 photosynthesis in terms of water economy would (like the corresponding smaller advantage of C_4 over C_3) have been smaller under atmospheric conditions of CO_2 and O_2 partial pressure believed to pertain in the Silurian than in those found today (Fig. 24.1; see Raven 1993). While global temperatures in the Silurian and early Devonian were higher than those of today, the configuration of the continents and the high sea level would have meant substantial rainfall in many areas and the likelihood that precipitation exceeded evaporation, further exacerbating the effects of high CO_2:O_2 on the relative advantage of CAM over C_3. The first time after the Silurian in which the CO_2:O_2 ratio in the atmosphere was believed to be as low as, or lower than, the extent value was the Late Carboniferous (Fig. 24.1), when the CO_2:O_2 ratio may have been lower than today's 35 Pa CO_2:21 kPa O_2 and thus have resembled the pre-industrial interglacial value of 28 Pa CO_2:21 kPa O_2, or even the full-glacial 18 Pa CO_2:21 kPa O_2, albeit with slightly higher absolute CO_2 and O_2 partial pressures (Figs. 24.1, 24.2). Other things being equal, the selective significance of the water-cost advantage of CAM over C_3 would have been quantitatively similar in the late Carboniferous to what is found today. According to the computations of Berner (1990) and Berner and Canfield (1989), and the measurements of Barnola et al. (1987; Figs. 24.1, 24.2), the next time that such low CO_2 partial pressures (and CO_2:O_2 ratios) occurred was in the

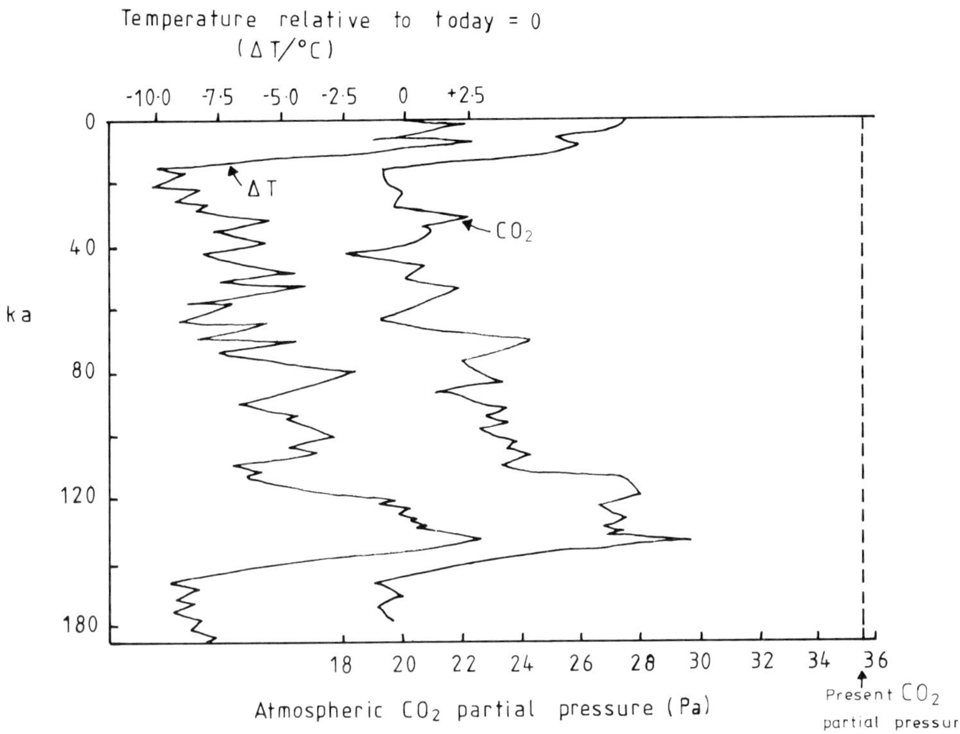

Fig. 24.2. Estimate of Late Pleistocene atmospheric partial pressure (after Barnola et al. 1987); *ka* 1000 years

glacial episodes (e.g. 115 to 15 ka) of the Pleistocene (2 to 0.01 Ma), with rather higher CO_2 partial pressures and CO_2:O_2 ratios in the interglacials (e.g. 130 to 115 ka). This is not to say that there was no selective significance of CAM relative to C_3 metabolism at other times in the last 420 million years, but rather that the selective significance would have been smaller. This point is particularly significant in the context of the possibility that CAM originated as a facultative response to exceptional, rather than typical global, conditions (see West-Eberhard 1986). It is also relevant to the next section on aquatic plants.

Modern desert forbs, grasses, deciduous shrubs and non-phreatophytic evergreens growing under present atmospheric concentrations of CO_2 and O_2 have upper limits for photosynthesis of about 50 °C, phreatophytes generally have upper limits below 50 °C, while CAM succulents have upper limits of some 59 °C (Smith and Nobel 1986). However, geological evidence shows that global average temperatures were higher than present throughout the late Silurian, Devonian and most of the Mesozoic (Frakes et al. 1992) and that, for at least 75% of land-plant history, low latitudes were characterized by high temperatures and seasonal aridity. Results obtained from Mesozoic climate-modelling experiments (Kutzbach and Gallimore 1989; Valdes and Sellwood

1992), which either directly or indirectly include higher than present CO_2 levels, suggest that the Late Jurassic and mid-Cretaceous midday low-latitude surface temperatures exceeded 50 °C (Valdes and Sellwood 1992, and in prep.) and 3-day averages exceeded 36 °C. Nighttime temperatures remained high due to extensive cloud cover. Moreover, in these models constantly moist soils were limited to high latitudes, and tropical regions were only seasonally or intermittently wet. To a first approximation these patterns of moisture regime compare well to the observed distributions of coals and evaporites (Zeigler et al 1987; Parrish et al. 1982), comparisons which lend credibility to the model results.

These low-latitude temperature and moisture regimes raise interesting questions regarding photosynthetic performance during times of extreme global warmth. For CAM plants, the "light" phase of CO_2 fixation can have an optimum of 40 to 45 °C (Smith and Nobel 1986), while that for the "dark" phase, involving net CO_2 fixation, is commonly much lower, i.e. 10 to 20 °C (with possible acclimation to 25 °C), at least in desert plants that are usually subject to relatively low night temperatures (Smith and Nobel 1986; Lüttge 1989). CAM epiphytes have rather higher temperature optima (25 to 30 °C) for the "dark" phase (Lüttge 1989), showing that there is adequate diversity of temperature optima in extant CAM plants to accommodate night temperatures of 30 °C. Interaction of these temperature effects with CO_2 partial pressure does not seem to have been investigated, but dark CO_2 fixation in CAM plants is essentially CO_2-saturated in today's 35 Pa CO_2.

The desert C_4 *Tidestromia oblongifolia* grown at 45 °C (optimum 40 to 45 °C) has a temperature optimum for light-saturated photosynthesis in 35 Pa CO_2 of 47 °C (Björkman 1975); since C_4 plants are CO_2-saturated by 35 Pa CO_2, the temperature optimum probably would not increase at higher CO_2 partial pressures. The desert C_3 *Nerium oleander*, when grown at 45 °C in 35 Pa CO_2, has a temperature optimum for photosynthesis at 35 Pa CO_2 of about 35 °C, and about 45 °C in 75 to 85 Pa (Björkman et al. 1978). The C_3 *Larrea divaricata* behaves similarly (Mooney et al. 1978). This indicates that C_3 as well as C_4 plants could cope with quite high temperatures, *provided* short-term effects of high CO_2 can be translated into long-term growth effects. While there may have been prima facie reasons for assuming a prolonged period of selection in favour of CAM during the Mesozoic, much depends on the long-term effects of elevated CO_2:O_2 ratios on C_3 plants at high temperatures.

Turning to the evidence as to the occurrence of terrestrial CAM photosynthesis in times past in relation to the selective significance of CAM relative to C_3 metabolism in different palaeoatmospheres, we consider (1) nucleotide-sequence evidence of the occurrence of major taxa which today exhibit terrestrial CAM, (2) morphological/anatomical evidence of the occurrence of major taxa which today exhibit terrestrial CAM, and (3) more generalized palaeotological evidence relating to taxon-independent indicators of terrestrial CAM, i.e. "succulence", vascular-bundle spacing, and $^{13}C/^{12}C$ ratio of organic matter.

Dealing first with the antiquity of the major taxa which nowadays have CAM performing members, the estimated 16 000 species of extant CAM plants (see p. 9) comprise several species of two genera of polypodiaceous ferns, the gnetophyte gymnosperm *Welwitschia mirabilis*, and members of a number of families of flowering plants. The very extensive morphological and anatomical evidence indicating the occurrence of these taxa is all from Cretaceous and later strata ($\leqslant$ 130 Ma: Crane and Lidgard 1990), indicating that CAM members of these taxa are very unlikely to have evolved during the low CO_2 and low CO_2:O_2 thought to have occurred in the late Carboniferous, i.e. ~300 Ma). However, the molecular (nucleotide-sequence) evidence indicates a much earlier origin of angiosperms (330 Ma: Martin et al. 1989 a,b; 200 Ma: Li et al. 1990; 230 to 350 Ma: Brandl et al. 1992), with less clear evidence on the gnetophytes (Hasebe et al. 1992) and the polypodiaceous ferns. These early timings of the origin of flowering plants have been challenged by those relying on morphological and anatomical data (Crane et al. 1989; Crane 1993). The "molecular" timings of as long ago as 350 Ma would mean that the flowering plants were subject to the low CO_2 and low CO_2:O_2 ratios of the late Carboniferous, with their relative significance for CAM evolution. At the moment the discrepancy between the "recent" origin indicated by fossil evidence, implying that the rate of accumulation of nucleotide changes was relatively high for flowering plants, and the "older" origin, indicated by nucleotide sequences, implying that pre-Cretaceous flowering plant fossils have somehow been missed, is unresolved.

The third, less taxon-dependent line of evidence mentioned above concerns the anatomical, morphological and $^{13}C/^{12}C$ ratio characteristics of CAM plants. Gibson (1982) discusses "succulence", epidermal features (usually glabrous, thick cuticle, stomata sunk up to 10 μm below the surface), the well-defined boundary to chlorenchyma, and the wide spacing of vascular bundles, which are frequent correlates of CAM. It should be noted that these features are not generally confined to CAM plants among terrestrial tracheophytes, and that some of them (e.g. the boundary of chlorenchyma) may not be readily distinguished in even anatomically well-preserved fossils. We also note that not all extant CAM vascular plants show the "typical" CAM anatomical features (e.g. *Welwitschia*). It is not currently possible to identify unequivocally a fossil CAM terrestrial tracheophyte on anatomical/morphological data alone.

Much of the anatomically better-preserved terrestrial tracheophyte remains from the Upper Carboniferous are from moist or wet tropical forest. This would not have been a suitable habitat for ground-dwelling, arid-habitat CAM plants, but would have been appropriate for CAM epiphytes. Views on the occurrence of bona fide vascular epiphyte fossils in the late Palaeozoic are somewhat divergent. Overall, it appears that the alleged vascular epiphytes only grew on their alleged phorophytes after they had fallen over (Benzing 1989; Lüttge 1989; Behrensmeyer et al. 1992). Furthermore, none of these (alleged) epiphytes has known CAM attributes.

A final possible line of evidence as to the occurrence of CAM terrestrial tracheophytes in times past comes from $^{13}C/^{12}C$ measurements. However, it is not only CAM which yields a small discrimination against ^{13}C relative to what occurs in C_3 metabolism: C_4 plants also have a small discrimination against ^{13}C (Troughton et al. 1974a; O'Leary 1981; Farquhar et al. 1989; Raven and Farquhar 1990; Griffiths 1992). Accordingly, whenever a high $^{13}C/^{12}C$ is found in terrestrially produced fossil organic C, or in fossil inorganic C derived from this organic C, external evidence must be sought to distinguish C_4 from CAM as the source. Troughton et al. (1974b) used the high-quality external evidence of specific identification of glacial and post-glacial (40 ka and 10 ka) specimens of *Opuntia polyacantha* with their $^{13}C/^{12}C$ measurements to show that the 10 ka specimens used CAM (high $^{13}C/^{12}C$), while the 40 ka specimens used C_3 metabolism (lower $^{13}C/^{12}C$); the authors relate this phenotypic plasticity (cf. Gerwick and Williams 1978) to the environments at 40 and 10 ka. Troughton et al. (1974b) also investigated specifically identified C_4 plants (see also Marino et al. 1992).

Of more potential use in identifying fossil CAM plants from strata lacking fossils identifiable as extant species are the sorts of investigations carried out by Tidwell and Nambudiri (1989, 1990), who investigated anatomically well-preserved late Miocene and Pliocene grasses. They found that the specimens with C_3 anatomy ($\leqslant$ 5.4 Ma) had a C_3 $^{13}C/^{12}C$ ratio, while those with Kranz anatomy ($\sim$ 10 Ma) had a C_4-like $^{13}C/^{12}C$ ratio, indicating that they were C_4 rather than C_3/C_4 intermediate (Raven 1991a). The relevance to identifying CAM plants is that a fossil terrestrial tracheophyte with a high $^{13}C/^{12}C$ which lacks Kranz anatomy is likely, even in the absence of identifiable CAM anatomical features *à la* Gibson (1982), to be a CAM plant. No examples of successful use of this procedure are known to the authors, despite a number of $^{13}C/^{12}C$ measurements on specimens of various fossil (Palaeozoic, Mesozoic) taxa (e.g. Table 24.1).

$^{13}C/^{12}C$ data uncoupled from plant anatomical evidence comes from the $^{13}C/^{12}C$ ratios of organic C and teeth-enamel inorganic C of grazing vertebrates (which closely reflect the C-isotope ratios of their plant food), and the $^{13}C/^{12}C$ ratio of soil carbonates/palaeocretes.

The grazing vertebrates with high $^{13}C/^{12}C$ values in organic C (collagen) are hadrosaurs of the genus *Anatosaurus* from the Late Cretaceous (Bocherens et al. 1988). This is consistent with a diet of terrestrial C_4 or CAM plants, or of aquatic plants. The latter are apparently ruled out by the arid environment inhabited by a large population of hadrosaurs, and their (coniferous) gut contents; however, access to the marine littoral and altered dietary inputs to organisms close to death cannot be ruled out. The data do not distinguish C_4 from CAM as the metabolic pathways employed by the most likely food plants.

The teeth-enamel, soil-carbonate and palaeocrete data come from work of Cerling and associates (see Cerling et al. 1993) and Wright and Vanstone (1991). Granted certain assumptions, the $^{13}C/^{12}C$ of $CaCO_3$ in palaeosols and

Table 24.1. Carbon-isotope ratios ($\delta^{13}C$ values) of organic parts of fossil vascular plants (unpubl. data of R.A. Spicer and J.A. Raven)

Taxon	Age	Part	$\delta^{13}C$(‰)
Sigillaria[a] (Arborescent lycophyte)	Upper Carboniferous	Leaf cushion	−23.7[d]
Weichselia[b] (Mantoniaceous pteropsid)	Lower Cretaceous	Leaf	−25.0[d]
Pseudofrenelopsis[b] (Cheirolepidaceous conifer)	Lower Cretaceous	Cuticle[c]	−25.6[d]
		Wood(1)[c]	−22.8[d]
		Wood(2)[c]	−22.6[d]

[a]Chosen as a (terrestrial) probable ancestor of the Isoetales, some extant members of which are secondarily aquatic CAM plants.

[b]Lived under seasonally arid conditions and experienced high temperatures, which may have reached 50 °C daily maximum.

[c]Cuticle would be expected to have a more negative $\delta^{13}C$ value than bulk plant organic C or wood (Farquhar et al. 1989; Rieley et al. 1991; Smith and Madhaven 1982).

[d]All the values are within the range for extant C_3 plants, albeit at the higher (less negative) end of the range (Troughton et al. 1974a). See also Bocherens et al. (1993).

teeth enamel can be related to that of respiratory CO_2 in the soil or vertebrate from which they were derived, which in turn closely reflects the $^{13}C/^{12}C$ of plant organic C. The work of Cerling and collaborators has been used to indicate C_4 or CAM in the late Miocene of East Africa, Pakistan and North America; C_4 rather than CAM is suggested from contextual information. The work of Wright and Vanstone (1991) may indicate C_4 or CAM photosynthesis in the Carboniferous.

This discourse on methods by which the timing of the evolution of terrestrial CAM, and the organisms carrying it out, has led to few positive results, but the indications obtained are sufficient to justify further work with molecular genetic and stable-isotopic methodology, as well as with more mainstream palaeobotanical techniques.

The environmental background to the selective significance of terrestrial CAM which has informed the foregoing discussion has been that of CO_2 partial pressure and CO_2:O_2 ratio in the atmosphere, and thus of variations in the potential water (transpiration) cost of growth of C_3 plants and in the incremental advantage in water cost of CAM relative to C_3. Other potential selective advantages of CAM in terrestrial habitats relate to the lower costs of growth in terms of various "catalytic" nutrient elements, such as N, Fe, Mn and Mo. While the "expended" resource water has a cost measured as water lost per unit C (or dry weight) gain, the "catalytic" resources N, Fe, Mn and Mo have costs expressed as quantity of N (or Fe, Mn, or Mo) required to yield unit rate of gain of C (or dry weight).

The only experimental data on these catalytic costs concerns N in relation to the higher CO_2 partial pressure, and CO_2:O_2 ratio, under which RUBISCO operates during CAM deacidification than pertains for RUBISCO in C_3 plants in the extant atmosphere. The prediction is that a similar rate of gain of C or dry weight could be supported by less RUBISCO per unit biomass. Hence, since RUBISCO can account for up to 0.25 of the N of photosynthetic organs of C_3 plants (Evans and Seeman 1989), significant savings of N could be envisaged in CAM plants relative to C_3 plants. This is similar to that suggested for C_4 plants. However, the evidence as to the occurrence of such economies of N in CAM relative to C_3 plants in today's atmosphere is equivocal (Lüttge 1987; Griffiths 1989). Any such economies would have been smaller or absent during long periods of terrestrial tracheophyte existence in which the CO_2 partial pressure (and the CO_2:O_2 ratio) were believed to be higher than they are today, i.e. Silurian-Lower Carboniferous and Permian-Pliocene.

The arguments as to the potentially lower cost of Fe, Mn and Mo in CAM relative to C_3 plants are analogous to those for C_4 plants, and to aquatics with CO_2-concentrating mechanisms (Raven 1988, 1990). No evidence as to the Fe, Mn and Mo costs of CAM plants relative to C_3 plants seems to be available.

The other resource which, like water, is expended or (more accurately) consumed during the growth of terrestrial plants is light as energy source. The units here are mol photons absorbed per mol C gain (or per unit dry weight gain). Computations of the photon cost of CAM suggest that it is similar to, or higher than, that of C_3 plants (including the costs of RUBISCO oxygenase and of the PCOC) in today's atmosphere (Table 24.2; see also Chap. 16 where energetics are compared on the basis of kJ energy per mol CO_2 fixed rather than mol photons used per mol CO_2 fixed) (Lüttge 1987; Griffiths 1989). CO_2 fixation by CAM would invariably have cost more than C_3, and the relative cost would have been higher in the past when the CO_2 partial pressure and CO_2:O_2 ratio were higher, yielding lower photon costs of growth of C_3 plants as RUBISCO oxygenase and the PCOC were suppressed, but with no effect on the photon costs of CAM plant growth (or of C_4 plant growth). This means that CAM plants would have been disadvantaged in terms of photon cost relative to C_3 plants over most of the last 420 Ma, with the exception of the Upper Carboniferous and at least the glacial episodes of the Pleistocene (Figs. 24.1 and 24.2). In general, then, CAM plants would have been disadvantaged (or more disadvantaged) in shade habitats over most of the last 420 Ma than they are today.

Overall, such advantages in resource costs of growth as CAM plants possess relative to C_3 plants today – predominantly the water cost – would have been lower over most of the last 420 Ma, while such disadvantages in photon cost would have been accentuated. The most obvious times at which the CAM advantages would have been highest, and its disadvantages would have been lowest, are the Upper Carboniferous and the Pleistocene.

A final consideration on terrestrial CAM relates to the possibility of CAM in poikilohydric terrestrial plants. Such data as are available for extant terrestrial poikilohydric plants (algae; lichens; bryophytes; gametophyte phase of pteridophytes) show that they do not have CAM photosynthesis (see also

Table 24.2. Photon costs of net CO_2 fixation by C_3 and CAM plants. Values not in parentheses for mol photons absorbed per mol CO_2 fixed assume the occurrence of a proton-motive Q cycle in both non-cyclic and cyclic photophosphorylation (Raven 1984, 1985) using 2 photons to move 1 electron from H_2O to $NADP^+$ and pump 3 H^+ into the intrathylakoid space, with 3 H^+ returning to the stroma per ADP phosphorylated; non-cyclic phosphorylation with a Q cycle reduces 2 mol $NADP^+$ and phosphorylates 4 mol ADP per 8 mol photons absorbed. Cyclic photophosphorylation with 1 photon to move 1 electron and pump 2 H^+ phosphorylates 0.67 mol ADP per mol photon absorbed. Values in parentheses for mol photons absorbed per mol CO_2 fixed assume the absence of a proton-motive Q cycle from both non-cyclic and cyclic photophosphorylation (Raven 1984, 1985), so that 2 photons would only pump 2 H^+ in non-cyclic photophosphorylation and 1 photon would only pump 1 H^+ in cyclic photophosphorylation. This means that 2 mol $NADP^+$ would be reduced and 2.67 mol ADP phosphorylated per 8 mol photons absorbed in non-cyclic photophosphorylation, and 0.33 mol ADP would be phosphorylated per mol photon absorbed in cyclic photophosphorylation. ATP requirements in the dark are met by oxidative phosphorylation assuming 5.33 mol ADP phosphorylated per mol C oxidized, with iterative procedures used to compute total photon cost when part of the energy for CO_2 fixation comes from oxidation of reduced C

CO_2 fixation pathway	ATP/CO_2	$NADPH/CO_2$	Photons absorbed per CO_2
C_3 with high CO_2:O_2 (sufficient to completely suppress RUBISCO oxygenase)	3	2	8 (9)
C_3 in present atmosphere yielding carboxylase: oxygenase ratio of 4 for RUBISCO[a]	4.46	2.875	11.5 (13.38)
CAM NAD(P)-ME[b]	7	2	12.965 (21.517)
CAM PEPCK[c,d]	6.25	2	12.139 (19.287)

[a] After Edwards and Walker (1983), adjusted for carboxylase: oxygenase of 4 rather than 3.5.

[b] Assumes basal 2 NADPH, 3 ATP per CO_2 refixed after malate decarboxylation in the light, 0.5 ATP per CO_2 for conversion of 3 C from starch via phosphorylase to vacuolar malic acid (2 H^+ per ATP used by the V-ATPase) during acidification, 2 ATP per CO_2 for pyruvate, phosphate dikinase, 1 ATP per CO_2 for glyceraldehyde 3-phosphate dehydrogenase, and 0.5 ATP per CO_2 for starch synthesis to regenerate the glucan working capital during deacidification (elaborated from Smith and Bryce 1992; see also Chap. 2).

[c] Assumes basal 2 NADPH, 3 ATP per CO_2 refixed after malate decarboxylation in light, 1 ATP per CO_2 for conversion of 3 C from vacuolar hexose via hexokinase to vacuolar malic acid (2 H^+ per ATP used by the V-ATPase) during acidification, 1 ATP per CO_2 for PEPCK, 1 ATP per CO_2 for glyceraldehyde 3-phosphate dehydrogenase, and 0.25 ATP per CO_2 to regenerate (via sucrose phosphate synthase and vacuolar invertase) the hexose working capital during deacidification (elaborated from Smith and Bryce 1992; see also Chaps. 2, 4 and 25).

[d] Note that if sucrose transport into the vacuole proceeds by H^+/sucrose antiport rather than facilitated diffusion, this would require 0.125 ATP per CO_2 to transport 0.25 sucrose into the vacuole (2 H^+ per ATP used by the V-ATPase, 1 H^+ per sucrose antiport), and the overall ATP requirement would be 6.375 ATP per CO_2.

Chap. 22). The data include gas-exchange quotients in light and in dark as a function of time of day, titratable acidity of homogenates, and $^{13}C/^{12}C$ ratios (Raven and Osborne, unpubl. data on titratable acidity of homogenates; Raven and Johnston 1991 and Proctor et al. 1992, and references therein). High $^{13}C/^{12}C$ ratios in terrestrial algae and lichens relate to restricted CO_2 diffusion to C_3 biochemistry or, more likely, to CO_2-concentrating mechanisms, rather than to CAM (or C_4) photosynthesis. While it could be argued that CAM is unlikely in terrestrial plants without functional stomata, the occurrence of CAM in aquatic plants means that it is not intrinsically impossible to have CAM in the absence of stomata. Such a poikilohydric CAM in land plants could help maximize CO_2 fixation (and dry-weight accumulation) in each wetting-drying cycle, although, as with higher plants, the water-cost advantage over C_3 metabolism is diminished at higher atmospheric CO_2 partial pressures (Raven et al. 1987; Callaghan et al. 1992; Raven 1992). Thus the search for CAM in fossil poikilohydric plants may not be fruitless.

24.3 Selective Significance of CAM Traits in Aquatic Plants in the Context of Climates and Palaeoclimates

For *aquatic* CAM plants, major selective significance is attributed to the contribution of CAM to inorganic C acquisition *per se*, since there is no trade-off between C acquisition and water loss in totally submerged aquatics (see also Chap. 18). A multiplicity of mechanisms in aquatic plants are associated with increased inorganic C supply to RUBISCO relative to diffusive CO_2 supply to C_3 biochemistry (Raven 1991b). Such mechanisms are especially diverse in freshwater macrophytes, even in sites where inorganic C supply is increased by input from terrestrial organic C followed by mineralization, or by input of groundwater enriched in inorganic C from soil metabolism. This constitutes a "CO_2 pump" from the atmosphere to the illuminated water body (Raven 1991 a,b).

The implication of these mechanisms for enhancing inorganic C acquisition is that C_3 physiology (diffusive supply of CO_2 to RUBISCO) would make C supply more rate-limiting for growth than is the supply of other resources. CAM (and other mechanisms enhancing CO_2 input) can enhance the rate of CO_2 input to RUBISCO relative to that of other resources, and might also serve to economize on the requirement of N, Fe, Mn and Mo to give a certain rate of C assimilation in the light, but might increase the requirement for light (absorbed mol photons per mol C assimilated): see Raven (1988, 1990, 1991a), Raven et al. (1987, 1988) and Surif and Raven (1989a, 1990).

The distribution of high-amplitude CAM in aquatic plants is limited to freshwater vascular macrophytes (i.e secondarily aquatic plants), and is commonly (*Isoetes*, *Littorella*) associated with the isoetid life-form and thus with CO_2 acquisition from sediment via the roots (see also discussion in Chap. 18). However, not all isoetids exhibit CAM (*Lobelia dortmanna*, *Subularia aquatica*,

Eriocaulon spp. do not), and aquatic CAM *Crassula* spp. are not isoetids (Raven et al. 1988). It is of interest that many of these aquatic CAM plants can also grow on land. In *Littorella* and in many *Isoetes* spp., leaves produced after the transition to land by falling water level develop stomata over at least part of the leaf and exhibit C_3 metabolism with normal stomatal behaviour (Raven et al. 1988; Keeley 1990; Nielsen et al. 1991). However, some *Isoetes* spp. (including *Stylites*) eschew atmospheric CO_2 when growing on land, maintaining the absence of functional stomata, root acquisition of CO_2, and CAM found in submerged specimens; the first two attributes are shared by terrestrial specimens of the non-CAM isoetid *Lobelia dortmanna* (Raven et al. 1988; Pedersen and Sand-Jensen 1992).

The occurrence of CAM and a range of other auxiliary CO_2-acquisition mechanisms in freshwater macrophytes, but not in microphytes such as phytoplankton which only have diffusive CO_2 entry or inorganic C active influx, may in some cases be a function of the life-form: the acid-alkaline zone mechanism of HCO_3^- use needs a certain minimum size, while root acquisition of CO_2 requires roots with gas spaces in the CO_2-rich sediment. However, a wider-ranging possible explanation of the more frequent occurrence of CO_2-acquisition processes (including CAM) in macrophytes than microphytes relates to their C:N, C:P and C:Fe ratios. These ratios are all higher in macrophytes than microphytes (Raven 1984; Duarte 1992), probably as a function of extra C in cell wall material rather than of chronic N, P, or Fe deficiency. This means that, for a given C:N:P:Fe availability in the bulk water, C is more likely to be rate-limiting for growth relative to N, P, or Fe in macrophytes than microphytes.

The most "obvious" role of CAM in submerged aquatics is seen in some *Isoetes* spp. (Keeley et al. 1983) in vernal pools where the bulk water phase undergoes large diel fluctuations in free CO_2 and pH, respectively decreasing and increasing in the daytime, due to photosynthesis by HCO_3^--using microphytes. CAM permits acquisition of CO_2 from bulk water at night, supplemented by root acquisition of CO_2 from sediments (see also Chap. 18). In other situations, the effectiveness of CAM relative to other methods of enhancing inorganic C acquisition is perhaps less readily assessed.

A number of freshwater macrophytes (including characean algae) show much lower amplitude CAM-like biochemical behaviour, insufficient to prevent net CO_2 evolution in the dark phase (Keeley 1990). Similar behaviour is found in a few marine macroalgae (Johnston and Raven 1986; Raven et al. 1985; Surif and Raven 1989b). In these cases CAM-like CO_2 acquisition can only contribute one or two percent to total C accumulation. For the intertidal Fucales, the CAM-like behaviour occurs when exposed to the atmosphere (as occurs in nature at low tide), as well as when submersed (Surif and Raven 1989b). The smaller abundance of auxiliary C-acquisition mechanisms in marine macrophytes than in freshwater macrophytes might relate to the more deterministic spatial and temporal inorganic C supply in marine habitats relative to the stochastic nature of inorganic C in many freshwaters.

We conclude this brief survey of extant aquatic CAM by pointing out that, like terrestrial CAM, plants using aquatic CAM may have lower N, Fe, Mn and Mo costs, but higher photon costs, than plants using diffusive CO_2 entry and C_3 biochemistry, at least in air-equilibrium media. Such an air-equilibrium CO_2 concentration is largely an ecological fiction in freshwaters (Raven 1991b), at least as far as major habitats of aquatic macrophytes are concerned. Generalizations about the relative significance of CAM in the diverse freshwater habitats in which they occur are thus difficult. These difficulties greatly complicate predictions concerning the effects of anthropogenic CO_2 increase on aquatic macrophytes, and especially the relative effects on CAM (and other auxiliary CO_2-acquisition-mechanism)-using plants and C_3 plants (Lemon 1986; Raven 1991b).

Such problems are compounded when we attempt to reconstruct the significance of aquatic CAM in past environments. The disequilibrium of CO_2 between the atmosphere and biologically active freshwaters noted above would have characterized earlier times, and changed atmospheric CO_2 partial pressures would have less predictable effects on inorganic C availability in freshwaters. Nevertheless, it is likely that CO_2 availability in freshwaters would generally have been higher through most of the Phanerozoic (570 to 0 Ma) than it is now, or in the Pleistocene or Upper Carboniferous.

Using similar arguments to those made for terrestrial C_3 plants, we note that extant aquatic CAM plants with a high amplitude of acidity change belong to the lycopsids and two families (Plantaginaceae, Crassulaceae) of angiosperms, with somewhat lower-amplitude CAM reported for three other angiosperm families, the Alismataceae, Hydrocharitaceae and Apiaceae (Chap. 18; Raven et al. 1988; Webb et al. 1988). While the lycopsids have existed since at least the early Devonian, the earliest Isoetaceae are Triassic (225 to 190 Ma) (Raven et al. 1988; Ash and Pigg 1991). The presence of stomata on the fossil specimens is not considered in the descriptions, so it is not clear when the Isoetaceae became (secondarily) aquatic. The timing of origin of aquatic (CAM) Plantaginaceae and Crassulaceae is equally obscure.

As for the possibility that aquatic CAM could be detected in fossil aquatic plants by anatomical or $^{13}C/^{12}C$ data, the likelihood seems smaller than in the case of terrestrial CAM. The anatomical distinctions between terrestrial vascular CAM and C_3 or C_4 plants noted by Gibson (1982) are not generally applicable to aquatic vascular plants, in that the cuticular features would not, for example, distinguish the non-CAM *Lobelia dortmanna* from the CAM *Isoetes* spp. Also, (functional) stomata are absent from all submerged organs of aquatic plants, succulence is not readily defined, and wide spacing of vascular bundles is characteristic of non-CAM as well as CAM aquatic vascular plants (Raven 1984). No distinguishing features of anatomy can be construed for non-vascular aquatic plants with low-amplitude CAM. $^{13}C/^{12}C$ ratios of freshwater aquatic macrophytes are generally not helpful in distinguishing CAM acquisition of inorganic C from other modes of inorganic C assimilation (Raven et al. 1986; Raven 1991b; Keeley and Sandquist 1992) in extant organisms, not least because of the wide temporal and spatial variations in the $^{13}C/^{12}C$ of the source inorganic C. Accordingly, the prospects for using $^{13}C/^{12}C$ to define aquatic CAM in fossil specimens are bleak.

To conclude this review of the evolution of aquatic CAM plants in relation to their environment and the selective significance of CAM, the timing of the evolution of aquatic CAM and the nature of the freshwater environment with respect to inorganic C availability relative to that of other resources during the Phanerozoic are still poorly understood. Furthermore, the prospects for improving our understanding from the anatomy of fossil aquatic plants and their $^{13}C/^{12}C$ ratio are less hopeful than for their terrestrial analogues.

24.4 Mechanistic Considerations of the Evolution of CAM

Initially we shall stand well back from present-day CAM plants and their mechanisms and consider what physical, chemical and biological possibilities and constraints occur on low-energy uptake of inorganic C in the dark phase with very substantial conversion of this inorganic C into organic C in the light using RUBISCO. Both inorganic and organic modes of diel CO_2 storage are considered. For terrestrial plants the possibilities for minimizing water loss as a result of diel storage of inorganic C are emphasized.

Our analysis of these possible CAM-like mechnisms is in the context of their resource costs in comparison with their resource-acquisition benefits. While the costs in terms of one or more potentially scarce resources of the hypothetical CAM-like mechanisms are frequently very large relative to the perceived resource-acquisition benefits, it is not possible to eliminate the possibility that these mechanisms would have selective advantage under certain conditions, especially if they are not in competition with the present-day CAM mechanism. We certainly do not wish to discourage further investigation of these possibilities; the absence of any of these possibilities in the real world might reflect the absence of the appropriate variant in the first place rather than their elimination as a result of excessive costs. We also note that recurrent, but perhaps rare, environmental extremes may have been important in the evolution of CAM. Here a mechanism such as that suggested by West-Eberhard (1986), in which alternative phenotypes are expressed under extreme conditions where the benefit of a solution that is usually too costly to be favoured is unusually high, could be operative.

Dealing first with *inorganic* C storage between dark CO_2 uptake and light CO_2 release and refixation, we consider the possibilities of storage in the gas phase, in solution, and as a solid.

Intercellular gas spaces, including the gas bladders of some Fucalean brown algae, are almost universal in CAM plants. In principle, external CO_2 could be pumped into an intercellular gas phase using modifications of mechanisms known for CO_2-accumulation mechanisms based on transmembrane active inorganic C flux or a C_4-C_3 cycle as in C_4 plants. To store quantities of CO_2

equal to the highest quantities stored as malic acid ($\sim$ 500 mol m^{-3}) in a tissue with a gas-phase volume equal to the vacuolar volume (cf. the gas-phase volume of 5% of tissue volume in extant CAM plants: Smith and Heuer 1981) requires a CO_2 partial pressure of 1.2 MPa ($\approx$ 12 atmospheres). While *total* gas pressures in excess of this have biological precedent (10 MPa in teleost gas bladders; Pelster and Scheid 1992), and pressure differences of 3 MPa can be withstood by plant cell-wall structures (Carpita 1985), CO_2 leakage would be a problem, as would CO_2 narcosis of metabolism (Amthor 1991) unless the gas-phase CO_2 could be kept out of equilibrium with cell water, a difficult proposition granted the CO_2-secretion mechanisms proposed. Gas-phase storage of CO_2 only plays an incidental and quantitatively minor role in CAM today, although CAM plants transiently show (Phase III) some of the highest intercellular CO_2 partial pressures known for plants (Cockburn et al. 1979; Spalding et al. 1979).

CO_2 storage in solution is also a possibility, although leakage and narcosis problems would limit its quantitative role if storage was as free CO_2. Storage as HCO_3^- in solution requires a very alkaline aqueous phase if high free-CO_2 levels are to be avoided; keeping the inorganic C in solution requires strict limits on the cation composition unless extreme supersaturation can be maintained. Substantial intracellular aqueous phases at a suitably high pH (pH 10) do not exist in plant cells; vacuoles are almost always at pH $\leqslant$ 6. Furthermore, the required reversible production and consumption of OH^- in the shoot of a terrestrial plant poses great difficulties (Raven 1986).

Extracellular phases of high pH could act as traps for atmospheric CO_2 in terrestrial plants. Nectaries of the holoparasite *Lathraea* can be quite alkaline (pH $\geqslant$ 9: Prÿs-Jones and Willmer 1992), but CO_2 storage is of little use to a non-photosynthetic plant. Surface fluids maintained at a high pH (up to 10) occur on leaves of several species of the Malvaceae (Elleman and Entwhistle 1982) and in *Tamarix* (Waisel 1991); in the latter case use as a "CO_2 store" has been proposed by Waisel (1991). The quantity of CO_2 that can be stored in solution has not been quantified, but the occurrence of leaf surface fluid is not a promising augury for reducing the water cost of C acquisition (Raven 1986).

An alternative potential store of CO_2 in solution is in oil accumulations in leaves. However, Meyer and Canny (1975) showed that the quantity of CO_2 at air-equilibrium concentrations in the oil depots in *Eucalyptus* leaves could only support a few minutes of light-saturated photosynthesis. This means that a CAM-like role of CO_2 storage in oil would not be useful, even if oil production had other fitness-enhancing functions, such as chemical defence and increasing the likelihood of fires, which might confer a competitive advantage over dominant, less fire-tolerant species (Gill et al. 1981). One possibility related to the water cost of growth is that volatilization of ethereal oils decreases the rate of water loss (e.g. Atay 1958; Heilbronn 1958); however, this can be dismissed as a significant role of these "secondary products" (Audus and Cheetham 1940; J.A. Raven, unpubl. calculations on C cost of using oil volatilization to replace H_2O volatilization in the thermal balance of photosynthetic organs).

Inorganic C storage as a solid is also a possibility. The secretion of alkaline solutions has already been mentioned for *Tamarix*; co-secretion of Ca^{2+} can give $CaCO_3$ precipitates on the surface of leaves of *Tamarix* and *Saxifraga*. $CaCO_3$ is also deposited inside aerial shoots (but outside cells) of various higher plants (e.g. Ulmaceae, Urticaceae, Moraceae) as cystoliths, where their role relates inter alia to acid-base regulation (Raven 1985). Extracellular $CaCO_3$ deposition is also widespread in freshwater and marine macrophytes (algae and vascular plants in freshwater; only algae in the sea), and Raven (1984) has presented a scheme in which diel changes in $CaCO_3$ (deposition at night, solution in the day) might act in a CAM-like fashion, with large H^+/Ca^{2+} exchange fluxes between the cell wall and the vacuole. However, in none of these cases is there any evidence that the $CaCO_3$ deposits act in a CAM-like acquisition mechanism.

As with CO_2 storage in an inorganic form, the possibilities for storage of CO_2 in a diel CAM cycle in an organic form include organic C in the gas phase, in solution (the actual method used in extant CAM plants), and as a solid. The gaseous possibility seems relatively unlikely; aside from the problems of gaseous storage of CO_2 mentioned earlier, the energetics of producing organic gases would be as great as reduction of CO_2 to the carbohydrate; accordingly, this is not a likely CAM-like mechanism.

Clearly the "dissolved organic C" option is an important one to consider, in that extant CAM does involve diel inorganic C storage as dissolved organic acids (mainly malic acid). This process involves a non-existent or a relatively small increment of energetic (ultimately photon) running costs over those of C_3 (or C_4) photosynthesis in the present atmosphere, and a small (relative to many "dissolved organic C" alternatives) capital cost of providing the 3 C organic skeletons to which the CO_2 is attached to form malic acid in terms of energy and C, but none in terms of N. The diel fluctuations in osmolality are greater in CAM plants (usually NAD(P)-ME, i.e. NAD(P)-malic enzyme as decarboxylating enzyme) with starch as the storage form of 3 C organic skeletons than in those with hexose or sucrose as the storage form (PEPCK, i.e. phospho*enol*pyruvate carboxykinase as decarboxylating enzyme) (Smith and Bryce 1992). The scheme suggested by Raven et al. (1988) for *aquatic* plants to explain why CAM could be missed in a survey of "titratable acidity of homogenates" by having an H^+-cation exchange across the plasmalemma and $2K^+$-malate^{2-} accumulation in the vacuole in Phase I of CAM has not been demonstrated in any CAM plant, and would in any case cost at least 2 mol ATP per net CO_2 fixed in the dark. It could not work in a terrestrial plant, at least as far as H^+-K^+ exchange with the environment is concerned (see Raven 1977).

An alternative to the secretion of organic acid into the vacuole is secretion into the apoplast, albeit at a greater energy expense due to the H^+: ATP stoichiometry of 1 at the plasmalemma rather than 2 at the tonoplast. Organic-acid secretion via leaf and stem glands of *Cicer arietinum* is *not* part of a CAM mechanism (Koundal and Sinha 1981, 1983; Winter 1981; Lauter and Munns 1986).

Net fixation of CO_2 in the dark to form a carboxyl compound need not involve PEPC; if a dicarboxylate C_4 compound is to be the stored product, it is possible that PC (pyruvate carboxylase) could be involved in higher plants (Wurtelle and Nikolau 1990), while PEPCK is involved in the low-amplitude CAM-like behaviour of fucoid algae (Johnston and Raven 1989); these two enzymes share O_2-independence with PEPC, although the HCO_3^--using PEPC has a much higher affinity expressed in terms of CO_2 than does PEPCK.

Other carboxylating enzymes that would yield stored organic compounds other than dicarboxylic 4C acids are present in plants. These enzymes include ACC and CPS.

ACC (acetyl CoA carboxylase) is involved in the synthesis of malonyl CoA, which is involved in almost all syntheses of long-chain fatty acids in phototrophs (Dennis and Turpin 1990); an exception is the fermentative production of wax esters in *Euglena* (see Raven 1984). It shares with PEPC the use of HCO_3^- with relatively high affinity, meaning, at the pH of the chloroplast stroma, a high affinity in terms of atmospheric CO_2, and O_2-independence. Malonic acid could be stored and decarboxylated to produce acetic acid. Problems arise for such CAM photosynthesis to the extent that the nature of the only feasible analogue of starch or hexose is long-chain fatty acids, so that large quantities of reductant would be generated in acetyl CoA generation in Phase I (2 NADH per acetyl unit) and required in malonyl CoA use in Phase III in fatty-acid resynthesis and CO_2 release. Compared to this major problem of disposing of the excess NADH, the ATP costs of reforming the malonyl CoA in Phase III, assuming storage of malonic acid in Phase I, and of transporting malonic acid into the vacuole in Phase I, are relatively trivial. This redox imbalance (excess in Phase I, deficit in Phase III) cannot be readily accommodated without additional capital inputs of an organic oxidant for Phase I and reductant for Phase III, e.g. an aldose/alditol couple, with associated running costs of active transport of alditol into the vacuole in Phase I and of aldose into the vacuole in Phase III, and a higher basal osmolality on which the increase in Phase I is superimposed. The capital and running costs of the putative ACC-based CAM, if a variant arose that could do this, would not be favoured by selection.

CPS (carbamoyl phosphate synthetase) converts HCO_3^-, NH_4^+ (or amido-N of glutamine) and ATP into carbamoyl phosphate and ADP (Atkinson 1992). Production of carbamoyl phosphate in Phase I, with vacuolar deposition until release of CO_2 in the cytosol in Phase III, would provide a CO_2-storage mechanism costing only 2 ATP per C per diel cycle. However, aside from the instability of carbamoyl phosphate, we have to consider the capital costs of N and P. With 500 mol CO_2 m^{-3} stored at the end of Phase I, 500 mol N m^{-3} and 500 mol P m^{-3} are needed as capital investment. With 450 kg C in the dry weight and a fresh : dry weight ratio of 10 for a succulent shoot, the "basal" C level is 3750 mol C m^{-3}. With molar C : N of 15, the "basal" N level is 250 mol C m^{-3}. Accordingly, the N capital for CO_2 storage is twice the basal, catalytic/structural N level. With a basal molar ratio of N : P of 20, the basal P level is 12.5 mol P m^{-3}. Thus, the P capital for CO_2 storage would be 40 times the basal catalytic/structural P level. While terrestrial CAM is associated with water as the limiting resource and aquatic CAM with C as the limiting resource, 3-fold and 40-fold increases in demand for N and P, respectively, could greatly alter this resource-limitation situation.

CPS involvement in producing a more stable CO_2-storing solute in Phase I could involve a urea cycle with (using NH_4^+ as the N donor for CPS) a running cost of 4 ATP per CO_2; an extra ATP (5 ATP per CO_2) is used if glutamine is the N donor for CPS (Atkinson 1992). While the P capital cost is less than for carbamoyl phosphate storage (since P is recycled in arginine synthesis), the N capital cost is higher, i.e. four times the basal N level. Furthermore, the recycling of N imposes osmotic problems and also extra ATP running costs of transporting organic- and NH_4^+-N across the tonoplast.

The capital and running costs of either of the hypothetical CPS-based CAM processes considered above appear, like those of the proposed ACC-based mechanism, to be adequate to explain why they would not be selected even if this occurred.

A final organic storage mechanism for CO_2 in a CAM-like process is one which the exemplary work of Cockburn and McAulay (1975) has eliminated for "real" CAM, i.e. CO_2 fixation in the dark catalyzed by RUBISCO. Such a mechanism may appear to be entirely counter-productive in terms of CO_2 gain and water loss, since the generation of NADPH and ATP needed for CO_2 fixation by RUBISCO in the dark by the oxidative pentose phosphate pathway, glycolysis, and the tricarboxylic acid cycle plus oxidative phosphorylation involves the generation of more CO_2 from organic C produced in the light at significant water cost in terrestrial plants than is fixed by RUBISCO in the dark. This is easily seen by considering that the 2.9 NADPH needed to fix 1 CO_2 (net) in C_3 photosynthesis in the extant atmosphere involves the production of 1.45 CO_2, while the 4.5 ATP needs ($ATP/CO_2 = 6$) conversion of a further 0.75 organic C into CO_2, i.e. a total of 2.2 CO_2. This would be higher if ATP generation in chloroplasts occurs at a lower ATP/CO_2 ratio.

However, there is hope for RUBISCO as a means of fixing CO_2 in the dark, albeit at substantial energy cost, and increasing net 24-h C gain. The suggestion is that aldoses are converted to alditols in the light, a process requiring no CO_2 input or water loss, using photoproduced NADPH, in parallel with the use of photoproduced NADPH and ATP in fixing exogenous CO_2 (with water loss in terrestrial examples). In the dark, oxidation of alditols to aldoses yields NAD(P)H, which could be used for the reductive aspects of RUBISCO-catalyzed CO_2 fixation, and also to generate ATP by oxidative phosphorylation. Based on pentitol (e.g. ribitol), the mechanism would, in the present atmosphere, consume at least 4.3 pentitol (22 C) in fixing 1 CO_2 in the dark, yielding 1 hexose and 3.3 pentose. In the light, 4.3 pentitol must be produced for the next night's CO_2 fixation, leaving 1 reduced C for biosynthesis. Further aldose-alditol cycling would be needed to support transtonoplast fluxes implicit in this proposal. At all events a large capital cost of organic C is involved.

We note that the aldose/alditol interconversions required here are the *reverse* of those suggested for redox balancing with ACC.

The conclusions as to the "alternative" CAM-like possibilities involving soluble organic compounds as CO_2 stores produced in Phase I are that "actual" CAM involves lower energetic running costs and/or capital costs [energy, carbon (and hence water in terrestrial plants)] and, for CPS, N and P. The same conclusion applies to the "honorary CAM" of dark activity of RUBISCO.

Overall, the conclusion is that the running and/or capital costs, or the toxicity, of the alternative CAM possibilities involving inorganic C storage in gaseous, soluble or solid inorganic forms, or insoluble organic forms, are greater than those of the actual CAM mechanism and would make the hypothetical alternatives to CAM relatively non-competitive with actual CAM.

Space does not permit a detailed consideration of the extent to which the structures and catalysts of homoiohydric CAM plants have homologues in C_3 homoiohydric plants (or the likely evolutionary intermediates, CAM-cycling homoiohydric plants: Gravatt and Martin 1992; Pilon-Smits 1992; see also Chaps. 12 and 23), or have parallels in C_4 homoiohydric plants (see Osmond 1978; Cockburn 1981, Osmond et al. 1982; Winter 1985; Lüttge 1987; Griffiths 1988, 1989, 1992; Nobel 1991). It appears that all the enzymes and membrane-associated transporters used in CAM in either the starch-NAD(P)-ME-PPDK or the hexose-PEPCK types have homologues in C_3 homoiohydric plants (Cockburn 1981, 1985; Nelson and Langdale 1992; Smith and Bryce 1992). However, a number of changes in the main site(s) of intracellular expression of enzymes (e.g. carbonic anhydrase) and in regulation (e.g. of vacuolar H^+, K^+ and malate^{2-}, and of PEPC) have occurred in the evolution of CAM from C_3 metabolism (Reed 1979; Tsuzuki et al. 1982; Holtum et al. 1984; Cushman and Bohnert 1989a,b; Bremberger and Lüttge 1992; Sanders et al. 1992; Smith and Bryce 1992).

Concerning the main sites of intracellular expression of enzymes, carbonic anhydrase in C_3 terrestrial plants is mainly expressed in the chloroplast stroma, although there is activity in the cytosol (Reed 1979). A major role for carbonic anhydrase in C_3 terrestrial plants is in enhancing the diffusion of inorganic C from the site at which atmospheric CO_2 dissolves in cell wall water to the site of CO_2 consumption by RUBISCO by enlisting HCO_3^- as an inorganic C species diffusing in parallel with CO_2 in the cytosol and in the plastid stroma. The importance of HCO_3^- is that it is present at an equilibrium concentration at least an order of magnitude greater than that of CO_2; the need for carbonic anhydrase relates to the slowness of the uncatalyzed interconversion of CO_2 and HCO_3^- relative to the CO_2 requirement for C_3 photosynthesis when all rates are expressed in terms of the volume of cytosol or stroma (Raven and Glidewell 1981; Raven and Newman 1994). Leaves of C_3 plants have carboxylases (PEPC, CPS) that use HCO_3^-, although whether they need carbonic anhydrase to supply their HCO_3^- from respiratory or exogenous CO_2 is not clear (Raven and Newman 1994).

CAM (and C_4) photosynthesis is characterized by the HCO_3^--consuming PEPC in the cytosol. Holtum et al. (1984) demonstrated, using an elegant mass-spectrometric technique, that the cytosolic carbonic-anhydrase activity exceeded the requirement for dark CO_2 fixation in all the CAM plants tested, although the work of Tsuzuki et al. (1982), using the Wilbur-Anderson technique, did not find significant cytosolic carbonic-anhydrase activity in some of the CAM plants they tested. Clearly this field is ripe for investigation by molecular techniques, which have recently yielded important information on the phylogeny of carbonic anhydrases in a number of non-CAM photolithotrophs (Fukuzawa et al. 1992).

As to PEPC, this enzyme is involved in a wide range of biosyntheses and in acid-base regulation in all C_3 terrestrial plants, and in the supply of C_4 dicarboxylates to plastids, mitochondria and N_2-fixing symbionts in root nodules in some of them. This "housekeeping" form of the enzyme is present constitutively in C_4 and CAM plants, and the nucleotide sequence of the gene for this form in the facultative CAM plant *Mesembryanthemum crystallinum* has been determined (Cushman and Bohnert 1989a). The gene sequence for the CAM-specific isoform of PEPC in *Mesembryanthemum crystallinum* is also known (Cushman and Bohnert 1989b), so the basis for the regulatory changes in evolution of the CAM isoform of PEPC (Rustin et al. 1988) should soon be deciphered.

As a final and, perhaps, trivial point on the evolution of terrestrial CAM, and which we do not wish to be taken as a belief on our part that phylogeny is recapitulated in ontogeny, it may be worth noting that even "obligate" CAM plants are C_3 as seedlings (Altesor et al. 1992; see Chap. 25).

As for the secondarily aquatic vascular plants (Raven 1984; Raven et al. 1988), the aquatic CAM species of *Crassula* are apparently derived from terrestrial CAM species of that genus. However, this is not likely to be the case for the aquatic CAM species of *Isoetes*, or for *Littorella uniflora*; there are no known terrestrial CAM members of the Isoetaceae or Plantaginaceae that take up CO_2 through stomata. *Littorella uniflora* is the only known CAM species of the Plantaginaceae; when grown with its leaves in a high-humidity atmosphere it shows CAM, but at lower humidities it develops functional stomata, and performs C_3 metabolism using CO_2 from the atmosphere (Raven et al. 1988). Most aquatic *Isoetes* spp. develop stomata and show C_3 photosynthesis when grown in air; others (e.g. *Isoetes triquetra*, *I. andina*, *I.* (=*Stylites*) *andicola*) continue to obtain CO_2 via the roots from the soil/sediment and to exhibit CAM (Raven et al. 1988). Overall, it appears that *Isoetes* spp. and *Littorella uniflora* evolved CAM after their ancestors became aquatic. This means that all of the "C_3 plus stomata" mechanistic precursors of CAM suggested by Cockburn (1981, 1985) as involved in the evolution of CAM were available to these secondarily aquatic vascular plants. Alas, too little is known of the mechanism of "CAM-like" processes in fucoids, etc., to comment usefully on their evolution (cf. Johnston and Raven 1986, 1989).

Acknowledgements. $\delta^{13}C$ values in Table 24.1 were obtained on a VG SIRA Series II Isotope Ratio Mass Spectrometer obtained on a NERC grant to JAR and Professor J.I. Sprent; the measurements were performed by Dr. C.M. Scrimgeour. Comments from Drs. M.J. West-Eberhard and C.J. Hutchison were most helpful.

References

Altesor A, Ezcurra E, Silva C (1992) Changes in photosynthetic metabolism during the early ontogeny of four cactus species. Acta Oecol 13: 777–785

Amthor JS (1991) Respiration in a future, high-CO_2 world. Plant Cell Environ 14: 13–20

Ash SR, Pigg KB (1991) A new Jurassic *Isoetites* (Isoetales) from the Wallowa Terrane in Hells Canyon, Oregon and Idaho. Am J Bot 78: 1636–1646

Atay S (1958) Über die Einwirkung der ätherischen Öle auf die Evaporation und Transpiration. Istanbul Univ Fen Fak Mecm Ser B 23: 143–170

Atkinson DE (1992) Functional roles of urea synthesis in vertebrates. Physiol Zool 65: 243–267

Audus LJ, Cheetham AN (1940) Investigations on the significance of ethereal oils in regulating leaf temperatures and transpiration rates. Ann Bot (New Ser) 4: 465–483

Barnola JM, Raynaud D, Korotkevich YS, Lorius C (1987) Vostok ice core provides 160 000-year record of atmospheric CO_2. Nature 329: 408–414

Behrensmeyer AK, Damuth JD, Di Michele WA, Potts R, Sues H-D, Wing SL (eds) (1992) Terrestrial ecosystems through time. University of Chicago Press, Chicago

Benzing DH (1989) The evolution of epiphytism. In: Lüttge U (ed) Vascular plants as epiphytes: evolution and ecophysiology. Springer, Berlin Heidelberg New York, pp 15–41

Berner RA (1990) Atmospheric carbon dioxide levels over phanerozoic time. Science 249: 1382–1386

Berner RA (1993) Palaeozoic atmospheric CO_2: importance of solar radiation and plant evolution. Science 249: 1382–1386

Berner RA, Canfield DE (1989) A new model for atmospheric oxygen over phanerozoic time. Am J Sci 289: 333–361

Björkman O (1975) Thermal stability of the photosynthetic apparatus in intact leaves. Carnegie Inst Washington Year Book 74: 748–751

Björkman O, Badger MR, Armond PA (1978) Thermal acclimation of photosynthesis: effect of temperature on photosynthetic characteristics of the photosynthetic apparatus in *Nerium oleander*. Carnegie Inst Washington Year Book 77: 262–282

Bocherens H, Fizet N, Cuif JP, Jaeger J-J, Michard J-G, Mariotti A (1988) Premières mèsures d'abondances isotopiques naturelles en ^{13}C et ^{15}N de la matière organique fossile de Dinosaure. Application à l'étude du regimes alimentaire du genre *Anatosaurus* (Ornitischia, Hadrosauridae). C R Hebd Acad Sci, Paris, D Serie II 306: 1521–1525

Bocherens H, Friis EM, Mariotti A, Pedersen KR (1993) Carbon isotope abundances in Mesozoic and Coenozoic fossil plants: palaeoecological implications. Lethaia 26: 347–358

Brandl R, Mann W, Sprinzl M (1992) Estimation of the monocot-dicot age through RNA sequences from the chloroplast. Proc R Soc Lond B 249: 13–17

Bremberger C, Lüttge U (1992) Dynamics of tonoplast proton pumps and other tonoplast proteins of *Mesembryanthemum crystallinum* L. during the induction of crassulacean acid metabolism. Planta 188: 575–580

Callaghan TV, Sonesson M, Somme L (1992) Responses of terrestrial plants and invertebrates to environmental change at high latitudes. Philos Trans R Soc Lond B 338: 279–288

Carpita NC (1985) Tensile strength of cell wall of living cells. Plant Physiol 79: 485–488

Cerling TE, Wang Y, Quade J (1993) Expansion of C_4 ecosystems as an indicator of global ecological change in the late Miocene. Nature 361: 344–345

Cockburn W (1981) The evolutionary relationship between stomatal mechanism, crassulacean acid metabolism and C_4 photosynthesis. Plant Cell Environ 4: 417–418

Cockburn W (1985) Variation in photosynthetic metabolism in vascular plants: CAM and related phenomena. New Phytol 101: 3–24

Cockburn W, McAulay A (1975) The pathway of carbon dioxide fixation in crassulacean plants. Plant Physiol 55: 87–89

Cockburn W, Ting IP, Sternberg LO (1979) Relationships between stomatal behavior and internal carbon dioxide concentration in crassulacean acid metabolism plants. Plant Physiol 63: 1029–1032

Crane PR (1993) Time for the angiosperms. Nature 366: 631–632

Crane PR, Lidgard S (1990) Angiosperm radiation and patterns of Cretaceous palynological diversity. In: Taylor PD, Larwood GP (eds) Major evolutionary radiations. Clarendon, Oxford, pp 377–407

Crane PR, Donoghue MJ, Doyle JA, Friis EM (1989) Angiosperm origins. Nature 342: 131–132

Cushman JC, Bohnert HJ (1989a) Nucleotide sequence of the *Ppc2* gene encoding a housekeeping isoform of phosphoenolpyruvate carboxylase from *Mesembryanthemum crystallinum*. Nucleic Acids Res 17: 6743–6744

Cushman JC, Bohnert HJ (1989b) Nucleotide sequence of the gene encoding a CAM specific isoform of phosphoenolpyruvate carboxylase from *Mesembryanthemum crystallinum*. Nucleic Acids Res 17: 6745–6746

Dennis DT, Turpin DH (eds) (1990) Plant physiology, biochemistry and molecular biology. Longman, Harlow

Duarte CM (1992) Nutrient concentration of aquatic plants: patterns across species. Limnol Oceanogr 37: 882–889

Edwards DI, Seldon PA (1992) The development of early terrestrial ecosystems. Bot J Scotland 46: 337–365

Edwards GE, Walker DA (1983) C_3, C_4: mechanisms, and cellular and environmental regulation, of photosynthesis. Blackwell, Oxford

Elleman CJ, Entwhistle PF (1982) A study of glands on cotton responsible for the high pH and cation concentration on the leaf surface. Ann Appl Biol 100: 553–558

Evans JR, Seeman JR (1989) The allocation of protein nitrogen in the photosynthetic apparatus: costs, consequences and control. In: Briggs WR (ed) Photosynthesis. A R Liss, New York, pp 183–205

Farquhar GD, Ehleringer JR, Hubick KT (1989) Carbon isotope discrimination and photosynthesis. Annu Rev Plant Physiol Plant Mol Biol 40: 503–537

Frakes LA, Francis JE, Syktus JI (1992) Climate modes of the Phanerozoic. Cambridge University Press, Cambridge

Fukuzawa H, Suzuki E, Komukai Y, Miyachi S (1992) A gene homologous to chloroplast carbonic anhydrase (*icfA*) is essential for photosynthetic carbon dioxide fixation by *Synechococcus* PCC 7942. Proc Natl Acad Sci USA 89: 4437–4441

Gerwick BG, Williams GJ III (1978) Temperature and water regulation of gas exchange of *Opuntia polyacantha*. Oecologia 35: 149–159

Gibson AC (1982) The anatomy of succulence. In: Ting IP, Gibbs M (eds) Crassulacean acid metabolism. American Society of Plant Physiologists, Rockville, pp 1–17

Gill AM, Groves RH, Nobel IR (eds) (1981) Fire and the Australian biota. The Australian Academy of Science, Canberra

Gravatt DA, Martin CE (1992) Comparative ecophysiology of five species of *Sedum* (Crassulaceae) under well-watered and drought-stressed conditions. Oecologia 92: 532–541

Griffiths H (1988) Crassulacean acid metabolism: a re-appraisal of physiological plasticity in form and function. Adv Bot Res 15: 43–92

Griffiths H (1989) Carbon dioxide concentrating mechanisms and the evolution of CAM in vascular epiphytes. In: Lüttge U (ed) Vascular plants as epiphytes: evolution and ecophysiology. Springer, Berlin Heidelberg New York, pp 42–86

Griffiths H (1992) Carbon isotope discrimination and the integration of carbon assimilation pathways in terrestrial CAM plants. Plant Cell Environ 15: 1051–1062

Hasebe M, Ito M, Kofuji R, Iwatsuki K, Veda K (1992) Phylogenetic relationships in Gnetophyta deduced from *rbc*L gene sequences. Bot Mag Tokyo 105: 385–391

Heilbronn A (1958) Über die Oberflächen-Aktivität ätherischer Öle und die biologische Bedeutung dieses Phänomens. Istanbul Univ Fen Fak Mecm Ser B 23: 131–141

Holtum JAM, Summons R, Roeske CA, Comins N, O'Leary MH (1984) Oxygen-18 incorporation into malic acid during nocturnal carbon dioxide fixation in crassulacean acid metabolism. A new approach to estimating in vivo carbonic anhydrase activity. J Biol Chem 259: 6870–6881

Jane JW, Urbauer JL, O'Leary MH, Cleland WW (1992) Mechanistic studies of phosphoenolpyruvate carboxylase from *Zea mays* with (Z)- and (E)-3-fluoro-phosphoenolpyruvate as substrate. Biochemistry 31: 6432–6440

Johnston AM, Rave JA (1986) Dark fixation studies on the intertidal macroalga *Ascophyllum nodosum* (Phaeophyta). J Phycol 22: 78–83

Johnston AM, Raven JA (1989) Extraction, partial purification and characterization of phosphoenolpyruvate carboxykinase from *Ascophyllum nodosum*. J Phycol 25: 568–576

Keeley JA (1990) Photosynthetic pathways in freshwater aquatic plants. Trends Ecol Evol 5: 330–333

Keeley JE, Sandquist DR (1992) Carbon: freshwater plants. Plant Cell Environ 15: 1021–1035

Keeley JE, Mathews RP, Walker CM (1983) Diurnal acid metabolism in *Isoetes howellii* from a temporary pool and a permanent lake. Am J Bot 70: 854–857

Koundal KK, Sinha SK (1981) Malic acid exudation and photosynthetic characteristics in *Cicer arietinum*. Phytochemistry 20: 1251–1252

Koundal KR, Sinha SK (1983) Evaluation of the significance of malic acid secretion in chickpea. Physiol Plant 58: 189–192

Kutzbach JE, Gallimore RG (1989) Pangean climates. Megamonsoons of the megacontinent. J Geophys Res 94: 3341–3357

Lauter DJ, Munns DN (1986) Water loss via the glandular trichomes of chickpea (*Cicer arietinum* L.). J Exp Bot 37: 640–649

Lemon ER (ed) (1986) CO_2 and plants. The responses of plants to rising levels of atmospheric carbon dioxide. American Association for the Advancement of Science, Washinton, DC

Li WH, Gouy M, Wolfe KH, Sharp PM (1989) Angiosperm origins. Nature 342: 131–132

Long SR (1991) Modification of the response of photosynthetic productivity to rising temperature by atmospheric CO_2 concentrations: has its importance been underestimated? Plant Cell Environ 14: 729–739

Lüttge U (1987) Carbon dioxide and water demand: crassulacean acid metabolism (CAM), a versatile ecological adaptation exemplifying the need for integration in ecophysiological work. New Phytol 106: 593–629

Lüttge U (1989) Vascular epiphytes: setting the scene. In: Lüttge U (ed) Vascular plants as epiphytes. Springer, Berlin Heidelberg New York, pp 1–14

Marino BD, McElroy MB, Salawitch RJ, Spaulding WG (1992) Glacial-to-interglacial variations in the carbon isotope composition of atmospheric CO_2. Nature 357: 461–466

Martin W, Gierl A, Saedler H (1989a) Molecular evidence for pre-Cretaceous angiosperm origins. Nature 339: 46–48

Martin W, Gierl A, Saedler H (1989b) Angiosperm origins. Nature 342: 132

Meyer CP, Canny MJ (1975) CO_2 storage in *Eucalyptus* oil glands: a hypothesis disproved. Aust J Plant Physiol 2: 647–658

Mooney HA, Björkman O, Collatz GJ (1978) Photosynthetic acclimation to temperature in the desert shrub *Larrea divaricata*. I. Carbon dioxide exchange characteristics of intact leaves. Plant Physiol 61: 406–410

Nelson T, Langdale JA (1992) Developmental genetics of C_4 photosynthesis. Annu Rev Plant Physiol Mol Biol 43: 25–47

Nielsen SL, Gacia E, Sand-Jensen K (1991) Land plants of amphibious *Littorella uniflora* (L.) Aschers maintain utilization of CO_2 from the sediment. Oecologia 88: 258–262

Nobel PS (1991) Achievable productivities of certain CAM plants: basis for high values compared with C_3 and C_4 plants. New Phytol 119: 183–205
O'Leary MH (1981) Carbon isotope fractionation in plants. Phytochemistry 20: 153–517
Osmond CB (1978) Crassulacean acid metabolism: a curiosity in context. Annu Rev Plant Physiol 29: 379–414
Osmond CB, Winter K, Ziegler H (1982) Functional significance of different pathways of photosynthesis. In: Lange OL, Nobel PS, Osmond CB, Ziegler H (eds) Physiological plant ecology II. Encyclopedia of plant physiology: New Series, vol 12B. Springer, Berlin Heidelberg New York, pp 479–547
Parrish JT, Ziegler AM, Scotese CR (1982) Rainfall patterns and the distribution of coals and evaporites in the Mesozoic and Cenozoic. Palaeogeogr Palaeoclimatol Palaeoecol 40: 67–101
Pedersen O, Sand-Jensen K (1992) Adaptations of submerged *Lobelia dortmanna* to aerial life form: morphology, carbon sources and oxygen dynamics. Oikos 65: 85–96
Pelster B, Scheid P (1992) Countercurrent concentration and gas secretion in the fish swim bladder. Physiol Zool 65: 1–16
Pilon-Smits (1992) Variation and evolution of crassulacean acid metabolism in *Sedum* and *Aeonium* (Crassulaceae). PhD Thesis, Rijksuniversiteit te Utrecht, Utrecht
Proctor MCF, Raven JA, Rice SK (1992) Stable carbon isotope discrimination measurements in *Sphagnum* and other bryophytes: physiological and ecological implications. J Bryol 17: 193–202
Prÿs-Jones OE, Willmer P (1992) The biology of alkaline nectar in the purple toothwort (*Lathraea clandestina*): ground level defences. Biol J Linn Soc 45: 373–388
Raven JA (1977) H^+ and Ca^{2+} in phloem and symplast: relation of relative immobility to the cytoplasmic nature of the transport paths. New Phytol 79: 465–480
Raven JA (1984) Energetics and transport in aquatic plants. AR Liss, New York
Raven JA (1985) Regulation of pH and generation of osmolarity in vascular land plants: costs and benefits in relation to efficiency of use of water, energy and nitrogen. New Phytol 101: 25–77
Raven JA (1986) Biochemical disposal of excess H^+ in growing plants? New Phytol 104: 175–206
Raven JA (1988) The iron and molybdenum use efficiencies of plant growth with different energy, carbon and nitrogen sources. New Phytol 109: 279–287
Raven JA (1990) Predictions of Mn and Fe use efficiencies of phototrophic growth as a function of light availability for growth and C assimilation pathway. New Phytol 116: 1–18
Raven JA (1991a) Plant responses to high CO_2 concentrations: relevance to previous high O_2 episodes. Palaeogeogr Palaeoclimatol Palaeoecol (Global and Planetary Change Section) 97: 19–38
Raven JA (1991b) Implications of inorganic C utilization: ecology, evolution and geochemistry. Can J Bot 69: 908–924
Raven JA (1992) Energy and nutrient acquisition by autotrophic symbioses. Symbiosis 14: 33–60
Raven JA (1993) The evolution of vascular land plants in relation to quantitative functioning of dead water-conducting cells and stomata. Biol Rev 68: 337–363
Raven JA, Farquhar GD (1990) The influence of N metabolism and organic acid synthesis on the natural abundance of C isotopes in plants. New Phytol 116: 505–529
Raven JA, Glidewell SM (1981) Processes limiting photosynthetic conductance. In: Johnson CB (ed) Physiological processes limiting plant productivity. Butterworths, London, pp 109–136
Raven JA, Johnston AM (1991) Photosynthetic carbon assimilation by *Prasiola stipitata* (Prasiolales, Chlorophyta) under emersed and submersed conditions: relationship to the taxonomy of *Prasiola*. Br Phycol J 26: 247–257
Raven JA, Newman JR (1994) Requirement for carbonic anhydrase activity in processes other than photosynthetic inorganic carbon assimilation. Plant Cell Environ 17: 123–130

Raven JA, Osborne BA, Johnston AM (1985) Uptake of CO_2 by aquatic vegetation. Plant Cell Environ 8: 417–425

Raven JA, MacFarlane JJ, Griffiths H (1986) The application of carbon isotope techniques. In: Crawford RMM (ed) Plant life in aquatic and amphibious habitats. Blackwell, Oxford, pp 129–149

Raven JA, Johnston AM, MacFarlane JJ, Surif MB, McInroy SG (1987) Diffusion and active transport of inorganic carbon species in freshwater and marine macroalgae. In: Biggins J (ed) Progress in photosynthesis research, vol 4. Nijhoff/Junk, Dordrecht, pp 333–340

Raven JA, Handley LL, McInroy S, McKenzie L, Richards JH, Samuelsson G (1988) The role of root CO_2 uptake and CAM in inorganic C acquisition by plants of the isoetid life form. A review, with new data on *Eriocaulon decangulare*. New Phytol 108: 1–20

Reed ML (1979) Intracellular location of carbonate dehydratase (carbonic anhydrase) in leaf tissue. Plant Physiol 63: 216–217

Rieley G, Collier RJ, Jones DM, Eglinton G, Eakin PA, Fallick AE (1991) Sources of sedimentary lipids deduced from stable carbon-isotope analyses of individual compounds. Nature 352: 425–427

Rustin P, Meyer CR, Wedding RT (1988) Identification of substrate and effector binding sites of phosphoenolpyruvate carboxylase from *Crassula argentea*. A possible role of phosphoenolpyruvate as substrate and activator. J Biol Chem 263: 17611–17614

Sanders D, Davies JM, Rea PA, Brosnan JM, Johannes E (1992) Transport of H^+, K^+ and Ca^{2+} at the vacuolar membrane of plants. In: Tobin AK (ed) Plant organelles. Cambridge University Press, Cambridge, pp 169–188

Smith BN, Madhaven S (1982) Carbon isotope ratios in obligate and facultative CAM plants. In: Ting IP, Gibbs M (eds) Crassulacean acid metabolism. American Society of Plant Physiologists, Rockville, pp 231–243

Smith JAC, Bryce JH (1992) Metabolite compartmentation and transport in CAM plants. In: Tobin AK (ed) Plant organelles. Cambridge University Press, Cambridge, pp 141–167

Smith JAC, Heuer S (1981) Determination of the volume of intercellular spaces in leaves and some values for CAM plants. Ann Bot (New Ser) 48: 915–918

Smith SD, Nobel PS (1986) Deserts. In: Baker NR, Long SP (eds) Photosynthesis in contrasting environments. Elsevier, Amsterdam, pp 13–62

Spalding MH, Stumpf DK, Ku MSB, Burris RH, Edwards GE (1979) Crassulacean acid metabolism and diurnal variations of internal CO_2 and O_2 concentrations in *Sedum praealtum* DC. Aust J Plant Physiol 6: 557–567

Surif MB, Raven JA (1989a) Exogenous inorganic carbon sources for photosynthesis in seawater by members on the Fucales and Laminariales (Phaeophyta): ecological and taxonomic implications. Oecologia 78: 97–105

Surif MB, Raven JA (1989b) Occurrence of diel changes in titratable acidity in plant cell contents: indication of CAM-like metabolism in plants native to Scotland and plants from elsewhere. Trans Bot Soc Edinb 45: 235–244

Surif MB, Raven JA (1990) Photosynthetic gas exchange under emersed conditions in eulittoral and normally submersed members of the Fucales and the Laminariales: interpretation in relation to C isotope ratio and N and water use efficiency. Oecologia 82: 68–80

Tidwell WD, Nambudiri EMV (1989) *Tomlinsonia thomassonii*, gen. et sp. nov., a permineralized grass from the Upper Miocene Ricardo Formation, California. Rev Palaeobot Palynol 60: 165–177

Tidwell WD, Nambudiri EMV (1990) *Tomlinsonia stichkania* sp. nov., a permineralized grass from the Pliocene to (?)Pleistocene China Ranch beds in Sperry Wash, California. Bot Gaz 151: 263–274

Troughton JH, Card KA, Hendy CH (1974a) Photosynthetic pathways and carbon isotope discrimination by plants. Carnegie Inst Washington Year Book 73: 768–780

Troughton JH, Wells PV, Mooney HA (1974b) Photosynthetic mechanisms and palaeoecology from carbon isotope ratios in ancient specimens of C_4 and CAM plants. Science 185: 610–612

Tsuzuki M, Miyachi S, Winter K, Edwards GE (1982) Localization of carbonic anhydrase in CAM plants. Plant Sci Lett 24: 211–218

Valdes PJ, Sellwood BW (1992) A palaeoclimatic model for the Kimmeridgian. Palaeogeogr Palaeoclimatol Palaeoecol 40: 67–101

Waisel Y (1991) The glands of *Tamarix aphylla*: a system for salt recretion or for carbon concentration. Physiol Plant 83: 506–510

Webb DR, Rattray MR, Brown JMA (1988) A preliminary survey for crassulacean acid metabolism (CAM) in submerged aquatic macrophytes of New Zealand. NZ J Mar Fresh Water Res 22: 231–235

West-Eberhard MJ (1986) Alternative adaptations, speciation, and phylogeny (a review). Proc Natl Acad Sci USA 82: 1388–1392

Winter K (1981) Carbon dioxide and water vapour exchange, malate content and $\delta^{13}C$ value in *Cicer arietinum* grown under two water regimes. Z Pflanzenphysiol 101: 421–430

Winter K (1985) Crassulacean acid metabolism. In: Barber J, Baker NR (eds) Photosynthetic mechanisms and the environment. Elsevier, Amsterdam, pp 329–387

Winter K, Ziegler H (1992) Induction of crassulacean acid metabolism in *Mesembryanthemum crystallinum* increases reproductive success under conditions of drought and salinity stress. Oecologia 92: 475–479

Wright VP, Vanstone SD (1991) Addressing the carbon dioxide content of ancient atmospheres using palaeocretes: theoretical and empirical constraints. J Geol Soc Lond 148: 945–957

Wurtelle ES, Nikolau BJ (1990) Plants contain multiple biotin enzymes: discovery of 3-methylcrotonyl-CoA carboxylase, propionyl CoA carboxylase and pyruvate carboxylase in the plant kingdom. Arch Biochem Biophys 278: 179–186

Zeigler AM, Raymond AL, Gierlowski TC, Horrell MA, Rowley DB, Lottes AL (1987) Coal, climate and terrestrial productivity: the present and early Cretaceous compared. In: Scott AC (ed) Coal and coal-bearing strata: recent advances. Geol Soc Spec Publ 32. Geological Society, London, pp 25–49

Part D

25 Crassulacean Acid Metabolism: Current Status and Perspectives

K. WINTER[1] and J.A.C. SMITH[2]

In this concluding chapter, we review some of the major findings discussed in this book and their implications for our understanding of the CAM pathway. In particular, we highlight areas of uncertainty as well as consensus and refer to a number of recent findings not covered in the individual chapters. We shall focus on the biochemistry and energetics of the CAM pathway, the developmental and environmental control of its expression in different species, factors determining the growth and productivity of CAM plants in the field, and finally its possible evolutionary origins.

25.1 Biochemistry and Energetics

25.1.1 Dark CO_2 Fixation and Vacuolar Storage

The last few years have brought considerable advances in our understanding of two crucial components in the CAM cycle: dark fixation of CO_2 by PEPC, and nocturnal malic-acid accumulation in the vacuole. Work on these processes has provided important insights into the control of carbon metabolism in photosynthetic tissues.

A large body of evidence now suggests that control of PEPC activity during the day-night CAM cycle is brought about primarily by phosphorylation of the enzyme at night and its dephosphorylation during the day (Chap. 3). These findings, like others from CAM research in the past, have stimulated corresponding studies with C_3 and C_4 plants (Jiao and Chollet 1991; Lepiniec et al. 1994; Duff and Chollet 1995). Characterization of the protein kinase and phosphatase responsible for reversible phosphorylation of PEPC is now a high priority (Li and Chollet 1994). Experiments with protein-synthesis inhibitors have already indicated that the PEPC kinase in CAM plants may be synthesized de novo each night, and several cDNA clones encoding protein kinases have recently been identified in *Mesembryanthemum crystallinum* (Baur et al. 1994, 1995). One of these (MK6) is a possible candidate for the PEPC

[1]Smithsonian Tropical Research Institute, P.O. Box 2072, Balboa Republic of Panama
[2]Department of Plant Sciences, University of Oxford, South Parks Road, Oxford OX1 3RB, UK

Ecological Studies Vol. 114
Winter/Smith (eds) Crassulacean Acid Metabolism

kinase in *M. crystallinum*, since the 40 kDa polypeptide that it encodes increases in abundance during CAM induction, and both protein and transcript levels are higher at night than during the day (Baur et al. 1995). Experiments to demonstrate the functional importance of these protein kinases could involve measurements of gas exchange for leaves in which protein synthesis has been suppressed using inhibitors. A powerful approach to understanding the regulation of PEPC activity and other aspects of the CAM cycle should become available through the generation of transgenic CAM plants in which key enzymes are over- or underexpressed.

During nocturnal CO_2 fixation in CAM plants, the rate of malate production in the cytoplasm must be closely linked to its removal into the vacuole across the tonoplast. Malic-acid accumulation can be energized by either the tonoplast H^+-ATPase or H^+-PP_iase, or a combination of the two. In *M. crystallinum*, the catalytic activity of the H^+-PP_iase declines as CAM is induced by salt stress, suggesting that this enzyme does not play a major role in energizing transport across the tonoplast (Bremberger et al. 1988). In *Kalanchoë*, the catalytic activity of the H^+-PP_iase is very high and of the same order as the tonoplast H^+-ATPase, even in mature leaves showing maximal CAM activity. As in C_3 plants, therefore, the relative contribution of these enzymes to tonoplast energization in vivo needs to be further investigated (Chap. 4).

In response to the inside-positive membrane potential established by the H^+ pump(s), malate^{2-} influx into the vacuole can occur passively by an electrophoretic mechanism. This transport pathway seems to be provided by a malate^{2-}-selective ion channel in the tonoplast that exhibits strong rectification, thereby favouring malate^{2-} influx over efflux (Chap. 4). However, we are still ignorant of the mechanism of malate efflux from the vacuole during the light period, and of the way in which the transport processes are regulated during the CAM cycle to prevent futile cycling of malate into and out of the vacuole. To maximize the energetic efficiency of malic-acid transport across the tonoplast, the influx process should be inactivated during the day and efflux downregulated at night. To determine whether such regulation occurs (and if so by what mechanism) will be an important question for future work in this area.

In a manner similar to malate influx, malate release from the vacuole during the day must be closely geared to metabolism so that it matches the rate of malate decarboxylation in the cytoplasm and the flux of carbon in gluconeogenesis. Factors such as ambient daytime temperature greatly influence the rate of malate decarboxylation, and may also affect the passive permeability of the tonoplast through the changes in membrane fluidity (Chap. 5). However, CAM plants can exhibit an essentially normal CAM rhythm under constant day-night temperature regimes (Winter 1980), suggesting that temperature-related changes in membrane fluidity are not prerequisites for the CAM rhythm, even though they can affect its amplitude at extreme temperatures.

The central vacuole may have an additional function in PEPCK-CAM plants like *Ananas* and *Clusia*, in which soluble sugars have an important role in carbohydrate storage during the CAM cycle (Chap. 2). It has been postulated in these plants that sucrose is transported into the vacuole during the day, where it is hydrolyzed by a vacuolar invertase, and that glucose and fructose are subsequently remobilized from the vacuole in the course of the night to provide PEP for dark CO_2 fixation (Smith and Bryce 1992). The transmembrane sugar fluxes involved must be quantitatively quite substantial (for glucose and fructose combined, up to 50% of malate). They could provide another model for understanding the integration of carbon metabolism and solute transport across the tonoplast. Application of in vivo fluorescence techniques, as shown in Chap. 6, may allow the coordination of these processes to be studied in real time during the decarboxylation phase and other parts of the CAM cycle.

25.1.2 Mitochondrial Metabolism and Carbon Fluxes in Phase III

One important area that clearly warrants more detailed study is the role of the mitochondrion in the CAM pathway. Labelling experiments on randomization of the C-1 and C-4 carbon atoms in malate by the mitochondrial enzyme fumarase have shown that a substantial proportion (between 30 and 100%) of the malate synthesized in the cytoplasm during the night passes through this organelle (Osmond et al. 1988), and that the extent of this label randomization is inversely proportional to the rate of malate synthesis (Kalt et al. 1990). However, we have no direct information on how much malate passes through the mitochondrion during the deacidification phase in the light period. This is linked to the still unresolved question of the relative contribution of the cytosolic NADP-malic enzyme and mitochondrial NAD-malic enzyme to malate decarboxylation in ME-CAM plants. During the induction of CAM by salinity in *Mesembryanthemum crystallinum*, the extractable activities of both NAD-ME and NADP-ME increase, suggesting that both participate in Phase III malate decarboxylation (Holtum and Winter 1982; Winter et al. 1982; Saitou et al. 1994). In *Kalanchoë daigremontiana*, on the other hand, the development of CAM in the course of leaf ontogeny is associated with an increase in NAD-ME activity but a significant decline in NADP-ME activity (Ingram and Smith 1995). Whether NAD-ME or NADP-ME activities might also be controlled by reversible covalent modification during the CAM cycle in a manner equivalent to PEPC has not yet been determined.

During the deacidification phase, malate entering the mitochondrion can be oxidatively decarboxylated by NAD-ME, but it is not clear how much of the resultant pyruvate (if any) subsequently enters the TCA cycle. Many representations of carbohydrate flow during the CAM cycle imply that the pyruvate produced by the NAD(P)-ME reaction is quantitatively converted to PEP (by the chloroplastic enzyme PPDK), so that the C_3 residue derived from malate decarboxylation is conserved in gluconeogenesis (Holtum and Osmond 1981). To operate in Phase III, this gluconeogenic pathway requires 3.5 ATP per

CO_2 assimilated (or 2.25 ATP per CO_2 in PEPCK-CAM plants: Chaps. 2 and 24). In principle, ATP could be generated through the cytochrome pathway in ME-CAM plants by mitochondrial oxidation of some of the NAD(P)H produced in the ME reaction (Edwards et al. 1982). This would supply about 2.5 ATP per CO_2 for each NADH produced by mitochondrical NAD-ME, or about 1.5 ATP per CO_2 for each NADPH produced by cytosolic NADP-ME, according to the current consensus on P:O ratios in oxidative phosphorylation (Hinkle et al. 1991; Nicholls and Ferguson 1992; Siedow 1995). However, these reducing equivalents would be lost to the glyceraldehyde 3-phosphate dehydrogenase (G3PDH) reaction involved in the gluconeogenic recovery of pyruvate, in which they are required in stoichiometric amounts. Similarly, any NADH derived from the NAD-MDH reaction in PEPCK-CAM plants that is oxidized via the cytochrome pathway would no longer be available for the G3PDH reaction.

One way in which the additional ATP demand of Phase III metabolism can be met without creating a shortfall in reducing equivalents was suggested by Edwards et al. (1982), who pointed out that a small proportion of the malate accumulated in Phase I could be completely oxidized through the TCA cycle during Phase III to the level of 4 CO_2. Based on the best current estimates of the relevant P:O stoichiometries (see above), this would generate 14.75 ATP per malate completely oxidized to CO_2 (rather than the 18 ATP per malate assumed by Edwards et al. 1982). However, reassimilation of the respired CO_2 resulting from pyruvate oxidation back into the starch or hexose 'working capital' via the PCR cycle – which is essential to sustain the CAM cycle – also requires input of energy (3 ATP plus 2 NADPH per CO_2 in the PCR cycle, and 0.167 ATP per CO_2 for conversion of triose phosphate to storage carbohydrate, i.e. 9.5 ATP plus 6 NADPH per pyruvate completely oxidized to CO_2). Thus, although pyruvate oxidation could in principle cover the extra ATP requirement of Phase III (i.e. gluconeogenic recovery of the three-carbon residue plus reassimilation of any respired CO_2 resulting from pyruvate oxidation), this pathway involves an additional demand for NADPH, which would presumably have to be met by photophosphorylation.

What would be the optimal ratio of pyruvate recovery in gluconeogenesis to pyruvate oxidation in respiration in terms of the overall energy costs of net CO_2 assimilation? This can be calculated if we assume (1) that the NADPH required for reassimilation in the PCR cycle of CO_2 derived from pyruvate oxidation is supplied by photophosphorylation together with ATP in the conventional ratio 1 NADPH:1.5 ATP (Bryce and Hill 1993), and (2) that the ATP cost of gluconeogenesis to the level of storage carbohydrate (3.5 ATP per malate in ME-CAM, or 2.25 ATP per malate in PEPCK-CAM) is quantitatively met by complete oxidation of part of the total malate to 4 CO_2 (producing 14.75 ATP per malate). For ME-CAM plants, it can be shown that if approximately 20% of the total malate is completely oxidized to CO_2, then the *additional* energy cost of Phase III metabolism (over and above that of assimilation through the PCR cycle of the CO_2 originally taken up from the atmosphere in Phase I and incorporated into malate) becomes 1.8 ATP plus 1.2 NADPH per net CO_2 assimilated (Fig. 25.1A). As the additional energy requirement can be met by photophosphorylation with a stoichiometry of 3 ATP:2 NADPH per O_2 evolved, this is consistent with the observation that the photon-use efficiency of photosynthetic O_2 evolution by CAM plants in Phase III can be as high as that of C_3 plants under non-photorespiratory conditions (i.e. saturating CO_2: Adams et al. 1986; Winter et al. 1986a; Björkman and Demmig 1987)[3]. In total, therefore, this scheme requires 4.8 ATP plus 3.2 NADPH per net CO_2 assimilated during Phase III, or 60% more ATP plus NADPH than is needed by the PCR cycle alone in the absence of photorespiration (Fig. 25.1A).

Using a similar argument for PEPCK-CAM plants, it can be shown that complete oxidation of 13% of the total malate would give a net energetic cost of Phase III metabolism of 1.2 ATP plus 0.8 NADPH per CO_2 in excess of the usual PCR cycle costs of 3 ATP plus 2 NADPH per CO_2 (Fig. 25.1B). Again, this energy cost can be met by photophosphorylation with the usual stoichiometry of 3 ATP:2 NADPH, so the photon costs of photosynthetic O_2 evolution are unaffected. These additional energetic costs for ME- and PEPCK-CAM plants will translate into photon-use efficiencies of *net* CO_2 assimilation for Phase III of the CAM cycle that are 37% or 29% lower, respectively, than those of C_3 plants under non-photorespiratory conditions. In a normal atmosphere, of course, the photon-use efficiency of net CO_2 assimilation in C_3 plants is considerably reduced by photorespiration (Chaps. 16 and 24) and is much more similar to that of CAM plants, a point to which we will return below. Here, it is important to note that only with the pathway stoichiometries illustrated in Fig. 25.1A,B would the extra energy cost of Phase III of the CAM cycle show an ATP:NADPH ratio of 1.5:1. The ratio of ATP:NADPH required would decrease or increase if a larger or smaller proportion of the malate pool, respectively, were completely oxidized to CO_2 via the cytochrome pathway, and this in turn would decrease the photon-use efficiency of *both* O_2 evolution and net CO_2 assimilation during Phase III.

Although oxidation of part of the total malate beyond the level of pyruvate can thus provide additional ATP, the respiratory CO_2 produced has to be reassimilated via the PCR cycle to avoid draining the CAM cycle of three-carbon residues, which provide the primary CO_2 acceptor in Phase I. What would be the energetic consequences for Phase III metabolism if no pyruvate

[3]The absolute photon requirement for production of ATP by photophosphorylation depends on whether cyclic electron flow contributes to ATP production in vivo, or whether the non-cyclic pathway can produce 3 ATP per $2e^-$ through engagement of a proton-motive Q cycle (Heber and Walker 1992; Nicholls and Ferguson 1992). If this applies to both C_3 plants and CAM plants, then the absolute photon requirement will not affect the *relative* costs of the two photosynthetic pathways, provided they utilize ATP and NADPH in the same ratio. If, on the other hand, the additional costs of the CAM pathway are calculated in terms of ATP alone (cf. Tables 16.3 and 24.2), then the involvement of Q-cycle activity will affect the relative costs of the two pathways.

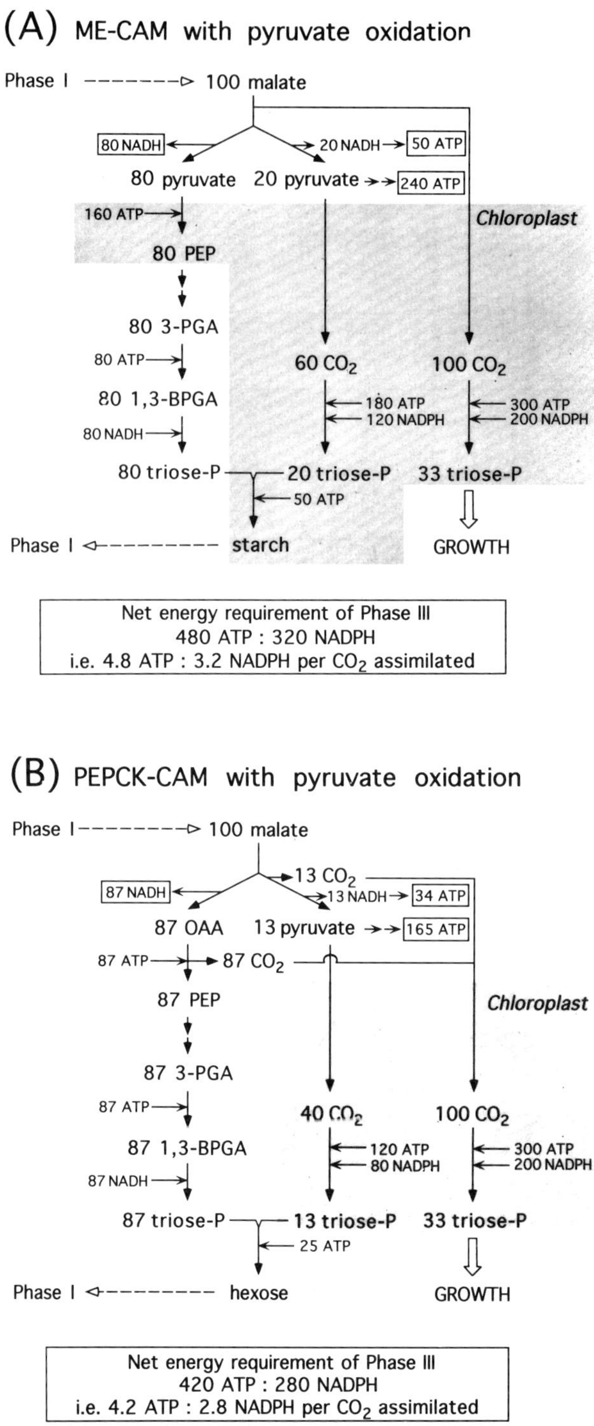
(A) ME-CAM with pyruvate oxidation
Phase I
100 malate
80 NADH
20 NADH
50 ATP
80 pyruvate
20 pyruvate
240 ATP
160 ATP
Chloroplast
80 PEP
80 3-PGA
80 ATP
60 CO2
100 CO2
80 1,3-BPGA
180 ATP
120 NADPH
300 ATP
200 NADPH
80 NADH
80 triose-P
20 triose-P
33 triose-P
50 ATP
Phase I
starch
GROWTH
Net energy requirement of Phase III
480 ATP : 320 NADPH
i.e. 4.8 ATP : 3.2 NADPH per CO2 assimilated
(B) PEPCK-CAM with pyruvate oxidation
Phase I
100 malate
13 CO2
87 NADH
13 NADH
34 ATP
87 OAA
13 pyruvate
165 ATP
87 ATP
87 CO2
87 PEP
Chloroplast
87 3-PGA
87 ATP
40 CO2
100 CO2
87 1,3-BPGA
120 ATP
80 NADPH
300 ATP
200 NADPH
87 NADH
87 triose-P
13 triose-P
33 triose-P
25 ATP
Phase I
hexose
GROWTH
Net energy requirement of Phase III
420 ATP : 280 NADPH
i.e. 4.2 ATP : 2.8 NADPH per CO2 assimilated

Fig. 25.1A,B

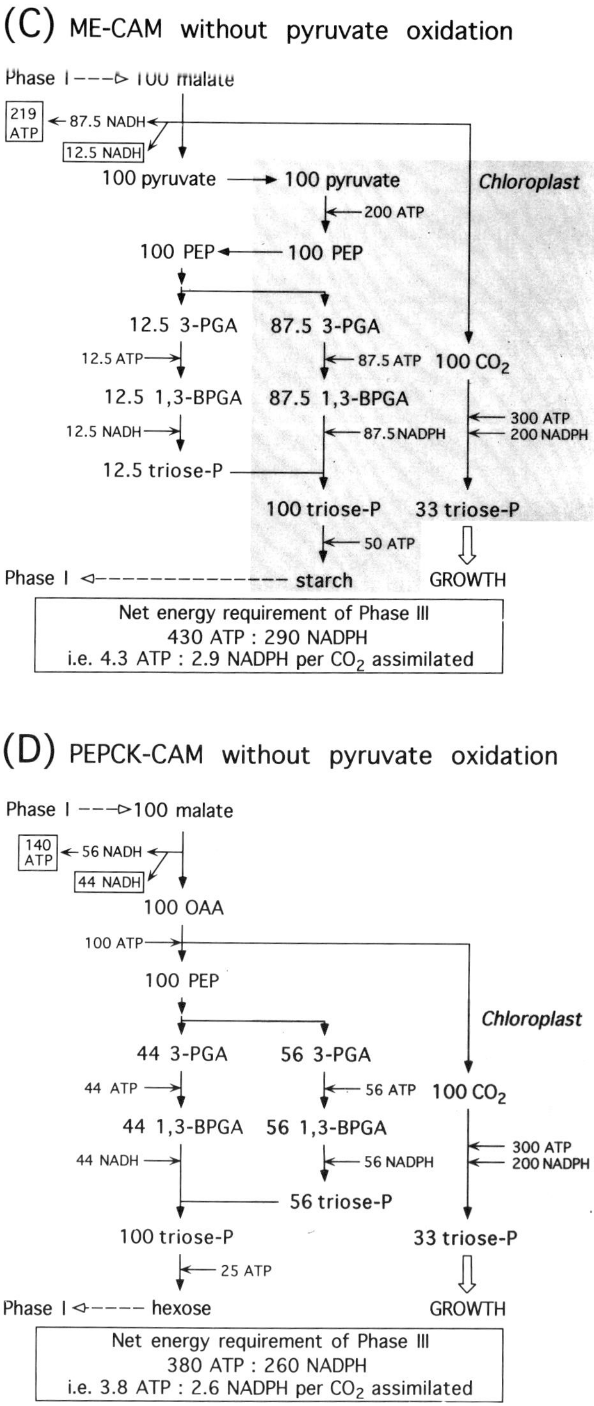
(C) ME-CAM without pyruvate oxidation
Phase I ---▷ 100 malate
219 ATP
87.5 NADH
12.5 NADH
100 pyruvate
100 pyruvate
Chloroplast
200 ATP
100 PEP
100 PEP
12.5 3-PGA
87.5 3-PGA
12.5 ATP
87.5 ATP
100 CO2
12.5 1,3-BPGA
87.5 1,3-BPGA
12.5 NADH
87.5 NADPH
300 ATP
200 NADPH
12.5 triose-P
100 triose-P
33 triose-P
50 ATP
Phase I ◁------------- starch
GROWTH
Net energy requirement of Phase III
430 ATP : 290 NADPH
i.e. 4.3 ATP : 2.9 NADPH per CO2 assimilated
(D) PEPCK-CAM without pyruvate oxidation
Phase I ---▷100 malate
140 ATP
56 NADH
44 NADH
100 OAA
100 ATP
100 PEP
Chloroplast
44 3-PGA
56 3-PGA
44 ATP
56 ATP
100 CO2
44 1,3-BPGA
56 1,3-BPGA
44 NADH
56 NADPH
300 ATP
200 NADPH
56 triose-P
100 triose-P
33 triose-P
25 ATP
Phase I ◁---- hexose
GROWTH
Net energy requirement of Phase III
380 ATP : 260 NADPH
i.e. 3.8 ATP : 2.6 NADPH per CO2 assimilated

Fig. 25.1C,D

Fig. 25A–D. Energetic costs of net CO_2 assimilation during Phase III in ME-CAM and PEPCK-CAM plants. These schemes illustrate the energetic implications of recovering the three-carbon residue resulting from malate decarboxylation back to the level of storage carbohydrate using two different mechanisms. ATP and NADH generated in these reactions are *boxed*, and *shaded areas* indicate reactions occurring in the chloroplast. For ME-CAM plants, the storage carbohydrate is starch, whereas in PEPCK-CAM plants it is predominantly soluble hexose synthesized via sucrose (Chap. 2); no additional energetic cost would be associated with compartmentalization of the storage carbohydrate, provided transport in PEPCK-CAM plants of soluble sugars into the vacuole (where they are assumed to be localized) occurs by facilitated diffusion (Chap. 24). In **A** and **B**, part of the total malate pool is oxidized completely to CO_2 via mitochondrial NAD-ME and the TCA cycle to produce ATP through the cytochrome pathway; in **C** and **D**, no pyruvate is oxidized to the level of CO_2, and the three-carbon residue resulting from malate decarboxylation is quantitatively converted in gluconeogenesis back to storage carbohydrate. **A**, **C**, ME-CAM plants (which possess both NAD-ME and NADP-ME); **B**, **D**, PEPCK-CAM plants (which possess significant NAD-ME activities). The flux stoichiometries in the pathways have in most instances been rounded to whole numbers for clarity, and the ATP and NADPH costs in Phase III per net CO_2 assimilated are shown in each *boxed caption* to one decimal place (exact numbers are given below). These stoichiometries derive from a number of assumptions. First, the net energetic cost of recovering in gluconeogenesis the three-carbon residue resulting from malate decarboxylation should be met entirely by photophosphorylation in the chloroplast, with the same overall ratio of ATP:NADPH required (1.5:1) as in the PCR cycle; this is mechanistically parsimonious in that it minimizes the photon cost of photosynthetic O_2 evolution (cf. Chap. 24), resulting in a photon-use efficiency of O_2 evolution virtually identical to that of C_3 plants under non-photorespiratory conditions (which has been observed experimentally: see main text). Second, oxidation of each NADH produced in the mitochondrial matrix by the NAD-ME reaction produces 2.5 ATP by oxidative phosphorylation; complete oxidation of each pyruvate produces 12.25 ATP; and oxidation of each NADPH produced in the cytosol by the NADP-ME reaction produces 1.5 ATP (for a discussion of consensus stoichiometries, see Hinkle et al. 1991; Taiz and Zeiger 1991; Nicholls and Ferguson 1992). Third, any alternative pathway activity (cyanide-resistant respiration) in the mitochondria is assumed not to be altered in Phase III compared with other times in the day-night cycle (see main text). Fourth, transfer of ATP produced in the mitochondria to the chloroplasts must be possible, but transfer of reducing equivalents (NADH/NADPH) between compartments is not required. Fifth, reduction of 1,3-bisphosphoglycerate (1,3-BPGA) to triose phosphate (triose-P) can occur via both an NAD-glyceraldehyde 3-phosphate dehydrogenase in the cytosol and an NADP-linked isoform in the chloroplast
A ME-CAM plants, with sufficient ATP being produced by complete oxidation of a proportion of the total malate (14.75 ATP per malate) to cover the ATP requirement for conversion of malate to starch in gluconeogenesis. It is assumed that malate oxidation to provide ATP by oxidative phosphorylation occurs exclusively via mitochondrial NAD-ME, and that 210 of the 290 ATP thereby produced are transferred to the chloroplast. (Consequently, 19.72% of the total malate can be completely oxidized to produce ATP, with the exact net energy requirement being 4.775 ATP plus 3.183 NADPH per CO_2 in Phase III.) If any 3-PGA derived from PEP were reduced to triose-P in the chloroplast rather than the cytosol, this would increase the net energy requirement per CO_2
B PEPCK-CAM plants, with requirements as for **A**. No transfer of ATP to the chloroplast is needed. (In this instance, 13.43% of the total malate is completely oxidized via mitochondrial NAD-ME to produce ATP, and the exact net energy requirement is 4.209 ATP plus 2.806 NADPH per CO_2 in Phase III.) Reduction of 3-PGA derived from PEP to triose-P in the chloroplast would increase the net energy requirement per CO_2
C ME-CAM plants, with pyruvate from the ME reaction being quantitatively conserved in gluconeogenesis. Photophosphorylation can support reduction in the chloroplasts of 87.5% of the 3-PGA produced from malate, with precisely the same proportion of NADH produced in the ME reaction being oxidized via the cytochrome pathway to give 219 ATP, most

were oxidized to CO_2 and the three-carbon residue were quantitatively converted back to storage carbohydrate through gluconeogenesis, as suggested by Holtum and Osmond (1981)? It has been noted by Gardeström and Edwards (1985) that this should provide an energetically more favourable way of regenerating the storage carbohydrate, although the precise energy costs depend upon the pathways by which the ATP equivalents are provided. If a greater proportion of the NADH derived from the oxidation of malate to pyruvate (in ME-CAM plants) or oxaloacetate (in PEPCK-CAM plants) is used to generate ATP by oxidative phosphorylation, then the reducing equivalents required for the G3PDH reaction must be made up by NADPH from photophosphorylation. The optimal proportion of NADH to be oxidized in the cytochrome pathway can be calculated as before by assuming that the ATP costs of gluconeogenesis are 3.5 ATP per CO_2 (ME-CAM) or 2.25 ATP per CO_2 (PEPCK-CAM), that oxidation of each NADH generated in the mitochondrial matrix produces 2.5 ATP, and that the NADPH supplied by photophosphorylation is produced together with ATP in the ratio 1:1.5. The calculations for ME-CAM plants (Fig. 25.1C) and PEPCK-CAM plants (Fig. 25.1D) show that the energy costs of the recovery of the three-carbon residue back to the level of storage carbohydrate are in fact 26% and 30% lower, respectively, than in the schemes involving complete oxidation of a proportion of the total malate to CO_2 (Fig. 25.1A,B). This would of course improve the photon-use efficiency of net CO_2 assimilation during Phase III. And as in the two previous cases, these additional net energy costs of Phase III metabolism can be entirely met by photophosphorylation in the chloroplasts.

These examples show that the net energy costs of Phase III metabolism can be minimized if the three-carbon residue resulting from malate decarboxylation is quantitatively recovered via gluconeogenesis, without any pyruvate being oxidized through the TCA cycle. Are there factors that might restrict

---

of which (206.5 ATP) must be transferred to the chloroplast. The stoichiometry shown assumes that the NADH used in oxidative phosphorylation is produced in the mitochondrial matrix by NAD-ME. (The exact net energy requirement is 4.3125 ATP plus 2.875 NADPH per CO_2 in Phase III.) An increase or decrease in the proportion of 3-PGA reduced in the chloroplast would increase the net energy requirement per CO_2. If reducing equivalents for mitochondrial oxidative phosphorylation were contributed by the NADP-ME reaction, this would increase the overall energy cost of Phase III metabolism, and more ATP and NADPH would have to be supplied by photophosphorylation

D PEPCK-CAM plants, with PEP from the PEPCK reaction being quantitatively conserved in gluconeogenesis. Photophosphorylation can support reduction in the chloroplast of 56.25% of the 3-PGA derived from malate, with precisely the same proportion of NADH produced in the MDH reaction being oxidized via the cytochrome pathway to give ATP. This stoichiometry assumes that oxidative phosphorylation utilizes matrix NADH produced by mitochondrial NAD-MDH. (The exact net energy requirement is 3.8438 ATP plus 2.5625 NADPH per CO_2 in Phase III.) An increase or decrease in the proportion of 3-PGA reduced in the chloroplast would again increase the net energy requirement per CO_2. In this example, there is export of 28 ATP from the chloroplast to the cytosol, which is the location of both PEPCK and the ATP (UTP)-requiring reaction (UDP-glucose pyrophosphorylase) in sucrose synthesis leading to hexose entry of pyruvate into the TCA cycle during Phase III? In C_3 plants at least, TCA cycle activity appears to be significantly reduced in the light period (Hanning and Heldt 1993; Villar et al. 1994). Entry of pyruvate into the TCA cycle will be strongly influenced by the activity of pyruvate dehydrogenase, which is known to be downregulated in the light in C_3 plants (Gemel and Randall 1992; Moore et al. 1993). So far, it has not been possible to obtain quantitative estimates of TCA cycle activity in illuminated tissue during Phase III of the CAM cycle. However, experiments with mitochondria isolated from leaves of *Kalanchoë daigremontiana* during the day and at night have shown that pyruvate dehydrogenase activity (assayed as the release of $^{14}CO_2$ from [1-^{14}C]-labelled pyruvate) is indeed considerably reduced in Phase III compared with Phase I (Smith and Bryce 1992).

Attempts to determine how much malate is oxidized in the mitochondria during Phase III have been made by measuring the rates of oxygen uptake in the dark by leaf discs of the ME-CAM plant *Kalanchoë daigremontiana* sampled at different times during the day-night cycle. By comparing the observed rates of O_2 uptake (in the dark) with those of malate consumption (in the light), Robinson et al. (1992) calculated that up to 58% of the malate decarboxylated in Phase III may be oxidized in the mitochondria by NAD-ME to pyruvate and CO_2 (or as little as 14.5% if the malate is completely oxidized to 4 CO_2). This is less than the proportions indicated in Fig. 25.1C and A (respectively), and would imply that significantly more ATP would have to be provided by photophosphorylation.

Using a novel oxygen-isotope-discrimination technique, Robinson et al. (1992) were also able to show that the flux through the alternative oxidase pathway was increased in Phase III tissue in parallel with the rate of malate decarboxylation, and that this accounted for up to 50% of mitochondrial respiration in the leaf discs. This confirmed a similar observation of Rustin and Queiroz-Claret (1985) that a large proportion of the respiratory activity in CAM-induced *K. blossfeldiana* was cyanide-resistant. What controls the activity of the alternative pathway in vivo is still not fully understood. According to the 'overflow' hypothesis, the pathway is engaged when the ubiquinone pool is sufficiently reduced and electron flow through the main respiratory chain is saturated (Moore and Siedow 1991). However, it has recently been shown that the activity of the alternative oxidase can be directly stimulated by pyruvate (Millar et al. 1993), which lowers the threshold level of reduced ubiquinone at which the alternative pathway is engaged (Umbach et al. 1994). Since pyruvate levels are known to increase in CAM plants during Phase III (Milburn et al. 1968), this would be consistent with an increase in activity of the alternative pathway during this part of the CAM cycle (Robinson et al. 1995).

A physiological role for the alternative pathway in Phase III metabolism is not readily apparent. Because electron transport through this pathway is non-phosphorylating, any cyanide-resistant respiration will inevitably increase the unit costs of net CO_2 assimilation. In general, the alternative pathway in plants seems to be engaged at times when very high rates of substrate oxidation are required, which is widely interpreted as a method of avoiding the restriction of TCA cycle metabolism by adenylate control (Lambers 1990). However, in Phase III, only small amounts of pyruvate may need to enter the TCA cycle and, in the limit, all the pyruvate produced by NAD-ME might be exported from the mitochondrion for gluconeogenesis (Fig. 25.1C). Another possibility is that for Phase III tissue incubated in the dark the energy demand is greatly reduced. Under such conditions, the energy-requiring conversion of pyruvate back to storage carbohydrate probably occurs at much reduced rates, for example as a result of inactivation of PPDK in the dark in ME-CAM plants (Sugiyama and Laetsch 1975). If malate decarboxylation continues, this would create a 'pyruvate overflow' that could cause a large increase in alternative pathway activity. In some instances, as in intact plants of the PEPCK-CAM species *Ananas comosus*, there appears to be no stimulation of respiratory O_2 uptake in the light in Phase III (Coté et al. 1989). Thus, it remains to be resolved whether the alternative pathway is an integral part of the CAM cycle under optimal physiological conditions. A possible role of the alternative oxidase is discussed in section 25.3.3 for conditions under which a significant proportion of the nocturnally accumulated malic acid is derived from recycling of respiratory CO_2.

25.2 Environmental and Developmental Control

25.2.1 *Mesembryanthemum crystallinum*

Considerable progress has been made in our understanding of the molecular processes underlying CAM induction in the halophyte *Mesembryanthemum crystallinum* (Chaps. 8, 9 and 10). This species has become one of the most intensively studied systems for investigation of the molecular basis of plant responses to salt and drought stress. As yet, the precise signal-transduction pathway underlying the induction of CAM in response to environmental and developmental signals has not been identified. In particular, we need to determine how a physical signal associated with water deficits (such as a change in cell turgor or cell volume) can lead to transcriptional activation of key genes in the CAM pathway. An important step towards understanding such transduction mechanisms will be to identify the promoter elements required for regulation of CAM-specific genes, such as the *Ppc1* gene encoding the CAM isoform of PEPC (Chaps. 9 and 10). There is considerable complexity in the ways in which gene expression can be regulated, which can include post-transcriptional mechanisms such as control of transcript stability (DeRocher and Bohnert 1993). Furthermore, the upregulation of certain genes in *M. crystallinum* in response to salinity is clearly related to ionic stress and is not associated with CAM induction per se. Important examples include genes involved in the production of compatible solutes (Chap. 9; Vernon et al. 1993).

The role of roots in stress perception and of root : shoot interactions in the induction and expression of CAM at the whole-plant level is another important area that requires more research (Winter and Gademann 1991). Several attempts have been made to alter the hormonal balance of intact *M. crystallinum* plants by external application of ABA and other growth regulators. However, these studies have led to equivocal results (see Chaps. 8, 9 and 10), possibly because of complications associated with the experimental treatments. For example, addition of ABA to culture solutions can lead to profuse microbial growth within 48 h, which may adversely affect water and nutrient uptake (K. Winter, unpubl. observ.; see also Chap. 10). Other observations are also hard to reconcile with a direct involvement of ABA in the stress-perception pathway, such as the finding that the ABA-biosynthesis inhibitor Lovastatin has no effect on CAM induction in response to salinity (Thomas et al. 1992).

Even in the absence of a stress treatment, some degree of CAM activity is commonly observed in older plants of *M. crystallinum*. This has given rise to a long, and still unresolved, debate about whether induction of CAM in response to salinity and drought simply represents acceleration of a normal developmental programme (Osmond 1978; Chaps. 8, 9 and 10). Even in constitutive CAM plants, a gradual shift from C_3 to CAM occurs during leaf development and maximal CAM activity is expressed only in mature leaves. The same is true for the photosynthetic organs of cacti and other stem succulents. But in well-watered *M. crystallinum*, CAM develops only slowly with plant age, so that fully expanded leaves of plants 6 to 8 weeks after germination show virtually no CAM activity. If such plants are then exposed to salt stress, a considerably greater CAM activity is induced than is ever observed in mature leaves of unstressed plants. This induction is much more pronounced in the expanded leaves than in developing leaves, but it is more a function of leaf age than of plant age. Salinity-induced CAM can be observed in the first (and only) true leaf pair of very young plants 2 to 3 weeks after germination, provided these leaves are close to maturity (K. Winter, unpubl. observ.).

These observations suggest a predominantly environmental component to CAM induction in *M. crystallinum*. Additional support for this view comes from the observed reversibility of the C_3 to CAM shift when plants are transferred from saline back to non-saline conditions (Winter 1974a; von Willert et al. 1976; Vernon et al. 1988). This too depends on leaf age, since the ability to revert to C_3 photosynthesis declines as the induced leaves become older (Winter 1974b). Experiments with detached leaves of unstressed plants operating in the C_3 mode have also shown that PEPC transcript levels increase over an 8-h period of dehydration, a response that is fully reversible when the leaves are rehydrated in water (Schmitt 1990). The relatively short life-span of leaves of *M. crystallinum* (several weeks) can make demonstration of CAM reversibility in mature leaves of intact plants problematic. But in C_3/CAM intermediate species of *Clusia*, for example, which show leaf longevities of many months, the reversibility of CAM activity caused by transient leaf water deficits can be more readily observed (Winter et al. 1992).

The importance of environmental conditions in determining the onset of CAM has also been shown for natural populations of *M. crystallinum* in areas with a mediterranean-type climate. Plants typically perform C_3 photosynthesis for the first 3 months of their annual life cycle in the winter and early spring when rainfall is frequent and temperatures are relatively low. In two independent field studies in Israel and California, it was found that CAM induction coincided with the onset of the dry season (Winter et al. 1978; Bloom and Troughton 1979). In California, the study included two populations that germinated at different times, yet there was a close correlation between CAM activity and decreasing soil moisture content independent of plant age (Bloom and Troughton 1979).

To demonstrate conclusively that there is no developmentally programmed ontogenetic shift from C_3 to CAM in *M. crystallinum*, it would be necessary to show that the plant can complete its life cycle without ever engaging in CAM. This is technically difficult, because even plants with an optimal water supply can experience transpirationally induced water deficits during the light period. Moreover, *M. crystallinum* is a halophilic species that requires moderate concentrations of NaCl for optimal growth and turgor maintenance (Winter and Lüttge 1976). Plants grown in the absence of NaCl are therefore particularly susceptible to turgor loss at times of high transpiration because of an inability to show effective osmotic adjustment. This is especially notable in older plants (K. Winter, unpubl. observ.), and may lead to CAM induction even in the absence of a specific stress treatment.

One way to determine the relative importance of developmental versus environmental factors in controlling CAM induction in *M. crystallinum* would be to keep plants well irrigated in their natural environment during the summer drought. Another approach would be to raise young plants for about 3 months in the C_3 mode under controlled conditions and then to transfer them to their natural habitat to complete their life cycle during the winter and early spring. If such plants under mesic conditions still developed significant CAM activity, this would be strong evidence for an ontogenetic component of CAM induction.

25.2.2 *Kalanchoë blossfeldiana*

One species in which CAM induction may be linked to acceleration of an underlying developmental programme is the photoperiod-sensitive cultivar *Kalanchoë blossfeldiana* cv. Tom Thumb. In plants maintained under long-day conditions, CAM activity develops gradually with increasing plant age. However, strong CAM induction occurs in response to a critical number of short days; this is also associated with flower induction and markedly increased leaf succulence. In contrast to *M. crystallinum*, the expression of CAM in a given leaf is not reversible on returning the plants to non-inductive long-day conditions (Queiroz and Brulfert 1982). Interestingly, the induction of CAM by drought in this species (under non-inductive photoperiods) is readily reversible by watering (Brulfert et al. 1988), which is more similar to the stress-induced CAM in *M. crystallinum*. In detached leaves, light-dependent increases in the maximum catalytic activity of PEPC, and in the amount of PEPC protein, are detectable within several hours (Brulfert et al. 1993). This would be a valuable system in which to explore further the developmental and environmental regulation of CAM induction at the molecular level (Brulfert et al. 1985).

25.2.3 C_3 to CAM Shifts in Other Species

In contrast to the extensive studies of CAM induction in *M. crystallinum* and *K. blossfeldiana*, there has as yet been relatively little work on the biochemical regulation of C_3 to CAM shifts in genera such as *Portulacaria* (Guralnick and Ting 1988), *Clusia* (Chaps. 19 and 20; Winter et al. 1992), *Peperomia* (Chap. 13; Ting et al. 1994) and *Sedum* (Chap. 11; Conti and Smirnoff 1994). Reversible increases in nocturnal CO_2 fixation in response to water deficits have been demonstrated in some of these species, although they are generally less striking than those in *M. crystallinum*. So far, these species have not been characterized at the genetic level, but there is evidence that CAM induction is associated with a distinct isoform of the PEPC protein in *S. telephium* (Groenhof et al. 1990), as observed in *M. crystallinum* (Chaps. 9 and 10; Slocombe et al. 1993).

CAM plants capable of such C_3 to CAM shifts differ with respect to their organic-acid content prior to CAM induction or enhancement of CAM activity (Ting 1985). Plants like *M. crystallinum*, or other members of the Aizoaceae with inducible CAM such as *M. nodiflorum* (Winter and Troughton 1978) and *Aptenia cordifolia* (Treichel 1975), show very low levels of organic acids when in the C_3 mode. In contrast, other species like *Sedum telephium* (Chap. 11), *Portulacaria afra* (Ting and Hanscom 1977) and *Clusia uvitana* (Chaps. 19 and 20) usually contain high levels of organic acids when fixing CO_2 predominantly during the day. In *C. uvitana*, increases in CAM activity can be triggered within 24 h by conditions that promote the diurnal consumption of organic acids for photosynthetic CO_2 reduction, for example when the stomata close in response to short-term water deficits or low ambient air humidity. Conversely, conditions that inhibit the diurnal consumption of organic acids in Phase III, such as low PPFD and decreased temperature, result in a switch to a more C_3-like gas-exchange pattern (Chap. 20; Zotz and Winter 1993a). Rapid changes in CAM activity have also been observed in *S. telephium* within several hours when diurnal malate concentrations are decreased by water deficits (Chap. 11). These short-term responses probably do not involve significant changes in the amount of PEPC protein, although in the longer term de novo synthesis of PEPC does occur (Winter et al. 1992). But they suggest that the expression of CAM activity may be modulated by daytime organic-acid concentrations (Zotz and Winter 1993a), and perhaps in particular by citrate levels (Borland et al. 1994, 1995).

The metabolic significance of the day-night changes in citric-acid concentrations seen in many CAM plants has not yet been determined. Nocturnal citrate accumulation does not lead to net carbon gain, since generation of acetyl coenzyme A for the citrate synthase reaction involves decarboxylation of the three-carbon precursor pyruvate (Lüttge 1988; Olivares et al. 1993). However, oxidation of storage carbohydrate during the dark period via 'citrate glycolysis' will produce ATP that can help to meet the energy cost of organic-acid accumulation in the vacuole (Lüttge 1988; Feng et al. 1994). The amount of ATP produced will depend on whether the pyruvate used to generate acetyl coenzyme A is produced from PEP by pyruvate kinase (Lüttge 1988) or via malate decarboxylation involving NAD-malic enzyme (ap Rees 1990). In principle, ATP could also be derived from complete oxidation of part of the nocturnally synthesized malate in the TCA cycle. The comparative benefits of these two mechanisms, taking into account the energy costs of reassimilating the CO_2 derived from citrate and malate decarboxylation in the light period and of gluconeogenesis, will be considered further in Section 25.3.3.

One proposal for the role of citrate accumulation is that CO_2 released by oxidation of citrate under conditions of water and high light stress may protect against photoinhibition (cf. Chap. 7). Refixation of CO_2 through the PCR cycle provides a sink for electrons, thereby preventing overreduction of the electron transport chain (Chap. 19; Haag-Kerwer et al. 1992). It has also been suggested that citrate accumulation may be important in increasing the pH-buffering capacity of the vacuole (Franco et al. 1992). However, the functional importance of these possible roles is difficult to test experimentally, and for the present the significance of nocturnal citrate accumulation in CAM remains enigmatic.

25.2.4 Field Studies

Few long-term studies have so far been made to assess the importance of light versus dark CO_2 fixation in inducible CAM plants growing in their natural environments. Most have involved measurements of seasonal changes in carbon-isotope ratios and day-night changes in organic-acid content throughout the year (Winter et al. 1978; Winter and Troughton 1978; Bloom and Troughton 1979). Long-term measurements of seasonal changes in CAM activity have also been made in situ for the constitutive CAM plants *Opuntia basilaris* (Szarek and Ting 1974), *O. inermis* (Osmond et al. 1979) and *Tillandsia usneoides* (Martin et al. 1981). In these studies, day/night changes in $^{14}CO_2$ uptake were used as an additional measure of CAM activity, although tracer-flux measurements do not allow direct estimates of diurnal and nocturnal carbon budgets. Such information is essential to assess the extent to which C_3 photosynthesis and CAM contribute at different times of the year to CO_2 acquisition in natural environments.

The most detailed in situ long-term investigation of 24-h net CO_2 exchange is that of *Clusia uvitana*, which was studied over an entire year in a semi-evergreen moist tropical forest on Barro Colorado Island, Panama (Chap. 20;

Zotz and Winter 1994a,b,c). Apart from providing detailed information on the responses of daytime and nocturnal CO_2 fixation to environmental conditions, this study revealed a close correlation between maximum instantaneous rates of net CO_2 uptake and total 24-h carbon gain, as was found previously for C_3 species (Zotz and Winter 1993b). This latter finding should facilitate estimation of the primary productivity of CAM plants in the field, an approach that may complement that based on the Environmental Productivity Index developed for cacti and agaves by Nobel (Chap. 16; Nobel 1991). More long-term studies of 24-h gas exchange are needed, with leaf-succulent CAM plants in particular, to evaluate (1) the relative importance of environmental and developmental factors in determining the balance between day and night carbon assimilation, and (2) the significance of this metabolic versatility for the performance and survival of such species in their natural habitats.

25.3 Growth and Productivity

25.3.1 Potential Productivity

Recent research by Nobel and collaborators, summarized in Chapter 16, has yielded some remarkable findings on the potential productivity of CAM plants. Two species of *Agave* (*A. mapisaga* and *A. salmiana*) and two species of *Opuntia* (*O. amyclea* and *O. ficus-indica*), cultivated under well-watered conditions in plantations in Mexico and Chile, show amongst the highest annual productivities reported for any plants to date! These high growth rates are at first sight surprising for plants commonly regarded as possessing in the CAM pathway a mere 'survival option'. How can such extremely high productivities of up to 50 Mg ha^{-1} $year^{-1}$ be rationalized?

In these perennial agaves and cacti, the surfaces of essentially all aboveground parts are photosynthetically active year-round, and the plants have very high shoot : root ratios (of the order of 10:1). Further, these particular CAM plants exhibited exceptionally high maximal rates of net nocturnal CO_2 uptake of up to 34 μmol m^{-2} s^{-1}, more than five-fold higher than those typically observed for CAM species. These productivities also refer to established stands of optimally spaced plants, and optimal light interception was maintained with *Opuntia* by periodic pruning (Nobel et al. 1992). While the growth rates of CAM plants under natural conditions will invariably be much lower, these findings–almost unimaginable a few years ago–suggest that the adaptive features of CAM plants that allow survival in stressful habitats do not necessarily compromise their potential for high photosynthetic productivities.

25.3.2 Elevated CO_2

Concern about global climatic change associated with rapidly increasing atmospheric CO_2 concentrations has led to numerous studies on the growth and photosynthetic responses of C_3 plants and, to some extent, C_4 plants to elevated CO_2. For CAM plants, in which primary CO_2 fixation via PEPC is commonly thought to be CO_2-saturated at present day ambient CO_2 concentration, there is a widespread expectation that nocturnal CO_2 fixation will be little affected in a high-CO_2 world. The results from short-term experiments have been consistent with this view. In studies with *Agave deserti*, *Ferocactus acanthodes* (Nobel and Hartsock 1986) and *Kalanchoë pinnata* (Winter and Engelbrecht 1994), a doubling of ambient CO_2 partial pressure had almost no effect on nocturnal carbon gain, although as anticipated there were strong increases in C_3-photosynthetic CO_2 uptake during the light period in Phase IV. In *K. pinnata* grown in open-top chambers under tropical conditions in Panama, doubling the CO_2 partial pressure led to a marked increase in dry-matter accumulation over a 3-month period (K. Winter and B. Engelbrecht, unpubl. observ.). Consistent with the lack of effect of elevated CO_2 on net dark CO_2 fixation in the short-term experiments, the nocturnal increase in organic-acid content was similar under the two conditions, suggesting that the increased dry-matter accumulation was largely the result of increased day-time CO_2 fixation via RUBISCO. However, in the C_3-CAM intermediate *Clusia uvitana* (Winter et al. 1992) and the constitutive CAM plant *Epiphyllum phyllanthus* (K. Winter, unpubl. observ.), high-CO_2-stimulated increases in carbon gain during the light period were accompanied by a *reduction* in nocturnal CO_2 uptake.

There is increasing evidence that CO_2 fixation via PEPC in fact may not always be completely CO_2-saturated under present-day atmospheric conditions. Summarizing data from the literature on 156 species grown under twice-ambient CO_2 concentrations, Poorter (1993) noted an average increase in dry-matter accumulation in C_4 plants of 22% (compared with 41% for C_3 plants). In long-term studies with *Opuntia ficus-indica*, a CAM plant showing CO_2 fixation almost entirely at night, doubling the CO_2 partial pressure increased cladode thickness and the thickness of the photosynthetic chlorenchyma tissue; these cladodes showed surprisingly large increases of 70% and 74% in the maximum rates of nocturnal CO_2 fixation and 24-h carbon gain, respectively (Nobel and Israel 1994; see also Cui et al. 1993; Luo and Nobel 1993; Cui and Nobel 1994; Israel and Nobel 1994; Nobel et al. 1994a,b). How much of these large increases are attributable to development of a greater amount of photosynthetic tissue per unit surface area of cladode, and how much to non-saturation of PEPC under normal atmospheric conditions, remains to be determined. Irrespective of the underlying mechanism, doubling of ambient CO_2 concentration resulted in an increased rate of annual dry-matter accumulation of up to 40%, reaching 65 Mg ha^{-1} $year^{-1}$ in field plots where optimal light interception was maintained by frequent pruning (Nobel and Israel 1994). It will be of considerable interest to determine how cacti and other CAM plants respond to elevated CO_2 in their natural habitats. The water-conserving feature of the CAM pathway and the importance of the CO_2-concentrating mechanism in suppressing photorespiration are also likely to favour the performance and survival of CAM plants in the face of the declining precipitation and increasing temperatures that will occur in many regions as a result of global climate change.

25.3.3 Light-Use and Energetics of the 24-h CAM Cycle

Ecologically, CAM plants are best represented in exposed terrestrial and epiphytic habitats where periodic drought stress is combined with high solar radiation. A number of studies have suggested a photoprotective role for the CO_2 cycled through the CAM pathway in Phase III, especially under conditions where stomatal opening at other times during the day-night cycle is restricted by water deficits (Osmond 1982; Griffiths et al. 1989; Osmond et al. 1989). Such CO_2 recycling in the absence of net CO_2 uptake is believed to provide an important sink for excitation energy, thereby reducing the likelihood of sustained photodamage. Conversion of violaxanthin to zeaxanthin in the xanthophyll cycle also appears to contribute significantly to dissipation of excess excitation energy (Chap. 7; Maxwell et al. 1994; Robinson and Osmond 1994).

The ecological range of CAM plants is sufficiently great that a number of species can be found in deeply shaded habitats, with dark CO_2 fixation being the main route for carbon gain even under these conditions (Winter et al. 1983; Pfitsch and Smith 1988; see also Medina et al. 1994a,b). For examples such as the bromeliads from the subfamily Bromelioideae, in which the vast majority of species are CAM plants and the ancestral forms presumably grew in relatively exposed environments, this may reflect a high degree of adaptive radiation within the group and the secondary colonization of more shaded habitats (Medina 1974; Smith 1989). Under low-light conditions, the photon costs of net CO_2 assimilation will be closely linked to growth rates, so the energy requirements of the CAM cycle over and above those of the C_3 pathway become of critical importance (Chaps. 2, 16 and 24). These were considered above (Sect. 25.1) in terms of the energy requirements of net CO_2 assimilation in Phase III, but for growth studies the relevant costs will be those of the entire 24-h CAM cycle.

To estimate total 24-h energy requirements for the CAM cycle, the Phase I costs of nocturnal malic-acid accumulation must be added to the minimum costs of Phase III metabolism, which were derived for ME-CAM and PEPCK-CAM plants in Figs. 25.1C and D, respectively. Based on known reaction stoichiometries, the dark-phase costs per malic acid accumulated in the vacuole that must be met by oxidative phosphorylation are predicted to be 0.5 ATP in ME-CAM plants and 1.0 ATP in PEPCK-CAM plants (Chap. 2). These costs assume not only that starch breakdown in ME-CAM plants is phosphorolytic rather than amylolytic, but also that the breakdown of storage carbohydrate occurs exclusively via glycolysis. However, some hexose-P metabolism might be routed through the oxidative pentose-phosphate (OPP) pathway, and indeed Lüttge and Smith (1988) noted that this pathway could reduce the ATP required from oxidative phosphorylation in Phase I by up to 20% in ME-CAM plants (but would not affect the ATP requirement in PEPCK-CAM plants). To set against this, the OPP pathway would need as 'working capital' a 20% *larger* pool of storage carbohydrate, because 1 CO_2 is released when 6-phosphogluconate is oxidatively decarboxylated to ribulose 5-phosphate; furthermore, this CO_2 would have to be reassimilated back to storage carbohydrate in Phase III (at a cost of 3.167 ATP and 2 NADPH per CO_2). Overall, therefore, operation of the OPP pathway would incur an additional energy cost for the CAM cycle, and experimental evidence suggests that carbohydrate oxidation in such tissues is likely to be predominantly glycolytic (ap Rees 1985). Another factor of importance for the ATP budget of Phase I is the role of the PP_i-dependent fructose 6-phosphate 1-phosphotransferase (PFP). The extent to which this enzyme is involved in the CAM cycle is not properly understood, but if PFP functioned in the glycolytic direction and could substitute for the ATP-dependent phosphofructokinase, this would reduce the total ATP requirement of Phase I metabolism. Recent evidence from potato is consistent with such a function of PFP (Hajirezaei et al. 1994; Scott et al. 1995), but much more work is needed with CAM plants before the role of this enzyme in carbohydrate metabolism is established.

As with the Phase III energy costs, the flux ratios observed during Phase I will be influenced by the way in which the ATP requirement for vacuolar malic-acid accumulation is met. In ME-CAM plants, ATP can be provided in equal amounts by substrate-level phosphorylation in glycolysis and by mitochondrial oxidative phosphorylation, but in PEPCK-CAM plants this ATP requirement must be met by oxidative phosphorylation alone (Lüttge and Smith 1988). The most plausible respiratory substrate would seem to be malate (and perhaps citrate) derived from the glycolytic breakdown of storage carbohydrate. If the energy budget is thereby balanced in Phase I, the marginal cost of the CAM cycle in Phase III – namely the regeneration of storage carbohydrate – can then be met by photophosphorylation as shown in Fig. 25.1.

If the balance of ATP required for malic-acid accumulation in Phase I is provided by respiration of malate in the mitochondria, it is possible to derive optimal flux stoichiometries for carbon flow during the complete 24-h cycle. These are shown in Figs. 25.2A and B for ME-CAM and PEPCK-CAM plants, respectively, again using the assumption that complete oxidation of 1 malate to 4 CO_2 can yield 14.75 ATP. These respired 4 CO_2 must be conserved in Phase I by refixation by PEPC into malate, but 3 of the 4 malate produced can be removed to the vacuole as malic acid (at a cost of 1 ATP per malic acid accumulated). Subsequently, the CO_2 released by decarboxylation of these 3 malate in Phase III must also be converted via the PCR cycle back to the level of the carbohydrate 'working capital' of the CAM cycle. That is, to sustain steady-state operation of the day-night cycle, any respiratory CO_2 produced from malate in Phase I must be quantitatively reassimilated back to storage carbohydrate in Phase III.

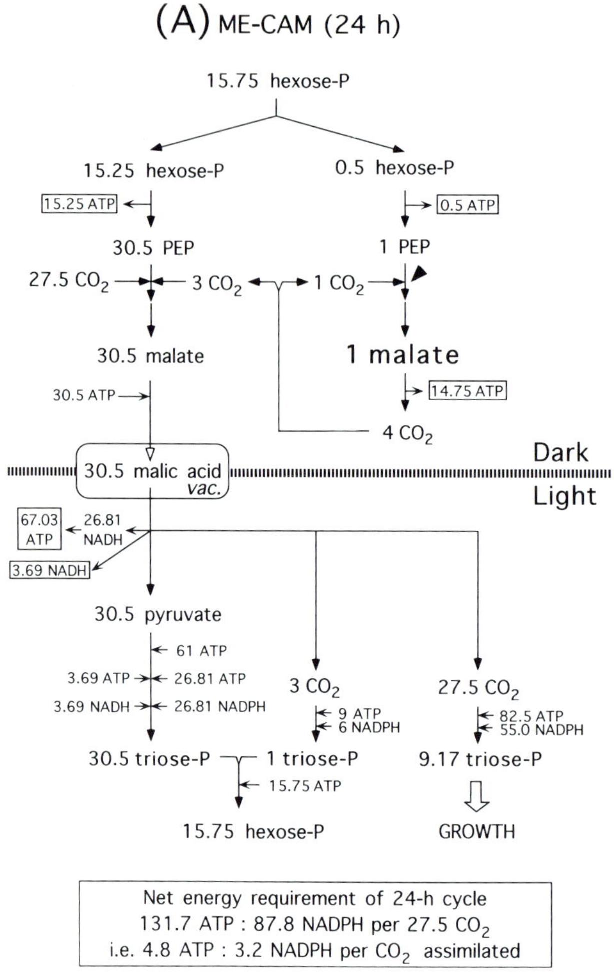

Fig. 25.2A,B. Energetic costs of net CO_2 assimilation through the CAM pathway over a complete 24-h cycle in **A** ME-CAM plants and **B** PEPCK-CAM plants. These schemes illustrate the energetic implications of Phase I metabolism at night (carbohydrate breakdown, malate synthesis and vacuolar malic-acid accumulation) combined with those of Phase III metabolism during the day. It is assumed that the energy budget during Phase I is balanced, with the energy required for malic-acid transport into the vacuole (1 ATP per 2 H^+ to drive the tonoplast H^+-ATPase, with the malate anion following passively) being met either by the production of ATP through substrate-level phosphorylation in glycolysis together with oxidative phosphorylation (in ME-CAM plants), or by oxidative phosphorylation alone (in PEPCK-CAM plants). Malate is shown as the sole substrate for mitochondrial respiration, yielding 14.75 ATP per malate oxidized to the level of 4 CO_2 via the TCA cycle. The flux stoichiometries are based on the complete oxidation of 1 malate to 4 CO_2, of which 1 CO_2 is then recycled through the respiratory pathway (a single CO_2 being required initially to prime this pathway, indicated by the *large arrow-head*), and 3 CO_2 are refixed to contribute to malate accumulating in the vacuole. During the light phase, the pyruvate **A** or PEP **B** resulting from malate decarboxylation is quantitatively conserved in gluconeogenesis,

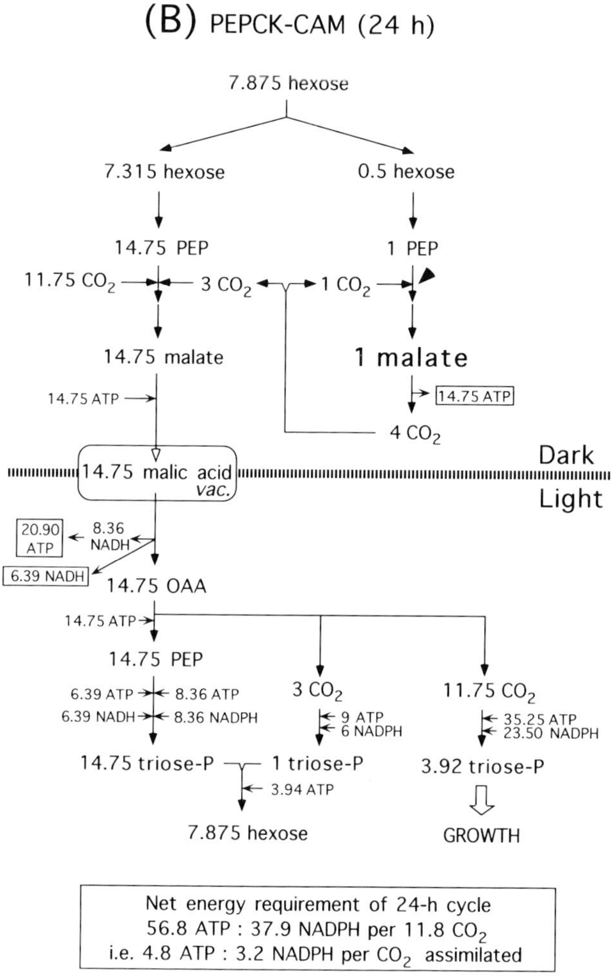

Fig. 25.2A,B (*Contd.*) since this represents the most energy-efficient pathway for the regeneration of storage carbohydrate. As in Fig. 25.1, it is assumed that the net energy costs of Phase III can be met entirely by the production of ATP and NADPH (in the ratio 1.5:1) by photophosphorylation in the chloroplasts. For the gluconeogenic reaction leading to triose-P, the ATP and NADH required in the cytosol are indicated on the *left-hand side* of the pathway, and the ATP and NADPH required in the chloroplasts on the *right-hand side*. For further assumptions, see legend to Fig. 25.1

For ME-CAM plants (Fig. 25.2A), the net energy requirement of the 24-h cycle is 4.8 ATP:3.2 NADPH per CO_2 assimilated from the atmosphere, which can be met entirely by photophosphorylation with the conventional ATP:NADPH ratio. These stoichiometries are also consistent with the experimental observation that the rate of dark respiration is sufficient to provide approximately half the ATP required for vacuolar malic-acid accumulation in Phase I (Lüttge et al. 1981). This would involve complete oxidation in the mitochondria of only about 3% of the malate synthesized via dark CO_2 fixation (Fig. 25.2A). In PEPCK-CAM plants (Fig. 25.2B), on the other hand, the ATP requirement of malic-acid accumulation in Phase I must be met entirely by respiration, which involves complete oxidation of just over 6% of the malate synthesized by dark CO_2 fixation. Thus, although the gluconeogenic pathway from three-carbon residue to storage carbohydrate in Phase III requires less energy in PEPCK-CAM plants (Fig. 25.1D), a greater proportion of the total malate fixed by PEPC in Phase I is respired to CO_2, which must also be reassimilated in the subsequent light period back into the carbohydrate working capital. Consequently, the net energy requirement of the 24-h cycle turns out to be exactly the same in PEPCK-CAM plants as in ME-CAM plants, i.e. 4.8 ATP:3.2 NADPH per net CO_2 assimilated from the atmosphere (Fig. 25.2B).

In comparing the energetics of these two types of CAM cycle, we should note that the pool of storage carbohydrate metabolized per CO_2 assimilated from the atmosphere is 17% larger in PEPCK-CAM plants (Fig. 25.2B) than in ME-CAM plants (Fig. 25.2A). Thus, PEPCK-CAM plants may have a fractionally higher energy requirement for generating the working capital of the CAM cycle compared with ME-CAM plants. However, these costs do not appear in the marginal costs of the CAM pathway, since the working capital is conserved over the 24-h cycle in the schemes of Fig. 25.2. In fact, PEPCK-CAM plants usually metabolize significant quantities of starch during the day-night rhythm (Sideris et al. 1948; Black et al. 1982; Medina et al. 1986; Fahrendorf et al. 1987; Popp et al. 1987; Ball et al.1991), so the CAM pathway may not be sustained entirely by hexose metabolism. However, the partial involvement of starch in place of hexose in the scheme of Fig. 25.2B would not affect the overall energy requirement of the 24-h cycle. On the other hand, PEPCK-CAM plants such as *Ananas* have particularly high activities of PFP (Carnal and Black 1989), which, as mentioned above, could reduce the ATP required from malate oxidation if it functioned in the glycolytic direction.

These schemes show that the *additional* energy cost of the entire CAM cycle, over and above the standard cost of 3 ATP:2 NADPH for CO_2 assimilation to the level of triose phosphate in the absence of photorespiration, is only 1.8 ATP plus 1.2 NADPH per net CO_2 assimilated. This additional energy demand of the 24-h cycle can be met entirely by photophosphorylation with the conventional ratio of 3 ATP:2 NADPH produced (this being a constraint imposed in deriving the flux ratios in Fig. 25.2). As discussed in Section 25.1.2, this ratio of ATP:NADPH serves to make the photon-use efficiency of photosynthetic O_2 evolution indistinguishable from that of C_3 plants in the absence of photorespiration. Consequently, when appropriate allowance is made for photorespiration, which results in a cost of net CO_2 fixation in C_3 plants of about 4.5 ATP:2.9 NADPH (Table 24.2), the predicted requirement for 4.8 ATP:3.2 NADPH per CO_2 in CAM plants (Fig. 25.2) represents a photon cost only about 10% higher than for C_3 plants under present atmospheric conditions.

Most previous energy-budget models have assumed that the marginal costs of the 24-h CAM cycle can be met by ATP alone, giving total energy requirements of 7 ATP:2 NADPH per CO_2 assimilated for ME-CAM plants, and of 6.25 ATP:2 NADPH for PEPCK-CAM plants. If calculated in terms of thermodynamic free energy content (50 kJ mol^{-1} for ATP, and 220 kJ mol^{-1} for NADPH), this actually represents a slightly *lower* energetic cost of the CAM pathway compared with net CO_2 fixation via C_3 photosynthesis (see Table 16.3). However, the additional ATP required for the CAM pathway, over and above the 3 ATP:2 NADPH supplied by the light reactions in C_3 photosynthesis, must be provided by auxiliary reactions such as cyclic or pseudocyclic electron transport. This inevitably means that the total photon costs of the CAM pathway per CO_2 assimilated are somewhat *higher* than those of C_3 photosynthesis (Table 24.2). In this respect, the pathways proposed in Fig. 25.2 are energetically favourable because the marginal costs of the 24-h cycle can be met entirely by ATP:NADPH produced in the light period in the ratio 3:2. At any rate, whichever energy-budget model proves to be most applicable to CAM plants growing in their natural environments, it would seem that the energetic cost of the CAM pathway is substantially smaller than often supposed.

Although these schemes suggest a way in which the energy costs of the CAM cycle can be minimized, further experiments will be needed to determine whether the flux ratios in plants growing under optimal conditions approach those shown in Fig. 25.2. However, a number of predictions follow from these schemes that should be readily testable. For example, if complete oxidation of malate provides just under half (ME-CAM) or all (PEPCK-CAM) of the ATP required in Phase I through oxidative phosphorylation, then the rates of dark respiration relative to net CO_2 uptake in ME-CAM plants should be less than half those in PEPCK-CAM plants. There is evidence that this may be the case from measurements of O_2 uptake in bromeliads, some of which at least are known to be PEPCK-CAM plants, since they show high rates of dark respiration relative to malate accumulation (Lüttge and Ball 1987). Application of the polarographic methods described in Chapter 1 for combined gas-phase measurements of O_2 and CO_2 exchanges may allow these fluxes to be quantified in a wider range of species under appropriate conditions. Another feature of these proposed pathways is that, even with the optimal flux stoichiometries shown in Fig. 25.2, a small proportion of the nocturnally accumulated malic acid derives from recycling of respiratory CO_2 (cf. p. 192). Under such conditions, the ratio of malic acid accumulated at night to net CO_2 uptake should be 1.09 in ME-CAM (Fig. 25.2A), or 1.26 in PEPCK-CAM (Fig. 25.2B); the equivalent values for titratable H^+ to CO_2 will be 2.18 for ME-CAM, or 2.52 for PEPCK-CAM, assuming that all the acid accumulated is malic acid. Ratios of titratable H^+ accumulation to net CO_2 uptake greater than 2.0 have been observed for several ME-CAM plants under well-watered conditions in some of the most detailed studies of these species (Medina and Osmond 1981; Medina 1982, 1984; Nobel and Hartsock 1983; Winter et al. 1986b). If some CO_2 is recycled into citrate (the accumulation of which is not believed to be associated with net CO_2 uptake: see Sect. 25.2.3), this will result in an increase in the ratio of titratable acid accumulated to net CO_2 taken up, since the observed H^+: citrate ratio in CAM is 3:1 (Lüttge 1988; Franco et al. 1992). Finally, the ratio of hexose equivalents used to net CO_2 taken up should be slightly greater than 0.5: the predicted ratio is 0.57 for ME-CAM (Fig. 25.2A), or 0.67 for PEPCK-CAM (Fig. 25.2B).

In some circumstances, a considerable proportion of the malic acid accumulated in Phase I may be derived from respired CO_2 refixed through PEPC. High degrees of recycling of respiratory CO_2 are thought to be particularly characteristic of environments in which uptake of atmospheric CO_2 has been restricted by partial stomatal closure (Griffiths 1988). In the limit, if the stomata are closed completely by severe drought stress, then all the nocturnally accumulated malic acid will be derived from recycling of respiratory CO_2: this corresponds to the situation known as 'CAM idling' (Szarek et al. 1973). A model for CO_2 recycling during idling is shown in Fig. 25.3, which is based on

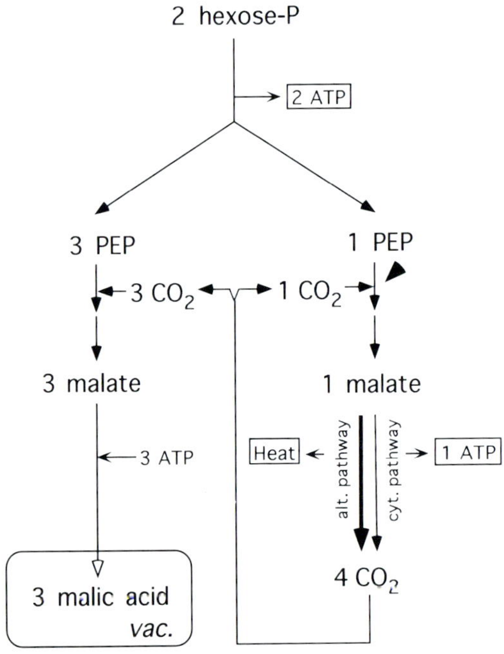

Fig. 25.3. Model for nocturnal malic-acid accumulation in ME-CAM plants based solely on recycling of respiratory CO_2 produced by mitochondrial oxidation of malate. The storage carbohydrate oxidized at night is assumed to be starch, which can be broken down by the phosphorolytic pathway via hexose-P to PEP, yielding 0.5 ATP per PEP by substrate-level phosphorylation (Chap. 2). Of the 4 CO_2 produced by complete oxidation of 1 malate via the TCA cycle, 1 CO_2 is recycled through the respiratory pathway (which must be primed by 1 CO_2, indicated by the *large arrow-head*), and 3 CO_2 are refixed into malate that accumulates in the vacuole. In this scheme, no net CO_2 uptake occurs from the atmosphere, and the cytochrome pathway is required only to produce 1 ATP per malate out of a possible 14.75 ATP; thus, most electron flow would be diverted through the alternative pathway, with energy dissipation occurring in the form of heat loss

the dark reactions of Fig. 25.2A considered in the absence of any uptake of atmospheric CO_2. In ME-CAM plants with starch breakdown occurring by the phosphorolytic pathway, oxidative phosphorylation would need to provide only one-third of the ATP required for vacuolar malic-acid accumulation (Fig. 25.3). An equivalent scheme for recycling in PEPCK-CAM plants would show that oxidative phosphorylation has to provide all the ATP needed for malic-acid accumulation. In both cases, however, the demand for ATP would be relatively low compared with the potential capacity of the cytochrome pathway to produce it, suggesting that much of the electron flow may be diverted through the alternative pathway. A significant fraction of the nocturnally accumulated malic acid can derive from refixation of respiratory CO_2 even in well-watered CAM plants, as for example in tropical environments with high night temperatures (Winter et al. 1986b; Lüttge and Ball 1987; Griffiths 1988). It would thus be interesting to determine whether the alternative oxidase also makes a proportionately greater contribution to respiratory O_2 uptake under such conditions compared with those under which recycling is minimal.

The alternative pathway is believed to draw off electrons from the mitochondrial electron transport chain at the level of ubiquinone, and so to bypass the 'coupling sites' of complexes III and IV (Lambers 1990; Siedow 1995). But even the alternative pathway could still produce more ATP than actually required by the scheme in Fig. 25.3 (1 ATP per malate oxidized to 4 CO_2), if the reducing equivalents resulting from malate oxidation feed into the H^+-translocating NADH dehydrogenase of complex I. However, mitochondria from CAM plants possess high activities of a rotenone-resistant NADH dehydrogenase, which does not translocate H^+ and is consequently non-energy-conserving (Arron et al. 1979; Day 1980). Together, this enzyme and the alternative oxidase might permit relatively high rates of malate oxidation in the absence of a large demand for ATP (Lance et al. 1985), such as may occur when a large proportion of the malic acid accumulating in Phase I is derived from recycling of respiratory CO_2 (Fig. 25.3).

Another factor influencing the ratio of net CO_2 uptake in Phase I to titratable H^+ accumulated will be the amount of CO_2 fixed into citric acid. Lüttge (1988) showed in his reassessment of this phenomenon that citric-acid accumulation occurs in a wide range of species, but that it most likely does not represent a pathway for net assimilation of CO_2 (see Sect. 25.2.3). Nevertheless, the H^+: citrate stoichiometry observed is 3:1, showing that the free acid accumulates in the vacuole, and under certain conditions the amount of citric acid accumulated can actually exceed that of malic acid (Chap. 19; Franco et al. 1992). If citrate is produced from malate via the mitochondrial NAD-ME and pyruvate dehydrogenase reactions, the reducing equivalents generated can be used to supply ATP by oxidative phosphorylation. Citrate thus synthesized via the first two reactions of the TCA cycle would be exported from the mitochondria to the cytosol and subsequently accumulated in the vacuole at a cost of 1.5 ATP per citric acid (assuming a stoichiometry for the tonoplast H^+-ATPase of 2 H^+:1 ATP).

The consequences of nocturnal citric-acid accumulation for the energetics of the CAM cycle are summarized for ME-CAM plants and PEPCK-CAM

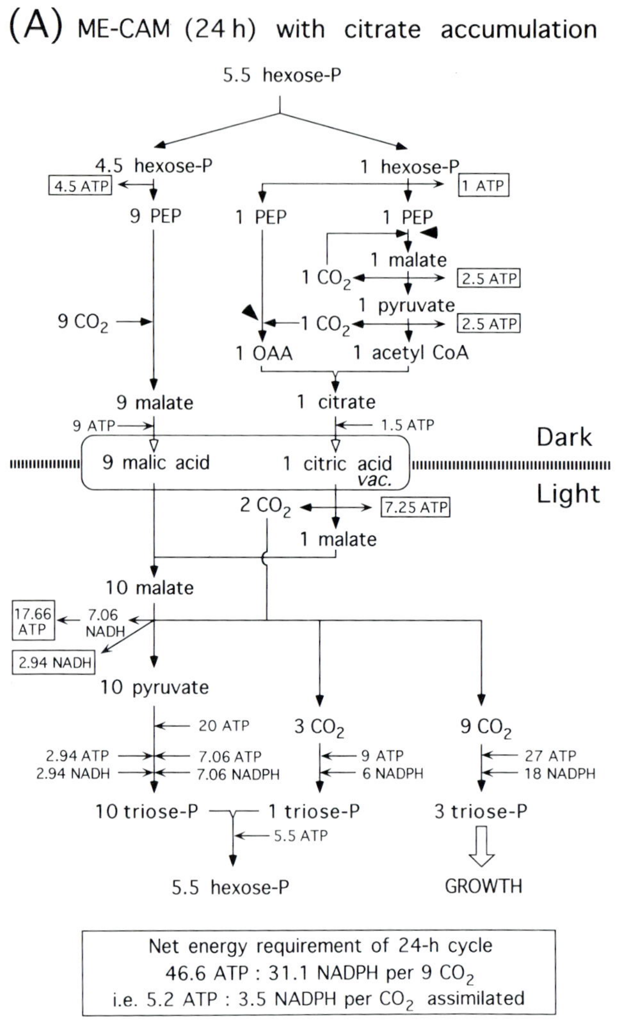

Fig. 25.4A,B. Energetic costs of net CO_2 assimilation through the CAM pathway over a complete 24-h cycle in **A** ME-CAM plants and **B** PEPCK-CAM plants showing nocturnal citric-acid accumulation. The ATP required for malic-acid and citric-acid transport into the vacuole (to drive the tonoplast H^+-ATPase with a stoichiometry of 2 H^+ per ATP, with the organic-acid anion following passively) is provided either by substrate-level phosphorylation in glycolysis together with oxidative phosphorylation (in ME-CAM plants), or by oxidative phosphorylation alone (in PEPCK-CAM plants). Malate is shown as the sole substrate for mitochondrial respiration, being partially oxidized to 2 CO_2 (which are refixed by PEPC) and generating citrate via the first two reactions of the TCA cycle. The flux stoichiometries are based on the accumulation of 1 citric acid in the vacuole in Phase I, which in this scheme accounts for all the nocturnally accumulated acid that derives from refixation of respiratory CO_2. Other assumptions are as in the legends to Figs. 25.1 and 25.2

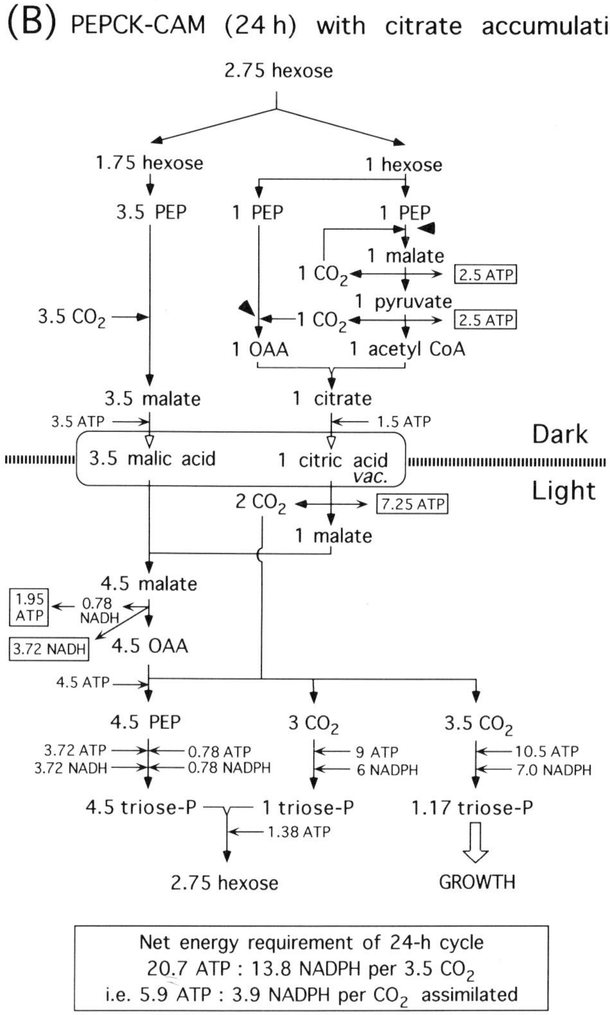

Fig. 25.4B

plants in Figs. 25.4A and B, respectively. Based on the vacuolar accumulation of 1 citric acid, these schemes show that the ATP supplied by partial oxidation of malate can be used to help balance the energy budget in Phase I. As in Fig. 25.2, the respiratory CO_2 refixed into citric acid in Phase I has to be assimilated back into the carbohydrate working capital of the CAM cycle in Phase III. This has a net energy cost and, for the ratio of net CO_2 assimilation to CO_2 recycling shown in Figs. 25.4A,B, results in a significantly higher energy requirement for the 24-h cycle than for malic-acid accumulation alone (Figs. 25.2A,B). (Similarly, the higher energy requirement for the PEPCK-CAM pathway shown in Fig. 25.4B compared with the ME-CAM pathway in Fig. 25.4A is simply the result of a higher ratio of recycled CO_2 to net CO_2 fixed.) These calculations support the notion that the CO_2 recycling involved in citric-acid accumulation may serve to prevent overreduction of the photosynthetic electron transport chain at times when CO_2 uptake from the atmosphere is restricted (Sect. 25.2.3). However, it is clear that much more work is needed to understand fully the metabolic significance of citrate metabolism in CAM plants.

25.3.4 CAM Versus C_3: Costs and Benefits

If the pathways proposed in Fig. 25.2 for the 24-h costs of CO_2 assimilation are mechanistically correct, then the CAM pathway presents a relatively small additional cost (of the order of 10%) for net CO_2 assimilation compared with the C_3 pathway. In practice, the comparative energetics of the two pathways will be strongly affected by ambient environmental conditions. For example, the energetics of C_3 photosynthesis depend considerably on the ratio of the oxygenation to carboxylation reactions by RUBISCO, which becomes much less favourable for carbon gain with increasing temperature. In CAM plants, on the other hand, photorespiration is suppressed by the CO_2-concentrating mechanism over a wide range of temperatures, although some carbon loss may occur during Phase III at high temperatures, especially at low light intensities.

Measured values for the photon requirement of net CO_2 uptake in leaves of C_3 plants are about 30% higher than those predicted theoretically (Ehleringer and Björkman 1977; Ehleringer and Pearcy 1983; Sharp et al. 1984). If the molar ratio of CO_2 taken up to O_2 evolved during photosynthesis is significantly lower than unity, this could indicate that the mean reduction level of assimilates formed is greater than that of carbohydrate, and/or that the rate of light-dependent nitrate reduction is appreciable (Björkman and Demmig 1987). However, these considerations would not be expected to affect the relative costs of the C_3 and CAM pathways. Nevertheless, accurate studies of the photon-use efficiency of 24-h carbon gain, based on measurements of complete day-night cycles, are badly needed to quantify the precise differences in photon costs of carbon assimilation in CAM and C_3 plants.

25.4 Evolutionary Origins

CAM has been traditionally viewed as a photosynthetic mechanism that helps to conserve water during carbon fixation in arid environments. These include habitats that are microclimatically arid, such as those of CAM epiphytes in tropical forests. However, given the properties of CAM as a 'CO_2 concentrating mechanism' (Osmond et al. 1982) and the occurrence of CAM in submerged aquatic species (Chap. 18), it is now clear that the increased water-use efficiency associated with CAM-CO_2 assimilation was possibly not the primary, and almost certainly not the only, driving force for the evolution of this photosynthetic pathway (Ehleringer and Monson 1993). Rather, reduced CO_2 availability may well have been the common, original selective pressure that resulted in the evolution of CAM in both aquatic and terrestrial habitats. Reduced CO_2 availability results from low concentrations and diffusivity of CO_2 in the aqueous media, and from daytime stomatal closure caused by drought stress in arid terrestrial and epiphytic habitats. Nighttime assimilation of respired CO_2 – as observed in present-day 'CAM-cycling' plants (Chap. 12) – could have been an initial evolutionary step that would have improved the plant's carbon balance by reducing nocturnal losses of respiratory CO_2 (Teeri 1982; Griffiths 1989, 1992; Monson 1989; Ehleringer and Monson 1993). Thus, improved carbon economy in environments that restrict daytime CO_2 availability may have been be the primary driving force in the evolution of CAM.

Carbon-isotope values of organic material from hadrosaurs and palaeocretes indicate the possible existence of CAM (or C_4!) plants in the Late Cretaceous and as early as the Carboniferous, respectively (Chap. 24). However, there is no direct prehistoric evidence on which to reconstruct the evolutionary origins of the CAM pathway. Fossil records for succulent plants are lacking, except for the recent discovery of a permineralized *Yucca* (Agavaceae) from the Middle Miocene (about 15 million years ago) (Tidwell and Parker 1990) and the recovery of 40000- and 10000-year-old *Opuntia polyacantha* (Cactaceae) in Nevada (Troughton et al. 1974). The present-day *Isoetes*, the most basal CAM taxon known (Chaps. 18 and 24), are considered remnants of an evolutionary reduction process (Mägdefrau 1968; Stewart and Rothwell 1993) that started with tall arborescent Devonian Lepidodendrales and included small Triassic *Isoetites* (about 200 million years ago) (Bock 1962) and Early Cretaceous *Nathorstiana* (about 120 million years ago). Both of these extinct genera were very similar in size and morphology to extant *Isoetes* and possibly had the capacity to perform CAM, and indeed there is fossil evidence for the existence of *Isoetes* by the Early Cretaceous (Collinson 1991). Since all present-day species of *Isoetes* appear to possess CAM (Chap. 18), it seems likely both that CAM is monophyletic within this taxon and that this represents the oldest clade of known CAM plants.

During the Cretaceous, there was probably also a high degree of speciation of the palaeotropic fern genus *Pyrrosia* in continental Asia (Griffiths 1989). The extant, largely epiphytic members of *Pyrrosia* are either poikilohydrous C_3 plants or succulent CAM plants, with the exception of the semi-succulent epiphytic *P. rupestris*, which may have lost CAM (Griffiths 1989). Furthermore, pollen evidence indicates that the Gnetales – of which *Welwitschia mirabilis* is an extant representative with very weak CAM – diversified considerably during the Mid-Cretaceous, particularly at low palaeolatitudes, but then suffered a rapid decline (Crane and Lidgard 1989; Stewart and Rothwell 1993). Global temperatures were probably also high during the Early and Mid-Cretaceous, and seasonal aridity could have been conducive to the development of CAM in a plant such as *Welwitschia*. On the other hand, atmospheric CO_2 concentrations (and CO_2:O_2 ratios) were probably much higher than present

(Chap. 24), which would have reduced the relative advantage of CAM over C_3 photosynthesis if restricted CO_2 availability were the primary selection pressure in CAM evolution.

The great majority of present-day CAM species are found in the monocotyledonous and dicotyledonous taxa of the angiosperms, which underwent their major diversification in the Early and Mid-Cretaceous (Crane et al. 1995). The taxonomic and biogeographic distribution of the CAM angiosperms and their assumed phylogenetic relationships clearly suggest that CAM is derived from C_3 ancestors and that CAM has independently arisen numerous times within different higher plant taxa. Further, the semi-arid or seasonally arid conditions of low-latitude areas during the Early Cretaceous (Crane et al. 1995) suggest that CAM may have arisen in some taxa at a relatively early stage in angiosperm evolution.

The Bromeliaceae were the first group in which the evolutionary relationships between CAM, life-form and habitat were considered in detail (Medina 1974; Griffiths and Smith 1983). Starting from a hypothetical terrestrial C_3 ancestor occupying exposed but possibly relatively moist habitats, both epiphytism and CAM are now believed to have arisen more than once during the evolution of the present-day bromeliads. There is also good circumstantial evidence that CAM is being progressively lost in some taxa. In the subfamily Bromelioideae, for example, which is almost wholly CAM, a number of species are characteristic of more mesic habitats such as upper montane rain forest and the forest understory, and some of these species also show only weak CAM activity (Smith 1989).

Extending the earlier studies of Teeri (1982), Pilon-Smits and collaborators (Chap. 23) have used $\delta^{13}C$ values, PEPC and NADP-ME activities, as well as data on leaf succulence, to study the expression of CAM in several genera of the Crassulaceae. Their work has provided detailed insights into phylogenetic and evolutionary relationships for the genus *Aeonium* in particular, in which adaptive radiation has led to a wide range of life-forms. These authors demonstrate how CAM has evolved from small undifferentiated species exhibiting little or no CAM to large woody species showing strong CAM, and how the expansion of the Sahara desert may have led to the disjunct present-day distribution of these large woody species over North Africa and the Canary and Cape Verde Islands.

A plausible scenario for the co-evolution of CAM and stem succulence in the Cactaceae has been depicted by Gibson and Nobel (1986), who suggest that the ancestral forms in this family may have been similar to the leaf-bearing *Pereskia*. This is the most basal taxon of extant Cactaceae (Barthlott and Hunt 1993), whose members assimilate carbon predominantly via C_3 photosynthesis by CO_2 uptake during the daytime. Gibson and Nobel (1986) describe how the breakup of Gondwanaland and the separation of Madagascar, Africa and South America in the Cretaceous (Axelrod and Raven 1978) may have led to the disjuntion of an ancestral group and the parallel evolution of similar vegetative characteristics and of the CAM pathway in three closely related families: the Cactaceae in the New World, the Didiereaceae endemic to Madagascar, and the Portulacaceae in Africa. Indeed, Rauh (1973) has specifically pointed out the similarities in growth-habit between *Alluaudia procera* and *A. humbertii* (Didiereaceae) from Madagascar and *Ceraria namaquensis* (Portulacaceae) from Namaqualand, South Africa. While the major radiation of new species in the Cactaceae is thought to have taken place some 30 to 35 million years ago from what is now southern Mexico or northern South America (Nobel 1994), evolutionary diversification within the Aizoaceae, another CAM family of the Caryophyllales, occurred much more recently. About 90% of the ca. 2000 present-day species of Aizoaceae are endemic to the Succulent Karoo, a semi-desert in southern Africa that opened up only 5 million years ago (Axelrod and Raven 1978; Ihlenfeldt 1994). The rapid speciation of the Aizoaceae provides an important example of the apparent ease and rapidity with which the CAM pathway facilitates new ecological opportunities for subsequent diversification, and also of the differing degrees of CAM activity that can be observed within a single family.

Since hot, dry climates – the preferred environments of CAM succulents – are apparently not conducive to the formation of fossils, there is only a limited prospect that new palaeobotanical discoveries might significantly advance our knowledge of the origins of the CAM pathway. However, we can expect that a considerable amount of new information will be gained from future studies of extant species at the molecular level. For example, the expected broadening of analyses of DNA sequences from the chloroplast gene *rbc*L (Chase et al. 1993), and other genes, to an ever-increasing number of taxa should substantially improve our knowledge of phylogenetic relationships between C_3 and CAM species and will provide a further impetus to efforts to reconstruct the origins of the CAM pathway. Further application of molecular techniques should also lead to a better understanding of the mechanisms underlying the evolution of CAM. PEPC and the other enzymes of the CAM cycle are all present in C_3 plants, but the induction of CAM in *M. crystallinum* is accompanied by the synthesis of a CAM-specific isoform of PEPC (Chap. 9). This suggests that the evolution of CAM is accompanied by the generation of CAM-specific genes within multigene families, perhaps as a result of modification of upstream sequence elements in the non-coding regions of the genes concerned (see also Hermans and Westhoff 1990 and Toh et al. 1994 on the molecular evolution of C_4-PEPC). Characterization of the gene families encoding key enzymes of the CAM pathway, such as PEPC, NADP-ME, NAD-ME, PPDK and PEPCK, in a wide range of species promises to be one of the most rewarding areas of CAM research in the years ahead. This work should help to reveal which modifications to C_3 photosynthetic metabolism are needed at the molecular level to define the fundamental operating principles of the CAM pathway.

References

Adams WW, Nishida K, Osmond CB (1986) Quantum yields of CAM plants measured by photosynthetic O_2 exchange. Plant Physiol 81: 297–300

ap Rees T (1985) The organization of glycolysis and the oxidative pentose phosphate pathway in plants. In: Douce R, Day DA (eds) Encyclopedia of plant physiology. New Series, vol 18. Higher plant cell respiration. Springer, Berlin Heidelberg New York, pp 391–417

ap Rees T (1990) Carbon metabolism in mitochondria. In: Dennis DT, Turpin DH (eds) Plant physiology, biochemistry and molecular biology. Longman, Harlow, pp 106–123

Arron GP, Spalding MH, Edwards GE (1979) Isolation and oxidative properties of intact mitochondria from the leaves of *Sedum praealtum*. A crassulacean acid metabolism plant. Plant Physiol 64: 182–186

Axelrod DI, Raven PH (1978) Late Cretaceous and Tertiary vegetation history of Africa. In: Werger MJA (ed) Biogeography and ecology of southern Africa. Junk, The Hague, pp 77–130

Ball E, Hann J, Kluge M, Lee HSJ, Lüttge U, Orthen B, Popp M, Schmitt A, Ting IP (1991) Ecophysiological comportment of the tropical CAM-tree *Clusia* in the field. II. Modes of photosynthesis in trees and seedlings. New Phytol 117: 483–491

Barthlott W, Hunt WR (1993) Cactaceae. In: Kubitzki K, Rohwer JG, Bittrich V (eds) The families and genera of vascular plants II. Flowering plants: dicotyledons; magnoliid, hamamelid and caryophyllid families. Springer, Berlin Heidelberg New York, pp 161–197

Baur B, Fischer K, Winter K, Dietz KJ (1994) cDNA sequence of a protein kinase from the halophyte *Mesembryanthemum crystallinum* L., encoding a SNF-1 homologue. Plant Physiol 106: 1225–1226

Baur B, Fischer K, Winter K, Dietz KJ (1995) Molecular cloning of a protein kinase preferentially expressed upon induction of crassulacean acid metabolism in the facultative CAM plant *Mesembryanthemum crystallinum* L. (submitted)

Björkman O, Demmig B (1987) Photon yield of O_2 evolution and chlorophyll fluorescence characteristics at 77 K among vascular plants of diverse origins. Planta 170: 489–504

Black CC, Carnal NW, Kenyon WH (1982) Compartmentation and the regulation of CAM. In: Ting IP, Gibbs M (eds) Crassulacean acid metabolism. American Society of Plant Physiologists, Rockville, pp 51–68

Bloom AJ, Troughton JH (1979) High productivity and photosynthetic flexibility in a CAM plant. Oecologia 38: 35–43

Bock W (1962) A study of fossil *Isoetes*. J Paleontol 36: 53–59

Borland AM, Griffiths H, Broadmeadow MSJ, Fordham MC, Maxwell C (1994) Carbon-isotope composition of biochemical fractions and the regulation of carbon balance in leaves of the C_3-crassulacean acid metabolism intermediate *Clusia minor* L. growing in Trinidad. Plant Physiol 106: 493–501

Borland AM, Griffiths H, Maxwell C, Fordham MC, Broadmeadow MSJ (1995) CAM induction in *Clusia minor* L. during the transition from wet to dry season in Trinidad: the role of organic acid speciation and decarboxylation. (submitted)

Bremberger C, Haschke H-P, Lüttge U (1988) Separation and purification of the tonoplast ATPase and pyrophosphatase from plants with constitutive and inducible crassulacean acid metabolism. Planta 175: 465–470

Brulfert J, Vidal J, Keryer E, Thomas M, Gadal P, Queiroz O (1985) Phytochrome control of phosphoenolpyruvate carboxylase synthesis and specific RNA level during photoperiodic induction in a CAM plant. Physiol Vég 23: 921–928

Brulfert J, Kluge M, Güçlü S, Queiroz O (1988) Interaction of photoperiod and drought as CAM inducing factors in *Kalanchoë blossfeldiana* Poelln., cv. Tom Thumb. J Plant Physiol 133: 222–227

Brulfert J, Güçlü S, Taybi T, Pierre JN (1993) Enzymatic responses to water stress in detached leaves of the CAM plant *Kalanchoë blossfeldiana* Poelln. Plant Physiol Biochem 31: 491–497

Bryce JH, Hill SA (1993) Energy production in plant cells. In: Lea PJ, Leegood RC (eds) Plant biochemistry and molecular biology. Wiley, Chichester, pp 1–26

Carnal NW, Black CC (1989) Soluble sugars as carbohydrate reserve for CAM in pineapple leaves. Plant Physiol 90: 91–100

Chase MW, Soltis DE, Olmstead RG, and 39 others (1993) Phylogenetics of seed plants: an analysis of nucleotide sequences from the plastid gene *rbc*L. Ann Mo Bot Gard 80: 528–580

Collinson ME (1991) Diversification of modern heterosporous pteridophytes. In: Blackmore S, Barnes SH (eds) Pollen and spores. Clarendon Press, Oxford, pp 119–150

Conti S, Smirnoff N (1994) Rapid triggering of malate accumulation in the C_3/CAM intermediate plant *Sedum telephium:* relationship with water status and phosphoenolpyruvate carboxylase. J Exp Bot 45: 1613–1621

Coté FX, André M, Folliot M, Massimino D, Daguenet A (1989) CO_2 and O_2 exchanges in the CAM plant *Ananas comosus* (L.) Merr. Plant Physiol 89: 61–68

Crane PR, Lidgard S (1989) Angiosperm diversification and paleolatitudinal gradients in Cretaceous floristic diversity. Science 246: 675–678

Crane PR, Friis EM, Pedersen KR (1995) The origin and early diversification of angiosperms. Nature 374: 27–33

Cui M, Nobel PS (1994) Gas exchange and growth responses to elevated CO_2 and light levels in the CAM species *Opuntia ficus-indica.* Plant Cell Environ 17: 935–944

Cui M, Miller PM, Nobel PS (1993) CO_2 exchange and growth of the crassulacean acid metabolism plant *Opuntia ficus-indica* under elevated CO_2 in open-top chambers. Plant Physiol 103: 519–524

Day DA (1980) Malate decarboxylation by *Kalanchoë daigremontiana* mitochondria and its role in crassulacean acid metabolism. Plant Physiol 65: 675–679

DeRocher EJ, Bohnert HJ (1993) Developmental and environmental stress employ different mechanisms in the expression of a plant gene family. Plant Cell 5: 1611–1625

Duff SMG, Chollet R (1995) In vivo regulation of wheat-leaf phospho*enol*pyruvate carboxylase by reversible phosphorylation. Plant Physiol 107: 775–782

Edwards GE, Foster JG, Winter K (1982) Activity and intracellular compartmentation of enzymes of carbon metabolism in CAM plants. In: Ting IP, Gibbs M (eds) Crassulacean acid metabolism. American Society of Plant Physiologists, Rockville, pp 92–111

Ehleringer J, Björkman O (1977) Quantum yields for CO_2 uptake in C_3 and C_4 plants. Plant Physiol 59: 86–90

Ehleringer JR, Monson RK (1993) Evolutionary and ecological aspects of photosynthetic pathway variation. Annu Rev Ecol Syst 24: 411–439

Ehleringer J, Pearcy RW (1983) Variation in quantum yield for CO_2 uptake among C_3 and C_4 plants. Plant Physiol 73: 555–559

Fahrendorf T, Holtum JAM, Mukherjee U, Latzko E (1987) Fructose 2, 6–bisphosphate, carbohydrate partitioning, and crassulacean acid metabolism. Plant Physiol 84: 182–187

Feng W, Ning L, Daley LS, Moreno Y, Azarenko A, Criddle RS (1994) Theoretical fitting of energetics of CAM path to calorimetric data. Plant Physiol Biochem 32: 591–598

Franco AC, Ball E, Lüttge U (1992) Differential effects of drought and light levels on accumulation of citric and malic acids during CAM in *Clusia.* Plant Cell Environ 15: 821–829

Gardeström P, Edwards GE (1985) Leaf mitochondria ($C_3 + C_4 +$ CAM). In: Douce R, Day DA (eds) Encyclopedia of plant physiology, New Series, vol 18, Higher plant cell respiration. Springer, Berlin Heidelberg New York, pp 314–346

Gemel J, Randall DD (1992) Light regulation of leaf mitochondrial pyruvate dehydrogenase complex. Role of photorespiratory carbon metabolism. Plant Physiol 100: 908–914

Gibson AC, Nobel PS (1986) The cactus primer. Harvard University Press, Cambridge

Griffiths H (1988) Crassulacean acid metabolism: a re-appraisal of physiological plasticity in form and function. Adv Bot Res 15: 43–92

Griffiths H (1989) Carbon dioxide concentration mechanisms and the evolution of CAM in vascular ephiphytes. In: Lüttge U (ed) Vascular plants as epiphytes. Springer, Berlin Heidelberg New York, pp 42–86

Griffiths H (1992) Carbon isotope discrimination and the integration of carbon assimilation pathways in terrestrial CAM plants. Plant Cell Environ 15: 1051–1062

Griffiths H, Smith JAC (1983) Photosynthetic pathways in the Bromeliaceae of Trinidad: relations between life–forms, habitat preference and the occurrence of CAM. Oecologia 60: 176–184

Griffiths H, Ong BL, Avadhani PN, Goh CJ (1989) Recycling of respiratory CO_2 during crassulacean acid metabolism: alleviation of photoinhibition in *Pyrrosia piloselloides*. Planta 179: 115–122

Groenhof AC, Smirnoff N, Bryant JA (1990) The appearance of a new molecular species of phospho*enol*pyruvate carboxylase (PEPC) and the rapid induction of CAM in *Sedum telephium* L. Plant Cell Environ 13: 437–446

Guralnick LJ, Ting IP (1988) Seasonal patterns of water relations and enzyme activity of the facultative CAM plant *Portulacaria afra* (L.) Jacq. Plant Cell Environ 11: 811–818

Haag-Kerwer A, Franco AC, Lüttge U (1992) The effect of temperature and light on the gas exchange and acid accumulation in the C_3-CAM plant *Clusia minor* L. J Exp Bot 43: 345–352

Hajirezaei M, Sonnewald U, Viola R, Carlisle S, Dennis D, Stitt M (1994) Transgenic potato plants with strongly decreased expression of pyrophosphate:fructose-6-phosphate phosphotransferase show no visible phenotype and only minor changes in metabolic fluxes in their tubers. Planta 192: 16–30

Hanning I, Heldt HW (1993) On the function of mitochondrial metabolism during photosynthesis in spinach (*Spinacia oleracea* L.) leaves. Partitioning between respiration and export of redox equivalents and precursors for nitrate assimilation products. Plant Physiol 103: 1147–1154

Heber U, Walker D (1992) Concerning a dual function of coupled cyclic electron transport in leaves. Plant Physiol 100: 1621–1626

Hermans J, Westhoff P (1990) Analysis of expression and evolutionary relationships of phosphoenolpyruvate carboxylase genes in *Flaveria trinerva* (C_4) and *F. pringlei* (C_3). Mol Gen Genet 224: 459–468

Hinkle PC, Kumar MA, Resetar A, Harris DL (1991) Mechanistic stoichiometry of mitochondrial oxidative phosphorylation. Biochemistry 30: 3576–3582

Holtum JAM, Osmond CB (1981) The gluconeogenic metabolism of pyruvate during deacidification in plants with crassulacean acid metabolism. Aust J Plant Physiol 8: 31–44

Holtum JAM, Winter K (1982) Activity of enzymes of carbon metabolism during the induction of crassulacean acid metabolism in *Mesembryanthemum crystallinum* L. Planta 155: 8–16

Ihlenfeldt HD (1994) Diversification in an arid world: the Mesembryanthemaceae. Annu Rev Ecol Syst 25: 521–546

Ingram J, Smith JAC (1995) Developmental regulation of NAD^+– and $NADP^+$–malic enzyme activities in the CAM plant *Kalanchoë daigremontiana*. (submitted)

Israel AA, Nobel PS (1994) Activities of carboxylating enzymes in the CAM species *Opuntia ficus-indica* grown under current and elevated CO_2 concentrations. Photosyn Res 40: 223–229

Jiao J, Chollet R (1991) Posttranslational regulation of phospho*enol*pyruvate carboxylases in C_4 and crassulacean acid metabolism plants. Plant Physiol 95: 981–985

Kalt W, Osmond CB, Siedow JN (1990) Malate metabolism in the dark after $^{13}CO_2$ fixation in the crassulacean plant *Kalanchoë tubiflora*. Plant Physiol 94: 826–832

Lambers H (1990) Oxidation of mitochondrial NADH and the synthesis of ATP. In: Dennis DT, Turpin DH (eds) Plant Physiology, biochemistry and molecular biology. Longman, Harlow, pp 124–143

Lance C, Chauveau M, Dizengremel P (1985) The cyanide–resistant pathway of plant mitochondria. In: Douce R, Day DA (eds) Encyclopedia of plant physiology, New Series, vol 18. Higher plant cell respiration. Springer, Berlin Heidelberg New York, pp 202–247

Lepiniec L, Vidal J, Chollet R, Gadal P, Cretin C (1994) Phosphoenolpyruvate carboxylase: structure, regulation and evolution. Plant Sci 99: 111–124

Li B, Chollet R (1994) Salt induction and the partial purification/characterization of phospho*enol*pyruvate carboxylase protein-serine kinase from an inducible crassulacean-acid-metabolism (CAM) plant, *Mesembryanthemum crystallinum* L. Arch Biochem Biophys 314: 247–254

Lüttge U (1988) Day-night changes of citric acid levels in CAM: phenomenon and ecophysiological significance. Plant Cell Environ 11: 445–451
Lüttge U, Ball E (1987) Dark respiration of CAM plants. Plant Physiol Biochem 25: 3–10
Lüttge U, Smith JAC (1988) CAM plants. In: Baker DA, Hall JL (eds) Solute transport in plant cells and tissues. Longman, Harlow, pp 417–452
Lüttge U, Smith JAC, Marigo G, Osmond CB (1981) Energetics of malate accumulation in the vacuole of *Kalanchoë tubiflora* cells. FEBS Lett 126: 81–84
Luo Y, Nobel PS (1993) Growth characteristics of newly initiated cladodes of *Opuntia ficus-indica* as affected by shading, drought and elevated CO_2. Physiol Plant 87: 467–474
Mägdefrau K (1968) Paläobiologie der Pflanzen. Fischer, Jena
Martin CE, Christensen NL, Strain BR (1981) Seasonal patterns of growth, tissue acid fluctuations, and $^{14}CO_2$ uptake in the crassulacean acid metabolism epiphyte *Tillandsia usneoides* L. (Spanish moss). Oecologia 49: 322–328
Maxwell C, Griffiths H, Young AJ (1994) Photosynthetic acclimation to light regime and water stress by the C_3-CAM epiphyte *Guzmania monostachia*: gas-exchange characteristics, photochemical efficiency and the xanthophyll cycle. Func Ecol 8: 746–754
Medina E (1974) Dark CO_2 fixation, habitat preference and evolution within the Bromeliaceae. Evolution 28: 677–686
Medina E (1982) Temperature and humidity effects on dark CO_2 fixation by *Kalanchoë pinnata*. Z Pflanzenphysiol 107: 251–258
Medina E (1984) On the temperature dependence of dark CO_2 fixation in CAM plants: acidity, growth and $\delta^{13}C$ values. In: Medina E (ed) Physiological ecology of CAM plants. CIET (Unesco-IVIC), Caracas, pp 52–72
Medina E, Osmond CB (1981) Temperature dependence of dark CO_2 fixation and acid accumulation in *Kalanchoë daigremontiana*. Aust J Plant Physiol 8: 641–649
Medina E, Olivares E, Diaz M (1986) Water stress and light intensity effects on growth and nocturnal acid accumulation in a terrestrial CAM bromeliad (*Bromelia humilis* Jacq.) under natural conditions. Oecologia 70: 441–446
Medina E, Lüttge U, Leal F, Ziegler H (1994a) Carbon and hydrogen isotope ratios in bromeliads growing under different light environments in natural conditions. Bot Acta 104: 47–52
Medina E, Ziegler H, Lüttge U, Trimborn P, Francisco M (1994b) Light conditions during growth as revealed by $\delta^{13}C$ values of leaves of primitive cultivars of *Ananas comosus*, an obligate CAM species. Func Ecol 8: 298–305
Milburn TR, Pearson DJ, Ndegwe NA (1968) Crassulacean acid metabolism under natural tropical conditions. New Phytol 67: 883–897
Millar AH, Wiskich JT, Whelan J, Day DA (1993) Organic acid activation of the alternative oxidase of plant mitochondria. FEBS Lett 329: 259–262
Monson RK (1989) On the evolutionary pathways resulting in C_4 photosynthesis and crassulacean acid metabolism (CAM). Adv Ecol Res 19: 57–110
Moore AL, Siedow JN (1991) The regulation and nature of the cyanide-resistant alternative oxidase of plant mitochondria. Biochim Biophys Acta 1059: 121–140
Moore AL, Gemel J, Randall DD (1993) The regulation of pyruvate dehydrogenase activity in pea leaf mitochondria. Plant Physiol 103: 1431–1435
Nicholls DG, Ferguson SJ (1992) Bioenergetics 2. Academic Press, London
Nobel PS (1991) Environmental productivity indices and productivity for *Opuntia ficus-indica* under current and elevated atmospheric CO_2 levels. Plant Cell Environ 14: 637–646
Nobel PS (1994) Remarkable agaves and cacti. Oxford University Press, New York
Nobel PS, Hartsock TL (1983) Relationships between photosynthetically active radiation, nocturnal acid accumulation, and CO_2 uptake for a crassulacean acid metabolism plant, *Opuntia ficus-indica*. Plant Physiol 71: 71–75
Nobel PS, Hartsock TL (1986) Short-term and long-term responses of crassulacean acid metabolism plants to elevated CO_2. Plant Physiol 82: 604–606
Nobel PS, Israel AA (1994) Cladode development, environmental responses of CO_2 uptake, and productivity for *Opuntia ficus-indica* under elevated CO_2. J Exp Bot 45: 295–303

Nobel PS, Garcia-Moya E, Quero E (1992) High annual productivity of certain agaves and cacti under cultivation. Plant Cell Environ 15: 329–335

Nobel PS, Cui M, Israel AA (1994a) Light, chlorophyll, carboxylase activity and CO_2 fixation at various depths in the chlorenchyma of *Opuntia ficus-indica* (L.) Miller under current and elevated CO_2. New Phytol 128: 315–322

Nobel PS, Cui M, Miller PM, Luo Y (1994b) Influences of soil volume and an elevated CO_2 level on growth and CO_2 exchange for the crassulacean acid metabolism plant *Opuntia ficus-indica*. Physiol Plant 90: 173–180

Olivares E, Faist K, Kluge M, Lüttge U (1993) $^{14}CO_2$ pulse-chase labelling in *Clusia minor* L. J Exp Bot 44: 1527–1533

Osmond CB (1978) Crassulacean acid metabolism: a curiosity in context. Annu Rev Plant Physiol 29: 379–414

Osmond CB (1982) Carbon cycling and stability of the photosynthetic apparatus in CAM. In: Ting IP, Gibbs M (eds) Crassulacean acid metabolism. American Society of Plant Physiologists, Rockville, pp 112–127

Osmond CB, Nott DL, Firth PM (1979) Carbon assimilation patterns and growth of the introduced CAM plant *Opuntia inermis* in eastern Australia. Oecologia 40: 331–350

Osmond CB, Winter K, Ziegler H (1982) Functional significance of different pathways of CO_2 fixation in photosynthesis. In: Lange OL, Nobel PS, Osmond CB, Ziegler H (eds) Physiological plant ecology II. Encyclopedia of plant physiology. New Series, vol 12B. Springer, Berlin Heidelberg New York, pp 479–547

Osmond CB, Holtum JAM, O'Leary MH, Roeske C, Wong OC, Summons RE, Avadhani PNC (1988) Regulation of malic-acid metabolism in crassulacean-acid-metabolism plants in the dark and light: in-vivo evidence from ^{13}C-labelling patterns after $^{13}CO_2$ fixation. Planta 175: 184–192

Osmond CB, Adams WW, Smith SD (1989) Crassulacean acid metabolism. In: Pearcy RW, Ehleringer JR, Mooney HA, Rundel PW (eds) Plant physiological ecology. Chapman and Hall, London, pp 255–280

Pfitsch WA, Smith AP (1988) Growth and photosynthesis of *Aechmea magdalenae*, a terrestrial CAM plant in a tropical moist forest, Panama. J Trop Ecol 4: 199–207

Poorter H (1993) Interspecific variation in the growth response of plants to an elevated ambient CO_2 concentration. Vegetatio 104/105: 77–97

Popp M, Kramer D, Lee H, Diaz M, Ziegler H, Lüttge U (1987) Crassulacean acid metabolism in tropical trees of the genus *Clusia*. Trees 1: 238–247

Queiroz O, Brulfert J (1982) Photoperiod-controlled induction and enhancement of seasonal adaptation to drought. In: Ting IP, Gibbs M (eds) Crassulacean acid metabolism. American Society of Plant Physiologists, Rockville, pp 208–230

Rauh W (1973) Über die Zonierung und Differenzierung der Vegetation Madagaskars. Akademie der Wissenschaften und der Literatur, Mainz

Robinson SA, Osmond CB (1994) Internal gradients of chlorophyll and carotenoid pigments in relation to photoprotection in thick leaves of plants with crassulacean acid metabolism. Aust J Plant Physiol 21: 497–506

Robinson SA, Yakir D, Ribas-Carbo M, Giles L, Osmond CB, Siedow JN, Berry JA (1992) Measurements of the engagement of cyanide-resistant respiration in the crassulacean acid metabolism plant *Kalanchoë daigremontiana* with the use of on-line oxygen isotope discrimination. Plant Physiol 100: 1087–1091

Robinson SA, Ribas-Carbo M, Yakir D, Giles L, Reuveni Y, Berry JA (1995) Beyond SHAM and cyanide: opportunities for studying the alternative oxidase in plant respiration using oxygen isotope discrimination. Aust J Plant Physiol 22: 487–496

Rustin P, Queiroz-Claret C (1985) Changes in the oxidative properties of *Kalanchoë blossfeldiana* leaf mitochondria during development of crassulacean acid metabolism. Planta 164: 415–422

Saitou K, Agata W, Masui Y, Asakura M, Kubota F (1994) Isoforms of NADP-malic enzyme from *Mesembryanthemum crystallinum* L. that are involved in C_3 photosynthesis and crassulacean acid metabolism. Plant Cell Physiol 35: 1165–1171

Schmitt JM (1990) Rapid concentration changes of phospho*enol*pyruvate carboxylase mRNA in detached leaves of *Mesembryanthemum crystallinum* L. in response to wilting and rehydration. Plant Cell Environ 13: 845–850

Scott P, Bettey M, Donath LN, Kruger NJ (1995) Effects of elevated fructose 2,6-bisphosphate levels on metabolism in transgenic potato tubers. (submitted)

Sharp RE, Matthews MA, Boyer JS (1984) Kok effect and the quantum yield of photosynthesis. Plant Physiol 75: 95–101

Sideris CP, Young NY, Chun HHQ (1948) Diurnal changes and growth rates as associated with ascorbic acid, titratable acidity, carbohydrate and nitrogenous fraction in the leaves of *Ananas comosus* (L.). Merr. Plant Physiol 23: 38–69

Siedow JN (1995) Bioenergetics: the mitochondrial electron transport chain. In: Levings CS, Vasil I (eds) The molecular biology of plant mitochondria. Kluwer, Dordrecht, pp 281–312

Slocombe SP, Whitelam GC, Cockburn W (1993) Investigation of phospho*enol*pyruvate carboxylase (PEPCase) in *Mesembryanthemum crystallinum* L. in C_3 and CAM photosynthetic states. Plant Cell Environ 16: 403–411

Smith JAC (1989) Epiphytic bromeliads. In: Lüttge U (ed) Vascular plants as epiphytes. Evolution and ecophysiology. Springer, Berlin Heidelberg New York, pp 109–138

Smith JAC, Bryce JH (1992) Metabolite compartmentation and transport in CAM plants. In: Tobin AK (ed) Plant organelles. Cambridge University Press, Cambridge, pp 141–167

Stewart WN, Rothwell GW (1993) Paleobotany and the evolution of plants. 2nd edn. Cambridge University Press, Cambridge

Sugiyama T, Laetsch WM (1975) Occurrence of pyruvate orthophosphate dikinase in the succulent plant, *Kalanchoë daigremontiana* Hamet. et. Perr. Plant Physiol 56: 605–607

Szarek SR, Ting IP (1974) Seasonal patterns of acid metabolism and gas exchange in *Opuntia basilaris*. Plant Physiol 54: 76–81

Szarek SR, Johnson HB, Ting IP (1973) Drought adaptation in *Opuntia basilaris*. Plant Physiol 52: 539–541

Taiz L, Zeiger E (1991) Plant physiology. Benjamin/Cummings, Redwood City

Teeri J (1982) Photosynthetic variation in the Crassulaceae. In: Ting IP, Gibbs M (eds) Crassulacean acid metabolism. American Society of Plant Physiologists, Rockville, pp 244–259

Thomas JC, McElwain EF, Bohnert HJ (1992) Convergent induction of osmotic stress-responses. Abscisic acid, cytokinin, and the effects of NaCl. Plant Physiol 100: 416–423

Tidwell WD, Parker LR (1990) *Protoyucca shadishii* gen. et sp. nov., an arborescent monocotyledon with secondary growth from the Middle Miocene of northwestern Nevada, U.S.A. Rev Palaeobot Palynol 62: 79–95

Ting IP (1985) Crassulacean acid metabolism. Annu Rev Plant Physiol 36: 595–622

Ting IP, Hanscom Z (1977) Induction of acid metabolism in *Portulacaria afra*. Plant Physiol 59: 511–514

Ting IP, Patel A, Sipes DL, Reid PD, Walling LL (1994) Differential expression of photosynthesis genes in leaf tissue layers of *Peperomia* as revealed by tissue printing. Am J Bot 81: 414–422

Toh H, Kawamura T, Izui K (1994) Molecular evolution of phospho*enol*pyruvate carboxylase. Plant Cell Environ 17: 31–43

Treichel S (1975) Crassulaceensäurestoffwechsel bei einem salztoleranten Vertreter der Aizoaceae: *Aptenia cordifolia*. Plant Sci Lett 4: 141–144

Troughton JA, Wells PV, Mooney HA (1974) Photosynthetic mechanisms in ancient C_4 and CAM plants. Carnegie Inst Washington Year Book 73: 812–816

Umbach AL, Wiskich JT, Siedow JN (1994) Regulation of alternative oxidase kinetics by pyruvate and intermolecular disulfide bond redox status in soybean seedling mitochondria. FEBS Lett 348: 181–184

Vernon DM, Ostrem JA, Schmitt JM, Bohnert HJ (1988) PEPCase transcript levels in *Mesembryanthemum crystallinum* decline rapidly upon relief from salt sress. Plant Physiol 86: 1002–1004

Vernon DM, Ostrem JA, Bohnert HJ (1993) Stress perception and response in a facultative halophyte: the regulation of salinity-induced genes in *Mesembryanthemum crystallinum*. Plant Cell Environ 16: 437–444

Villar R, Held AA, Merino J (1994) Comparison of methods to estimate dark respiration in the light in leaves of two woody species. Plant Physiol 105: 167–172

von Willert DJ, Treichel S, Kirst GO, Curdts E (1976) Environmentally controlled changes of phosphoenolpyruvate carboxylase in *Mesembryanthemum*. Phytochemistry 15: 1435–1436

Winter K (1974a) NaCl-induzierter Crassulaceen-Säurestoffwechsel bei der Salzpflanze *Mesembryanthemum crystallinum*. Oecologia 15: 383–392

Winter K (1974b) CO_2-Gaswechsel von an hohe Salinität adaptiertem *Mesembryanthemum crystallinum* bei Rückführung in glykisches Anzuchtmedium. Ber Dtsch Bot Ges 86: 467–476

Winter K (1980) Carbon dioxide and water vapor exchange in the crassulacean acid metabolism plant *Kalanchoë pinnata* during a prolonged light period. Plant Physiol 66: 917–921

Winter K, Gademann R (1991) Daily changes in CO_2 and water vapor exchange, chlorophyll fluorescence, and leaf water relations in the halophyte *Mesembryanthemum crystallinum* during the induction of crassulacean acid metabolism in response to high NaCl salinity. Plant Physiol 95: 768–776

Winter K, Engelbrecht B (1994) Short-term CO_2 responses of light and dark CO_2 fixation in the crassulacean acid metabolism plant *Kalanchoë pinnata*. J Plant Physiol 144: 462–467

Winter K. Lüttge U (1976) Balance between C_3 and CAM pathway of photosynthesis. In: Lange OL, Kappen L, Schulze ED (eds) Water and plant life. Springer, Berlin Heidelberg New York, pp 323–334

Winter K, Troughton JH (1978) Carbon assimilation pathways in *Mesembryanthemum nodiflorum* L. under natural conditions. Z Pflanzenphysiol 88: 153–162

Winter K, Lüttge U, Winter E, Troughton JH (1978) Seasonal shift from C_3 photosynthesis to crassulacean acid metabolism in *Mesembryanthemum crystallinum* growing in its natural environment. Oecologia 34: 225–237

Winter K, Foster JG, Edwards GE, Holtum JAM (1982) Intracellular localization of enzymes of carbon metabolism in *Mesembryanthemum crystallinum* exhibiting C_3 photosynthetic characteristics or performing crassulacean acid metabolism. Plant Physiol 69: 300–307

Winter K, Wallace BJ, Stocker GC, Roksandic Z (1983) Crassulacean acid metabolism in Australian vascular epiphytes and some related species. Oecologia 57: 129–141

Winter K, Osmond CB, Hubick KT (1986a) Crassulacean acid metabolism in the shade. Studies on an epiphytic fern, *Pyrrosia longifolia*, and other rainforest species from Australia. Oecologia 68: 224–230

Winter K, Schröppel-Meier G, Caldwell MM (1986b) Respiratory CO_2 as carbon source for nocturnal acid synthesis at high temperatures in three species exhibiting crassulacean acid metabolism. Plant Physiol 81: 390–394

Winter K, Zotz G, Baur B, Dietz KJ (1992) Light and dark CO_2 fixation in *Clusia uvitana* and the effects of plant water status and CO_2 availability. Oecologia 91: 47–51

Zotz G, Winter K (1993a) Short-term regulation of CAM activity in a tropical hemiepiphyte, *Clusia uvitana*. Plant Physiol 102: 835–841

Zotz G, Winter K (1993b) Short-term photosynthesis measurements predict leaf carbon balance in tropical rain-forest canopy plants. Planta 191: 409–412

Zotz G, Winter K (1994a) Annual carbon balance and nitrogen-use efficiency in tropical C_3 and CAM epiphytes. New Phytol 126: 481–492

Zotz G, Winter K (1994b) A one-year study on carbon, water and nutrient relationships in a tropical C_3 hemi-epiphyte, *Clusia uvitana* Pittier. New Phytol 127: 45–60

Zotz G, Winter K (1994c) Predicting annual carbon balance from leaf nitrogen. Naturwissenschaften 81: 449

26 Taxonomic Distribution of Crassulacean Acid Metabolism

J.A.C. SMITH[1] and K. WINTER[2]

The list presented in Table 26.1 summarizes the available information on the taxonomic occurrence of the CAM pathway. Details of the methods used to detect CAM activity can be found in the original publications: these have included nocturnal increases in malic-acid content, nocturnal CO_2 uptake, or carbon–isotope composition (see p. 5). The latter has been particularly useful in taxonomic surveys, since it can be applied to dried plant material from herbarium collections, and modern mass spectrometers can provide adequate signals from as little as 1 mg of plant dry matter.

As has been shown in several chapters of this Volume, the expression of CAM activity can be strongly influenced by environmental conditions, most notably water availability. In some species, carbon assimilation by the CAM pathway may make only a very small, or barely significant, contribution to biomass production, and then only under conditions of water shortage. Examples include *Guzmania monostachia* (Bromeliaceae: Medina et al. 1977; Smith et al. 1985). *Sedum* spp. (Crassulaceae: Chaps. 11,15 and 23; Kluge 1977; Teeri 1982), *Carpobrotus edulis* (Aizoaceae: Winter 1973) and *Othonna opima* (Asteraceae: von Willert et al. 1992; Flach and Eller 1994). Similarly, there has been much debate about whether to classify the gnetophyte *Welwitschia mirabilis* as a CAM plant (von Willert 1985): small amounts of malic acid can accumulate in the leaves of this species at night, but these are generally associated with very low rates of net nocturnal CO_2 *loss*, rather than net nocturnal CO_2 uptake (Ting and Burk 1983; Winter and Schramm 1986). All these species have been included in Table 26.1, since they demonstrate the ability to fix carbon through the CAM pathway. However, on the continuous spectrum of photosynthetic types from full CAM to C_3, these species are clearly very close to normal C_3 plants.

The taxonomic arrangement of Table 26.1 follows Tryon and Tryon (1982) for the ferns and fern allies, Foster and Gifford (1974) for the gymnosperm, Dahlgren et al. (1985) for the monocotyledons (except that the ending -anae instead of -iflorae is used for the superorders: Thorne 1992; Chase et al. 1993), and Cronquist (1981) for the dicotyledons. The sequence of orders and

[1]Department of Plant Sciences, University of Oxford, South Parks Road, Oxford OX1 3RB, UK
[2]Smithsonian Tropical Research Institute, P.O. Box 2072, Balboa, Republic of Panama

Ecological Studies Vol. 114
Winter/Smith (eds) Crassulacean Acid Metabolism

Table 26.1. List of genera containing species capable of crassulacean acid metabolism

Class, superorder, or subclass	Order	Family (genera/species)	Genus	Reference
Class Lycopodiopsida				
	Isoetales	Isoetaceae (1/150)	*Isoetes*[a]	Chapter 18
			Stylites (= *Isoetes*)	Chapter 18
Class Filicopsida				
	Polypodiales	Polypodiaceae (29/660)	*Drymoglossum* (= *Pyrrosia*)	1
			Pyrrosia	1
		Vittariaceae (6/120)	*Vittaria*	2
Class Gnetopsida				
	Gnetales	Welwitschiaceae (1/1)	*Welwitschia*	3, 4
Class Liliopsida (Monocotyledonae)				
Alismatanae	Alismatales	Alismataceae (16/100)	*Sagittaria*[a]	Chapter 18
		Hydrocharitaceae (17/130)	*Vallisneria*[a]	5
Lilianae	Asparagales	Agavaceae (8/300)	*Agave*	6
			Hesperaloë	7
			Polianthes	8
			Yucca	9
		Dracaenaceae (9/230)	*Sansevieria*	10
		Asphodelaceae (15/800)	*Aloë*	11
			Astroloba (= *Apicra*)	45
			Bulbine	12
			Gasteria	11
			Haworthia	11
			Poellnitzia (= *Haworthia*)	13
	Liliales	Orchidaceae (736/19 500)	*Aerangis*	14–17
			Aeranthes	
			Aerides	
			Angraecum	
			Arachnis	
			Ascocentrum	
			Brassavola	
			Bulbophyllum	
			Cadetia	
			Calanthe	
			Cattleya	
			Caularthron	
			Chiloschista	
			Coelogyne	
			Cymbidium	
			Cyrtopodium	
			Dendrobium	
			Dimerandra	
			Encyclia	

Table 26.1 (*Contd.*)

Class, superorder, or subclass	Order	Family (genera/species)	Genus	Reference
			Epidendrum	
			Eria	
			Flickingeria	
			Gussonea (= *Microcoelia*)	
			Jumellea	
			Laelia	
			Lanium	
			Lissochilus (= *Eulophia*)	
			Luisia	
			Maxillaria	
			Microcoelia	
			Micropera	
			Mobilabium	
			Oberonia	
			Oeonia	
			Oncidium	
			Ornithidium (= *Maxillaria*)	
			Paphiopedilum	
			Phalaenopsis	
			Pholidota	
			Plectorrhiza	
			Pleurothallis	
			Polyradicion	
			Rhinerrhiza	
			Robiquetia	
			Saccolabiopsis	
			Saccolabium	
			Sarcochilus	
			Schoenorchis	
			Schomburgkia	
			Solenangis	
			Sophronitis	
			Taeniophyllum	
			Thrixspermum	
			Thunia	
			Trachoma	
			Trichoglottis	
			Vanda	
			Vanilla	
Commelinanae	Commelinales	Commelinaceae (40/605)	*Callisia*	18, 19
			Tradescantia	
			Tripogandra	
Bromelianae	Bromeliales	Bromeliaceae (51/2500)	*Abromeitiella*	20–22
			Acanthostachys	
			Aechmea	
			Ananas	
			Araeococcus	
			Billbergia	

Table 26.1 (*Contd.*)

Class, superorder, or subclass	Order	Family (genera/species)	Genus	Reference
			Canistrum	
			Cryptanthus	
			Deuterocohnia	
			Dyckia	
			Encholirium	
			Guzmania	
			Hechtia	
			Hohenbergia	
			Neoglaziovia	
			Neoregelia	
			Nidularium	
			Orthophytum	
			Portea	
			Puya	
			Quesnalia	
			Streptocalyx	
			Tillandsia	
			Wittrockia	
Class Magnoliopsida (Dicotyledonae)				
Magnoliidae	Piperales	Piperaceae (5/3000)	*Peperomia*	Chapter 13; 23
Caryophyllidae	Caryophyllales	Aizoaceae (127/2500)	*Aizoanthemum*	Chapters 8–10; 12, 24, 25
			Amoebophyllum	
			Anisocalyx (= *Drosanthemum*)	
			Aptenia	
			Argyroderma	
			Aridaria	
			Aspazoma	
			Astridia	
			Bergeranthus	
			Brownanthus	
			Carpobrotus	
			Carruanthus	
			Cephalophyllum	
			Cheiridopsis	
			Conophytum	
			Disphyma	
			Drosanthemum	
			Eberlanzia	
			Faucaria	
			Fenestraria	
			Hereroa	
			Jacobsenia	
			Lampranthus	
			Lapidaria	
			Leipoldtia	
			Lithops	
			Malephora	
			Mesembryanthemum	
			Mestoklema	
			Meyerophytum	
			Mitrophyllum	

Table 26.1 (*Contd*)

Class, superorder, or subclass	Order	Family (genera/species)	Genus	Reference
			Monilaria	
			Nananthus	
			Opophythum (= *Mesembryanthemum*)	
			Platythyra	
			Prenia	
			Psilocaulon	
			Rhombophyllum	
			Ruschia	
			Ruschianthemum	
			Sceletium	
			Sphalmanthus	
			Stoeberia (= *Ruschia*)	
			Tetragonia	
			Titanopsis	
			Trichodiadema	
			Vanheerdia	
		Didiereaceae (4/11)	*Alluaudia*	26
			Alluaudiopsis	26
			Decaryia	26
			Didierea	26
		Cactaceae (122/1600)	all genera	6, 27
		Portulacaceae (29/450)	*Anacampseros*	28
			Calandrinia	29, 30
			Ceraria	12
			Portulaca[b]	31, 32
			Portulacaria	33
			Talinum	Chapter 12
Dilleniidae	Violales	Cucurbitaceae (118/825)	*Seyrigia*	34
			Xerosicyos	34
		Passifloraceae (18/630)	*Adenia*	28
	Theales	Clusiaceae (45/1010)	*Clusia*	Chapters 19, 20
Rosidae	Rhamnales	Vitaceae (13/735)	*Cissus*	Chapter 14
			Cyphostemma	28
	Geraniales	Oxalidaceae (6/890)	*Oxalis*	11
		Geraniaceae (14/775)	*Pelargonium*	35
			Sarcocaulon	Chapter 22
	Rosales	Crassulaceae (35/1500)	*Adromischus*	Chapters 21, 23; 11, 36
			Aeonium	
			Cotyledon	
			Crassula[c]	
			Cremnophila (= *Sedum*)	
			Diamorpha	
			Dudleya	
			Echeveria	
			Graptopetalum	

Table 26.1. (*Contd.*)

Class, superorder, or subclass	Order	Family (genera/species)	Genus	Reference
			Greenovia	
			Hasseanthus (= *Sedum*)	
			Hylotelephium	
			Kalanchoë	
			Lenophyllum	
			Monanthes	
			Pachyphytum	
			Parvisedum (= *Sedum*)	
			Rochea (= *Crassula*)	
			Sedum	
			Sempervivum	
			Tylecodon (= *Cotyledon*)	
			Umbilicus	
			Villadia	
	Euphorbiales	Euphorbiaceae (317/8000)	*Euphorbia*	28, 37
			Monadenium	37
			Pedilanthus	8
			Synadenium	37
	Apiales	Apiaceae (365/2800)	*Lilaeopsis*[a]	5
Asteridae	Asterales	Asteraceae (1160/19085)	*Kleinia*	35
			Notonia (= *Kleinia*)	38
			Othonna	12
			Senecio	39
	Gentianales	Apocynaceae (215/2100)	*Pachypodium*	40
		Asclepiadaceae (250/2000)	*Caralluma*	41
			Ceropegia	42
			Cynanchum	17
			Dischidia	15
			Folotsia (= *Cynanchum*)	17
			Frerea (= *Desmidorchis*)	43
			Hoodia	44
			Hoya	39
			Huernia	45
			Sarcostemma	15
			Stapelia	39
			Trichocaulon	28
	Rubiales	Rubiaceae (550/9000)	*Hydnophytum*	15
			Myrmecodia	15
	Scrophulariales	Gesneriaceae (126/2850)	*Codonanthe*	46
	Plantaginales	Plantaginaceae (3/255)	*Littorella*[a]	Chapter 18
	Lamiales	Lamiaceae (= Labiatae) (258/6970)	*Plectranthus*	11

[a]Aquatic CAM only.
[b]C_4 photosynthesis with very weak CAM.
[c]Includes species with aquatic CAM.

families is arranged according to the results of the phylogenetic analysis of Chase et al. (1993), but within each family the genera are listed in alphabetical order. Estimates of the number of genera and species in each family are given below the family name in parentheses (genera/species) and were taken from Thorne (1992) except in the following cases: Isoetaceae (Chap. 18), Polypodiaceae and Vittariaceae (Kubitzki 1990), Orchidaceae (species: Dressler 1993), Bromeliaceae (species: Kress 1989), Piperaceae and Aizoaceae (Kubitzki 1993), Cactaceae (Gibson and Nobel 1986), Portulacaceae (Kubitzki 1993), Euphorbiaceae (Webster 1987, 1994), Apocynaceae (Mabberley 1987), Asclepiadaceae (Kress 1989) and Plantaginaceae (Mabberley 1987). References to original sources or important reviews are given for individual genera, except for the Orchidaceae, Bromeliaceae, Aizoaceae and Crassulaceae, where the references are to detailed reviews of the whole families.

Table 26.1 records CAM activity for 328 genera in 33 families. (Included in this number are all 122 genera of the Cactaceae, not all of which have been investigated experimentally, but for which there is very strong circumstantial evidence for the occurrence of the CAM pathway: Gibson and Nobel 1986; Nobel 1988.) The taxa described in Table 26.1 include one family in the Lycopodiopsida, two in the Filicopsida and one in the Gnetopsida, as well as eight families of Liliopsida (monocotyledons) and 21 families of Magnoliopsida (dicotyledons). Interestingly, no CAM species have been recorded for one of the six subclasses of the Magnoliopsida, the Hamamelidae, a group containing many deciduous tress. Rao et al. (1979) reported substantial day-night changes in malate content in thin-leaved species in the Celastraceae, Ebenaceae, Flacourtiaceae and Sapindaceae (one species of each); however, these plants all showed C_3-type $\delta^{13}C$ values and more work is needed to evaluate whether the CAM cycle genuinely operates in these species. CAM activity has been confirmed in aquatic macrophytes in six families, although further work is needed for two of these, the Hydrocharitaceae and Apiaceae, in which the observed nocturnal acid accumulation is extremely low (Webb et al. 1988). For some of the larger families listed in Table 26.1, and most especially the Orchidaceae, the number of genera currently known to contain CAM species probably represents only a fraction of the true total.

←

Table 26.1 (*Contd.*) **References.** 1, Kluge et al. (1989); 2, Carter and Martin (1994); 3, Dittrich and Huber (1974); 4, Winter and Schramm (1986); 5, Webb et al. (1988); 6, Nobel (1988); 7, Ravetta and McLaughlin (1993); 8, Reddy and Das (1978); 9, Eickmeier (1978); 10, Virzo De Santo et al. (1981–1982); 11, Kluge and Ting (1978); 12, von Willert et al. (1992); 13, Troughton et al. (1974); 14, Avadhani et al. (1982); 15, Winter et al. (1983); 16, Goh and Kluge (1989); 17, M. Kluge, unpubl. observ.; 18, Martin et al. (1990); 19, Martin et al. (1994); 20, Medina et al. (1977); 21, Smith (1989); 22, Martin (1994); 23, Holthe et al. (1992): 24, von Willert et al. (1977); 25, K. Winter, unpubl. observ.; 26, Winter (1979); 27, Gibson and Nobel (1986); 28, Mooney et al. (1977); 29, Mooney et al. (1974); 30, Winter et al. (1981); 31, Koch and Kennedy (1980); 32, Ku et al. (1981); 33, Ting and Hanscom (1977); 34, De Luca et al. (1977); 35, Schütte et al. (1967); 36, Teeri (1982); 37, McWilliams (1970); 38, Milburn et al. (1968); 39, Nuernbergk (1961); 40, von Willert et al. (1980); 41, Lange et al. (1975); 42, Ziegler et al. (1981); 43, Lange and Zuber (1977); 44, Schulze et al. (1976); 45, Dittrich et al. (1973); 46, Guralnick et al. (1986).

References

Avadhani PN, Goh CJ, Rao AN, Arditti J (1982) Carbon fixation in orchids. In: Arditti J (ed) Orchid biology, reviews and perspectives, vol II. Cornell University Press, Ithaca, pp 173–193

Carter JP, Martin CE (1994) The occurrence of crassulacean acid metabolism among epiphytes in a high-rainfall region of Costa Rica. Selbyana 15: 104–106

Chase MW, Soltis DE, Olmstead RG, and 39 others (1993) Phylogenetics of seed plants: an analysis of nucleotide sequences from the plastid gene *rbc*L. Ann Mo Bot Gard 80: 528–580

Cronquist A (1981) An integrated system of classification of flowering plants. Columbia University Press, New York

Dahlgren RMT, Clifford HT, Yeo PF (1985) The families of the monocotyledons. Structure, evolution, and taxonomy. Springer, Berlin Heidelberg New York

De Luca P, Alfani A, Virzo De Santo A (1977) CAM, transpiration, and adaptive mechanisms to xeric environments in the succulent Cucurbitaceae. Bot Gaz 138: 474–478

Dittrich P, Huber W (1974) Carbon dioxide metabolism in members of the Chlamydospermae. In: Avron M (ed) Proc 3rd Int Congr Photosynthesis. Elsevier, Amsterdam, pp 1573–1578

Dittrich P, Campbell WH, Black CC Jr (1973) Phosphoenolpyruvate carboxykinase in plants exhibiting crassulacean acid metabolism. Plant Physiol 52: 357–361

Dressler RL (1993) Phylogeny and classification of the orchid family. Cambridge University Press, Cambridge

Eickmeier WG (1978) Photosynthetic pathway distributions along an aridity gradient in Big Bend National Park, and implications for enhanced resource partitioning. Photosynthetica 12: 290–297

Flach BMT, Eller BM (1994) Diurnal pattern of transpiration, water uptake and water budget of succulents with different CO_2 fixation pathways. Bot Acta 107: 46–53

Foster AS, Gifford EM Jr (1974) Comparative morphology of vascular plants, 2nd edn. WH Freeman, San Francisco

Gibson AC, Nobel PS (1986) The cactus primer. Harvard University Press, Cambridge

Goh CJ, Kluge M (1989) Gas exchange and water relations in epiphytic orchids. In: Lüttge U (ed) Vascular plants as epiphytes: evolution and ecophysiology. Springer, Berlin Heidelberg New York, pp 139–166

Guralnick LJ, Ting IP, Lord EM (1986) Crassulacean acid metabolism in the Gesneriaceae. Am J Bot 73: 336–345

Holthe PA, Patel A, Ting IP (1992) The occurrence of CAM in *Peperomia*. Selbyana 13: 77–87

Kluge M (1977) Is *Sedum acre* L. a CAM plant? Oecologia 29: 77–83

Kluge M, Ting IP (1978) Crassulacean acid metabolism. Analysis of an ecological adaptation. Springer, Berlin Heidelberg New York

Kluge M, Avadhani PN, Goh CJ (1989) Gas exchange and water relations in epiphytes and tropical ferns. In: Lüttge U (ed) Vascular plants as epiphytes: evolution and ecophysiology. Springer, Berlin Heidelberg New York, pp 87–108

Koch K, Kennedy RA (1980) Characteristics of crassulacean acid metabolism in the succulent C_4 dicot, *Portulaca oleracea* L. Plant Physiol 65: 193–197

Kress WJ (1989) The systematic distribution of vascular epiphytes. In: Lüttge U (ed) Vascular plants as epiphytes: evolution and ecophysiology. Springer, Berlin Heidelberg New York, pp 234–261

Ku SB, Shieh YJ, Reger BJ, Black CC (1981) Photosynthetic characteristics of *Portulaca grandiflora*, a succulent C_4 dicot. Plant Physiol 68: 1073–1080

Kubitzki K (ed) (1990) The families and genera of vascular plants, vol I. Kramer KU, Green PS (eds) Pteridophytes and gymnosperms. Springer, Berlin Heidelberg New York

Kubitzki K (ed) (1993) The families and genera of vascular plants, vol II. Kubitzki K, Rohwer JG, Bittrich V (eds) Flowering plants: dicotyledons, magnoliid, hamamelid and caryophyllid families. Springer, Berlin Heidelberg New York

Lange OL, Zuber M (1977) *Frerea indica*, a stem succulent CAM plant with deciduous C_3 leaves. Oecologia 31: 67–72
Lange OL, Schulze ED, Kappen L, Evenari M, Buschbom U (1975) CO_2 exchange pattern under natural conditions of *Caralluma negevensis*, a CAM plant of the Negev desert. Photosynthetica 9: 318–326
Mabberley DJ (1987) The plant-book. Cambridge University Press, Cambridge
Martin CE (1994) Physiological ecology of the Bromeliaceae. Bot Rev 60: 1–82
Martin CE, Gravatt DA, Loeschen VS (1994) Crassulacean acid metabolism in three species of Commelinaceae. Ann Bot 74: 457–463
Martin CE, Loeschen VS, Coke LB (1990) Crassulacean acid metabolism in selected terrestrial succulents in southeastern Jamaica, including two species in the Commelinaceae. Oecologia 84: 99–102
McWilliams EL (1970) Comparative rates of dark CO_2 uptake and acidification in the Bromeliaceae, Orchidaceae, and Euphorbiaceae. Bot Gaz 131: 285–290
Medina E, Delgado M, Troughton JH, Medina JD (1977) Physiological ecology of CO_2 fixation in Bromeliaceae. Flora 166: 137–152
Milburn TR, Pearson DJ, Ndegwe NA (1968) Crassulacean acid metabolism under natural tropical conditions. New Phytol 67: 883–897
Mooney H, Troughton JH, Berry JA (1974) Arid climates and photosynthetic systems. Carnegie Inst Washington Year book 73: 793–805
Mooney HA, Troughton JH, Berry JA (1977) Carbon isotope ratio measurements of succulent plants in southern Africa. Oecologia 30: 295–305
Nobel PS (1988) Environmental biology of agaves and cacti. Cambridge University Press, Cambridge
Nuernbergk EL (1961) Endogener Rhythmus und CO_2 Stoffwechsel bei Pflanzen mit diurnalem Säurerhythmus. Planta 56: 28–70
Ravetta DA, McLaughlin SP (1993) Photosynthetic pathways of *Hesperaloe funifera* and *H. nocturna* (Agavaceae): novel sources of specialty fibers. Am J Bot 80: 524–532
Reddy AR, Das VSR (1978) The decarboxylating systems in fourteen taxa exhibiting CAM pathway. Z Pflanzenphysiol 86: 141–146
Schulze E-D, Ziegler H, Stichler W (1976) Environmental control of crassulacean acid metabolism in *Welwitschia mirabilis* Hook. fil. in its range of natural distribution in the Namib desert. Oecologia 24: 323–334
Schütte KH, Steyn R, Van der Westhuizen M (1967) Crassulacean acid metabolism in South African succulents: a preliminary investigation into its occurrence in various families. J South Afr Bot 33: 107–110
Smith JAC (1989) Epiphytic bromeliads. In: Lüttge U (ed) Vascular plants as epiphytes: evolution and ecophysiology. Springer, Berlin Heidelberg New York, pp 109–138
Smith JAC, Griffiths H, Bassett M, Griffiths NM (1985) Day-night changes in the leaf water relations of epiphytic bromeliads in the rain forests of Trinidad. Oecologia 67: 475–485
Teeri J (1982) Photosynthetic variation in the Crassulaceae. In: Ting IP, Gibbs M (eds) Crassulacean acid metabolism. American Society of Plant Physiologists, Rockville, pp 92–111
Thorne RF (1992) Classification and geography of the flowering plants. Bot Rev 58: 225–348
Ting IP, Burk JH (1983) Aspects of carbon metabolism in *Welwitschia*. Plant Sci Lett 32: 279–285
Ting IP, Hanscom III Z (1977) Induction of acid metabolism in *Portulacaria afra*. Plant Physiol 59: 511–514
Troughton JH, Card KA, Hendy CH (1974) Photosynthetic pathways and carbon isotope discrimination by plants. Carnegie Inst Washington Year book 73: 768–780
Tryon RM, Tryon AF (1982) Ferns and allied plants. Springer, New York
Virzo De Santo A, Fioretto A, Alfani A (1981–1982) The adaptive significance of growth form, leaf morphology, and CAM in the genus *Sansevieria*. Delpinoa N.S. 23–24: 307–323
von Willert DJ (1985) *Welwitschia mirabilis* – new aspects in the biology of an old plant. Adv Bot Res 11: 157–191

von Willert DJ, Thomas DA, Lobin W, Curdts E (1977) Ecophysiologic investigations in the family of the Mesembryanthemaceae. Occurrence of a CAM and ion content. Oecologia 29: 67–76

von Willert DJ, Brinckmann E, Scheitler B, Schulze ED, Thomas DA, Treichel S (1980) Ökophysiologische Untersuchungen an Pflanzen der Namib-Wüste. Naturwissenschaften 67: 21–28

von Willert DJ, Eller BM, Werger MJA, Brinckmann E, Ihlenfeldt HD (1992) Life strategies of succulent plants in deserts, with special reference to the Namib desert. Cambridge University Press, Cambridge

Webb DR, Rattray MR, Brown JMA (1988) A preliminary survey for crassulacean acid metabolism (CAM) in submerged aquatic macrophytes in New Zealand. N Z J Mar Freshw Res 22: 231–235

Webster GL (1987) The saga of the spurges: a review of classification and relationships in the Euphorbiales. Bot J Linn Soc 94: 3–46

Webster GL (1994) Synopsis of the genera and suprageneric taxa of Euphorbiaceae. Ann Mo Bot Gard 81: 33–144

Winter K (1973) NaCl-induzierter Crassulaceensäurestoffwechsel bei einer weiteren Aizoacee: *Carpobrotus edulis*. Planta 115: 187–188

Winter K (1979) $\delta^{13}C$ values of some succulent plants from Madagascar. Oecologia 40: 103–112

Winter K, Schramm MJ (1986) Analysis of stomatal and nonstomatal components in the environmental control of CO_2 exchange in leaves of *Welwitschia mirabilis*. Plant Physiol 82: 173–178

Winter K, Osmond CB, Pate JS (1981) Coping with salinity. In: Pate JS, McComb AJ (eds) The biology of Australian plants. University of Western Australia Press, Nedlands, pp 88–113

Winter K, Wallace BJ, Stocker GC, Roksandic Z (1983) Crassulacean acid metabolism in Australian vascular epiphytes and some related species. Oecologia 57: 129–141

Ziegler H, Batanouny KH, Sankhla N, Vyas OP, Stichler W (1981) The photosynthetic pathway types of some desert plants from India, Saudi Arabia, Egypt, and Iraq. Oecologia 48: 93–99

Subject Index

A complete list of taxa in which CAM activity has been detected is given in Table 26.1, and the individual taxa are indexed here only if mentioned in the main text.

Printing: Saladruck, Berlin
Binding: Buchbinderei Lüderitz & Bauer, Berlin

Ecological Studies

Volumes published since 1989

Volume 77
Air Pollution and Forest Decline: A Study of Spruce (*Picea abies*) on Acid Soils (1989)
E.-D. Schulze, O. L. Lange, and R. Oren (Eds.)

Volume 78
Agroecology: Researching the Ecological Basis for Sustainable Agriculture (1990)
S. R. Gliessman (Ed.)

Volume 79
Remote Sensing of Biosphere Functioning (1990)
R. J. Hobbs and H. A. Mooney (Eds.)

Volume 80
Plant Biology of the Basin and Range (1990)
B. Osmond, G. M. Hidy, and L. Pitelka (Eds.)

Volume 81
Nitrogen in Terrestrial Ecosystems: Questions of Productivity, Vegetational Changes, and Ecosystem Stability (1991)
C. O. Tamm

Volume 82
Quantitative Methods in Landscape Ecology: The Analysis and Interpretation of Landscape Heterogeneity (1990)
M. G. Turner and R. H. Gardner (Eds.)

Volume 83
The Rivers of Florida (1990)
R. J. Livingston (Ed.)

Volume 84
Fire in the Tropical Biota: Ecosystem Processes and Global Challenges (1990)
J. G. Goldammer (Ed.)

Volume 85
The Mosaic-Cycle Concept of Ecosystems (1991)
H. Remmert (Ed.)

Volume 86
Ecological Heterogeneity (1991)
J. Kolasa and S. T. A. Pickett (Eds.)

Volume 87
Horses and Grasses: The Nutritional Ecology of Equids and Their Impact on the Camargue (1992)
P. Duncan

Volume 88
Pinnipeds and El Niño: Responses to Environmental Stress (1992)
F. Trillmich and K. A. Ono (Eds.)

Volume 89
Plantago: A Multidisciplinary Study (1992)
P. J. C. Kuiper and M. Bos (Eds.)

Volume 90
Biogeochemistry of a Subalpine Ecosystem: Loch Vale Watershed (1992)
J. Baron (Ed.)

Volume 91
Atmospheric Deposition and Forest Nutrient Cycling (1992)
D. W. Johnson and S. E. Lindberg (Eds.)

Volume 92
Landscape Boundaries: Consequences for Biotic Diversity and Ecological Flows (1992)
A. J. Hansen and F. di Castri (Eds.)

Volume 93
Fire in South African Mountain Fynbos: Ecosystem, Community, and Species Response at Swartboskloof (1992)
B. W. van Wilgen et al. (Eds.)

Volume 94
The Ecology of Aquatic Hyphomycetes (1992)
F. Bärlocher (Ed.)

Volume 95
Palms in Forest Ecosystems of Amazonia (1992)
F. Kahn and J.-J. DeGranville

Volume 96
Ecology and Decline of Red Spruce in the Eastern United States (1992)
C. Eagar and M. B. Adams (Eds.)

Ecological Studies

Volumes published since 1989

Volume 97
The Response of Western Forests to Air Pollution (1992)
R.K. Olson, D. Binkley, and M. Böhm (Eds.)

Volume 98
Plankton Regulation Dynamics (1993)
N. Walz (Ed.)

Volume 99
Biodiversity and Ecosystem Function (1993)
E.-D. Schulze and H.A. Mooney (Eds.)

Volume 100
Ecophysiology of Photosynthesis (1994)
E.-D. Schulze and M.M. Caldwell (Eds.)

Volume 101
Effects of Land Use Change on Atmospheric CO_2 Concentrations: South and South East Asia as a Case Study (1993)
V.H. Dale (Ed.)

Volume 102
Coral Reef Ecology (1993)
Y.I. Sorokin

Volume 103
Rocky Shores: Exploitation in Chile and South Africa (1993)
W.R. Siegfried (Ed.)

Volume 104
Long-Term Experiments With Acid Rain in Norwegian Forest Ecosystems (1993)
G. Abrahamsen et al. (Eds.)

Volume 105
Microbial Ecology of Lake Plußsee (1993)
J. Overbeck and R.J. Chrost (Eds.)

Volume 106
Minimum Animal Populations (1994)
H. Remmert (Ed.)

Volume 107
The Role of Fire in Mediterranean-Type Ecosystems (1994)
J.M. Moreno and W.C. Oechel

Volume 108
Ecology and Biogeography of Mediterranean Ecosystems in Chile, California and Australia (1994)
M.T.K. Arroyo, P.H. Zedler, and M.D. Fox (Eds.)

Volume 109
Mediterranean-Type Ecosystems. The Function of Biodiversity (1995)
G.W. Davis and D.M. Richardson (Eds.)

Volume 110
Tropical Montane Cloud Forests (1995)
L.S. Hamilton, J.O. Juvik, and F.N. Scatena (Eds.)

Volume 111
Peatland Forestry. Ecology and Principles (1995)
E. Paavilainen and J. Päivänen

Volume 112
Tropical Forests: Management and Ecology (1995)
A.E. Lugo and C. Lowe (Eds.)

Volume 113
Arctic and Alpine Biodiversity. Patterns, Causes and Ecosystem Consequences (1995)
F.S. Chapin III and C. Körner (Eds.)

Volume 114
Crassulacean Acid Metabolism. Biochemistry, Ecophysiology and Evolution (1995)
K. Winter and J.A.C. Smith (Eds.)

Volume 115
Islands. Biological Diversity and Ecosystem Function (1995)
P.M. Vitousek, L.L. Loope, and H. Adsersen (Eds.)

Volume 116
High Latitude Rainforests and Associated Ecosystems of the West Coast of the Americas: Climate, Hydrology, Ecology and Conservation (1995)
R.G. Lawford, P. Alaback, and E.R. Fuentes (Eds.)

Volume 117
Global Change and Mediterranean-Type Ecosystems (1995)
J. Moreno and W.C. Oechel (Eds.)